Health Risk Assessment for Asbestos and Other Fibrous Minerals

Health Risk Assessment for Asbestos and Other Fibrous Minerals

Edited by

Andrey Korchevskiy
Chemistry & Industrial Hygiene, Inc., Lakewood, CO, USA

James Rasmuson
Chemistry & Industrial Hygiene, Inc., Lakewood, CO, USA

Eric Rasmuson
Chemistry & Industrial Hygiene, Inc., Lakewood, CO, USA

Copyright © 2024 by John Wiley & Sons, Inc. All rights reserved, including rights for text and data mining and training of artificial intelligence technologies or similar technologies.

Published by John Wiley & Sons, Inc., Hoboken, New Jersey.
Published simultaneously in Canada.

No part of this publication may be reproduced, stored in a retrieval system, or transmitted in any form or by any means, electronic, mechanical, photocopying, recording, scanning, or otherwise, except as permitted under Section 107 or 108 of the 1976 United States Copyright Act, without either the prior written permission of the Publisher, or authorization through payment of the appropriate per-copy fee to the Copyright Clearance Center, Inc., 222 Rosewood Drive, Danvers, MA 01923, (978) 750-8400, fax (978) 750-4470, or on the web at www.copyright.com. Requests to the Publisher for permission should be addressed to the Permissions Department, John Wiley & Sons, Inc., 111 River Street, Hoboken, NJ 07030, (201) 748-6011, fax (201) 748-6008, or online at http://www.wiley.com/go/permission.

Trademarks: Wiley and the Wiley logo are trademarks or registered trademarks of John Wiley & Sons, Inc. and/or its affiliates in the United States and other countries and may not be used without written permission. All other trademarks are the property of their respective owners. John Wiley & Sons, Inc. is not associated with any product or vendor mentioned in this book.

Limit of Liability/Disclaimer of Warranty: While the publisher and author have used their best efforts in preparing this book, they make no representations or warranties with respect to the accuracy or completeness of the contents of this book and specifically disclaim any implied warranties of merchantability or fitness for a particular purpose. No warranty may be created or extended by sales representatives or written sales materials. The advice and strategies contained herein may not be suitable for your situation. You should consult with a professional where appropriate. Further, readers should be aware that websites listed in this work may have changed or disappeared between when this work was written and when it is read. Neither the publisher nor authors shall be liable for any loss of profit or any other commercial damages, including but not limited to special, incidental, consequential, or other damages.

For general information on our other products and services or for technical support, please contact our Customer Care Department within the United States at (800) 762-2974, outside the United States at (317) 572-3993 or fax (317) 572-4002.

Wiley also publishes its books in a variety of electronic formats. Some content that appears in print may not be available in electronic formats. For more information about Wiley products, visit our web site at www.wiley.com.

Library of Congress Cataloging-in-Publication Data applied for:
ISBN: 9781119438434 (HB)

Cover Design: Wiley
Cover Image: © Brasil2/Getty Images

Set in 9.5/12.5pt STIXTwoText by Straive, Pondicherry, India

Contents

List of Contributors *xv*
Preface *xvii*

Part I Hazard Identification *1*

1 **Mineralogical Characteristics and Risk Assessment of Elongate Mineral Particles (EMPs): Asbestos, Fiber, and Fragment** *3*
Ann G. Wylie
Introduction *3*
Nomenclature *6*
Source Specificity: Chemical and Physical Properties *8*
Source Specificity: Dimension *11*
Structural Groupings of Common Elongate Minerals *13*
Establishing the Chemical Composition of Minerals *15*
Mineral Intergrowths and Associations *16*
Bioreactivity of Mineral Surfaces: Chemical Factors *17*
   The Specificity of Mineral Surfaces: The Example of Quartz *17*
   General Considerations of Solubility *18*
   Formation of Reactive Oxygen Species (ROS) *20*
   Coatings *21*
   Surface Charge *22*
   EMP Surfaces: Chain Silicates and Zeolites *23*
Physical Factors *24*
   Specific Surface Area *24*
   Enthalpy and Other Thermodynamic Properties *26*
   Density and Aerodynamic Diameter *26*
   Stiffness and Tensile Strength *28*

　　　　The Effects of Heat *30*
　　　　Dimensionality: General Considerations *30*
　　　Establishing Measurement Protocols *32*
　　　　Optical vs. Electron Microscopy Methods *32*
　　　　Stratified Counting *34*
　　　　Sample Preparation for TEM: Direct vs. Indirect Preparation *34*
　　　　Frequency Distributions of Length and Width *35*
　　　　Lung Burden *37*
　　　　Dimensionality and Carcinogenicity *38*
　　　Discussion *39*
　　　References *40*

2　Toxicology of Mineral Fibers and Implications for Risk Assessment *52*
　　Brooke T. Mossman
　　Introduction *52*
　　Use of Rodent Models to Analyze the Toxicity to Disease Potential
　　of Naturally Occurring and Synthetic Fibers *53*
　　　　Inhalation Studies *53*
　　　　Intratracheal Instillation and Oropharyngeal Aspiration Studies *54*
　　　　Intrapleural Injection Studies *54*
　　　　Intraperitoneal Injection Studies *54*
　　　　Comparative Results on Effects of Asbestos and Other Naturally
　　　　Occurring Fibers in Rodent Studies *54*
　　In vitro Models of Toxicity *66*
　　　　Advantages and Disadvantage of *In vitro* Models *66*
　　　　Contributions of *In vitro* Models to Understanding Mechanisms of
　　　　Cytotoxicity and Carcinogenesis by Mineral Fibers *67*
　　Properties of Mineral Fibers Important in Toxicity
　　and Carcinogenic Effects *68*
　　A Systems Biology Approach to Understanding Connections and Interactions
　　Between Adverse Outcomes in Mineral Fiber-Induced Diseases *71*
　　References *72*

**3　Health Outcomes of Asbestos Exposure – A Pathology
　　and Diagnostic Perspective** *82*
　　Bruce Case
　　Introduction *82*
　　Nonmalignant Change in Structure or Function *83*
　　Nonmalignant Asbestos-Related Disease *84*
　　　　Pleural *84*
　　　　　　Asbestos Effusion *84*

Pleural Plaques and Localized Pleural Thickening (LPT)   *84*
Diffuse Pleural Thickening   *88*
Rounded Atelectasis   *89*
Lung   *89*
Asbestosis   *89*
Malignant Diseases Attributable to Asbestos Exposure   *92*
General Comments   *92*
Asbestos-Related Lung Cancer   *94*
Mesothelioma – Accelerating Knowledge   *96*
References   *102*

Part II   Exposure Assessment   *109*

4   Principles of Exposure Assessment for Elongate Mineral Particles (EMPs)   *111*
Eric Rasmuson, James Rasmuson, and Andrey Korchevskiy
General Principles and Methods   *113*
Gathering Information   *113*
Evaluating the Quality of Data   *114*
Measurement Techniques   *116*
Comparison of the Results of Different Analytical Methodologies   *120*
Proximity to the Emission Source   *121*
Adjusting Results for Censored Data   *122*
Correlation of EMP Exposures and Lung Burden Analysis   *122*
References   *123*

5   Asbestos Exposure Measurements: Principles of Current and Historical Data Interpretation   *127*
Garry Burdett
Aim and Background   *127*
Causes of Asbestos-Related Lung Disease and Their Relationship to Exposure Assessment   *128*
Exposure Measurement   *130*
Historic Methods of Asbestos Exposure Measurement   *131*
Gravimetric Methods   *131*
Impaction Sampling and Microscopic Particle Counting   *132*
Impinger Sampling and Microscopic Particle Counting   *132*
Thermal Precipitator (TP) Sampling and Microscopic Particle Counting   *133*

Direct Reading Instruments for Particle and Fiber Counting   *134*
Early Sampling Strategies   *135*
Development of the Current Analytical Methods for Fiber Counting   *136*
Membrane Filter Sampling and Phase Contrast Microscopy Fiber Counting (MF-PCM)   *136*
Membrane Filter Sampling and Electron Microscopy (EM) Analysis   *137*
Limitations of Current Indices of Exposure Assessment   *139*
Variability of the MF-PCM Index Over Time   *140*
Sampling Method   *140*
Sample Preparation   *141*
Microscope Equipment and Set-Up   *142*
Fiber Definition   *143*
Counting Procedures and Performance   *144*
Effect of Changes to the MF-PCM Counts Over Time   *145*
Conclusion   *146*
Acknowledgements   *147*
References   *147*

6   Asbestos Exposure Modeling Using Advanced Tools Including Computational Fluid Dynamics (CFD)   *153*
Daniel Hall, James Rasmuson, and Cassidy Strode
Introduction   *153*
Validation and Application of CFD Air Dispersion Modeling   *155*
Overview of CFD General Methodology   *157*
CFD Simulation Set-Up   *159*
Geometry Creation and Set-Up   *159*
Mesh Creation   *160*
Parameter Set-Up   *160*
Computational Solve   *162*
Post-processing   *162*
Complementary Modeling Software Tools   *163*
Other Software Tools   *164*
Indoor and Outdoor Modeling Examples   *164*
First Example – Indoor CFD Modeling   *164*
Preliminary Outdoor CFD Wind Simulation – Effect on Indoor Ventilation   *166*
Indoor CFD Simulations   *168*
Mill Ventilation   *168*
Other Model Parameters   *169*

      Source Descriptions   *170*
      Reheat Furnace Brick Removal Source   *170*
      Pipe Insulation Removal Source   *171*
      CFD Results   *172*
   Second Example – Outdoor CFD, AERMOD, and CALPUFF Models   *174*
      Model Geometry   *177*
      Receptor Descriptions   *177*
      Source Descriptions   *177*
      Fugitive Plant Emission – Manufacturing, Finishing, Fiber Warehouse, Tray Loading, and Stripping Station   *180*
      Baghouse Source Emission Rates   *182*
      Pipe Storage and Shipping Yard Source Emission Rate   *183*
      Crusher Source Emission Rate   *183*
      Meteorology   *184*
      CFD Results   *186*
      EPA Outdoor Dispersion Models   *188*
      Geophysical Set-Up   *188*
      CALMET Set-Up   *189*
      CALPUFF Processor   *189*
      CALPUFF Results   *191*
      AERMOD Model   *191*
      Geophysical Set-Up   *191*
      Meteorology Set-Up   *191*
      AERMOD Set-Up   *193*
      AERMOD Results   *194*
      Comparison of CFD, CALPUFF, and AERMOD Results   *194*
   Discussion and Conclusions   *194*
   References   *197*

Part III   Dose-Response Assessment   *201*

7   Asbestos Dose–Response Assessment: The Peto Model and Its Application in the US EPA and Berman and Crump Studies   *203*
Andrey Korchevskiy
   Rationale and Meaning of the Peto Model   *203*
   Utilization of the Peto Model by the US EPA   *212*
   Berman and Crump Meta-analysis Based on Peto Model   *218*
   References   *228*

8 The Hodgson and Darnton Approach to Quantifying the Risks of Mesothelioma and Lung Cancer in Relation to Asbestos Exposure 233
Lucy Darnton

Introduction *233*
Overview of the Hodgson and Darnton Approach *234*
Metrics and Data Requirements *235*
   Lung Cancer *235*
   Mesothelioma *236*
   Other Data Issues *236*
Summary of Cohorts Included in the Original and Updated Meta-Analyses *237*
   Crocidolite Cohorts *238*
   Amosite Cohorts *239*
   Other Amphiboles: Vermiculite Miners and Associated Workers, Libby, Montana, USA *241*
   Chrysotile Cohorts *242*
Summary of Original and Updated Meta-Analyses *245*
   Mesothelioma *245*
   Lung Cancer *250*
Nonlinear Exposure–Response Relationship *256*
   Pleural Mesothelioma *257*
   Peritoneal Mesothelioma *259*
   Lung Cancer *260*
   Summary *262*
Application of Hodgson and Darnton for Risk Assessment *262*
Conclusions *264*
References *266*

9 Prediction of Mesothelioma Mortality in the Context of Country-wide Risk Evaluation 270
Lucy Darnton

Conclusions *284*
References *284*

10 Implications of Exposure Measurement Methodologies for Dose–Response Assessment in Asbestos Worker Cohorts 286
Garry Burdett

Electron Microscopy Fiber Size Distribution for Different Cohorts and Their Relationship to PCM Fiber Counts *287*
   TEM Fiber-Size-Distribution in Cohorts from Mines and Mills *288*
   TEM Size Distributions from Manufacturing Cohorts *289*

SEM Size Distributions from Manufacturing Cohorts   *292*
EM Determinations of Asbestos Fiber Types in Asbestos Industry Cohorts   *293*
   Natural Occurrence   *294*
   Mixed-Use   *295*
   External Sources   *296*
   Lung Burden Analysis   *296*
Conversions of Historic Cohort Measurement Indices to MF-PCM Fiber Counts   *296*
   Conversion from Impinger Counts to MF-PCM   *297*
   Conversions from Other Particle Counting Methods   *299*
   Conversions from Gravimetric Measurement   *299*
Crocidolite Cohort Exposures   *302*
   Wittenoom Occupational   *302*
   Wittenoom Environmental   *305*
   South African Mines and Mills   *306*
   Massachusetts Cigarette Filter Manufacturing   *309*
   UK Gas Mask Workers   *310*
   Other Cohorts Exposed to Crocidolite   *311*
   Crocidolite Summary   *311*
Amosite Cohort Exposures   *311*
   South African Amosite Mining   *311*
   Patterson, New Jersey   *314*
   Tyler, Texas   *315*
   Uxbridge   *315*
   Amosite Summary   *316*
Chrysotile Mining and Milling Cohort Exposures   *317*
   Quebec, Canada   *318*
   Balangero, Italy   *318*
   Qinghai, China   *319*
   Uralasbest, Russia   *321*
   Chrysotile Mining Summary   *322*
Chrysotile Textiles   *322*
   South Carolina Textile Workers   *324*
   North Carolina Textile Workers   *325*
   Chongqing Chrysotile Cohort   *327*
   Chrysotile Textiles Summary   *328*
   Other Chrysotile Cohorts   *328*
Discussion and Outlook   *330*
Acknowledgement   *333*
References   *333*

11 Mathematical Modeling of Cancer Potency for Various Fibrous Minerals *344*
Andrey Korchevskiy, James Rasmuson, and Eric Rasmuson
References *360*

12 Theoretical and Practical Aspects of Asbestos Dose–Response Assessment *366*
Andrey Korchevskiy and James Rasmuson

General Considerations and Model of Asbestos Dose–Response Assessment *366*
   Linear Model *367*
   Nonlinear Model *368*
Relationship Between Different Estimation of Mesothelioma and Lung Cancer Potency Factors *371*
Life Tables and Life Expectancy of the Exposed Population *374*
Linearity and Nonlinearity of the Dose–Response Curves *375*
Threshold and Benchmark Dose Response in Asbestos Risk Assessment *376*
Community and Occupational Risk Assessment *378*
Peritoneal Mesothelioma *378*
Other Types of Cancer *380*
Inhalation Unit Risk (IUR) for Asbestos Fibers *383*
Asbestos Dose–Response and Tobacco Smoking *385*
Other Factors Impacting the Dose–Response Relationship for Elongate Mineral Particles *387*
References *388*

Part IV Risk Characterization *393*

13 Risk Characterization for Occupational and Environmental Exposure to Asbestos: Case Studies *395*
James Rasmuson, Andrey Korchevskiy, and Eric Rasmuson
References *408*

14 Asbestos in Soil: Risk Characterization for Occupational and Environmental Exposures *412*
Andrey Korchevskiy and Robert Strode
References *424*

15 Asbestos in Brakes: Risk Assessment for Exposure Patterns with Nonlinear Dynamics *427*
Andrey Korchevskiy, Robert Strode, and Arseniy Korchevskiy
Ambient Air Emissions from the Brakes in Street Canyons *428*
Fibers in Car Brakes: Chaotic Behavior of Emissions in a Self-regulated Community *433*
   Diagnosing the Chaotic Trends *439*
References *441*

Index *443*

List of Contributors

Garry Burdett
Research and Measurement Scientist (Retired), UK Health and Safety Executive

Bruce Case
Department of Pathology
Faculty of Medicine and Health Sciences
McGill University, Montreal
Canada

Lucy Darnton
Science Division
Epidemiology and Predictive Modelling
Bootle, Merseyside
UK

Daniel Hall
Chemistry & Industrial Hygiene, Inc.
Lakewood, CO
USA

Andrey Korchevskiy
Chemistry & Industrial Hygiene, Inc.
Lakewood, CO
USA

Arseniy Korchevskiy
Chemistry & Industrial Hygiene, Inc.
Lakewood, CO
USA

Brooke T. Mossman
Department of Pathology and Laboratory Medicine
University of Vermont College of Medicine
Burlington, VT
USA

Eric Rasmuson
Chemistry & Industrial Hygiene, Inc.
Lakewood, CO
USA

James Rasmuson
Chemistry & Industrial Hygiene, Inc.
Lakewood, CO
USA

Cassidy Strode
Chemistry & Industrial Hygiene, Inc.
Lakewood, CO
USA

Robert Strode
Summit Exposure and Risk Sciences LLC, Silverthorne, CO
USA

Ann G. Wylie
Department of Geology
University of Maryland
College Park, MD
USA

Preface

In the twenty-first century, asbestos remains a serious occupational and environmental hazard. First, asbestos is still produced in different countries and widely used in China, Latin America, and other regions. Second, asbestos is present in numerous buildings, equipment, and the environment as a legacy from the times when it was most actively utilized. Third, naturally occurring asbestos (NOA) is present in soil and rocks as a contaminant presenting a possible hazard for various human activities worldwide. It has been long recognized that human exposure to multiple types of unregulated fibrous minerals, which sometimes do not fit the traditional definitions of asbestos, may have similar potential health consequences as commercial asbestos itself. Some examples of asbestiform minerals include erionite, fluoro-edenite, and balangeroite, among others. Also, various minerals, when affected by mechanical force, can produce so-called cleavage fragments that may resemble asbestos fibers by chemical composition, but they typically do not have the same size distribution, rigidity, lung penetration rate, and biopersistence as asbestos fibers.

In quantitatively and qualitatively estimating the degree of potential adverse effects from exposure to asbestos and other fibrous minerals, health risk assessment allows for evaluation and prioritization of potential control and other related actions. This book is intended to summarize approaches for quantitative risk assessment in applications to asbestos and its analogs, using the most recent mineralogical assessment, epidemiological data, and toxicological concepts. The book also aims to fill in voids in the scientific literature on asbestos and other fibrous mineral risk assessment methodologies and to demonstrate practical applications of risk quantification and characterization.

The science of health risk assessment involves integration of epidemiological, toxicological, industrial hygiene, medical, and environmental evaluation expertise to identify hazards, measure or estimate exposure levels, determine dose–response relationships for various health effects, and characterize risks of specific diseases. Thus, authors of the book chapters have multidisciplinary backgrounds. Our intention was to invite among the most advanced scientists working in the area of asbestos research, aiming to update the major works published in this area and suggest future research needs.

The past decades brought better understanding of the significant differences in various characteristics of fibrous minerals. Erionite fibers from Cappadocia, Turkey, became a confirmed factor in the local mesothelioma epidemic in a magnitude never observed for other types of fibers. Chrysotile fibers were convincingly demonstrated to have biopersistence in human lungs in the range of several months versus amphiboles with half-life in lungs of several decades. It was shown that asbestiform mineral fibers can be distinguished from non-asbestiform varieties, or cleavage fragments, by characterizing size distribution. Many epidemiological studies found deviations in cancer potency not only between mineral types of fibers but for varieties of the same mineral type (for example, between textile and non-textile chrysotile). While all types of asbestiform mineral fibers are potentially carcinogenic, the difference in potency factors for various minerals can reach the magnitude of several thousands. The quantitative risk assessment is a valuable tool to apply the knowledge about characteristics of fibers to specific exposure levels, not to achieve universal "100% safety" goals but to compare realistic predictions of the health outcomes for various scenarios and types of fibrous and nonfibrous agents and help determine priorities for interventions.

It should be noted that differences in cancer potency between various types of mineral fibers are still rarely considered in the regulations. For many of the fibrous minerals (like erionite), there are no established regulatory exposure limits. However, quantitative risk assessment becomes a tool of choice for many regulators internationally. The readers of this book will see how different approaches to asbestos risk assessment can be reconciled if better understanding of the available mineralogical, toxicological, epidemiological, and other relevant information about asbestos would be employed by regulators. We will demonstrate that attempts to equalize the regulatory limits for all types of asbestos would sometimes cause under-protection of workers in various industries. Application of quantitative risk assessment should help environmental and occupational health professionals evaluate health hazards related to various mineral types of fibers realistically, based on the most recent scientific results and approaches.

In our book, we followed the structural paradigm of the risk assessment procedure as described by the National Research Council (National Academies of Sciences) (NRC/NAS).[1,2] Based on the paradigm, Stage 1 of risk assessment is considered "planning", and Stage 2 consists of the following:

- Hazard Identification
- Exposure Assessment
- Dose–Response Assessment
- Risk Characterization

Accordingly, this book is also organized into four sections representing the four steps of the risk assessment process by NRC/NAS. Each section contains chapters describing relevant scientific achievements and developed methodologies for the risk assessment.

An attempt has been made to focus our narrative on important and complex issues such as defining terms of asbestos, asbestiform, cleavage fragments, and elongate mineral particles; major mechanisms of asbestos toxicity along with potential health effects of exposure; pathological indications of asbestos exposure and disease; historical asbestos testing methods and comparability with current techniques; exposure assessment methodology and validation; comparison and utilization of various meta-analytical data for risk assessment purposes; mathematical modeling of risk; and practical adaptation of scientific results for retrospective, current, and perspective risk characterization. Several case studies of asbestos risk assessment will be demonstrated.

This book was not intended to cover every possible aspect of asbestos risk assessment. Further studies are needed to improve our understanding of specific mechanisms for asbestos carcinogenicity. There is a need for further epidemiological studies in the cohorts exposed to commercial and naturally occurring asbestos. The issues of the role of fiber size distribution in carcinogenic potential of asbestos as well as deep exploration of the differences between asbestiform and non-asbestiform minerals in their toxic effects probably deserve a separate book to be developed. There are obvious limitations in the scope of this book: for example, we focused on the carcinogenic asbestos risks, though additional steps can be suggested to develop similar methodology for nonmalignant health outcomes such as pleural plaque and pulmonary asbestosis.

---

1 National Research Council (US) Committee on Improving Risk Analysis Approaches Used by the U.S. EPA. (2009). *Science and Decisions: Advancing Risk Assessment*. Washington (DC): National Academies Press (US). https://doi.org/10.17226/12209.
2 The National Academies of Science, Engineering, Medicine. (2017). *Using 21st Century Science to Improve Risk-Related Evaluations*. Washington (DC): National Academies Press (US); Jan 5. https://doi.org/10.17226/24635

We anticipate that dose–response science for asbestos will continue its development in the future. Further studies are needed to fully clarify the issues of possible nonlinearity for asbestos risks as a function of exposure. At some level of low cumulative exposure, a threshold model for asbestos risk may appear to be quite realistic for noncarcinogenic, but also for carcinogenic outcomes. Scientific exploration of the risk assessment paradigm requires understanding of uncertainties and limitations in the methodology, which may define areas of future research.

The editors are grateful to the contributors of the book chapters, but also to the wide group of scientists who helped our better understanding of the risk assessment methodology. We especially appreciate our discussions with Professor Julian Peto, Dr. D. Wayne Berman, John Hodgson, Dr. Victor Roggli, and many others. It should be noted that our coauthors, Dr. Ann Wylie, Dr. Bruce Case, Dr. Brooke Mossman, Dr. Garry Burdett, Lucy Darnton, Dan Hall, and Arseniy Korchevskiy, not only worked on their own chapters but also helped us organize all the process of work and review various materials included to the book. At the same time, the editors acknowledge that the opinions of the authors for different chapters may not always reflect the position of the editors. Also, the authors are responsible for the accuracy of their references and quotations, as well as for the quality of data used in their studies.

The editors appreciate the work of Debbie Vaughan who processed the entire book, helping with the editing, proofreading, and organization of the chapters.

The editors especially thank Wiley for the support in the preparation of this book and for recognizing the potential scientific need.

The editors also dedicate the book to the memory of Sue Rasmuson, the wife, the mother, and the friend, who was the inspiration for this project when it only started. We also are grateful to our entire families for their support and love.

Andrey Korchevskiy, PhD, DABT, CIH
James Rasmuson, PhD, CIH, DABT, FAIHA
Eric Rasmuson, MS, MHS, DABT, CIH

Part I

Hazard Identification

1

Mineralogical Characteristics and Risk Assessment of Elongate Mineral Particles (EMPs): Asbestos, Fiber, and Fragment

Ann G. Wylie

Department of Geology, University of Maryland, College Park, MD, USA

Introduction

This chapter describes the characteristics of elongate mineral particles (EMPs). Within the regulatory and public health communities, this broad category is understood to include particulates of any mineral that satisfiye the shape criterion of a minimum aspect ratio (length/width) of 3 when viewed microscopically. Mineral particles that fall into this group may form directly in nature as fiber or acicular single crystals or as brick-like particles that form when humans break rock. Because minerals are naturally occurring, EMP does not apply to man-made materials such as carbon nanotubes, mineral wool, fiberglass, silicon carbide, and whiskers. These elongate particles (EPs) are much simpler in chemical and physical properties than minerals generally, and as such, experiments with them often provide information that is useful in understanding the risk arising from the inhalation of EMP aerosols. Conversely, risks from the inhalation of fabricated materials are also understood in the context of demonstrated risks from asbestos.

The physical properties of "asbestos" that gave it utility in commerce also made it a deadly carcinogen. They derive from a specific growth form referred to as the *asbestiform habit*. The term asbestiform means that the flexible fibers visible in hand specimen are actually bundles of loosely bound, very narrow, single crystals called *fibrils*, which are aligned parallel to their common fiber axis direction, tightly packed, but randomly or semi-randomly aligned in the directions perpendicular. Fibers composed of these bundles of tiny fibrils can possess extraordinary tensile strength. The sizes of asbestos fibrils vary among occurrences but fibril widths as small as 0.01 µm have been reported from commercially mined asbestos.

Health Risk Assessment for Asbestos and Other Fibrous Minerals, First Edition.
Edited by Andrey Korchevskiy, James Rasmuson, and Eric Rasmuson.
© 2024 John Wiley & Sons, Inc. Published 2024 by John Wiley & Sons, Inc.

Narrow fiber (width <0.2μm) also characterizes the so-called natural occurrences of asbestos (NOA) where its disturbance is associated with asbestos-related disease. It is abundant in lung burden of asbestos-exposed individuals. (Examples can be found in Wylie (2016) and Wylie et al. (2020).) Fibrils wider than 0.2μm may also be present in NOA, increasing surface area and fiber mass, while decreasing tensile strength. Anthophyllite asbestos from Finland and Russia, which possesses the lowest tensile strength among the commercially mined large asbestos occurrences, has a geometric mean width of 0.34μm and a small component of fiber with width ≤0.15μm (Korchevskiy and Wylie 2022a).

Fibrils wider than 0.2μm are common among museum specimens of asbestiform amphiboles and serpentines as well as other sources throughout the world. A good example of this is the NIST tremolite-asbestos standard, which lacks very narrow fiber altogether (Beard et al. 2007). I have observed many samples labeled asbestos that include single crystals of several micrometers or more in width. These are glassy and brittle; in my writing, I use the term byssolite from Dana and Ford (1949), meaning a stiff fibrous variety of asbestos to describe this type of brittle fiber observed by polarized light microscopy. While the term fiber mineralogically implies macroscopic flexibility, microscopically the distinction between a brittle and flexible, highly elongated naturally formed EMP may not be possible. Normally other properties serve to identify and distinguish glassy byssolite from composite asbestos fibers.

Members of mineral groups other than amphibole and serpentine may be asbestiform in habit of growth, but their habit does not result in fibers of sufficient tensile strength and length to compete with asbestos for utility. Minerals such as sepiolite, palygorskite, brucite, and talc can occur in an asbestiform habit, and the asbestiform variety of erionite, referred to as woolly erionite in the mineralogical literature, has a demonstrated carcinogenic potency in animals (Wagner et al. 1985).

Some rock such as that found in the blueschist belts in California is composed of formerly asbestiform fiber that has been "lithified" in most places so fibrils are no longer able to disaggregate and be readily released from the rock. This process of lithification also forms jade, in which high tensile strength fiber of amphibole asbestos is "cemented" into an extraordinarily tough gemstone. In the case of the blueschist, release of fibrils may occur only where the lithification was incomplete.

Also included among EMPs are mineral particles formed when rock is crushed. Many minerals have planes of weakness inherent in their structure that leads to rock particles with regular shapes such as prisms, plates, and rhombohedra when broken. These shaped particles are called *cleavage fragments*. Amphibole and pyroxene, two common rock-forming minerals, have two planes of weakness in their structures that produce elongated fragments when broken. Other planes of weakness (called partings) that reflect defects in the atomic structure may also control shape. Elongation up to 15 : 1 is not uncommon among the longer fragments of amphibole. EMPs formed by fragmentation can be very common wherever amphibole, pyroxene, or any mineral with prismatic cleavage or parting is reduced to dust by excavation, road construction, mining, etc.

Sometimes it is implied or stated directly that **parallel sides** are also expected to be found on EMPs because mineral fiber has this characteristic, as do prismatically shaped cleavage fragments, but parallel sides are not normally required in exposure methods recommended by National Institute of Occupational Safety and Health (NIOSH).

Another categorization of EMPs relates to the fate of airborne particles once inhaled. In general, **inhalable** means particles whose mean aerodynamic diameter is less than 20 µm, **thoracic** means particles with a mean aerodynamic diameter of <10 µm, and **respirable** are particles with a mean aerodynamic diameter of <4 µm applied to mineral particles (Brown et al. 2013).

Not all minerals have well-defined cleavage or any cleavage at all. Mineral particles without geometrically controlled shapes are simply referred to under the more general term **fragment.** The term **nonasbestiform** is used to describe all mineral fragments, whether they are formed by cleavage, parting, or irregular fracture, as well fibrous material with habit like byssolite that microscopically are demonstrably not composed of sub-light fibrils. It also applies to all mineral occurrences that are not fibrous.

The pathogenic potential of EMPs derives from physical properties, such as their shape and size, as well as chemical factors that affect the potential for biopersistence and host–mineral interactions. In addition, for certain site-specific cancers, such as pleural and peritoneal cancer, factors governing translocation following inhalation may also be important. Physical properties that affect access to disease sites include shape, density, stiffness, size, surface area, and surface charge, which in turn derive from chemical composition, mineral habit, structure, and thermal and mechanical history. Once a mineral fiber is "in place" in the lung following deposition of an inhaled particle, other properties governing particle surface–host interactions such as solubility rates and products and surface chemical reactivity come into play. Governing the magnitude and sensitivity of potency factors are mechanisms for removing and coating particles. Because of the multiplicity of factors that may play a role in EMP pathogenicity, unraveling the chain of cause and effect is complex.

Most of what we know about the carcinogenic potential of EMPs derives from human exposure from mining, utilization of asbestos in commerce, and the resulting wave of mesothelioma and other asbestos-related diseases. There are also human exposures to asbestos encountered during daily living, e.g. whitewash, stone working, etc., and to miners of other minerals such as vermiculite (Libby) and stone (Italy). In addition, exposure to woolly erionite in Turkey mimics that for asbestos in terms of potential to cause mesothelioma. Despite multiple associations, there is yet very little agreement about what in the exposure distinguishes it from other forms of rock and mineral dust as a mesotheliomagenic aerosol. However, one need only examine the mesothelioma rates resulting from exposure to crocidolite (Berry et al. 2012) and chrysotile (Darnton 2023) to see that there can be a difference in potency for mesothelioma among asbestos occurrences.

Experiments with both cell cultures and animals have extended our understanding of EMP potency. They confirm that to produce asbestos-related diseases, durable mineral fiber must be part of the exposure, and while that fiber does not have to be "asbestos" (as defined by statute), it must have dimensional characteristics and durability that are very similar. The question has been, however, how best to describe that similarity.

The NIOSH in 2011 released a Roadmap for Research covering what they define formally as EMPs (NIOSH 2011). This new name responds to the fact that NIOSH occupational monitoring methods for asbestos do not require identification of the particle being counted but instead rely on a particle's length and shape for designation as a "fiber." As such, asbestos monitoring today and in the past is not specific for asbestos. It is an EMP exposure measurement that includes all elongated particles of a certain length that attain a length-to-width ratio of at least 3. During asbestos mining and fabricating, conditions with abundant airborne asbestos, most EMPs would be asbestos, but the assumption that "fiber" in the exposure is real mineral fiber **is not valid** in the general mining environment, and its application severely complicates risk assessment for miners and anyone exposed to mined materials. Fragments of many common minerals that are not asbestos can be 3 : 1 in aspect ratio without any known association with asbestos-related diseases. If these particles are present, additional testing becomes necessary to determine their identity and measure their dimensions.

Perhaps, the most difficult aspect of working with minerals is the fact that each occurrence of a specific mineral is formed during a specific geologic event, affecting specific rocks of variable composition, at a specific time in Earth history. As such, each occurrence differs from every other occurrence of the same mineral in some way. Examples are provided in the discussion that follows. Mineralogists always include the locality with any mineral report, and anyone working with minerals is urged to follow suit. In that way, the chemical and physical properties of **that specific mineral occurrence** can be related to disease outcome, and others can study the same material.

Nomenclature

Communication, scientific discovery, and public health can be compromised when terms used by one group are understood to mean something else by another. Terms used by mineral scientists who understand one meaning and by a health scientist who understand another make public health decisions about inhalable mineral particle risks, the establishment of meaningful permissible exposure limits, and monitoring exposures difficult. For these reasons, this section was included for the reader's reference.

A *mineral name* specifies a naturally occurring substance with a chemical composition represented by a formula of major elements and an ordered and defined internal atomic arrangement. The nomenclature of minerals is designated by the International Mineralogical Association (IMA). Some mineral names incorporate ranges in chemical composition by defining compositional end members in a *solid solution series*, for example $(Mg,Fe)_2SiO_4$, the chemical formula for the mineral olivine. The formula indicates that the mineral olivine may range in composition from no Fe to no Mg or anything in between (most common). Chemical variability of major elements will affect the details of the atomic structure (lattice dimensions and interfacial angles) and the mineral's physical properties. All minerals contain trace elements that are not represented by the end-member formulas.

Minerals are sometimes grouped together because of similar structure or chemical composition. *Amphiboles* are a group of minerals formed from double chains of Si–O tetrahedra on which common cations such as Fe, Mg, Ca, Al, Na, and K attach and within which reside hydroxyl anions. *Phyllosilicates*, or sheet silicates, are all minerals built on a framework of sheets of Si–O tetrahedra. *Polymorphs* are two or more minerals with the same chemical composition, but different atomic arrangement and different names. Antigorite and chrysotile are polymorphs in the serpentine mineral group of phyllosilicates.

The term *mineral fiber* has two meanings, which are often confused.

1) Mineral fibers form in nature as single crystals or bundles of single crystals that are highly elongate in hand specimens and flexible. In the case of asbestos, the very thin individual single crystals that occur in bundles are referred to as fibrils. The term bundle reflects this *fibrillar structure of asbestos*. (See, for example, Franco et al. 1977.) Mineral fibers and fibrils acquire their shape when they form in nature. Their surfaces are growth faces.

2) To assess the concentration of an exposure to asbestos within the asbestos industries, NIOSH and OSHA recommend counting airborne particles defined by length and aspect ratio (length/width) and call them fibers; other federal agencies have followed suit. They also established and adopted analytical protocols reflecting this dimensional approach to defining fiber. Different protocols apply different dimensional designations for "fiber," such as length $>5\,\mu m$ and aspect ratio ≥ 3 (the original Occupational Safety and Health and Mining Safety and Health fiber designations) or length $>0.5\,\mu m$ and aspect ratio $\geq 5\,\mu m$ (the AHERA protocol for clearance following asbestos removal) and aspect ratio ≥ 3 largely parallel asides and $L > 10\,\mu m$ (US EPA method 100.2 for water).

Protocol fibers may have attained their shape by growth, i.e. are mineral fibers, or they may have attained their shape through fracture, e.g. cleavage fragments produced during grinding or blasting rock.

A *mineral's habit* is its external appearance or form in hand specimens as it occurs in nature. Common mineral habits include terms like *platy, globular,* or *prismatic.* Minerals with a *fibrous* habit are, or appear to be, formed of fiber. Occasionally, minerals in particular habits are given a *variety name,* such as *crocidolite,* the variety name for the mineral riebeckite occurring in an *asbestiform habit,* or *asbestos,* the variety name for an asbestiform variety of an amphibole or serpentine mineral (these terms were defined earlier).

In this chapter, I use the terms *fiber* and *fibril* in its mineralogical sense, *cleavage fragment or fragment* to define fragmented EMPs and *EMP* or *protocol fiber* to designate a particle of any origin defined by a 3 : 1 aspect ratio and straight sides. I will also use the term "short" fiber or "short" EMP to refer to particles that are ≤5 µm in length and "long" fiber or "long" EMP to refer to particles with $L > 5$ µm.

The fate of fiber once inhaled may be referred to in a number of ways. The term *retention* normally refers to the retained dose. It is normally expressed as fibers per gram of dry lung. The term *biopersistence* refers to *in vivo* behavior. It may be expressed by the *half-life,* the length of time for 50% of the mineral to be removed without consideration of the method of removal, for example as derived from rat inhalation studies. The term *biodurability* refers specifically to chemical and physical processes *in vivo.* It may also be expressed as the half-life or sometimes the time for complete dissolution and may be estimated from experiments with synthetic extracellular lung and intracellular macrophage fluids. Translocation and splitting of fiber once inhaled are physical processes that affect retention, biopersistence, and biodurability.

Source Specificity: Chemical and Physical Properties

A mineral name specifies a naturally occurring crystalline substance. However, despite the same name, there will be variability in the physical and chemical characteristics among every occurrence; each occurrence is unique in some way(s). For example, variations among localities of the amphibole mineral actinolite include habit, color, chemical composition, luster, tensile strength, indices of refraction, lattice dimensions, and the minerals found with it. These variables normally occur within defined limits that are characteristic of the occurrence.

A mineral's formula often indicates a solid solution between compositional end members, in particular, Fe-rich vs. Mg-rich or Si-rich vs. (Al + Na)-rich end, and nomenclature of end-member compositions are specified by the International Mineralogical Association (IMA). (See, for example, amphibole nomenclature in

Hawthorne et al. (2012).) Substitutions of major elements affect lattice dimensions, X-ray diffraction patterns, electron diffraction patterns, energy-dispersive X-ray analysis, and indices of refraction and other optical properties, normally in predictable ways. Trace metals such as Ni and Co are not normally represented in the formula of silicates, although they may occur in measurable amounts, and less is known about their impacts on physical and surface properties. At each occurrence, however, chemical compositions, both major and trace element, normally vary within rather narrow limits.

X-ray and electron diffraction are used to establish and compare the lattice symmetry and lattice dimensions of mineral particles. While single crystal X-ray diffraction provides the most detailed information, X-ray diffraction patterns gathered from powdered samples are normally sufficient for mineral identification of the major minerals making up the powder; components with abundance <5% may be difficult to identify due to incomplete pattern development and masking by more abundant minerals. Electron diffraction can be very useful in comparing diffraction patterns from zone axes of very small particles to those of known minerals with similar habit and chemical composition to confirm identification. Lattice dimensions vary with composition and are, therefore, also location specific within defined limits.

Habit is also a significant locational variable. The same mineral may form a tough fine-grained rock, brittle prismatic crystals, or flexible asbestiform fibers in different locations. Habit terms, such as massive, prismatic, fibrous, and asbestiform, are generally qualitative because they describe the way minerals physically appear or feel, properties acquired when the mineral formed and evident in hand-sized specimens. Tensile strength and stiffness also vary among occurrences, even among those that are asbestiform.

Other qualitative terms used to describe minerals are also location specific. These include luster, the way the surface reflects/scatters light and color. Asbestiform minerals may be described as having a silky luster, or looking like flax, which are terms describing the effects of light on a material composed of composite fibrils. The same mineral from a different location in a different habit may be described as glassy. Colors vary somewhat with luster, with asbestiform minerals of the same chemical composition being somewhat lighter in appearance than more massive material. The color of a mineral often varies among locations because of trace elements. For example, quartz (SiO_2) can be pink, purple, green yellow, black white, or colorless. However, trace elements rarely affect the color of the streak, the finely powdered mineral, and streak color normally is given as diagnostic since it does not vary with location.

The optical properties of minerals vary with chemical composition and habit and as such are also location specific within well-defined limits. These include the commonly detected interference figure, principal vibration directions, indices of refraction, birefringence, and extinction angle. Verkouteren and Wylie (2002) describe several types of anomalous properties they found for members of the tremolite-ferroactinolite solid solution series developed because of a fibrous habit.

The case of crocidolite provides an example of the importance of place in discussing the toxic potential of a mineral. The two largest and most well-studied occurrences of crocidolite are the Northern Cape and Northwest (Cape Crocidolite Mines) South Africa and the Hamersley Range Australia. These two occurrences are remarkably similar mineralogically. They are of about the same geologic age, occur in the same type of rock, (a banded iron formation), and were formed by the same geologic processes. In another South African district, the Northern Province (Transvaal), both amosite and crocidolite were mined, also from a banded Iron formation of similar age. The conditions that formed the asbestos there, however, were slightly different from those in the Cape. Not only did both amosite and crocidolite form, sometimes together, but ferroactinolite asbestos is also known from this area. The differences in geologic history resulted in an average fibril width wider in the Transvaal asbestos deposits than occurs in the Cape. In Bolivia, there is a fourth occurrence of crocidolite that is mined for roofing materials. Here the geologic conditions of formation were again different. The crocidolite has lower iron content and fibrils are wider than crocidolite from the Transvaal. A good description of the fiber characteristics from four crocidolite occurrences can be found in Shedd (1985).

By 1971, it was known that mesothelioma was much more common among workers from the Cape than the Transvaal, and Timbrell et al. (1971) proposed the narrower width of Cape crocidolite as the explanation. A review of 123 cases of mesothelioma by Rees et al. (1999) confirmed that mesothelioma cases in South Africa are dominated by exposure to Cape crocidolite to such a degree that the paucity of cases from the Transvaal (and Mpumalanga chrysotile) cannot be explained by numbers of exposed miners, dust levels, or degree of environmental contamination. One must look to the site-specific properties of the minerals for an explanation.

While this discussion has emphasized the variability among sources of the same mineral, it is also true that characteristics of minerals from the same source are likely to be similar. For a specific occurrence, the chemical composition, optical, and physical properties of the mineral found there normally will be relatively uniform if the habit is the same. For example, a particular massive amphibole occurrence at Kakanui, New Zealand, is so uniform that it is used as an international chemical standard. The uniformity of chemical and physical properties of minerals from the same location reflects the fact that most rocks are in chemical equilibrium or have been formed by the same processes from the same sources. There are some exceptions, of course, e.g. where strong chemical and physical gradients were present, where veins result from dissolution and reprecipitation of the same mineral, or for some mineral fiber occurrences.

Source Specificity: Dimension

While it is obvious that many mineral properties are location specific, it may be surprising that the dimensions of aerosolized particles derived from disturbance and/or crushing rock are also source specific. Their shapes are influenced by the mineral structure, the processes that formed it, the mineral habit, and the geologic events that have occurred since the minerals were first formed. They are comparatively little affected by the methods of fragmentation, although abundances of characteristic dimensional sets may be.

Figure 1.1 provides the frequency of width of grunerite cleavage fragment EMPs in an aerosol generated by mining in Homestake Mine, South Dakota, where grunerite forms part of the matrix that contains gold (Wylie et al. 2015). The data from Beard et al. (2007) were generated from a single-hand specimen of grunerite from the same locality, crushed in the laboratory. The aerosol data were derived from occupational air monitoring filters in the mine and as such represent a larger sampling of crushed material than that used by Beard et al. The characterizations were done by two different laboratories; Wylie et al. used the scanning electron microscope (SEM), Beard et al. used the transmission electron microscope (TEM), and the analyses were done more than 20 years apart. Both curves have three modes, slightly offset, and more prominently developed in the single crushed sample. The modes develop because of preferential fracture of the grunerite along planes which spacing reflects the thermal and mechanical history of the mineral *in situ*. Some variation in the position of these planes of weakness occurs within a single deposit, hence the broadening and offset of the modes in the airborne particles.

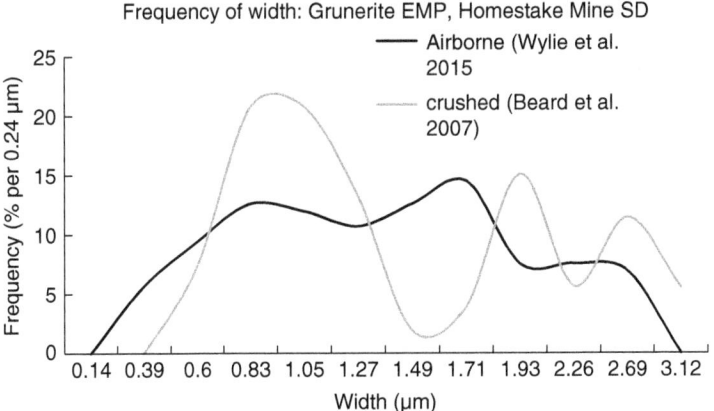

Figure 1.1 Frequency of width of grunerite EMPs from Homestake Gold Mine, SD. Contrast the broad range and wide modes of width of asbestiform winchite and richterite from Libby MT.

12 | Mineralogical Characteristics and Risk Assessment of Elongate Mineral Particles (EMPs)

For asbestiform minerals, the frequency of width behaves in ways similar to fragments in that there may be multiple modes. Figure 1.2a and b shows the frequency distributions of width (a) and length (b) of amphibole associated with the vermiculite deposit at Libby Montana. One curve was generated from data

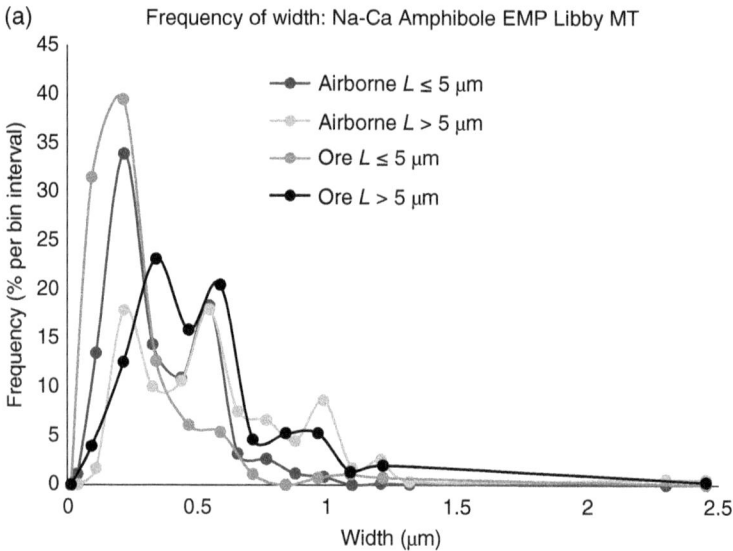

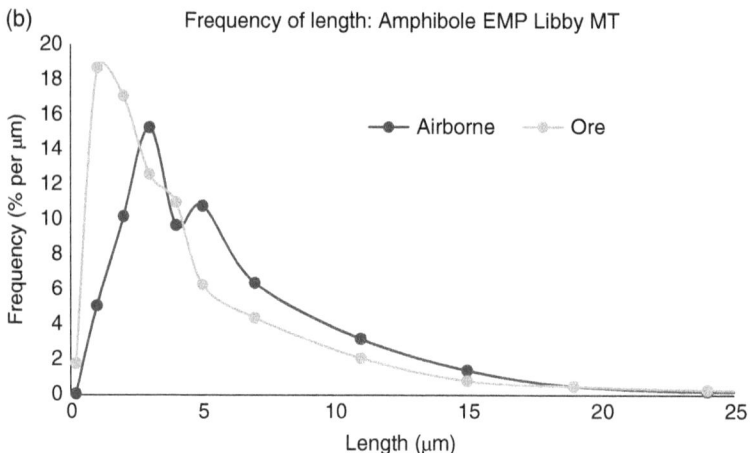

Figure 1.2 Frequency of width (a) and length (b) of airborne and ore extracted Na–Ca amphibole from Libby, MT. Airborne particles were measured and collected from town dust by the US EPA with TEM (US EPA 2006). Amphibole EMP was extracted from ore in the mine and mill by Atkinson et al. (1981) and measured with TEM. *Source:* Atkinson et al. (1981) / Environmental Health Perspectives.

Table 1.1 Proportion of thin EMPs with length >5 μm from four occurrences of asbestos.

Mineral occurrence	Number of samples	% total EMP (SD), width ≤0.15 μm	% total EMP (SD), width ≤0.25 μm
Crocidolite, Cape SA	6	51.6 (3.3)	79.5 (7.8)
Crocidolite, Hamersley, Australia	4	46.7 (3.8)	71.1 (9.4)
Amosite, Transvaal SA	5	8.0 (0.8)	25.3 (1.5)
Na–Ca amphiboles, Libby MT	4	2.1 (0.3)	18.4 (1.1)

Source: Adapted from Wylie et al. (2020) / John Wiley & Sons.

collected from air monitoring filters in the town of Libby, where the source of fiber was primarily the mine waste used for gravel road aggregate (US EPA 2006). The other curve was generated from fibers extracted from the vermiculite ore and mill products (Atkinson et al. 1981). The similarity of these two data sets demonstrates that source characteristics are retained in the dimensions of aerosol-sized particles just as they are in cleavage fragment populations.

Another example of source specificity of mineral dimensions was published by Wylie et al. (2020). Table 1.1 summarizes their data on the proportion of EMPs >5 μm with width ≤0.15 and ≤0.25 μm from four or more characterizations of each of the four occurrences of asbestos. Standard deviations are relatively small (<15%) and the proportions are in remarkable agreement despite differences in protocols and sample types (bulk and air). The similarities are consistent with the fact that dimensional properties of asbestos reflect the conditions of formation, which varied little within these four localities.

Structural Groupings of Common Elongate Minerals

The more common minerals described in reference mineralogical texts as fibrous, acicular, or asbestiform are found in Table 1.2. The most common insoluble elongate minerals are from the silicate and oxide groups. Extensive mineralogical literature describing details of their occurrence, atomic structure, chemical variability, optical properties, and mineral associations is available for most minerals forming EMPs. The interested reader is referred to the series *Rock-Forming Minerals* published by the Geological Society of London, the series *Reviews in Mineralogy* published by the Mineralogical Society of America, *Asbestos and other Fibrous Materials* by Skinner et al. (1988), Mineralogy of asbestos and fibrous erionite (Wylie 2017), and the website https://www.mindat.org.

Table 1.2 Common minerals that occur in a fibrous, acicular, and/or asbestiform habit and may also form EMPs when fragmented.

Mineral types and characteristics

A. Silicates
  I. Inosilicates (Chain silicates)
      i. Pyroxenes and pyroxenoids including enstatite, spodumene, aegirine, wollastonite, and pectolite.
      ii. Amphibole is a group composed of structurally similar minerals with a wide compositional variability expressed as the proportion of more than 100 root end-member compositions of which 77 occur naturally and have a specific mineral name. In addition, there is the possibility of adding 11 prefixes (such as fluoro or potassic) if warranted by the composition (Hawthorne et al. 2012). The following amphiboles have been identified as fibers and/or asbestos: tremolite, actinolite, anthophyllite, ferro-actinolite, fluoro-edenite, magnesio-cummingtonite, cummingtonite, grunerite, winchite, potassic-winchite, richterite, magnesio-riebeckite, riebeckite, edenite, glaucophane, arfvedsonite, and eckermannite. Fiber forming Al-rich amphiboles are very rare.
      iii. Biopyriboles including jimthompsonite, clinojimthompsonite, and chesterite.
      iv. Other less common chain structures including balangeroite and carlosturanite
  II. Phyllosilicates (sheet silicates)
      i. Fibrous clays including palygorskite, sepiolite, saponite, halloysite, and metahalloysite
      ii. *Talc* and minnesotaite
      iii. Serpentine including *chrysotile* and asbestiform antigorite
      iv. Greenalite
      v. Stilpnomelane
      vi. Muscovite and illite
      vii. Vermiculite
      viii. Pyrophyllite
  III. Tectosilicates (framework silicates)
      ix. Zeolites including *erionite*, mordenite, natrolite, offretite, ferrierite and scolecite, mesolite
      x. *Quartz*
  IV. Other
      i. *Aluminosilicates* including kyanite and sillimanite
      ii. Epidote
      iii. Tourmaline
B. Nonsilicates
   Oxides such as *rutile*, cassiterite, and crocoite, pyrolusite; certain phosphates (vivianite, pyromorphite, apatite), some sulfates (gypsum, basanite, halotrichite, ulexite, celestite, alunite, calcanthite, and crocoite) and carbonates like aragonite and its isostructural group (witherite, strontianite, cerussite), hydroxides such as brucite, manganite, and goethite, and even some sulfides such as millerite, galena, argentite, and jamesonite.

Silicate minerals are by far the most important mineral group. Si and O alone comprise almost 75% of the Earth's crust (Rudnick and Gao 2003) and silicates, Si and O in combination with other elements, form about 95% of all crustal rock. Si and O form a tetrahedral unit with O at the four apices and Si in the middle in fourfold coordination. These tetrahedra polymerize in ways that characterize five silicate mineral groups. Among these, EMPs from the *phyllosilicates* (sheets of tetrahedra), *inosilicates* (both single and double silicate chains as well as other chain forms), and *tectosilicates* (frameworks of tetrahedra) have been identified in dusts from occupational and environmental settings, and members of these groups have been studied for toxicity. Silicates from other groups may also form EMPs, notably the alumino-silicates sillimanite and kyanite, and even quartz, but these are less common. Oxides may also form EMPs, for example, rutile, TiO_2. Table 1.2 lists the more common minerals; Skinner et al. (1988) provide a more comprehensive list in their Appendix A.

Establishing the Chemical Composition of Minerals

Standard methods for determining mineral composition have been established by the mineralogical community, and these should be used and relied upon whenever possible. Today, the most commonly recognized approach is to use electron beam techniques (a microprobe) on polished surfaces to generate an energy-dispersive X-ray (EDX) spectrum that is then compared to standards of similar composition. The minimum size for analysis is a volume about 1 μm in diameter and as such can often be limited to an individual mineral grain or fiber. By comparison to recognized, well-characterized standards, analyzed under the same conditions, a quantitative chemical analysis, normally expressed as oxide abundances, is obtained. Water is not normally analyzed so its content may be assumed. A high-quality analysis results in a formula for the mineral occurrence, with the atomic proportions of each element specified, that is electrically neutral, and in which all sites specified by the structure are filled. For example, in amphiboles, which are represented by the general formula $A_{0-1}X_2Y_5Z_8O_{22}(OH)_2$ in which X and Y are in octahedral coordination and Z is a tetrahedral coordination, Si + Al in the tetrahedral sites must sum to at least eight atoms per formula unit (apfu) (excess Al may be assigned to an octahedral site) so that a sum of 7.5 apfu would indicate a poor analysis. Quantitative chemical composition of particles <1 μm can also be obtained from analytical TEM as long as standards are used. Some silicates such as zeolites can be damaged by the electron beam, making chemical analyses by TEM problematic (Harper et al. 2017). Without standards, a simple EDX spectrum may be an indicator of composition, but it is not a quantitative chemical analysis.

Minerals from some locations, especially fibers, may display some compositional variability resulting from changing physio-chemical conditions during formation.

Fibers are precipitated from aqueous fluids, and the chemical environment in which they form is often complex. It is not unusual to see some compositional variability, within limits, among fibers associated with hydrothermal alteration of igneous rocks found at Libby, MT, where compositional variability places fiber in a number of amphibole compositional boxes through small variations primarily in Na and Ca contents. Elsewhere, there may be different forms of fiber from a single location. At Balangero, Italy, asbestiform chrysotile is intergrown with asbestiform balangeroite.

Chemical composition may be a factor in risk analysis, since many characteristics of minerals with the same atomic structure vary with composition. Among mineral groups, there are also systematic compositional variables, such as the proportion of Si, which increases as the degree of polymerization of the Si–O tetrahedron increases from inosilicates through tectosilicates. Formation of reactive oxygen species (ROS) can be promoted by iron in silicate minerals. An application of composition to risk is provided by the work of Korchevskiy et al. (2019), who developed an empirical model for mesothelioma potency of EMPs based on chemical composition and dimensionality.

Mineral Intergrowths and Associations

Mineral samples are unlikely to be "pure" unless they come from hand-picked single crystals, and even then, inclusions of other minerals may be present. Minerals normally occur with other minerals, each forming an individual grain with grain size varying among and between minerals in the same sample and among occurrences. Sometimes, minerals are intergrown on a submicroscopic scale and particles of two or more minerals may exist in a single aerosol particle.

The importance of understanding mineral intergrowths in an exposure comes from the experience at Balangero, Italy, where chrysotile is intergrown with fiber from an uncommon single-chain silicate mineral called balangeroite. This asbestiform mineral makes up only a small percentage of the mineral fiber in the aerosol, but in combination with the long chrysotile fiber, it may be responsible for the unexpectedly high incidence of mesothelioma among minerals and millers of the chrysotile at this location (Turci et al. 2009; Korchevskiy and Wylie 2023). Jimthompsonite and other fibrous biopyriboles may be closely associated with talc or anthophyllite and with each other. They have been reported with anthophyllite from southwest Finland (Schumacher and Czank 1987) and near talc mines in Vermont (Veblen and Burnham 1978a, 1978b). In industrial mineral products, intergrowths of minerals in the ore may persist after mining and beneficiation. For example, magnetite is intergrown with the fibers of amosite, crocidolite, and chrysotile, and it travels with asbestos in building materials. Other examples include amosite and crocidolite occurring together in the same seam in some Transvaal mines, an iron-rich mica intergrown with crocidolite from Australia (Ahn and Buseck 1991), offretite intergrown with erionite, and talc intergrown with asbestiform anthophyllite (Watson 1999).

Sub-light mineral intergrowths may affect the physical and chemical properties of mineral particles. For example, intergrowths of talc and anthophyllite, two hydrated Mg silicates with similar chemical composition but different atomic structures, may result in fibers with densities and indices of refraction intermediate between the two (Greenwood 1998). Sheet silicates or biopyriboles may form between fibrils of asbestos, and intergrowths of triple, quadruple chains, and single chains have been found in some fibers. Because mineral intergrowths affect properties attributed to the pure mineral, they may also affect toxicity, depending on the sensitivity of the biological response.

EMPs in certain metal mines may be closely associated with other minerals that carry toxic metals. In particular, common sulfides of As, Cd, Cu, Pb Ni, etc. are widespread and may be carcinogenic. A review is provided by Entwistle et al. (2019).

Bioreactivity of Mineral Surfaces: Chemical Factors

Once a particle has been deposited in the lung, chemical processes and reactions take place at the host–particle interface that may affect toxicity. As reviewed by Hochella (1993), surface properties include surface composition, surface atomic structure, surface micro topography, and surface charge. These are altered by dissolution as the surface structure is changed and cations are released, by oxidation and the formation of ROS, and by deposition of coatings. These processes and reactions do not necessarily progress uniformly on all surfaces of the mineral. Their rate influences biodurability which in turn influences biopersistence, a main contributor to EMP toxicity. A recent publication, *Mineral fibres: Crystal chemistry, chemical physical properties, biological interaction and toxicity* (Gualtieri 2017) provides many examples of studies on the chemical reactivity of mineral surfaces *in vivo*.

The Specificity of Mineral Surfaces: The Example of Quartz

IARC (2012) points out in its review of crystalline silica as a lung carcinogen that the data support more and less pathogenic occurrences. However, IARC concluded that studies have failed to identify specific pathogenic risk factors to explain observed variance in disease outcomes among exposures. It is thought by many that it is the ability of the quartz surface to generate free radicals and ROS that leads to cell death, oxidative stress, inflammation, progressive lung fibrosis (silicosis), and, ultimately, cancer of the lung. This ability varies among quartz surfaces. For example, fracture surfaces of quartz have been shown to be more biologically active than crystal faces (Turci et al. 2016), and epidemiological studies have demonstrated that the highest toxicity is associated with exposure to cutting, grinding, or abrading quartz-bearing rock. In some environments, mantling of quartz particles by other minerals may be important. For example, work by Stone et al. (2004) suggests that coal dust and clay may mask the reactivity of aerosolized quartz. Quartz has simple chemistry, a highly symmetrical lattice, and fracture that does not normally favor a particular crystallographic orientation. There is no indication that the shape of quartz particles plays a role in their toxicity although both size and surface specificity have been demonstrated to do so (Inoue et al. 2021). Most elongate silicate particles are more complex in both chemistry and structure than quartz, and variability in mineral–host interactions would be expected to be somewhat surface specific.

General Considerations of Solubility

Biopersistence is considered to be a major factor in EMP toxicity, and **biodurability**, the resistance to dissolution and fragmentation, is a component of **biopersistence**. The chemical environment surrounding mineral fiber once deposited in lungs is that of extracellular lung fluid (pH 7.4), intracellular (pH 4.5) fluids, and gastric fluids (pH 1.2). The chemistry of extracellular and intracellular fluids can be approximated to study mineral solubility *in vitro*, but lung fluids are complex in time and composition, and studies on mineral–host interactions *in vivo* are just beginning. While the relative rates can be reasonably relied upon, the absolute rates *in vivo* may be different. For example, dissolution rates of common minerals during weathering have been shown to be lower than predicted by laboratory studies. Several different approaches have been used to measure dissolution rates and biodurability, including *in vitro* and *in vivo* studies. These are reviewed by Utembe et al. (2015) and Turci et al. (2017). Dissolution rates may be expressed on a surface area or mass basis. Innes et al. (2021) provide a comprehensive review of the use of simulated lung fluids in determining solubility *in vivo*.

The range in dissolution rates among silicate and oxide groups is related to atomic structure, chemical composition, and the composition of the fluid. Studies on the dissolution rates of common silicates (Table 1.2) show that they vary over four orders of magnitude, and, in general, are least soluble in the pH range of 6–8 (Brantley 2003). In general, the solubility of silicate minerals varies inversely as the degree of polymerization of the Si–O tetrahedra, with the highest solubility rates among those with nonpolymerized tetrahedra like olivine and lowest among tectosilicates like quartz (maximum polymerization). Dissolution rates (expressed as K_{dis}, the dissolution coefficient) based on surface area may be expressed as ng/cm^2/h. According to Eastes and Hadley (1996), no fiber with K_{dis} higher than 100 g/cm^2/h produced fibrosis or tumors in animal inhalation studies.

Experiments on mineral fiber solubility in synthetic lung fluids have been carried out by a number of researchers. The results are expressed as the length of time required for half of the particles of a specified size to be dissolved or the time required for full dissolution. Table 1.3 provides some examples of experimentally

Table 1.3 Dissolution rates of some common asbestos and other minerals.

Mineral particle	Dimension (μm)	Fluid	Half-life	Sources
Chrysotile	0.25 fiber	Extracellular	94–177 d	1
Amphibole	0.25 fiber	Extracellular	49–245 yr	1
Erionite	0.25 fiber	Extracellular	181 yr	1
Brazilian Chrysotile	$L > 20$	Rat lung	1.3 d	2
	$5 < L < 20$	Rat lung	2.4 d	2
	$L < 5$	Rat lung	23 d[a]	2
Coalinga Chrysotile	$L > 20$	Rat lung	0.31 d	3
	$5 < L < 20$	Rat lung	7 d	3
	$L < 5$	Rat lung	64 d[a]	3
Fibrogenic Tremolite	$L > 20$	Rat lung	∞	3
	$5 < L < 20$	Rat lung	∞	3
	$L < 5$	Rat lung	152 d	3
Crocidolite	All lengths	Rat lung	>6 mo	4
Amosite	$L > 20$	Rat lung	466 d	5
Wollastonite	All lengths	Rat lung	21 d	4
Rock wool	All lengths	Rat lung	32 d[a]	6
	$L > 20$	Rat lung	10 d	6
			Complete dissolution	
Chrysotile	1 × 10	Gastric	33 h	7
	1 × 10	Extracellular	19 mo	7
Chrysotile	$d = 1$ sphere	Extracellular	7 mo	8
Tremolite asbestos	1 × 10	Gastric	9 mo	7
	1 × 10	Extracellular	4 yr	7
Crocidolite Cape	15 × 0.25	Extracellular	6 yr	8
Talc	$d = 1$ sphere	pH 2–8	8 yr	9
Olivine	$d = 1$ sphere	Extracellular	4.8 yr	9
Quartz	$d = 1$ sphere	Extracellular	5,000 yr	9

[a] Increase explained by disaggregation.
Source: Gualtieri et al. (2018).

derived solubility half-lives. Despite these experiments, however, many researchers have reported no signs of dissolution of amosite and crocidolite fibers in lung tissue, despite their residence there for 20 years or more.

Even when atomic structures are similar, e.g. within the monoclinic amphibole group, chemical composition plays a role in solubility. Cations in octahedral or higher coordination are more likely to be leached than Si and Al that occupy tetrahedral sites, with larger ions such as Ca leached more readily than smaller cations such as Mg. Based on the work of Brantley (2003) on solubility of silicates under ambient conditions on Earth surface, it is likely that the half-life in the lung of the same-sized particle of wollastonite (Ca-rich) is less than the half-life of enstatite (Mg-rich), and the half-life of sepiolite (Mg-rich) is less than the half-life of palygorskite (Al-rich), reflecting compositional influences. However, experimental data for solubility from synthetic lung fluid experiments are lacking for most minerals and there are no standard protocols for conducting the experiments.

While pH is a primary variable affecting estimates of solubility rates, other fluid characteristics may play a role. For example, experiments by Werner et al. (1995) on crocidolite showed that while crocidolite would be almost insoluble in neutral pH fluids, Fe chelators present in lung fluid alter the dissolution mechanisms and increase the rate of Fe^{+2} release from the surface such that the half-life of a crocidolite 1 µm particle would be on the order of 10 years.

The dissolution rate is not uniform across particle surfaces because solubility can be affected by atomic configuration on the surface or structural defects. Cleaved surfaces have been shown to be more readily attacked by acid than crystal faces, likely due to the higher density of defects on surfaces formed by fragmentation. Crawford (1980) demonstrated the presence of a partially crystalline leached layer preferentially found on the {100} surface of crocidolite after it was immersed in blood for two weeks; this surface is parallel to pervasive defects in the crocidolite structure.

Despite these differences among surfaces, and other factors affecting solubility, overall solubility rate is a factor in biopersistence and therefore one factor among many in toxicity. For this reason, fibers of low solubility have repeatedly been shown to be noncarcinogenic, demonstrating the importance of this variable.

Formation of Reactive Oxygen Species (ROS)

Beginning with early work with cell cultures in 1985 (Goodglick and Kane 1986; Mossman et al. 1986), the role in fiber toxicity of ROS generated by oxidation of metals in silicates has been proposed as an important process affecting toxicity. A review of the role of ROS in the pathogenesis of asbestos-induced disease is provided by Shukla et al. (2003) and Cox (2023) and in silicosis by Fubini and Hubbard (2003). Of particular focus is the role of iron (reviewed by Fubini and Mollo 1995), given its ubiquity in rock. The IARC model for cancer associated with fibril exposure includes the generation of free radicals from interactions involving Fe^{2+}. However, the amount of iron in the structure does not seem to be a major factor as even small amounts of iron have been shown to activate these processes. Fe^{3+} is less active than Fe^{2+}, and the degree of availability of Fe^{2+} would depend on bond strengths with coordinating oxygen atoms. For example, Fe^{2+} in the oxide magnetite does not appear to be available to form ROS (IARC 2012). The strength of the Fe-oxygen bonds in magnetite is high, rendering it effectively insoluble under most Earth surface conditions. Among iron-bearing silicates with their complex crystal structures, Fe^{2+} and other transition metals will normally reside in octahedrally coordinated sites where they are susceptible to oxidation and would be expected to participate in the generation of ROS if on the surface. However, recent publications suggest that the role of Fe^{3+} in fiber carcinogenicity might be more pronounced than considered previously (Ghio et al. 1992; Toyokuni 1996; Korchevskiy et al. 2019).

A quantitative mineral parameter that can be used to predict the relative activity of ROS at the mineral surface, both initially and over time, is lacking. In developing such parameters(s), relationship to chemical composition, atomic structural and mineral intergrowths should be considered, as well as the surfaces itself (e.g. {100} vs. {010} vs. {110}). Over time, surfaces will be altered and so will the fluids in contact with them, altering the chemical processes and reactions that take place at the mineral–host interface, complicating quantitative modeling for risk assessment.

Coatings

Surface modifications begin once a mineral fiber is inhaled and these modifications may involve coatings. The formation of "asbestos bodies," ferruginous material composed mostly of the iron-containing protein ferritin coating long fibers of asbestos in lung tissue is the best example. Ferruginous bodies are not found on all lung fibers uniformly. Morgan and Holmes (1985) reported that they are less common on fibers <0.2 μm in width or less than 10 μm in length, and for this reason they are absent on most fibrils of chrysotile and crocidolite and on erionite from Turkey[1]; however, they are almost always found on fibers greater than 40 μm, suggesting a short residence time for uncoated short, narrow fibers. (The absence of coatings on fibers less than 0.2 μm in width may relate to the fact that such fibers can more readily translocate from the lung, and fibers less than 10 μm are small

---

1 Later work by Kliment et al. (2009) reported ferruginous bodies around 8 or 10 analyzed erionite fibers from exposure in the United States; no widths provided.

enough to be removed by macrophages which have difficulty removing fibers about >15 µm in length.) These observations and the increase in proportion of coated fibers as length increases above 10 µm suggest that the formation of asbestos bodies may signify a long residence time in the lung. Work by Gandolfi et al. (2016) on the formation of asbestos bodies on erionite in rat lungs showed that at any width, none were found on erionite. In addition, asbestos bodies that do form are found on only a small proportion of lung burden fiber in both humans and rats. These observations suggest that more than size governs their formation. When present, they isolate the fiber and limit host–fiber interactions.

Experiments have demonstrated reduced toxicity from chrysotile after coating with organosilane (Feuerbacher et al. 1980) and reduced membranolytic activity after reaction with phosphorus oxychloride, creating a fiber referred to as chrysophosphate (Langer and Nolan 2018). Dissolution creates an amorphous surface layer on minerals, which reduces the solubility from the high rates typical of fresh surfaces and has been shown to reduce cytotoxicity (Chamberlain and Brown 1978). Other coatings might enhance metal availability and promote redox reactions such as the adhesion of nitric oxide. Some EMPs may attract organic molecules such as polycyclic aromatic hydrocarbons, which themselves are toxic.

Surface Charge

A mineral surface in an aqueous environment will normally carry a charge. Ions in the liquid will be attracted to the surface to balance the charge, producing an electrical double layer. The ions on the solid surface may be tightly bound or may diffuse on the surface and even enter the liquid. Surface charge may be positive or negative, and it may be different on different crystallographic surfaces. Surface charge may change with time after deposition in the body or following phagocytosis, and net surface charge will be affected by shape.

It has long been known that surface charge affects the toxicity of nanoparticles by influencing cellular interactions such as particle uptake by macrophages and fibrogenic potential. For example, cell death has been attributed to the high positive surface charge of quartz, and a high negative charge developed during phagocytosis of amphibole may be toxic to the macrophage (Fröhlich 2012). The positive surface charge of quartz and chrysotile enhances hemolysis of red blood cells (Light and Wei 1977) and is important in binding certain environmental carcinogens such as polycyclic aromatic hydrocarbons to the surface of chrysotile.

Most silicate EMPs carry a net negative charge along surfaces parallel to length and a positive charge at particle terminations. As such, the particles are slightly polar and attract water molecules to them, rendering them hydrophilic. Dissolution of cations enhances the net negative charge. Because of the distribution, the net negative charge will increase with aspect ratio. However, it has been demonstrated that amphibole single crystals and amphibole cleavage fragments carry the same charge if their aspect ratios are the same, reflecting the fact that cleavage surfaces and crystal faces are structurally parallel and provide the same chemical surface (Schiller et al. 1980). Sheet silicates may carry a negative charge parallel to the flat cleavage surfaces {001} but positive on the edges.

Among silicates, a negative charge is generally associated with surfaces parallel to the Si–O chains and/or sheets, while the positive charge is associated with surfaces that are perpendicular or fracture surfaces that break the tetrahedral framework. Quartz fragments carry a positive charge because all surfaces disrupt the tetrahedral framework.

Chrysotile carries a positive charge because of its rare tubular structure in which the brucite-like hydroxide ($Mg(OH)_2$) layer forms the outer surface. *In vivo*, the hydroxide layer is readily leached, decreasing the magnitude of the positive charge of chrysotile with time. It may be noteworthy that in similar experiments, the negative charge on crocidolite increased, as positively charged ions were leached increasing the hemolytic activity of crocidolite (Light and Wei 1977).

Surface charge of particles in fluids is frequently represented by the zeta potential, a measure of electrophoretic mobility. Most of the work on differences in surface change among asbestos and other EMPs has used zeta potential measurement. The zeta potential will be affected by measurement conditions including temperature, time, and the chemical composition of the fluids in which particles are suspended; many studies have examined these variables. Another type of surface charge can be defined by surface conductivity and another by static surface charge, which may affect agglomeration of particles (Hartkamp et al. 2018).

EMP Surfaces: Chain Silicates and Zeolites

Amphibole, pyroxene, pyroxenoid, biopyribole, and other chain silicates form crystals that have well-developed plane-bounding surfaces parallel to the c crystallographic axis. The most common are designated {100}, {010}, and {110}. These surfaces are parallel to the Si–O tetrahedral chains and expose octahedrally coordinated cations. In these mineral groups, cleavage and parting surfaces developed by breaking, grinding, or abrading rock have the same crystallography orientation as these common crystal faces. Cleavage in these cases does not break the covalently bonded chains parallel to the faces. These are broken only by fracture perpendicular to the axis of elongation. For these mineral groups, surface chemical reactions include leaching of the octahedrally coordinated cations, most commonly, K, Na, Ca, Mg, and Fe.

The surfaces of asbestos fibrils are not necessarily parallel to crystal faces. Fibrils may be irregular in shape and follow the normal low-index planes for only short distances. Asbestos surfaces are also frequently associated with sheet silicates as Cressey et al. (1982) demonstrated for amosite and anthophyllite asbestos and Ahn and Buseck (1991) demonstrated for crocidolite. For these fibers, it may not be amphibole at all forming the surface. Amphibole asbestos fibrils are frequently twinned on {100} and wider amphibole fibrils when viewed microscopically are frequently oriented with {100} approximately perpendicular to the electron or light beam, especially common for amosite.

Zeolites are tectosilicates, built from a framework of interconnected Si and Al tetrahedral; some may be fibrous. Zeolites are widespread in volcanic terranes in Turkey, Greece, Italy, and the western United States. Erionite is known to occur in an asbestiform habit where it is known as *woolly erionite* (Shedd et al. 1982). The structure of zeolites is open and contains channel ways that range in size from less than 10 to more than 1,000 Å in natural and synthetic zeolites; water and other molecules can pass through the mineral structure along these channels and porosities range from about 20% to 40% (Zoltai and Stout 1984, pp. 315–320). This structure loosely binds the cations that fill the interstices between the tetrahedra so that they are readily exchanged with cations in the surrounding fluids and may be leached. Although zeolites do not contain iron in their formulas, iron in fluids may enter the structure by ion exchange. (Ion exchange reactions involving low-temperature hydrous fluids are absent in the inosilicate group of minerals.) Iron may be present on surfaces of zeolites in thin layers of clay.

Physical Factors

Specific Surface Area

Specific surface area is measured as area per unit volume or unit mass. It has been correlated with lung tumor and other lung abnormalities in rats induced by chemically inert TiO_2 (Oberdörster 1996) and with the degree of fibrosis in humans from three types of asbestos (Timbrell et al. 1988). Korchevskiy and Wylie (2022b) report a high correlation between mesothelioma potency (R_M) and specific surface area. Oberdörster (1994) demonstrated that ultrafine TiO_2 particles (20 nm) with high specific surface area were preferentially retained, especially in the alveolar space, over larger TiO_2 particles (250 nm) and cause more inflammation. Wylie et al. (2022a) showed that a high specific surface area distinguishes the asbestiform habit. When asbestos exposures are measured in terms of particle number or mass, because of differing fiber sizes, surface areas can vary significantly for the same exposure as Timbrell et al. (1988) demonstrated. For populations with a range of size, the specific surface area will be concentrated in the small particles, just as the mass is concentrated in the large.

The specific surface area of a material can be measured directly with BET methods which rely on measurement by pressure change of the amount of gas adsorbed onto the surfaces of all the particles in a powdered sample of known weight. This measurement includes surface roughness and may even include the interior surface of chrysotile tubes. For this reason, measurements by this method are normally higher than surface area calculated from measurements of length and width. In the case of asbestiform minerals, surface area increases with "fiberization," meaning disaggregation into component fibrils. For example, surface area of raw amphibole asbestos as shipped from the South African mines may be <20,000 cm^2/g, increasing to 90,000 cm^2/g when used in manufacturing (Hodgson 1979). If disaggregation of fiber bundles takes place after sample characterization or following inhalation, surface area of the particles in the aerosol sample could be much less than in the eventual dose.

Surface area of individual particles can be calculated from length, width, thickness, and an approximation of a geometric form; the specific surface area is estimated from the calculated volume or mass of the form. For mineral aerosols, the instrument that enables the most precise measurements of very small particles for these calculations is the TEM. Surface area calculated in this way has many uncertainties because particles are rarely precise geometric forms with smooth plane bounding surfaces, and the thickness of the particles must be assumed. Measurement of the thickness of Cape crocidolite and Transvaal amosite by SEM and TEM showed that the thickness-to-width ratio (t:w) approaches 1 : 1 for the single fibrils, decreasing to 0.3 : 1 for fibers of 1 μm wide as described by Wylie et al. (1982), who derived the relationship between width and thickness for both crocidolite and amosite as shown in Eq. (1.1):

$$\text{Log thickness} = 0.692 \log \text{width} - 0.493 \tag{1.1}$$

Fibrils of crocidolite with approximately equal dimensions in an irregular cross section were shown by Alario et al. (1977).

Cross sections of cleavage fragments have not been measured directly in the same way as fibers of asbestos; therefore, some assumptions must be made to establish an estimate of thickness. It is commonly assumed that width/thickness = 2 based on general observations on fragments whose thickness can be observed by rolling grains under the microscope. This relationship would be expected from the geometry of an amphibole particle which, if shaped ideally by cleavage planes, would have a width that is larger than the thickness in a ratio of 1.73. It is likely that the true relationship is somewhere between 1.7 and 2. In some of our works, we used the estimation of width/thickness = 1.9 as a midpoint between the ratio of 1.7 and 2 (Korchevskiy and Wylie 2022a).

Enthalpy and Other Thermodynamic Properties

The change in heat content (ΔH) that accompanies the formation of one mole of a mineral from its elements with all components in their standard states is defined as enthalpy; it is also called the "standard heat of formation." In general, the larger the absolute value of ΔH, the more stable will be the compound at standard conditions. Puzyn et al. (2011) describe the results of a study correlating the change in enthalpy of formation expressed in kcal/mol, on a set of metal oxides with cytotoxicity to bacteria. They found that the larger the enthalpy of formation, the lower the cytotoxicity. Simple oxides of Zn, Cu, Ni, and Co had the highest toxicity with lower toxicity among dioxides like TiO_2 and SiO_2. The enthalpy of formation, calculated under the same conditions, is an indication of the relative stability of minerals, which in turn may govern the availability of metals on the surface to participate in chemical reactions.

The utility of application of this measure to the toxicity of silicates has not been explored, although enthalpy data are available for many minerals. The range of enthalpies reported by Puzyn et al. (2011) for simple oxides and dioxides is 596.70 (NiO) to 1,686.38 (SiO_2) kcal/mol (2,496.59 to 7,055.81 kJ/mol). For reference, values reported for chrysotile are about 4,360 kJ/mol (Ogorodova et al. 2006). Values for other minerals of interest include; Transvaal amosite 176 and Cape crocidolite 431 kJ/mol (Bennington et al. 1978). Enthalpy should be independent of mineral habit.

Giese and van Oss (1993) review surface thermodynamic properties of silicates that may affect their interactions with biological materials. They point out that surface thermodynamic properties may help to understand both physical and chemical interactions that take place at the mineral–host interface. They discuss surface tension, interfacial tension and surface free energy, hydrophobic and hydrophilic surfaces, and zeta potential and its relation to surface charge. As they point out, relatively few studies have investigated host–mineral interactions with respect to surface thermodynamic properties. One example is the application of atomic force microscopy to examine the interfacial angles between adhered bacteria and mineral surface by Lower et al. (2000).

Density and Aerodynamic Diameter

Common silicates and oxides range in density from about 2–5 g/cm³. Mineralogists describe density by the term specific gravity (SG), which is a dimensionless parameter defined as the density of the mineral divided by the density of the same volume of water. SG is equivalent to density measured in units of g/cm³. Aerodynamic diameter depends on density, width, length, and shape (Asgharian et al. 2018). However, because of the dominance of width on settling velocity of EMPs, the influence of density on aerodynamic diameter can be illustrated by comparing spherical mineral particles of the same measured width or calculated aerodynamic diameter separately by the formula in Eq. (1.2):

$$\text{Aerodynamic Diameter} \approx (\text{Measured width})(\sqrt{\text{Specific gravity}}) \quad (1.2)$$

For EMPs, the aerodynamic diameter and measured width is not such a simple relationship. There are a number of formulae given in the literature. Timbrell (1965) provided an empirical formula from his work with asbestos of the form:

$$AD = 66W\left(AR/(2+4AR)\right)^{2.2}(SG)^{0.5}$$

where AD is the aerodynamic diameter, AR is the aspect ratio, W is the measured width, and SG is the specific gravity.

Table 1.4 gives the aerodynamic diameter of particles of a variety of different silicate EMPs all with a measured width of 0.2 μm and an aspect ratio of 20 but with different densities. Aerodynamic diameter is calculated from the formula of Timbrell (1965).

Table 1.5 provides examples of how length might affect the AD based on formulae for aerodynamic diameter developed by Timbrell (1965) and by Stöbler (1971), as shown in Eq. (1.3):

$$AD = 1.3(\text{density})^{1/2}(\text{width})^{5/6}(\text{length})^{1/6} \quad (1.3)$$

Dusts are categorized as suspended dusts (with aerodynamic diameter 0.001–1 μm), settled dusts (1–100 μm), and heavy dusts (100–1,000 μm). Because

Table 1.4 Aerodynamic diameter, density, and width of fibers with an aspect ratio of 20 : 1.

Mineral (SG)	Measured Width (μm)	Aerodynamic Diameter (μm)	Aerodynamic Diameter (μm)	Measured Width (μm)
Erionite (2.1)	0.20	0.85	1.0	0.31
Chrysotile (2.53)	0.20	0.93	1.0	0.28
Quartz (2.65)	0.20	0.96	1.0	0.27
Tremolite (3.0)	0.20	1.02	1.0	0.26
Ferroactinolite (3.1)	0.20	1.030	1.0	0.25
Riebeckite (3.4)	0.20	1.08	1.0	0.24
Grunerite (3.5)	0.20	1.10	1.0	0.24
Rutile (4.23)	0.20	1.21	1.0	0.21
Magnetite (5.17)	0.20	1.34	1.0	0.19

Table 1.5 Effect of length on aerodynamic diameter Stobler and Timbrell equations.

			Aerodynamic Diameter (μm)	
Diameter (μm)	Density (g/cm³)	Length (μm)	Stobler	Timbrell
0.2	3.0	1	0.59	0.88
0.2	3.0	10	0.86	1.04
0.2	3.0	100	1.27	1.07

most asbestos fibrils are so narrow, they are moved readily as suspended dusts that deeply penetrate the lung. However, as inhalation and translocation pathways approach the width of fibers, aerodynamics may have little bearing on the fate of asbestos fibers.

Stiffness and Tensile Strength

Stiffness is defined as resistance to bending or deflection. Among commercially produced asbestos, amosite and crocidolite are stiff in comparison to chrysotile. Some asbestos occurrences are described as "harsh" or "soft," reflecting stiffness of the fiber, which may be due to inherent stiffness, fibril size, and/or the presence of inter-fibril adhesion.

It has been proposed that stiffness is a factor in limiting the ability of macrophages to remove fiber from the lung and for this reason they are more likely to remove chrysotile than amphibole fiber (Donaldson et al. 2010). Stiffness may also be a factor in the translocation of fiber once inhaled with stiff fibers migrating more readily because of their ability to penetrate vessel walls and tissue. Zhu et al. (2016) showed that stiffness along with length were factors in vesicle damage from carbon nanotubes. Stiffness can be measured by the critical buckling load which is approximately equal to

$$CBL \approx EW^4/L^2$$

where E is Young's modulus (GPa), W is width, and L is length in nanometers (Kane et al. 2018). Young's modulus has been shown to increase as width decreases, and measurements vary among samples of the same mineral due to differences in type and abundance of structural defects.

Tensile strength and stiffness are mechanical properties, not chemical properties. For that reason, while chrysotile is as strong as amphibole asbestos as measured by tensile strength or Young's modulus, it is not chemically stable under the conditions within the lung or within macrophages. In addition, because of the very narrow width of the fibrils that make up chrysotile, chrysotile fibrils longer than a few micrometers are not stiff. When Mg is leached from the chrysotile fibril, a Si-rich structure with unknown mechanical properties remains.

Tensile strength is the ability to withstand pulling or stretching forces. For asbestos, it describes resistance to breaking perpendicular to length. It arises from the nature of the atomic structure, the presence of defects, and the cross-sectional area of the fiber. In general, the fewer the surface defects and the smaller the cross section, the greater the tensile strength will be. For example, O'Hanley (1986) demonstrated in the case of tremolite that the asbestiform fiber bundles had higher tensile strength than cleavage fragments of the same width, reflecting the higher tensile strength of narrow fibrils. The surfaces of fibers formed in nature have fewer defects than cleavage fragment surfaces because during crushing, breakage will occur preferentially along planes with a higher density of defects (Walker and Zoltai 1979). Ahn and Buseck (1991) propose that the enhanced tensile strength of crocidolite may arise from planar defects parallel to {110} and {100} planes that extend along the fiber axis by mitigating the propagation of cracks and by providing sites for interplanar slip.

Tensile strength for asbestos can be 100 times greater than the tensile strength of nonasbestiform varieties of the same mineral. However, as with so many other properties, those that affect tensile strength, such as surface defects, the cross-sectional area of the fibril and the abundance of planar defects, are location specific. Hodgson (1979) provides tensile strength measurements for crocidolite as 35×10^3 kg/cm^2, for amosite as 24×10^3 kg/cm^2, chrysotile as 31×10^3 kg/cm^2, and tremolite and actinolite as $<5 \times 10^3$ kg/cm^2. Table 1.6 provides averages from Aveston (1969). Although sometimes generalized to all occurrences, these

Table 1.6 Asbestos tensile strength (Aveston 1969) and fiber size.

Asbestos type and source	Tensile strength		EMPB
	kg/cm$^2 \times 10^4$	GPa	(%)
Chrysotile Thetford	3.86 (0.20)	375	69.2
Crocidolite Cape	2.26 (0.14)	219	64.7
Ferroactinolite Cape SA	1.94 (0.31)	188	29.7[a]
Amosite Transvaal	1.55 (0.13)	150	20.4
Anthophyllite Finland	1.22 (0.46)	118	26.6
Tremolite Pakistan	0.42 (0.04)	41	15.8[b]

[a] Proportion ≤0.27.
[b] Sample from India of similar texture.
Source: Adapted from Wylie et al. (2020) / John Wiley & Sons.

measurements are specific for Cape crocidolite, Transvaal amosite, and SA chrysotile; the source of tremolite, actinolite, and anthophyllite asbestos were not specified, but based on discussion in the paper, were likely were likely from Italy/Pakistan/Korea, Transvaal South Africa, and Finland, respectively. Generalization of low tensile strength to all occurrences of asbestiform anthophyllite, tremolite, and actinolite may not be justified, however, because low tensile strength in most cases can be explained by a small proportion of narrow fiber, as illustrated in Table 1.6 by EMPB ($W < 0.25$ µm).

The Effects of Heat

Many experiments have shown that heating minerals to temperatures below their melting point affects their behavior *in vitro*. Thermal modifications of chrysotile have been shown to reduce cytotoxicity (Valentine et al. 1983) although heating to temperatures above the stability of chrysotile resulted in increased cytotoxicity (Hayashi 1974). Heating amphibole to the point of water loss reduces the free radical activity on its surface (Areán et al. 2000). When asbestos is in contact with high temperature, e.g. as an insulator around a boiler, its color and indices of refraction may be altered due to oxidation and loss of adsorbed water. Its tensile strength can also be significantly altered (Hodgson, 1979).

Dimensionality: General Considerations

The importance of dimensionality in explaining disease outcomes in animals has been established by the work of Wagner, Pott, Stanton, and Timbrell, among others, who have documented the carcinogenicity to animals of narrow, long durable fiber whether delivered by inoculation, implantation, or inhalation. Specific dimensions of high potency fibers have been suggested by various authors, and they generally include widths <0.25 µm and lengths >5 µm. Very long lengths were identified by the meta-analysis of Berman and Crump (2008b) as high potency. Similarly, extensive work has been undertaken to understand cell–mineral interactions *in vitro*, recently reviewed by Cox (2023). The carcinogenicity of EMPs was recently reviewed by Wylie and Korchevskiy (2023), who, using an epidemiological approach based on the work of Hodgson and Darnton (2000) and Darnton (2023), concluded that dimensionality is the main driver of durable mineral carcinogenicity in humans and in particular, narrow widths ($W < 0.2$) of long lengths ($L > 5$ µm) are the primary potency predictors for mineral fiber.

A number of different dimensional characteristics of EMP populations have been used as parameters for assessing carcinogenic risk. These include frequencies of width, length, aspect ratio, and specific surface area as a whole or of a subset defined by limiting one or more of these parameters. The parameters have been examined separately and in combination, and on both a geometric and an arithmetic basis (reviewed in Korchevskiy and Wylie 2021). Mean lengths and mean widths may obscure modal frequencies that reflect the sizes of single fibrils. Dimensions may vary over four orders of magnitude in a single material and as such require carefully developed protocols specific for research objectives.

Asbestos aerosols share certain characteristics derived from their formation in nature in the asbestiform habit, which enable fibers to disaggregate and disperse readily when disturbed, forming aerosols containing abundant, very thin *fibrils* whose dimensions often persist through mineral processing, industrial applications, inhalation, and translocation in the body. The length of single fibrils varies among deposits, but most fibrils are short ($L < 4\,\mu m$); fibrils over 100 µm in length and fibers with $L > 5$ µm and widths are as small as 0.02 µm have been measured in lung burden. Fibrils are the building blocks of asbestos fibers and they dominate occupational and environmental exposure in mining in sheer particle number. Long fibrils are highly potent for mesothelioma (Korchevskiy and Wylie 2022a).

In bulk samples, single fibrils will occupy little of the overall mass with the rest being bound in larger bundles or composed of accessory minerals such as magnetite or mica. When a fiber sample is implanted, inoculated, or digested by animals, dose is frequently reported as the number of EMPs per unit mass with the EMP category constrained by a particular dimensional set, e.g. length > 0.5 µm. Because mass is concentrated in the largest particles, error in mass-based abundances can be introduced if only the smaller particles are measured to characterize a bulk material on fiber/mass basis. Frequently, bulk samples are aerosolized, and aerosol-sized particles are collected to be used as a proxy for the bulk material.

An optically visible asbestos fiber is a bundle of fibrils. Single fibrils of chrysotile (< 0.15 µm) and amphibole (< 0.05 µm) can be too narrow to be readily visible so composite fibers will dominate occupational monitoring by phase contrast (optical) microscopy (PCM). Composite fiber has the potential to disaggregate into fibrils following inhalation, thus changing the dose. Disaggregation potential varies among asbestos deposits and industrial applications, although it may be an important factor in asbestos potency. Nonasbestiform EMPs may accompany asbestos, especially in asbestos collected from noncommercial sources, along with brittle, glassy, wide fibrils, called byssolite. EMP fragments formed by crushing minerals have wider widths for the same length fiber of asbestos and they lack the potential for disaggregation.

While the exposure to most carcinogens is measured by wt%, or ppm, during occupational air monitoring for asbestos, exposure is measured in terms of the number of mineral particulates visible by PCM in a set that is defined by a range of lengths and aspect ratios per unit volume of air and the visibility of narrow fibers. However, in the total exposure EMPs outside the designated set can be much more abundant than those in the set. Furthermore, the lengths and widths of these nonprotocol fibers may vary in their relative abundances from one exposure to the other, especially among industrial applications of asbestos. The issues can complicate the comparison of dose–response variation among occupational cohorts. However, as Hodgson and Darnton (2000) and Darnton (2023) point out, at the same level of exposure measured by PCM, the mortality from mesothelioma among miners and millers varies by several orders of magnitude among different asbestos exposure occurrences. Dimensionality is one explanation.

Establishing Measurement Protocols

The measurement protocols used to obtain length and width frequencies vary in ways that may affect the usefulness of the measurements for modeling of risk. For example, differences in counting criteria, in assumptions about fiber thickness, and in the assumed size of visible fibers will affect populations and inter-population comparisons.

Optical vs. Electron Microscopy Methods

Optical microscopy is limited in the study of aerosolized particles by resolution, normally represented by $d = 0.61\ \lambda/NA$ where d = resolving power, NA = numerical aperture, and λ = wavelength of illumination; the resolution for PCM is somewhat enhanced and the formula for resolution $d = 0.4756\ \lambda/NA$ (Longhurst 1967). A resolution of about $0.34 \pm 0.1\ \mu m$ is the lower limit for PCM (NA = 0.75) as configured for occupational air monitoring and $0.4 \pm 0.1\ \mu m$ for polarized light microscopy with an objective of NA = 0.85. Although visible by PCM, particles narrower than about $0.5\ \mu m$ may not be visible by PLM under crossed polars unless the retardation is high.

Depending on the index of refraction contrast, fibers narrower than the optical resolution of the microscope may still be visible. Experiments have shown fibers of amphibole with widths as small as 0.1–$0.15\ \mu m$ are visible by PCM (Rooker et al. 1982; Kenny et al. 1987; Pang et al. 1988). Recently, NIOSH revised 7400 Method (version 3) to indicate that amphibole fibers as narrow as $0.05\ \mu m$ and chrysotile as narrow as $0.15\ \mu m$ will be visible to the careful observer. Several TEM methods of analysis recommend equating the results to PCM by removing EMPs with widths ≤$0.25\ \mu m$ from the population, e.g. NIOSH 7402 (NIOSH 1994), resulting in a count NIOSH calls PCME (phase contrast microscopy equivalent). PCME is unlikely to be valid for amphiboles and many other EMPs for which visibility and resolution are close. Given the size of aerosol particles, optical microscopy is of limited use in measurements of width.

Studies in the literature on length and width of aerosolized particles have relied almost exclusively on TEM, and this continues to be the instrument of choice. Measurements with uncertainties of <0.01 μm are possible, although most laboratories normally use magnifications of about 20,000 for measurements of width introducing an uncertainty of ±0.05 μm (Wylie 2016). Although resolution and magnification of the SEM may be high, most experimental protocols do not rely on it for width measurements when fibers are very narrow due to the complication of the coating used to dissipate charge and a low mass contrast with the substrate. Visibility of chrysotile by SEM has been shown to be particularly problematic. A comparison of optical and electron microscopy methods as it applies to lung tissue is provided by Dodson et al. (2008). The Field Emission Scanning Electron Microscopy (FESEM) has not been used extensively for particle measurements, but it could have similar problems with low-density materials like chrysotile and with very thin fibers.

A careful and accurate characterization of widths by TEM may be important given that very narrow fiber, once inhaled, can readily translocate from the lung to other sites including the parietal and peritoneal pleura and its abundance correlates with toxicity (Wylie and Korchevskiy 2023). Recent studies suggest that fiber <0.2 μm may be most likely to migrate following inhalation. Fibers of widths near 0.02 μm may be encountered so measurements of width must be made at high magnification to measure narrow widths with uncertainties of <0.01 μm.

In all EMP populations, the most abundant particles are the smallest, so in terms of characterizing particle populations by number, TEM is recommended. It is frequently assumed that the mass proportion of a component in a dust is the same for all particle sizes, but as Turcotte (1986) described, the relative proportions of a component in a particular particle bin size vary among components. He represented particle number vs. frequency by the equation $N \approx r^{-D}$, where D is the fractal dimension of the material, N is the number of particles, and r is the radius. Among natural materials, he reports D values ranging from 1.44 to 3.54. For this reason, mass calculations based on particles visible by TEM are valid for the particle sizes examined but they may or may not be valid for the sample as a whole, depending on the range in D of the components. For example, Militello et al. (2019) showed different asbestos concentrations in different particle size ranges of materials separated by sieving and grinding. Campbell et al. (1980) reported that the longest 21 fibers of COF-25 chrysotile in a population of 1,072 fibers composed more than 41% of the mass calculated from particle dimensions, while 529 of the smallest fibers made up only 2% of the mass. For this reason, SEM or even optical microscopy may be necessary to determine mass proportions of a component in a bulk sample.

Stratified Counting

Since lung burden studies of exposed individuals were first undertaken, long, narrow fiber has been considered particularly important in the development of lung cancer and mesothelioma (reviewed by Barlow et al. 2017). However, given the high frequency of short particles, many population studies are focused on the frequency of the most common particle sizes, so they poorly characterize the less abundant long fiber. For example, if in a population of 1,000, one fiber longer than 50 µm is found, then to have confidence that this one fiber size represents 0.1% ± 0.05 of the population, one would need to count 4,000 fibers, not 1,000 (Van der Plas and Tobi 1965). Chatfield (2018) reports that characterizations of chrysotile that fail to count sufficient numbers of long fiber have resulted in data that show Coalinga and UICC-B chrysotile to be indistinguishable when in fact if more fibers with $L > 10$ µm are included in the measurements, the differences in the samples are clear.

To address the large uncertainty in abundance and nature of long fibers and to be sure that they are properly accounted for in exposure assessments, a stratified approach can be taken. This approach rests on counting a small proportion of a preparation at high magnification, measuring all fibers in a known area of the dispersed sample, and then decreasing magnification and increasing area scanned to measure larger particles. (High magnification may be necessary to measure the width of these large particles precisely.) Based on the assumption of uniform distribution of fibers over the largest area scanned, the number and sizes of fiber found at the highest magnifications can be scaled to provide a full distribution of the population. Some approaches transfer a TEM grid to the SEM to find fibers that may extend over even larger areas crossing grid lines and openings. One detailed approach for stratified counting is provided by Bernstein and Kunzendorf (2018).

Sample Preparation for TEM: Direct vs. Indirect Preparation

Particles captured on air monitoring filters must be removed from the filter to enable TEM examination of the particles. Two methods are in use: a direct transfer method and an indirect transfer method. The direct transfer method dissolves the filter and captures the fiber on a transparent coating. The indirect method removes the particles by oxidizing the filter and by sonicating and redistributing them on TEM grids. Both methods are described in Chesson and Hatfield (1990).

A direct method preserves the particle size distributions collected on the filter. An indirect method has been shown to break up fiber bundles, creating additional fibers without a change in mass, in some cases by three orders of magnitude (Chesson and Hatfield 1990)… For asbestos, increases in fiber number are accompanied by decreases in mean widths and lengths, although size distributions of nonasbestiform minerals such as amphibole cleavage fragments should be only minimally affected by the choice of techniques. In addition, the indirect method was developed to measure mass, not fiber number and may be useful if that is the major objective. Data from indirect transfers are often expressed in ng/cm^3.

Frequency Distributions of Length and Width

Frequency distributions of length and width of EMPs are normally log normal on a first-order basis. This arises from the very high abundances of the narrowest and shortest particles in comparison to those that are long and wider. The fractal model of fragmentation describes a linear log–log relationship between particle number and size or mass. For minerals formed as fiber, however, superimposed on this log–log distribution can be modes in both length and width that arises from the growth habit of the fiber. Fibrils form under chemical and physical conditions that promote rapid growth in one direction but are limited in the other two. Physical and chemical conditions that result in uniform fibers may change over the time of formation, or from place to place in an occurrence, resulting in abundances of widths or lengths that in detail are not always a continuum. Figure 1.3 illustrates a bimodal distribution of the width of a highly dispersed sample of actinolite asbestos EMPs. Figure 1.2 illustrates the same for amphibole EMP fibers from Libby, MT. Figure 1.4 demonstrates a bimodal distribution of length in a fibrous tremolite EMP sample from India.

For EMPs formed by crushing, modes may develop from defects, intergrowths, or twinning that in addition to cleavage provide planes of weakness called parting that are not part of the mineral structure. The combination of cleavage and

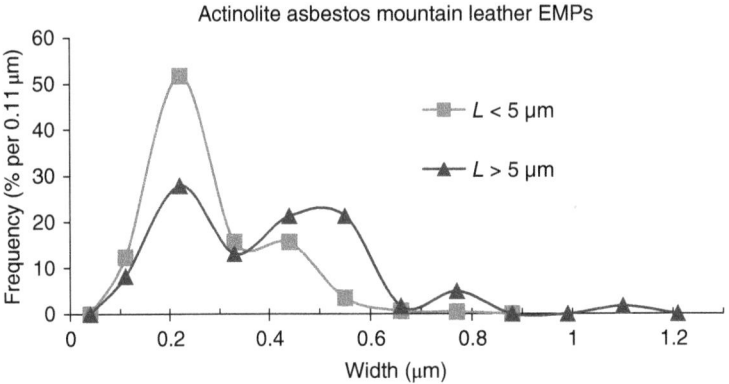

Figure 1.3 Frequency of width. Long and short fiber distributions are multimodal. Widths were measured with a precision of ±0.055 μm.

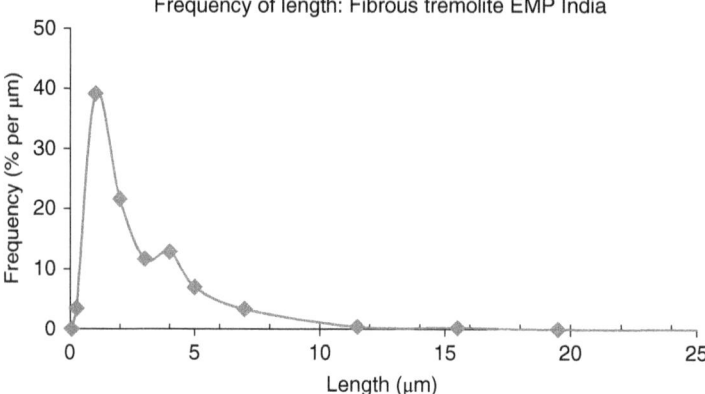

Figure 1.4 Frequency distribution of length. Two modes are defined. Lengths were measured with a precision of ±0.1 μm by TEM.

parting may result in more than one modal width. Asbestos and cleavage fragment EMP populations differ in the position of the most abundant modal width, which is always the smallest mode in asbestos populations.

Figure 1.5 shows the frequency of length of the particles from Homestake, SD whose width distribution is given in Figure 1.1. Comparing Figures 1.4 and 1.5 demonstrates a smaller number of short EMPs in the cleavage fragment population. This reflects the fact that a higher proportion of short particles of fragmented amphibole do not meet the 3 : 1 aspect ratio for inclusion in the population. Cleavage fragment populations rarely contain elongate particles with widths <0.25 μm and even for wider particles aspect ratios >3 : 1 are uncommon in particles with lengths about <3 μm. In general, for amphibole EMPs of all types, the aspect ratio increases with length. Because asbestos fiber has narrower widths

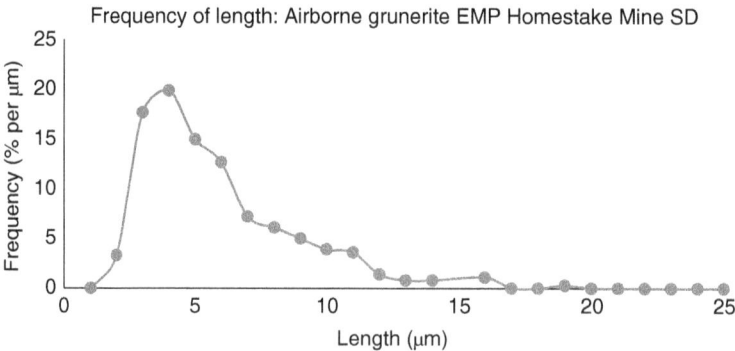

Figure 1.5 Frequency of length of airborne grunerite. 357 cleavage fragments make up this distribution.

than cleavage fragments of the same length, the impact of a 3 : 1 counting criterion eliminates few particles from asbestos populations and has little impact on frequency distributions.

Lung Burden

Mineral particulates found in lung burden reflect exposure history, intensity of exposure, duration of exposure, disaggregation of fiber bundles, dissolution of fibers, and elimination following deposition. In EMP populations produced by crushing, the individual particles will not normally disaggregate following deposition, since they are not composed of individual fibrils but the potential for fiber bundles of asbestos to break up after inhalation has been clearly demonstrated by the work of Coffin et al. (1983).

Figure 1.6a and b compares the frequency of width and length of fibers of SA Cape crocidolite longer than 5 μm in mine air and in the lungs of miners working in that mine (Pooley and Clark 1980). In both populations, the fibers are so narrow that aerodynamic differences among the fibers cannot explain the differences in the curves. Fibers in the lung are longer and narrower than the airborne fibers in the exposures. Longer fibers in lung tissue are usually explained by their higher biopersistence, and by the selective removal of short fibers. Narrower widths are most likely due to disaggregation.

The lung burden initially deposited is a function of the exposure intensity and duration, and, as the curves shown in Figure 1.6 illustrate, the characteristics of the fibers found in the lung can be expected to change over time, both in dimension and in concentration. As Korchevskiy and Wylie (2022b) demonstrated through mechanistic modeling of deposition and fiber elimination, the biopersistence of fiber in the lung is mineral specific (measured by the elimination coefficient reflecting proportion eliminated per year) with crocidolite and amosite exhibiting the highest biopersistence and chrysotile the least. Modeled results were close to published values derived from human and animal studies and confirm that some amphibole fiber can be expected to remain in the human lung for a lifetime; they predict that chrysotile would remain in the lung for two years or less.

Pooley (2018), through a detailed examination of fiber burden location in an individual exposed exclusively to amosite, has shown that amosite fiber 7–11 μm in length and less than 0.175 in width preferentially is found in the margins of the lung nearest the pleura and that the aerodynamic diameter calculated for the fibers also decreased toward the lung periphery (consistent with a decreasing width). He also reported that the concentration of fiber is greatest toward the center of the lung where the lung burden fiber is somewhat shorter and slightly wider but all fiber he reported had widths <0.25 μm. The regularity of the distribution by length and width demonstrates a filtering/removing mechanism that is extremely sensitive to small changes in dimension.

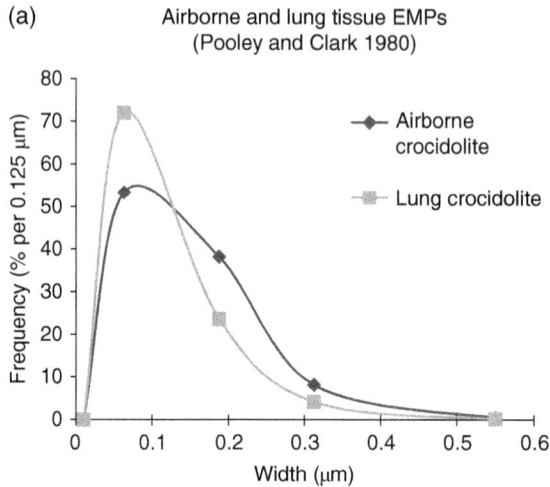

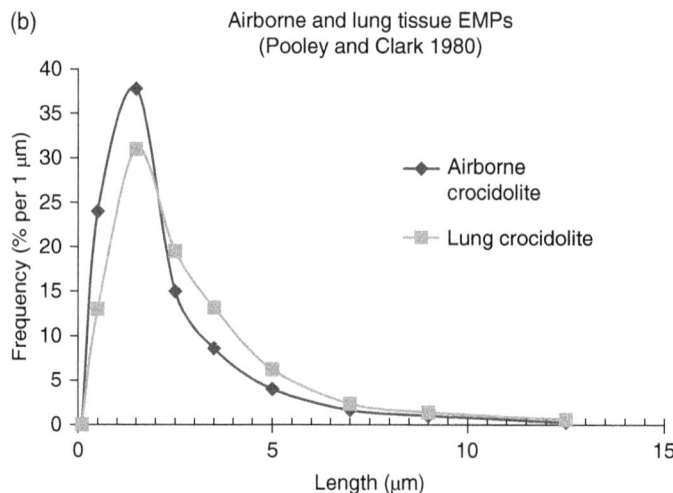

Figure 1.6 Frequency distributions of width (a) and length (b) for crocidolite for SA. Pooley and Clark (1980) provide width data in 0.125 μm intervals with the largest interval designated as >0.375 μm. The largest bin interval is arbitrarily plotted at 0.55 μm. Similarly, the largest length interval is given as >10 μm. This bin was arbitrarily plotted as 12.5 μm. *Source:* Pooley and Clark (1980) / Licensed Under CC BY 4.0.

Dimensionality and Carcinogenicity

Dimensionality has been used to correlate with the incidence of animal tumors following inhalation, cell toxicity and proliferation, and human disease. One of the earliest and perhaps the most important studies explaining tumor incidence following implantation in rats was published by Stanton et al. (1981) who put forward the hypothesis that dimensions of durable fibers define toxicity rather than chemical composition or other properties. Studies by Pott (1978) as well as others have supported this hypothesis. A major study of asbestiform and nonasbestiform tremolite by Davis et al. (1985, 1991) is important because it correlated dimension of the specific tremolite samples with tumor development in rats. The Davis study and others were reviewed by Addison and McConnell (2008) who affirmed the importance of dimensionality in animal experiments. Chatfield (2018) developed an algorithm for identifying pathogenic durable mineral fiber by dimension. He correlated the fraction of particles meeting these criteria with the frequency of mesothelioma developed in rates after intraperitoneal installation of a number of tremolite samples as reported by Davis et al. (1991).

Cell studies have demonstrated the importance of dimension. Not only do dimensions play a role in toxicity to cells, it has been shown that the rate and direction of cellular movement can be controlled by nanoscale topographic features and that the effects extend to features on a scale of a few hundred nanometers, the size of the smallest fibrils (Driscoll et al. 2014; Sun et al. 2015). Cell studies have also substantiated the importance of physio-chemical interactions (Mossman 2018; Gu et al. 2023).

More recently, Wylie et al. (2020) correlated the proportion of amphibole asbestos EMP in the exposure with $W < 0.15$ and $W < 0.25\,\mu m$ with mesothelioma potency (R_M) as given by Hodgson and Darnton (2023) and Garabrant and Pastula (2018) with a high degree of conformance to both power and linear models as discussed below. The widths chosen were based on a review by Lippmann (2014), who proposed that mesothelioma results from exposure to populations containing fiber with $W < 0.15\,\mu m$. Cleavage fragment populations which lack these narrow EMPs are not implicated in mesothelioma (Gamble and Gibbs 2008). Following this publication, Korchevskiy and Wylie and others have published a series of papers correlating dimensional parameters of width, specific surface area, and habit with potency for mesothelioma and lung cancer derived from epidemiological studies by Darnton and Hodgson and Berman and Crump (2008a) summarized recently in Wylie and Korchevskiy (2023).

Discussion

The characteristics of asbestos aerosols that relate to its toxicity can be best explored by close examination of fibrous material (aerosol-sized particle from powder, hand sample, or air) obtained directly from primary sources where excess asbestos-related disease or diseases have resulted from known dust inhalation. Unfortunately, the possible sources of such dusts have grown considerably as awareness of asbestos-related diseases and their diagnosis has spread around the world. Comparison among occurrences of the same mineral-asbestos can be particularly informative, such as is the case with the Cape asbestos province, South Africa and the Hamersley range, Australia, or Canadian chrysotile exposure in mining vs. exposures in the textile mills of South Carolina. For chrysotile, the single fibrils are similarly narrow; their chemical compositions are almost indistinguishable. For the crocidolite occurrences, the mortality rates are similarly high; for the two environments of chrysotile exposure, mortality rates are vastly different. Other occurrences such as chrysotile from Africa, crocidolite and amosite from Transvaal SA, anthophyllite-asbestos from Finland, Na–Ca amphibole-asbestos from Libby, MT, fluoro-edenite-asbestos from Italy, tremolite-asbestos from New Caledonia, Cyprus, Greece, and Turkey, wooly erionite from Turkey, etc., are highly variable in mineralogy, chemical composition, and fibril dimension. Together, they constitute mineral-specific asbestos exposures with chemical and physical properties that differ and from which, perhaps risk potency factors for mesothelioma, lung cancer, and asbestosis can be isolated.

IARC (2012) points out that there are heterogeneities in the disease outcome for mesothelioma that cannot be explained without considering mineral components in the aerosol, including the size, shape, composition and structure, and abundances. For example, the expected mortality from mesothelioma from exposure to Canadian chrysotile-asbestos is estimated to be almost two orders of magnitude lower than that for the same levels of exposure to Cape and Wittenoom crocidolite (Garabrant and Pastula 2018), despite the fact that fibrils from these three occurrences of asbestos have a similar morphology and possess similarly high tensile strength. Hodgson and Darnton (2000) found differences in dose response of lung cancer between the Canadian miners and millers and the textile workers in South Carolina of nearly 100-fold.

Careful quantitative measures of physical and chemical properties of mineral exposures are necessary to explain the observed variability in toxicity. They are essential for predicting toxicity of new exposures.

References

Addison, J. and McConnell, E.E. (2008). A review of carcinogenicity studies of asbestos and non-asbestos tremolite and other amphiboles. *Regulatory Toxicology and Pharmacology* 522: S187–199. https://doi.org/10.1016/j.yrtph.2007.10.001.

Ahn, J.H. and Buseck, P.R. (1991). Microstructures and fiber-formation mechanisms of crocidolite asbestos. *American Mineralogist* 76: 1467–1478.

Alario, F. M., Hutchison, J.L., Jefferson D.A., et al. (1977). Structural imperfection and morphology of crocidolite (blue asbestos). *Nature* 266: 520–521.

Areán, C.O., Barceló, F., Fenoglio, I., et al. (2000). Free radical activity of natural and heat-treated amphibole asbestos. *Journal of Inorganic Biochemistry* 83: 211–216. https://doi.org/10.1016/S0162-0134(00)00191-4.

Asgharian, B., Owens, T.P., Kuempel, D., et al. (2018). Dosimetry of inhaled elongate mineral particles in the respiratory tract: the impact of shape factor. *Toxicology and Applied Pharmacology* 361: 27–35. https://doi.org/10.1016/j.taap.2018.05.001.

Atkinson, G.R., Rose, D., Thomas, K., et al. (1981). Collection analysis and characterization of vermiculite samples for fiber content and asbestos contamination. Midwest Research Institute report for the US Environmental Protection Agency, Project 4901-A32 under EPA Contract 68-0d1-5915. Document Display | NEPIS | US EPA.

Aveston, A. (1969). The mechanical properties of asbestos. *Journal of Material Science* 4: 625–633

Barlow, C.A., Grespin, M., and Best, E.A. (2017). Asbestos fiber length and its relation to disease risk. *Inhalation Toxicology* 29(12–14): 541–554. https://doi.org/10.1080/08958378.2018.1435756.

Beard, M.E., Ennis, J.T., Crankshaw, O.S., et al. (2007). Preparation of non-asbestiform amphibole minerals for method evaluation and health Studies Summary Report and appendices, Prepared for Martin Harper, NIOSH, Morgantown WV by RTI International (Personal communication Frank Hearle CDC/NIOSH/OD).

Bennington, K.O., Ferrante, M.J., and Stuve, J.M. (1978). Thermodynamic data on the amphibole asbestos minerals amosite and crocidolite. Report of Investigations: 8265. U.S. Department of the Interior, Bureau of Mines, Washington.

Berman, D.W. and Crump, K.S. (2008a). Update of potency factors for asbestos-related lung cancer and mesothelioma. *Critical Reviews in Toxicology* 38: 1–47.

Berman, D.W. and Crump, K.S. (2008b). A meta-analysis of asbestos-related cancer risk that addresses fiber size and mineral type. *Critical Reviews in Toxicology* 38 (Suppl 1): 49–73. https://doi.org/10.1080/10408440802273156.

Bernstein, D.M., Rogers, R., and Smith, P. (2004). The biopersistence of Brazilian chrysotile asbestos following inhalation. *Inhalation Toxicology* 16(11–12): 745–761. https://doi.org/10.1080/08958370490490176.

Bernstein, D.M., Chevalier, J., and Smith, P. (2005). Comparison of Calidria chrysotile asbestos to pure tremolite: final results of the inhalation biopersistence and histopathology examination following short-term exposure. *Inhalation Toxicology* 17(9): 427–449. https://doi.org/10.1080/08958370591002012.

Bernstein, D.M. and Kunzendorf, P. (2018). Standardized methods for preparation and bi-variate length and diameter counting/sizing of aerosol and tissue digestion fiber samples. *Toxicology and Applied Pharmacology* 361: 174–184. https://doi.org/10.1016/j.taap.2018.04.026.

Berry, G., Reid, A., Aboagye-Sarfo, P., et al. (2012). Malignant mesotheliomas in former miners and millers of crocidolite at Wittenoom (Western Australia) after more than 50 years follow-up. *British Journal of Cancer* 106(5): 1016–1020. https://doi.org/10.1038/bjc.2012.23.

Brantley, S.L. (2003). 5.03 – Reaction kinetics of primary rock-forming minerals under ambient conditions. In: *Treatise on Geochemistry* (eds. H.D. Holland and K.K. Turekian) 73–117. Elsevier. https://doi.org/10.1016/B0-08-043751-6/05075-1.

Brown, J.S., Gordon, T., Price, O., et al. (2013). Thoracic and respirable particle definitions for human health risk assessment. *Particle and Fibre Toxicology* 10: 12–24. https://doi.org/10.1186/1743-8977-10-12.

Campbell, W.J., Wylie, A.G., and Huggins, C.W. (1980). Chemical and physical characterization of amosite, chrysotile, crocidolite, and non-bibrous tremolite for oral ingestion studies by the national institute of environmental health sciences. Bureau of Mines Report of Investigations 8452. U.S. Department of the Interior. NIOSH/00193277.

Chamberlain, M. and Brown, R.C. (1978). The cytotoxic effects of asbestos and other mineral dusts in tissue culture cell line. *British Journal of Experimental Pathology* 59(2): 183–189.

Chatfield, E.J. (2018). Measurement of elongate mineral particles: what should we measure and how do we do it? *Toxicology and Applied Pharmacology* 361: 36–46. https://doi.org/10.1016/j.taap.2018.08.010.

Chesson, J. and Hatfield, J. (1990). Comparison of airborne asbestos levels determined by transmission electron microscopy (TEM) using direct and indirect transfer techniques. EPA 560/5-89-004. March, 1990.

Coffin, D.L., Palekar, L.D., and Cook, P.M. (1983). Correlation of in vitro and in vivo methods by means of mass, dose and fiber distribution for amosite and fibrous ferroactinolite. *Environmental Health Perspectives* 51: 49–53. https://doi.org/10.1289/ehp.835149.

Cox, L.A., Bogen, K.T., Conolly, R., et al. (2023). Mechanisms and shapes of causal exposure-response functions for asbestos in mesotheliomas and lung cancers. *Environmental Research* 230: 115607. https://doi.org/10.1016/j.envres.2023.115607.

Crawford, A. (1980). Electron microscopy applied to studies of the biological significance of defects in crocidolite asbestos. *Journal of Microscopy* 120(2): 181–192. https://doi.org/10.1111/j.1365-2818.1980.tb04134.x.

Cressey, B.A., Whittaker, E.J.W., and Hutchison, J.L. (1982). Morphology and alteration of asbestiform grunerite and anthophyllite. *Mineralogical Magazine* 46(338): 77–87. https://doi.org/10.1180/minmag.1982.046.338.13.

Darnton, L. (2023). Quantitative assessment of mesothelioma and lung cancer risk based on phase contrast microscopy (PCM) estimates of fibre exposure: an update of 2000 asbestos cohort data. *Environmental Research* 230: 114753. https://doi.org/10.1016/j.envres.2022.114753.

Dana, E.S. and Ford, W.E. (1949). *A Textbook of Mineralogy*, 4th Edition. New York: Wiley.

Davis, J.M.G., Addison, J., McIntosh, C., et al. (1985). Variations in the carcinogenicity of tremolite dust samples of differing morphology. *Annals of the New York Academy of Sciences* 643: 473–490. https://doi.org/10.1111/j.1749-6632.1991.tb24497.x.

Davis, J.M., Jones, A.D., and Miller, B.G. (1991). Experimental studies in rats on the effects of asbestos inhalation coupled with the inhalation of titanium dioxide or quarts. *International Journal of Experimental Pathology* 72(5): 501–525. PMCID: PMC2002319.

Dodson, R.F., Hammar, S.P., and Poye, L.W. (2008). A technical comparison of evaluating asbestos concentration by phase-contrast microscopy (PCM), scanning electron microscopy (SEM) and analytical transmission electron microscopy (TEM) as illustrated from data generated from a case report. *Inhalation Toxicology* 20(7): 723–732. https://doi.org/10.1080/08958370701883250.

Donaldson, K., Murphy, F.A., Duffin, R., et al. (2010). Asbestos, carbon nanotubes and the pleural mesothelium: a review of the hypotheses regarding the role of long fiber retention in the parietal pleura, inflammation, and mesothelioma. *Particle and Fibre Toxicology* 7: 5. https://doi.org/10.1186/1743-8977-7-5.

Driscoll, M.K., Sun, X., Guven, C., et al. (2014). Cellular contact guidance through dynamic sensing of nanotopography. *ACS Nano* 8(4): 3546–3555. https://doi.org/10.1021/nn406637c.

Entwistle, J.A., Hursthouse, A.S., Marinho Reis, P.A., et al. (2019). Metalliferous mine dust: human health impacts and the potential determinants of disease in mining communities. *Current Pollution Reports* 5: 67–83. https://doi.org/10.1007/s40726-019-00108-5.

Eastes, W. and Hadley, J.G. (1996). A mathematical model of fiber carcinogenicity and fibrosis in inhalation and intraperitoneal experiments in rats. *Inhalation Toxicology* 8: 323–342.

Feuerbacher, D.G., Dimataris, G.T., Mace, M.L., et al. (1980). Comparative cytotoxicity and mutagenicity of organosilane reacted chrysotile asbestos. (Abstract). *Annual Meeting Program*, Clay Mineral Society, Waco, TX 34.

Fröhlich, E. (2012). The role of surface charge in cellular uptake and cytotoxicity of medical nanoparticles. *International Journal of Nanomedicine* 7: 5577–5591. https://doi.org/10.2147/IJN.S36111.

Fubini, B. and Mollo, L. (1995). Role of iron in the reactivity of mineral fibers. *Toxicological Letters* 82–83: 951–960. https://doi.org/10.1016/0378-4274(95)03531-1.

Fubini, B. and Hubbard, A. (2003). Reactive oxygen species (ROS) and reactive nitrogen species (RNS) generation by silica in inflammation and fibrosis. *Free Radical Biology and Medicine* 34(12): 1507–1516. https://doi.org/10.1016/s0891-5849(03)00149-7.

Gamble, J.F. and Gibbs, G.W. (2008). An evaluation of the risks of lung cancer and mesothelioma from exposure to amphibole cleavage fragments. *Regulatory Toxicology and Pharmacology* 52(1 Suppl): S154–186. https://doi.org/10.1016/j.yrtph.2007.09.020.

Gandolfi, N.B., Gualtieri, A.F., Pollastri, S., et al. (2016). Assessment of asbestos body formation by high resolution FEG-SEM after exposure of Sprague-Dawley rats to chrysotile, crocidolite, or erionite. *Journal of Hazardous Materials* 306: 95–104. https://doi.org/10.1016/j.jhazmat.2015.11.050.

Garabrant, D.H. and Pastula, S.T. (2018). A comparison of asbestos fiber potency and elongate mineral particle (EMP) potency for mesothelioma in humans. *Toxicology and Applied Pharmacology* 361: 127–136. https://doi.org/10.1016/j.taap.2018.07.003.

Ghio, A.J., Zhang, J., and Piantadosi, C.A. (1992). Generation of hydroxyl radical by crocidolite asbestos is proportional to surface [Fe^{3+}]. *Archives of Biochemistry and Biophysics* 298(2): 646–650. https://doi.org/10.1016/0003-9861(92)90461-5.

Giese, R.F. and van Oss, C.J. (1993). The surface thermodynamic properties of silicates and their interactions with biological materials. In: *Reviews in Mineralogy, Volume 28: Health Effects of Mineral Dusts* (eds. G. Guthrie and B. Mossman) 327–346, Washington DC: Mineralogical Society of America. https://doi.org/10.1515/9781501509711-011.

Goodglick, L.A. and Kane, A.B. (1986). Role of reactive oxygen metabolites in crocidolite asbestos toxicity to mouse macrophages. *Cancer Research* 46(11): 5558–5566.

Greenwood, W. (1998). Mineralogical Characteristics of Fibrous Talc. MS Thesis. Department of Geology, University of Maryland.

Gu, A., Bull, A., Perry, J.K., et al. (2023). Excitable systems: a new perspective on the cellular impact of elongate mineral particles. *Environmental Research* 230: 115353. https://doi.org/10.1016/j.envres.2023.115353.

Gualtieri, A.F. Ed. (2017). *Mineral Fibres: Crystal Chemistry, Chemical Physical Properties, Biological Interaction and Toxicity*. EMU Notes in Mineralogy, Volume 18. London: The European Mineralogical Union and the Mineralogical Society of Great Britain and Ireland.

Gualtieri, A.F., Pollastri, S., Gandolfi, N.B., et al. (2018). In vitro acellular dissolution of mineral fibres: a comparative study. *Scientific Reports* 8(1): 7071. https://doi.org/10.1038/s41598-018-25531-4.

Harper, M., Dozier, A., Chouinard, J., et al. (2017). Analysis of erionite from volcanoclastic sedimentary rocks and possible implications for toxicological research. *American Mineralogist* 102(8): 1718–1726. https://doi.org/10.2138/am-2017-6069.

Hartkamp, R., Biance, A., Fu, L., et al. (2018). Measuring surface charge: why experimental characterization and molecular modeling should be coupled. *Current Opinion in Colloid and Interface Science* 37: 101–114. https://doi.org/10.48550/arXiv.1808.08799.

Hawthorne, F.C., Oberti, R., Harlow, G.E., et al. (2012). IMA Report: nomenclature of the amphibole supergroup. *American Mineralogist* 97: 2031–2048. https://doi.org/10.2138/am.2012.4276.

Hayashi, H. (1974). Cytotoxicity of heated chrysotile. *Environmental Health Perspectives* 9: 267–270. https://doi.org/10.1289/ehp.749267.

Hochella, M.F. (1993). Surface chemistry, structure, and reactivity of hazardous mineral dust. In: *Reviews in Mineralogy, Volume 28: Health Effects of Mineral Dusts* (eds. G. Guthrie and B. Mossman) 275–308, Washington DC: Mineralogical Society of America. https://doi.org/10.1515/9781501509711-011.

Hodgson, A.A. (1979). Chemistry and physics of asbestos. In: *Asbestos. Properties, Applications and Hazards, Volume 1* (eds. S.S. Chissick and L. Michaels), 67–114, Wiley.

Hodgson, J.T. and Darnton, A. (2000). The quantitative risks of mesothelioma and lung cancer in relation to asbestos exposure. *Annals of Occupational Hygiene* 44(8): 565–601. PMID: 11108782.

IARC (2012). Monograph 100C: Asbestos (chrysotile, amosite, crocidolite, tremolite, actinolite, anthophyllite) 219–309. https://monographs.iarc.fr/wp-content/uploads/2018/06/mono100C-11.pdf.

Inoue, M., Sakamoto, K., Suzuki, A., et al. (2021). Size and surface modification of silica nanoparticles affect the severity of lung toxicity by modulating endosomal ROS generation in macrophages. *Particle and Fibre Toxicology* 18(1): 21. https://doi.org/10.1186/s12989-021-00415-0.

Innes, E., Yiu, H.H.P., McLean, P., et al. (2021). Simulated biological fluids – a systematic review of their biological relevance and use in relation to inhalation toxicology of particles and fibres. *Critical Reviews in Toxicology* 51(3): 217–248. https://doi.org/10.1080/10408444.2021.1903386.

Jurinski, B. (1998). Geochemical Investigations of Respirable Particulate Matter. PhD Thesis. Virginia Polytechnic Institute and State University, Blacksburg Virginia.

Jurinski, J.B. and Rimstidt, J.D. (2001). Biodurability of talc. *American Mineralogist* 86: 392–399. https://doi.org/10.2138/am-2001-0402.

Kane, A.B., Hurt, R.H., and Gao, H. (2018). The asbestos-carbon nanotube analogy: an update. *Toxicology and Applied Pharmacology* 361: 68–80. https://doi.org/10.1016/j.taap.2018.06.027.

Kenny, L.C., Rood, A.P., and Blight, B.J.N. (1987). A direct measurement of the visibility of amosite asbestos fibers by phase contrast optical microscopy. *The Annals of Occupational Hygiene* 31(2): 261–264. https://doi.org/10.1093/annhyg/31.2.261.

Kliment, C.R., Clemens, K., and Oury, T.D. (2009). North American erionite-associated mesothelioma with pleural plaques and pulmonary fibrosis: a case report. *International Journal of Clinical and Experimental Pathology* 2(4): 407–410.

Korchevskiy, A., Rasmuson, J.O., and Rasmuson, E.J. (2019). Empirical model of mesothelioma potency factors for different mineral fibers based on their chemical composition and dimensionality. *Inhalation Toxicology* 31(5): 180–191. https://doi.org/10.1080/08958378.2019.1640320.

Korchevskiy, A.A. and Wylie, A.G. (2021). Dimensional determinants for the carcinogenic potency of elongate amphibole particles. *Inhalation Toxicology* 33(6–8): 244–259. https://doi.org/10.1080/08958378.2021.1971340.

Korchevskiy, A.A. and Wylie, A.G. (2022a). Dimensional characteristics of the major types of amphibole mineral particles and the implications for carcinogenic risk assessment. *Inhalation Toxicology* 34(1–2): 24–38. https://doi.org/10.1080/08958378.2021.2024304.

Korchevskiy, A. and Wylie, A.G. (2022b). Asbestos exposure, lung fiber burden, and mesothelioma rates: mechanistic modelling for risk assessment. *Computational Toxicology* 24: 100249. https://doi.org/10.1016/j.comtox.2022.100249.

Korchevskiy, A.A. and Wylie, A.G. (2023). Toxicological and epidemiological approaches to carcinogenic potency modeling for mixed mineral fiber exposure: the case of fibrous balangeroite and chrysotile. *Inhalation Toxicology* 35(7–8): 185–200. https://doi.org/10.1080/08958378.2023.2213720.

Kudo, Y., Kohyama, N., Satoh, T., et al. (2006). Behavior of rock wool in rat lungs after exposure by nasal inhalation. *Journal of Occupational Health* 48(6): 437–445. https://doi.org/10.1539/joh.48.437.

Langer, A.M. and Nolan, R.P. (2018). Chrysotile and chrysophosphate chemical modification and the chrysotile surface and its effect on biological behavior. *Toxicology and Applied Pharmacology* 361: 118–126. https://doi.org/10.1016/j.taap.2018.10.018.

Light, W.G. and Wei, E.T. (1977). Surface charge and asbestos toxicity. *Nature* 265: 537–539. https://doi.org/10.1038/265537a0.

Lippmann, M. (2014). Toxicological and epidemiological studies on effects of airborne fibers: coherence and public [corrected] health implications. *Critical Reviews in Toxicology* 44(8): 643–695. https://doi.org/10.3109/10408444.2014.928266.

Longhurst, R.S. (1967). *Geometrical and Physical Optics*. 2nd Edition. Wiley: London.

Lower, S.K., Tadanier, C.J., and Hochella, M.F. (2000). Measuring interfacial and adhesion forces between bacteria and mineral surfaces with biological force microscopy. *Geochimica et Cosmochimica Acta* 64(18): 3133–3139. https://doi.org/10.1016/S0016-7037(00)00430-0.

Macdonald, J.L. and Kane, A.B. (1997). Mesothelial cell proliferation and biopersistence of wollastonite and crocidolite asbestos fibers. *Fundamental and Applied Toxicology* 38(2): 173–183. https://doi.org/10.1093/toxsci/38.2.173.

Militello, G.M., Sanguineti, E., Gonzalez, A.Y., et al. (2019). The concentration of asbestos fibers in bulk samples and its variation with grain size. *Minerals* 9(9): 539–558. https://doi.org/10.3390/min9090539.

Moolgavkar, S.H., Brown, R.C., and Turim, J. (2001). Biopersistence, fiber length, and cancer risk assessment for inhaled fibers. *Inhalation Toxicology* 13(9): 755–772. https://doi.org/10.1080/089583701316941294.

Morgan, A. and Holmes, A. (1985). The enigmatic asbestos body: its formation and significance in asbestos-related disease. *Environmental Research* 38(2): 283–292. https://doi.org/10.1016/0013-9351(85)90092-1.

Mossman, B.T., Marsh, J.P., and Shatos, M.A. (1986). Alteration of superoxide dismutase activity in tracheal epithelial cells by asbestos and inhabitation of cytotoxicity by antioxidants. *Laboratory Investigation* 54(2): 204–212. PMID: 3945053.

Mossman, B.T. (2018). Mechanistic in vitro studies: what they have told us about carcinogenic properties of elongated mineral particles (EMPs). *Toxicology and Applied Pharmacology* 361: 62–67. https://doi.org/10.1016/j.taap.2018.07.018.

NIOSH (1994). Asbestos by TEM: Method 7402. https://www.cdc.gov/niosh/nmam/pdf/7402.pdf (accessed 24 January 2024).

NIOSH (2011). Asbestos fibers and other elongate mineral particles: state of the science and roadmap for research revised. Department of Health and Human Services 174 pages. https://www.cdc.gov/niosh/docs/2011-159/pdfs/2011-159.pdf.

Oberdörster, G., Ferin, J., and Lehnert, B.E. (1994). Correlation between particle size, in vivo particle persistence, and lung injury. *Environmental Health Perspectives* 102(Suppl 5): 173–179. https://doi.org/10.1289/ehp.102-1567252.

Oberdörster, G. (1996). Significance of particle parameters in the evaluation of exposure-dose-response relationship of inhaled particles. *Inhalation Toxicology* 8(Suppl): 73–89. PMID: 11542496.

Ogorodova, L.P., Kiseleva, I.A., Korytkova, E.N., et al. (2006). The enthalpies of formation of natural and synthetic nanotubular chrysotile. *Russian Journal of Physical Chemistry* 80(7): 1021–1024. https://doi.org/10.1134/S003602440607003X.

O'Hanley, D.S. (1986). The origin and mechanical properties of asbestos. PhD dissertation, Department of Geology, University of Minnesota.

Oze, C. and Solt, K. (2010). Biodurability of chrysotile and tremolite asbestos in simulated lung and gastric fluids. *American Mineralogist* 95(5–6): 825–831. https://doi.org/10.2138/am.2010.3265.

Pang, T.W.S., Schonfeld, F.A., and Patel, K. (1988). The precision and accuracy of a method for the analysis of amosite asbestos. *American Industrial Hygiene Association Journal* 49(7): 351–356. https://doi.org/10.1080/15298668891379882.

Pooley, F.D. and Clark, N. (1980). A comparison of fibre dimensions in chrysotile, crocidolite and amosite particles from sampling of airborne dust and from post mortem lung tissue. *IARC Scientific Publication* 30: 78–86, Lyon, France.

Pooley, F.D. (2018). Characterization of lung burden E.M.P.S. *Toxicology and Applied Pharmacology* 361: 18–20. https://doi.org/10.1016/j.taap.2018.10.019.

Pott, F. (1978). Some aspects on the dosimetry of the carcinogenic potency of asbestos and other fibrous dusts. *Staub-Reinhalt Luft* 38: 486–489.

Puzyn, T., Rasulev, B., Gajewicz, A., et al. (2011). Using nano-QSAR to predict the cytotoxicity of metal oxide nanoparticles. *Nature Nanotechnology* 6(3): 175–178. https://doi.org/10.1038/nnano.2011.10.

Rees, D., Goodman, K., Fourie, E., et al. (1999). Asbestos exposure and mesothelioma in South Africa. *South African Medical Journal* 89(6): 627–634. PMID: 10443212.

Rooker, S.J., Vaughan, N.P., and LeGuen, J.M. (1982). On the visibility of fibers by phase contrast microscopy. *American Industrial Hygiene Association Journal* 43(7): 505–515. https://doi.org/10.1080/15298668291410125.

Rudnick, R.L. and Gao, S. (2014). Composition of the continental crust. Chapter 1 *Treatise on Geochemistry*. Vol. 41–51, 64. Elsevier.

Schiller, J.E., Payne, S.L., and Khalafalla, S.E. (1980). Surface charge heterogeneity in amphibole cleavage fragments and asbestos fibers. *Science* 209(4464): 1530–1532. https://doi.org/10.1126/science.209.4464.1530.

Schumacher, J.C. and Czank, M. (1987). Mineralogy of triple- and double-chain pyriboles from Orijarvi, southwest Finland. *American Mineralogist* 72(3–4): 345–352.

Shedd, K.B., Virta, R.L., and Wylie, A.G. (1982). Size and shape characterization of fibrous zeolites by electron microscopy. *Report of Investigations – United States, Bureau of Mines*. 8674: 1–20.

Shedd, K.B. (1985). Fiber dimensions of crocidolites from Western Australia, Bolivia, and the Cape and Transvaal Provinces of South Africa. *Report of Investigations – United States, Bureau of Mines*. 8998: ii–33.

Shukla, A., Gulumian, M., Hei, T.K., et al. (2003). Multiple roles of oxidants in the pathogenesis of asbestos-induced diseases. *Free Radical Biology and Medicine.* 34(9): 1117–11229. https://doi.org/10.1016/S0891-5849(03)00060-1.

Skinner, H.C.W., Ross, M., and Frondel, C. (1988). *Asbestos and Other Fibrous Materials: Mineralogy, Crystal Chemistry and Health Effects*. New York: Oxford University Press. https://doi.org/10.1093/oso/9780195039672.001.0001.

Stanton, M.F., Layard, M., Tegeris, A., et al. (1981). Relation of particle dimension to carcinogenicity in amphibole asbestoses and other fibrous minerals. *Journal of the National Cancer Institute* 67: 965–975.

Stöbler, W. (1971). A note on the aerodynamic diameter and the mobility of non-spherical aerosol particles. *Journal of Aerosol Science* 2: 453–456.

Stone, V., Jones, R., Rollo, K., et al. (2004). Effect of coal mine dust and clay extracts on the biological activity of the quartz surface. *Toxicology Letters* 149(1–3): 255–259. https://doi.org/10.1016/j.toxlet.2003.12.036.

Sun, X., Driscoll, M.K., Guven, C., et al. (2015). Asymmetric nanotopography biases cytoskeletal dynamics and promotes unidirectional cell guidance. *Proceedings of the National Academy of Sciences* 112(41): 12557–12562.

Timbrell, V. (1965). Human exposure to asbestos: dust controls and standards. The inhalation of fibrous dusts. *Annals of the New York Academy of Sciences* 132(1): 255–273. https://doi.org/10.1111/j.1749-6632.1965.tb41107.x.

Timbrell, V., Griffiths, D.M., and Pooley, F.D. (1971). Possible biological importance of fibre diameters of South African amphiboles. *Nature* 232(5305): 55–56. https://doi.org/10.1038/232055a0.

Timbrell, V., Ashcroft, T., Goldstein, B., et al. (1988). Relationships between retained amphibole fibres and fibrosis in human lung tissue specimens. *Annals of Occupational Hygiene* 32: 323–340. https://doi.org/10.1093/annhyg/32.inhaled_particles_VI.323.

Toyokuni, S. (1996). Iron-induced carcinogenesis: the role of redox regulation. *Free Radical Biology and Medicine* 20(4): 553–566. https://doi.org/10.1016/0891-5849(95)02111-6.

Turci, F., Tomatis, M., Compagnoni, R., et al. (2009). Role of associated mineral fibres in chrysotile asbestos health effects: the case of balangeroite. *Annals of Occupational Hygiene* 53(5): 491–497. https://doi.org/10.1093/annhyg/mep028.

Turci, F., Pavan, C., Leinardi, R., et al. (2016). Revisiting the paradigm of silica pathogenicity with synthetic quartz crystals: the role of crystallinity and surface disorder. *Particle and Fibre Toxicology* 13(1): 32. https://doi.org/10.1186/s12989-016-0136-6.

Turci, F., Tomatis, M., and Pacella, A. (2017). Surface and bulk properties of mineral fibres relevant to toxicity. In: *Mineral Fibres: Crystal Chemistry, Chemical-Physical Properties, Biological Interaction and Toxicity* (ed. A.F. Gualiteri) 171–214. London: European Mineralogical Union Notes in Mineralogy. https://doi.org/10.1180/EMU-notes.18.

Turcotte, D.L. (1986). Fractals and fragmentation. *Journal of Geophysical Research* 91(B2): 1921–1926. https://doi.org/10.1029/JB091iB02p01921.

US EPA. (2006). US EPA produced access database, Libby Montana, airborne particles in the matter of United States of America vs. W R Grace et al., CR-05-070M-DMW (Montana) 2005-2006 (R.J. Lee personal communication).

Utembe, W., Potgieter, K., Stefaniak, A.B., et al. (2015). Dissolution and biodurability: important parameters needed for risk assessment of nanomaterials. *Particle and Fibre Toxicology* 12: 11–23. https://doi.org/10.1186/s12989-015-0088-2.

Valentine, R., Chang, M.J., Hart, R.W., et al. (1983). Thermal modification of chrysotile asbestos: evidence for decreased cytotoxicity. *Environmental Health Perspectives* 51: 357–368. https://doi.org/10.1289/ehp.8351357.

Van der Plas, L. and Tobi, A.C. (1965). A chart for judging the reliability of point counting results. *American Journal of Science* 263(1): 87–90. https://doi.org/10.2475/ajs.263.1.87.

Veblen, D.R. and Burnham, C.W. (1978a). New biopyriboles from Chester, Vermont; I, descriptive mineralogy. *American Mineralogist* 63(9–10): 1000–1009.

Veblen, D.R. and Burnham, C.W. (1978b). New biopyriboles from Chester, Vermont; II, The crystal chemistry of jimthompsonite, clinojimthompsonite, and chesterite, and the amphibole-mica reaction. *American Mineralogist* 63(10–11): 1053–1073.

Verkouteren, J.R. and Wylie, A.G. (2002). Anomalous optical properties of fibrous tremolite, actinolite, and ferro-actinolite. *American Mineralogist* 87(8–9): 1090–1095. https://doi.org/10.2138/am-2002-8-905.

Wagner, J.C., Skidmore, J.W., Hill, R.J., et al. (1985). Erionite exposure and mesotheliomas in rats. *British Journal of Cancer* 51(5): 727–730. https://doi.org/10.1038/bjc.1985.108.

Walker, J.S. and Zoltai, T. (1979). A comparison of asbestos fibers with synthetic crystals known as "whiskers". *Annals of the New York Academy of Sciences* 330(1): 687–704. https://doi.org/10.1111/j.1749-6632.1979.tb18772.x.

Watson, M. (1999). Effects of Intergrowths on the Physical Characteristics of fibrous Anthophyllite. MS Thesis. University of Maryland, College Park.

Werner, A.J., Hochella, M.F., Guthrie, G.D., et al. (1995). Asbestiform riebeckite (crocidolite) dissolution in the presence of Fe chelators: implications for mineral-induced disease. *American Mineralogist* 80(11–12): 1093–1103. https://doi.org/10.2138/am-1995-11-1201.

Wylie, A.G., Shedd, K.B., and Taylor, M.E. (1982). *Measurement of the Thickness of Amphibole Asbestos Fibers with the Scanning Electron Microscopy and the Transmission Electron Microscope*. Washington, D.C.: Microbeam Analysis Society-Electron Microscope Society of America.

Wylie, A.G., Virta, R.L., Shedd, K.B., et al. (2015). Size and shape characteristics of airborne amphibole asbestos and amphibole cleavage fragments. *Digital Repository at the University of Maryland*. https://doi.org/10.13016/M2HP87.

Wylie, A.G. (2016). Amphibole dusts: fibers, fragments and mesothelioma. *The Canadian Mineralogist* 54(6): 1403–1435. https://doi.org/10.3749/canmin.1500109.

Wylie, A.G. (2017). Mineralogy of asbestos and fibrous erionite. In: *Current Cancer Research: Asbestos and Mesothelioma* (ed. J. Testa), 11–41. Heidelberg, Germany: Springer. https://doi.org/10.1007/978-3-319-53560-9_2.

Wylie, A.G., Korchevskiy, A., Segrave, A., et al. (2020). Modeling mesothelioma risk factors from amphibole fiber dimensionality: mineralogical and epidemiological perspective. *Journal of Applied Toxicology* 40(4): 515–524. https://doi.org/10.1002/jat.3923.

Wylie, A.G., Korchevskiy, A.A., Van Orden, D.R., et al. (2022a). Discriminant analysis of asbestiform and non-asbestiform amphibole particles and its implications for toxicological studies. *Computational Toxicology* 23: 100233. https://doi.org/10.1016/j.comtox.2022.100233.

Wylie, A.G. and Korchevskiy, A.A. (2023). Dimensions of elongate mineral particles and cancer: a review. *Environmental Research* 230: 114688. https://doi.org/10.1016/j.envres.2022.114688.

Zhu, W., von dem Bussche, A., Yi, X., et al. (2016). Nanomechanical mechanism for lipid bilayer damage induced by carbon nanotubes confined in intracellular vesicles. *Proceedings of the National Academies of Science* 113(44): 12374–12379. https://doi.org/10.1073/pnas.1605030113.

Zoltai, T. and Stout, J.H. (1984). *Mineralogy: Concepts and Principles*. Minneapolis: Burgess.

2

Toxicology of Mineral Fibers and Implications for Risk Assessment

Brooke T. Mossman

Department of Pathology and Laboratory Medicine, University of Vermont College of Medicine, Burlington, VT, USA

Introduction

After the discovery that various asbestos types were associated with the development of malignant mesotheliomas (MMs) and lung cancers in occupational settings, concerns continued to increase regarding the characterization of health hazards associated with exposures to asbestos and potentially other mineral fibers in the workplace and environment. Thereafter, research began on mechanisms of fiber toxicology, defined broadly as alterations in normal cell function and structure that over time may result in disease. Research initially focused on whether certain asbestos fibers produced tumors or pulmonary fibrosis (asbestosis), a nonmalignant hyperproliferative disease of the lung interstitium, in rodents. However, it soon became apparent that a combination of animal and *in vitro* studies were needed to characterize interactions between minerals and cells and the evolution of cellular changes over time. *In vitro* studies also facilitated the assessment of dose–response relationships resulting in the observation that cell repair occurs at low concentrations of mineral fibers and cell injury at high concentrations.

This chapter describes the historical use of animal (*in vivo*) and *in vitro* models designed to understand the mechanisms of MMs, lung cancers and pulmonary fibrosis. These data can be used in risk assessment and also have enabled preventive and therapeutic approaches to these diseases. There are distinct advantages and disadvantages in use of either animal or *in vitro* models that are detailed below. Moreover, where possible, results should be correlated with human pathology and epidemiology using a systems biology approach for relevant adverse outcomes.

Health Risk Assessment for Asbestos and Other Fibrous Minerals, First Edition.
Edited by Andrey Korchevskiy, James Rasmuson, and Eric Rasmuson.
© 2024 John Wiley & Sons, Inc. Published 2024 by John Wiley & Sons, Inc.

Use of Rodent Models to Analyze the Toxicity to Disease Potential of Naturally Occurring and Synthetic Fibers

With few exceptions, animal studies have been generally conducted in rats as these rodents develop malignant and fibrotic lesions similar in histopathology to that observed in humans. Exposures to mineral fibers range from acute observations (hours to days) to subchronic (weeks to months) or chronic (many months to the life span of two to three years). These experiments at high concentrations of fibers have provided an understanding of initial cell responses to fibers and the evolution of disease. However, since the life span of rodents is generally less than three years, the differential long-term clearance and dissolution of fibers that may occur over decades in the human lung cannot be mimicked. For example, chrysotile fibers are rapidly cleared from and/or break down in human lungs becoming amorphous silica structures, whereas amosite and crocidolite fibers persist (Giacobbe et al. 2021; Gualtieri 2023). Chronic animal experiments are also expensive and time consuming, thus dose–response studies with minerals are rare. Historically, rodent studies have compared fibers at equal weight concentrations, which may reflect different numbers and surface areas of fibers, more important parameters in dosimetry.

Inhalation Studies

Inhalation studies are the "gold standard" for animal experiments as airborne exposure is the natural route of exposure to particulates in occupational settings (McClellan et al. 1992; Mossman et al. 2011). Rodents are either exposed unrestrained in inhalation chambers designed for multiple animals per group or restrained in individual head or nose-only chambers. The expense of these experiments due to sophisticated generation systems and maintenance of animals over time has prohibited dose–response experiments using multiple concentrations of fibers. Characterization of the delivered aerosol preparations, including the fiber size distribution and inclusion of short fibers and nonfibrous particles, are often not included in the published literature. In addition, fiber "overload," a situation in which the host lung defenses are overwhelmed at a high lung burden of fibers (Vincent et al. 1985; Morrow 1988), may occur. A progressive reduction of particle clearance from the deep lung due to impairment of macrophage function then results in inflammation and a nonspecific tumorigenic response to many low-toxicity particles (Oberdörster 1995). Historically, animal inhalation studies have been performed at a chamber concentration of 10 mg/m^3 air, greater than an order of magnitude higher fiber concentrations than those encountered by humans (Mossman et al. 2011).

Intratracheal Instillation and Oropharyngeal Aspiration Studies

These methods allow the delivery of a bolus of minerals suspended in liquid to the lung in single or multiple doses. However, normal clearance mechanisms in the upper respiratory tract are bypassed, and the distribution of fibers may be nonuniform throughout the lungs. Since animals are anesthetized during these procedures and deaths may occur, multiple long-term applications of dusts are not feasible.

Intrapleural Injection Studies

This method is technically difficult but allows injection of fibers in liquid into the pleural space and their direct interaction with visceral and parietal pleural mesothelial cells to induce MMs. Often, the procedure is associated with the death of rodents; thus, it is not feasible for multiple injection or dose–response studies. Because this methodology bypasses normal clearance mechanisms in both the respiratory tract and pleura, its significance to normal mesothelial cell carcinogenesis is uncertain.

Intraperitoneal Injection Studies

This alternative route for studying mesothelial cell carcinogenesis involves direct injection of fibers in liquid into the peritoneal cavity. The rodent death rate is minimal, small amounts of materials administered over time in dose–response studies are feasible, and the test is more sensitive than other tumorigenicity assays. However, it has been criticized as giving rise to a number of false positive results as compared to other rodent tumorigenicity models (Mossman et al. 2011; Drummond et al. 2016). Moreover, the injection of a bolus of material directly into the peritoneum does not mimic what may occur in humans after inhalation, and the human peritoneal mesothelial cell may not respond to fibers as do pleural mesothelial cells (Dragon et al. 2015). In addition, it is still unclear in humans whether or not inhaled mineral fibers get to or are retained at sites of peritoneal mesothelial cells.

Comparative Results on Effects of Asbestos and Other Naturally Occurring Fibers in Rodent Studies

Chronic or life span studies using rodents have generally focused on the development of MMs, lung cancers and pulmonary fibrosis as revealed by pathology of organs at death. Correlations between the tumorigenic potential of various naturally occurring and synthetic fibers in inhalation assays as compared to other models have been described in an excellent review by Drummond et al. (2016).

This article summarizes results in comparative rodent models as reported in the literature using asbestos, many synthetic vitreous fibers, titanium dioxides, and man-made inorganic fibers such as potassium octatitanate (PO). This synthetic fiber type and wollastonite are described as examples of low toxicity or noncarcinogenic fibers that have been evaluated in a number of animal models. For example, a low dose (2.2 ± 0.7 mg/m^3 or 111 ± 34 fibers/ml) of PO in an inhalation study over a year showed mild fibrosis and no malignancies (Yamato et al. 2003), whereas a multidose two-year study at 0, 20, 60, and 200 WHO fibers/cc (defined by the World Health Organization as having a length greater than or equal to 5 µm, a diameter of less than 3 µm, and a length to diameter, i.e. aspect ratio, of greater than or equal to 3) showed no pulmonary neoplasms or MMs and only modest alveolar fibrosis (Ikegami et al. 2004). After analysis of these data, Yamato et al. (2003) proposed that a threshold dose for fibrotic changes existed between approximately 1.5 and 2.4 mg of dust. In contrast, Adachi et al. (2001) reported that 20% of animals developed mesotheliomas after injection of 5 mg PO intraperitoneally. Because of the overall lack of conformity between results with the use of PO and other particles and fibers in intraperitoneal injection models and inhalation studies, Drummond et al. (2016) suggest that the intraperitoneal injection model "should not be used to positively determine that a dust or fiber is carcinogenic by inhalation" and conclude "We would argue against the use of intraperitoneal tests for human health risk assessment except for perhaps the purpose of exoneration of a material from classification as a carcinogen."

Based upon these recommendations and those of others illustrating the importance of inhalation studies in replicating the route of human exposures to dusts in the workplace (Health Effects Institute-Asbestos Research 1991; McClellan et al. 1992; Mossman et al. 2011), emphasis on rodent inhalation models is recommended for purposes of human risk assessment. Table 2.1 presents the results of acute and subchronic inhalation studies suggesting concentrations in animals below which adverse outcomes are not observed or disappear over time. Special attention is given to the results of dose–response studies demonstrating thresholds, that is, concentrations below which markers of adverse health effects are not observed. Table 2.2 illustrates the results of chronic inhalation studies using asbestos fibers. Because chronic inhalation studies are less abundant using other naturally occurring fibers, Table 2.3 summarizes both subchronic and chronic inhalation studies as well as instillation studies using wollastonite, erionite or Libby amphibole as these naturally occurring fibers also have been studied in human populations.

In acute inhalation studies, emphasis historically has been on chrysotile asbestos as it is the most common asbestos type used commercially worldwide (see Table 2.1). In some studies, brief (one to five hours) exposures were to high concentrations of chrysotile fibers (4–13 mg/m^3 air), and rodents were evaluated over

Table 2.1 Acute and subchronic rodent inhalation studies providing information on risk assessment of asbestos fibers.

Reference	Fiber	Exposure duration	Length (μm)	Diameter (μm)	Mass/Unit air (number of fibers)	Outcome
(Chang et al. 1988) (Rats)	Chrysotile (Jeffrey mine)	1 h (follow-up at 2 d and 1 mo)	NR	NR	13 mg/m³	>Type I and II epithelial cells (morphometry) at 2D only
(Brody and Overby 1989) (Rats)	Chrysotile (NR)	5 h (follow-up at 19 and 33 h, 2, 8, 14, 30 d	NR	NR	10 mg/m³	>³H-thymidine-containing epithelial cells at 19 and 33 h only
(McGavran et al. 1990) (Mice)	Chrysotile (NR)	5 h (follow-up at 0, 19, 24, 31, 48 h; 8 d, 4 wk	NR	NR	4 mg/m³	>³H-thymidine containing epithelial cells at 31 and 48 h only
(Brass et al. 1999) (Two strains mice)	Chrysotile (CA)	5 h (follow-up at 0, 48 and 33 h, 2, 8, 14, 30 d	NR	NR	10 mg/m³	>BrdU-containing epithelial cells at 48 h and 1 wk only
(Mossman et al. 2011) (also see Quinlan et al. 1994, 1995) and (Shukla et al. 2004)	NIEHS Chrysotile • LD • HD NIEHS Crocidolite • LD • HD	5 d, 4 wk (follow-up at 4 wk post-4 wk exposure)	(See Campbell et al. 1980)	(See Campbell et al. 1980)	LD = 0.18 (32f ≥ 5 μm/ml air) HD = 8.2 (2,457f ≥ 5 μm/ml air) LD = 0.16 (60f ≥ 5 μm/ml air) HD = 8.3 (2,800f ≥ 5 μm/ml air)	>BrdU-containing lung epithelial/mesothelial cells at HD only. Only crocidolite at HD for 20+20 D caused mesothelial cell proliferation, gene expression and fibrosis

Study	Fiber Type	Exposure Duration	# fibers >20/cm³	Aspect ratio	Concentration	Findings
(Bernstein et al. 2014)	Brake dust powder + Chrysotile (Jeffrey mine)	5d (follow-up at 0, 7, 32, 91 d)	189	29.8	3.48 mg/m³ (1,007 WHO fibers/cm³)	Progressive interstitial fibrosis only in crocidolite group. Estimated clearance times of >20 µm fibers were 30–32 d for brake dusts ± chrysotile vs. 1,000 d for crocidolite
	Brake dust powder	5d (follow-up at 0, 7, 32, 91, d)	3.6	17.0	1.52 mg/m³ (46 WHO fibers/cm³)	
	Crocidolite (South African)	5d (follow-up at 0, 7, 32, 91, d)	93	17.6	6.34 mg/m³ (1,496 WHO fibers/cm³)	
(Bernstein et al. 2015)	Same fiber groups as above	5d (follow-up 365 D)	189	29.8	3.48 mg/m³ (1,007 WHO fibers/cm³)	Only crocidolite caused > fibrosis in lung and visceral pleura. Long (>20 µm) crocidolite fibers retained in lung and pleura for 365 d)
			3.6	17.0	1.52 mg/m³ (46 WHO fibers/cm³)	
			93	17.6	6.34 mg/m³ (1,496 WHO fibers/cm³)	
(Bernstein et al. 2018)	Brake dust	28 d (follow-up at 1, 14, 28, d and 14 and 28 d postexposure)				Chrysotile at HD caused some inflammation but no fibrosis. Crocidolite persisted in lung and pleura to cause inflammation and fibrosis at HD.
	• Low (LD)					
	• Medium (MD)					
	• High (HD)					
	Chrysotile					
	• Low (LD)		42			
	• High (HD)		62			
	Crocidolite					
	• Low (LD)		36			
	• High (HD)		55			

NR = Not reported.

Table 2.2 Chronic rodent inhalation studies with asbestos fibers.

Reference	Fiber	Exposure Duration	Length (μm)	Diameter (μm)	Mass/Unit/Air (number of fibers)	Rodents with disease MM[b]	Carcinomas	Fibrosis	Total pulmonary tumors[a]
(Wagner et al. 1974) (Rats)	Amosite	1 d to 24 mos.	NR[c]	NR	10 mg/m³	1	11	+	Increased with duration of exposure in all groups
	Anthophyllite					2	16	+	
	Crocidolite					4	16	+	
	Chrysotile (Canadian)					4	17	+	
	Chrysotile (Rhodesian)					0	31	+	
(Davis et al. 1978) (Rats)	Chrysotile (UICC A)[d]	12 mos. +Life span	NR (PCOM)[e]	NR	2 mg/m³ (390f > 5 μm/ml)			3.8%	21.4%
					10 mg/m³ (950f ≥5 μm/ml)			9.0%	37.5%
	Crocidolite (UICC)		NR	NR	5 mg/m³ (430f ≥5 μm/ml)			0.8%	6.9%
					10 mg/m³ (860f ≥5 μm/ml)			1.4%	2.5%
(Muhle et al. 1987) (Rats)	Chrysotile (UICC)	12 mos. + 12 mos. Follow-up	2–14	0.28–1.6	6 mg/m³ (cumulative exposure = 6,600 mg/h/m³)	NI[f]		42%	
	Crocidolite (South African)	12 mos. + 12 mos. Follow-up	0.72–45	0.17–0.46	2.2 mg/m³ (cumulative exposure = 2,200 mg/h/m³)	NI	1(2%)	36%	

Study	Fiber type	Duration	Length	Diameter	Concentration				
(Smith et al. 1987) (Rats, Hamsters)	Crocidolite (UICC)	24 mos.	95% <5	2.5 ± 0.2	7 mg/m³ (3,000 fibers/cm³)	1.8%	NS	53%	3.5% bronchoalveolar
(Hesterberg et al. 1995)	Chrysotile (Rats)	24 mos.	>5	<3	10 mg/m³ (1.1 ± 1.1 × 10⁴ WHO[g] fibers/cm³)	1.4%	NS	+	18.9%
	Chrysotile (Hamsters)	18 mos.	>5	<3	10 mg/m³ (3,000 ± 1,400 × 10⁴ WHO fibers/cm³)	0	0	0	0
	Crocidolite (Rats)	24 mos.	>5	<3	10 mg/m³ (.16 ± 0.1 × 10⁴ WHO fibers/cm³)	0.9%	NS	+	14.2%
(Cullen et al. 2000)	Amosite	12 mos. + 12 mos. follow-up	>0.4 <20	>.1 <0.9	1,000 fibers/cm³	4.8%	16.7%	+	38.1%

[a] May include benign and malignant neoplasms.
[b] MM = Malignant mesothelioma.
[c] NR = Not reported.
[d] UICC = Union International Contre le Cancer reference sample.
[e] PCOM = Phase contrast optical measuring (precludes fiber size distributions <5μm in length).
[f] NI = Not identified.
[g] WHO fibers as defined by the World Health Organization as having a length > 5 μm, a diameter of <3 μm, and a length to diameter, that is, aspect ratio of ≥ 3.

Table 2.3 Subchronic and chronic rodent inhalation and instillation using naturally occurring fibers.

Reference	Fiber	Exposure duration	Approximate length (μm)	Diameter (μm)	Mass/Unit/Air (number of fibers)	Rodents exhibiting adverse outcomes			
						MMs	Carcinomas	Fibrosis	Other Comments
(Warheit et al. 1994) (Rats)	Wollastonite (calcium silicate)	5 D + 6 mos. Follow-up		2.6–4.3	115 mg/m^3 (800 f/cc)			Mild	Rapid clearance noted
(Tatrai et al. 2004) (Rats?)	Wollastonite (calcium silicate)	1 mg IT (1, 3, 6 mos. Follow-up)	10–20	≤0.1				Mild	Nonprogressive mild inflammation and fibrosis
(Wagner et al. 1985) (Rats)	Erionite Karain	12 mos. (IH) or 1 × IP Lifetime Follow-up	IP = 10% > 4		IP = 20 mg (2.4 × 10^8 f/mg)	38/40			UICC crocidolite at 20 mg/m^3 caused 1 carcinoma
	Oregon		40% > 4		20 mg (2.9 × 10 f/mg)	40/40			
	Oregon	IH	50% > 5		10 mg/m^3	27/28			
(Cyphert et al. 2015) (Rats)	Libby Amphibole	IT 1 × or bimonthly for 13 wk + 20 mos. Follow-up	5.8 ± 4.5	0.29	Mass doses (0.15, 0.5, 1.5 or 5 mg/rat)	Tumors observed at all concentrations. No dose response.			Libby amphibole had > surface area/fiber
	Amosite (UICC)		7.5 ± 36.8	0.28					

(Cyphert et al. 2016) (Rats)	Libby Amphibole	1 × IT + 15 mos. follow-up	Mean 1.9	Max 27.3	(Mean Aspect Ratio) 19.5	0.5 or 1.5 mg/rat	0	12.5 mg	+	No adverse effects on lung function at lower concentrations of all dusts
	Sumas Mountain Chrysotile	1 × IT + 15 mos. follow-up	2.0	17.5	18.8	0.5 or 1.5 mg/rat	0	0	+	
	El Dorado Tremolite	1 × IT + 15 mos. follow-up	0.9	6.4	7.8	0.5 or 1.5 mg/rat	0	12.5 mg	+	No inflammation or mesothelial cell proliferations observed at either concentration of all dusts
	Ferroactinolite cleavage fragments	1 × IT + 15 mos. follow-up	1.1	9.4	8.4	0.5 or 1.5 mg/rat	0	0	0	

IH = inhalation; IP = intrapleural injection; IT = intratracheal/installation.

brief periods of time (Chang et al. 1988; Brody and Overby 1989; McGavran et al. 1990; Brass et al. 1999). Cell proliferation was determined by morphometry or labeling with ^{3}H-thymidine or bromo-deoxyuridine (BrdU) as a marker of unscheduled DNA synthesis. In comparison to sham rodents in clean air, early increases in proliferation of Type I and Type II epithelial cells in the distal lungs, endothelial cells, and epithelial cells at broncho-alveolar bifurcations were observed initially, but commenced with time, presumably due to repair mechanisms.

In a subsequent series of studies at low (LD = 0.15–0.18 mg/m^3 air) and high doses (HD = 8.00–8.45 mg/m^3 air) of crocidolite asbestos or chrysotile asbestos, several markers of cell proliferation were measured in rats after 5, 10, or 20 days of inhalation with additional follow-up of animals maintained for 20 days post exposure in clean air (Quinlan et al. 1994, 1995; BéruBé et al. 1996; Shukla et al. 2004). As summarized in Mossman et al. (2011), crocidolite at the high dose caused early bronchial epithelial cell proliferation and sustained mesothelial cell proliferation that occurred with increased gene expression in lung tissue of the early response protooncogene, *cjun*, and *odc* (*ornithine decarboxylase*) that encodes proteins linked to tumor promotion. Elevated expression of hydroxyproline, an indicator of pulmonary fibrosis, also was observed in the lungs of high-dose crocidolite-exposed rats, but appeared in both asbestos groups after an additional 20 days without fiber exposure. In contrast, high exposures to chrysotile fibers caused transient epithelial and mesothelial cell proliferation at five days without elevations in hydroxyproline or gene expression. No changes in hydroxyproline, markers of cell proliferation or gene expression were observed at lower concentrations of either crocidolite or chrysotile asbestos.

Results support the hypothesis that early increases in cell proliferation observed with high-dose chrysotile reflect injury and repair as opposed to sustained hyperplasia associated with tumor development. They also indicate the presence of a threshold below, which results are not seen in exposures to crocidolite asbestos. In perspective, no adverse effects in abnormal cell division or early markers of tumor development were observed at 600-fold higher levels than current US governmental workplace standards (0.1 f/ml air). These data and human exposure models, epidemiological studies and lung burden analyses support a possible threshold model for tumor development (Cox et al. 2023; Weill 2023).

The experiments by Bernstein et al. (Bernstein et al. 2014, 2015, 2018) in Table 2.1 are noteworthy as they also show differential effects of chrysotile and crocidolite asbestos exposures in rats after brief exposures of five days. In these studies, comparable doses and sizes of fibers were used, and the characterization of aerosols was meticulous. These and other chronic experiments by this group described below are the only published inhalation studies that address use and no adverse effects of a commercial, chrysotile asbestos-containing product, that is, brake dust.

In follow-up studies for as long as 365 days, progressive interstitial fibrosis and visceral pleural thickening were only noted after exposures to South African crocidolite as compared to brake dust alone or brake dust supplemented with chrysotile fibers. Estimated clearance times of crocidolite asbestos from the lungs and pleura were protracted, that is, greater than 1,000 days, whereas estimated clearance times of chrysotile or brake dust was 30–32 days. Long (>20 µm) fibers of crocidolite were preferentially trapped at sites of pleural mesothelial cells for as long as 365 days.

In more recent work, rats were exposed to brake dust, chrysotile asbestos, or crocidolite asbestos for 28 days with additional follow-up times up to 28 days post exposure (Bernstein et al. 2018). These dose–response studies evaluated three concentrations of brake dust in comparison to two concentrations of chrysotile or crocidolite asbestos. Only HD crocidolite resulted in persistent inflammation and fibrosis in the lungs and pleura.

Table 2.2 presents the results of chronic inhalation studies using asbestos fiber types in chronological order, beginning with the classical studies of Wagner et al. (Wagner and Skidmore 1965; Wagner et al. 1974). Early studies focused on the development of MMs and lung cancers, but it became apparent that rats also developed pulmonary fibrosis that often killed animals at an earlier time point, precluding the development of malignancies. In addition, benign tumors and early events in carcinoma development were noted using histopathology. These studies showed that pulmonary tumors were more plentiful than MMs and increased with the duration of inhalation. However, no trends were apparent when comparing different asbestos types, and studies have been criticized as overwhelming the defense and repair mechanisms of animals.

Long (greater than 5 µm), thin asbestos fibers have been historically linked to lung cancer and MM development in humans (Bernstein 2023; Wylie and Korchevskiy 2023; Weill 2023). Studies by Davis et al. (1978) explored dose–response effects of UICC chrysotile (2 and 10 mg/m^3 air), UICC amosite (10 mg/m^3 air), and UICC crocidolite (5 and 10 mg/m^3 air), but it was noted that the dimensions of these reference samples could not be accurately determined using phase contrast light microscopy. A modest dose-response was observed with chrysotile, whereas the results with crocidolite were less striking and not dose-related. These results were later attributed to the fact that the UICC reference sample of crocidolite asbestos consisted of very short fibers (Health Effects Institute-Asbestos Research 1991). It was also noted that most fiber preparations used in animal experiments were heterogeneous in size and consisted primarily of fibers less than 5 µm in length. Moreover, milling of asbestos and other dusts reduced or eliminated their carcinogenic potential.

Muhle et al. (1987) conducted experiments with UICC chrysotile (6 mg/m^3) or crocidolite (South African) (2.2 mg/m^3) in rats after exposures for 12 weeks with a 12-week follow-up period. Although pulmonary fibrosis was observed in 36–42%

of the rats, only one crocidolite rat developed an adenocarcinoma. In comparative studies using intraperitoneal injections, 55% of crocidolite-exposed rats showed malignant tumors compared to 84% of chrysotile-exposed animals. In contrast to the chrysotile fiber preparation, these authors noted that crocidolite fibers were all less than 5 µm in length. Smith et al. (1987) also used UICC crocidolite fibers of less than 5 µm in length (95%) in a 24-month inhalation exposure study. Fibrosis in approximately half the animals (Osborne-Mendel rats), one MM, and two bronchoalveolar tumors were reported. In summary, all of these studies indicated the importance of fibers >5 µm in causing tumorigenicity. It was noted in several papers that the fibrogenic potential of asbestos minerals also was linked to fiber length (Health Effects Institute-Asbestos Research 1991).

Hesterberg et al. (1995) performed a series of inhalation studies using chrysotile and crocidolite asbestos as positive controls for carcinogenicity in studies with a variety of man-made synthetic vitreous fibers. In these experiments, rats were exposed to fibers for two years (exposure to crocidolite had to be stopped at 10 months because of excess morality of the exposed animals), and hamsters were exposed for 18 months to 10 mg/m^3 air of chrysotile (both species), or crocidolite (rats only). In both the chrysotile and crocidolite groups, fibrosis and a single MM were reported, but total pulmonary tumors (14–18.9% of rats) were not specified by type. Fibrosis was observed in hamsters without the development of malignancies indicating species differences in response to minerals.

Cullen et al. (2000) subsequently exposed rats in an inhalation chamber to amosite asbestos fibers for up to 12 months and followed animals for an additional 12 months. In a parallel intraperitoneal injection study, animals received a single injection of 10^9 fibers of amosite. In contrast to the inhalation study, where 4.8% of animals developed MMs, 81% of the animals developed MMs after intraperitoneal injections.

In a 90-day subchronic inhalation protocol and life span study adhering to a design advocated by the European Union for demonstration of fibrotic and carcinogenic disease potential, Bernstein et al. (2021) studied three concentrations of brake dusts, two concentrations of chrysotile asbestos and high doses of amosite asbestos, crocidolite asbestos, and fine titanium dioxide. Because of the complexity of these studies, the reader is referred to the original publications for details. Rats were monitored for inflammation, pleural and pulmonary fibrosis, tumors, and retention of particles until death due to disease or other causes. In these experiments, brake dusts at all concentrations were cleared rapidly from the lung and pleura with no inflammation nor adverse tissue changes (fibrosis and tumors). Clearance with very slight inflammation and fibrosis was observed in chrysotile groups, whereas tumors, fibrosis, and retention of amosite and crocidolite asbestos occurred. These experiments confirm that long amphibole fibers impacted pleural sites of MM and pleural fibrosis, and were preferentially retained there.

Table 2.3 summarizes the results of inhalation and instillation studies with other naturally occurring fibers used occupationally (wollastonite, Libby amphibole) or occurring in the environment (erionite). Wollastonite is a calcium silicate fiber mined in New York State with no reported adverse effects in workers (Ross et al. 1993). This fiber has been used as a nonpathogenic negative control material in many *in vitro* studies and has been studied in rats by inhalation, intratracheal instillation, and intraperitoneal injection. Warheit et al. (1994) exposed rats for five days to 115 mg/m^3 air (800 fibers/cc) and documented its rapid clearance from the lungs. A subchronic study using intratracheal instillation with UICC crocidolite as a positive control showed that lungs of wollastonite-exposed animals examined at one, three, and six months exhibited mild inflammation and fibrosis which did not progress over time, whereas inflammation by crocidolite was more striking (Tatrai et al. 2004). Pott et al. (1987) have also injected wollastonite intraperitoneally into rats in five 20 mg injections. No tumors were observed for 28 months after dosing the animals. These and other fibers and particles including titanium dioxides, PO fibers, and synthetic glass fibers have been studied extensively in several animal models and have a low toxicity rating (Drummond et al. 2016). It should be noted, however, that fine and ultrafine titanium dioxide particles were found to be possibly carcinogenic based on animal experiments. In humans, however, carcinogenicity of titanium dioxide remains questionable (Le et al. 2018).

Erionite is a fibrous zeolite found in parts of Oregon, Nevada, and South Dakota and is also present in many other locations in the world (Turkey, Mexico, and others). It has not been used commercially but was incorporated into family dwellings in parts of Turkey where familial MMs have developed (Baris et al. 1987). Both samples of Oregon and Karain, Turkey erionite have been studied by Wagner et al. (1985) in rats after inhalation or intrapleural injection. A single injection of 20 mg erionite from either Karain or Oregon gave rise to MMs in 90–100% rats injected, and inhalation of Oregon erionite caused MMs in 27 of 28 animals. These proportions of tumors are markedly higher than has been observed using various types of asbestos fibers.

A complex naturally occurring fiber, Libby amphibole (LA)-containing vermiculite, has been mined in Libby, Montana where an increased incidence of MMs and pleural fibrosis has been reported (Larson et al. 2020). This fiber type has been studied after intratracheal instillations into rats over a range of concentrations (Cyphert et al. 2015, 2016). In initial experiments where UICC amosite asbestos was included as a positive control for tumorigenicity, tumor induction was observed over a 20-month period at all concentrations of either amosite asbestos or LA. Curiously, no dose–response relationship existed in these studies but the authors comment "...there is a possibility of greater long-term pathological changes with repeated lower LA dose exposures, which more accurately simulates chronic environmental exposures." Subsequently, this hypothesis was addressed in intratracheal instillation studies using two concentrations (0.5 and 1.5 mg/rat) of LA, a Sumas Mountain chrysotile (stated as containing 2–27% chrysotile fibers), an El Dorado tremolite, and a ferroactinolite cleavage fragment preparation. These studies examined baseline lung function, pulmonary mechanics, histopathology, inflammatory cells in bronchoalveolar lavage fluid (BALF), and markers of lung injury in BALF. In contrast to their earlier study, a dose response was indicated in all mineral groups in that parameters of lung injury, inflammation, and abnormal lung function were not observed at the lower concentrations of all minerals tested. However, only two lung carcinomas were observed in rats exposed to LA or El Dorado tremolite, respectively, and these occurred at the lower concentration of minerals. Although quantitative data are not presented, these authors state, "No apparent evidence for mesothelial cell proliferation was noted in any groups" (Cyphert et al. 2016).

In vitro Models of Toxicity

Advantages and Disadvantage of *In vitro* Models

In vitro models of toxicity have included studies with isolated DNA, isolated red blood cells (RBCs), isolated rodent and human cells, and organ cultures, that is, explants. They are less expensive overall in maintenance than animals, and reference samples of mineral fibers with defined properties can be introduced into cultures using various dose parameters, that is, equal fiber numbers, weights, or surface areas, etc. Accordingly, dose–response experiments over a range of times can be conducted. Mineral exposed cells or tissues can also be implanted or injected into syngeneic or immuno-compromised rodents to assay their tumorigenic potential. Although isolated normal cell cultures have a finite lifespan *in vitro*, that is, hours or days, organ cultures and immortalized cells can persist for months. Most importantly, the cell is where disease begins; thus, information from *in vitro* experiments has been fundamental to understanding mechanisms of disease causation by asbestos and other mineral fibers.

There are also a number of disadvantages to reliance on *in vitro* experiments alone for risk assessment. First, these short-term models do not allow assessment of fiber durability, an important parameter in toxicity and tumor development. For example, a number of false positive results have been observed in rodent cells with nondurable fibers, that is, synthetic vitreous fibers that are not disease-related in man. As in animal studies, high concentrations of fibers have been used; thus, their relevance and extrapolation to human exposures are often questionable (International Agency for Research on Cancer 2002). Many published studies do not identify the type, source, and characteristics of mineral fiber samples and employ only one concentration of fibers. These gaps and the absence of negative controls, that is, nonpathogenic dusts, in many experiments render interpretation of data difficult. Data from some *in vitro* studies also do not correlate with observations in animals and humans. For example, endpoints such as genotoxicity, chromosomal fragmentation, and aneuploidy by fibers in general have not been observed in mesothelial cells of rodents after inhalation of asbestos or other fiber types.

Contributions of *In vitro* Models to Understanding Mechanisms of Cytotoxicity and Carcinogenesis by Mineral Fibers

Many *in vitro* studies have examined cytotoxicity or cell death as an endpoint after addition of mineral fibers to cells although the relationship between cytotoxicity and carcinogenicity is questionable as dead cells cannot give rise to tumors. When compared on an equal mass basis, chrysotile asbestos is more toxic than various types of amphibole asbestos or synthetic vitreous fibers. However, trends are reversed if toxicity is measured as fiber numbers per cell (Mossman et al. 1990). The cytotoxic effects of chrysotile are largely attributed to its positive surface charge due to Mg^{++} that is leached after its cellular uptake into acidic lysosomes (Jaurand et al. 1977, 1984). Dissolution of chrysotile also occurs due to loss of Mg^{++}.

A marker of carcinogenesis that has been studied in many *in vitro* models is mutagenesis, as this is one characteristic of many agents that initiate cancer. Unlike soluble carcinogens, mineral fibers, including asbestos (Reiss et al. 1982; Sincock et al. 1982; Daniel 1983; Denizeau et al. 1985) and erionite (Kelsey et al. 1986) fail to test positively in many *in vitro* models of genotoxicity and mutagenesis. For these reasons, asbestos was classified as an agent that did not directly interact with DNA (Williams 1979; Shelby 1988). It was subsequently revealed in a unique hamster-human cell model that gene mutations by chrysotile asbestos occurred only at lethal concentrations of fibers. This was attributed to the observation that mutations by chrysotile consisted of large deletions in DNA that were incompatible with cell survival (Hei et al. 1992).

Aneuploidy and chromosomal damage by asbestos and other fiber types were reported in several *in vitro* models, but dose–response studies are limited (reviewed in Health Effects Institute-Asbestos Research 1991). In these studies, genetic aberrations were not seen at lower concentrations of asbestos (reviewed in Mossman 2018). Later studies suggested the importance of many cellular repair mechanisms including DNA repair enzymes and antioxidant enzymes (reviewed in Mossman et al. 2013). Studies using tracheobronchial epithelial cells as models of lung cancer development showed that human lung epithelial cells were more resistant to DNA damage by asbestos as compared to mesothelial cells and rodent cell cultures (Lechner et al. 1985; Kodama et al. 1993). Other research demonstrated that crocidolite and chrysotile asbestos fibers did not interact directly with nor form adducts with the DNA of rodent tracheal epithelial cells (Eastman et al. 1983; Mossman et al. 1983). These changes and breakage of DNA were, however, observed with the addition of soluble polycyclic aromatic hydrocarbons to cells.

The results above indicate that effects of mineral fibers are unlike those of chemical carcinogens that act directly on DNA and/or are metabolized by cells. In contrast, carcinogenic mineral fibers appear to stimulate chronic inflammation that induces sustained cell proliferation and abnormal cell function (reviewed in Mossman et al. 2013; Cox 2018). In support of these observations, physical or foreign body carcinogenesis was first noted in experiments where foreign bodies such as plastic gave rise to tumors after their implantation under the skin of animals (Brand et al. 1976). The hypothesis that MMs and lung cancers are induced after foreign body perturbations that stimulate chronic inflammation by carcinogenic mineral fibers is an attractive one supported by data from both cell cultures and animal experiments. For example, erionite and crocidolite asbestos initiate an autocrine pathway of inflammation in human mesothelial cells *in vitro* that begins with inflammasome activation (Hillegass et al. 2013; Sayan and Mossman 2016; Cox 2018). Long ($>5\,\mu m$) fibers may induce chronic inflammation in the lung and pleura because they persist at sites of tumor development (Murphy et al. 2012). Moreover, long thin fibers at high concentrations are trapped at stomata between mesothelial cells that are important in normal clearance of fibers by the lymphatic system. Thus, they cannot be cleared effectively (Donaldson et al. 2010; Murphy et al. 2011; Schinwald et al. 2012).

In support of the hypothesis that asbestos fibers act epigenetically instead of directly on DNA, human mesothelial cell transformation to malignancy has been reported using epigenetic modification (Pacaud et al. 2014). In these studies, global DNA hypomethylation of a human mesothelial cell line (MET5A cells) caused their transformation to malignancy that was demonstrated after their injection into immune-compromised mice.

Properties of Mineral Fibers Important in Toxicity and Carcinogenic Effects

Mineral fibers are complex in their chemical, physical, and crystallographic characteristics; thus, it has been difficult to address which of these properties are linked to toxic and carcinogenic effects of fibers. However, comparisons and correlations between animal and *in vitro* data have been fundamental to understanding the complexity of mineral fibers and certain properties of mineral fibers important in disease development. For example, the importance of fiber length in tumorigenicity has been studied by many laboratories using a variety of rodent models as summarized above. Links between fiber length and toxic cell responses have been attributed to physical interactions of long fibers (>15 μm) with proteins in the cytoplasm regulating the mitotic spindle of aberrantly dividing cells (Jensen and Watson 1999; MacCorkle et al. 2006). In addition to rodent studies described above in Tables 2.1–2.3, classical rodent studies by Stanton et al. (1981) also support the conclusion that long (>8 μm), thin (<0.25 μm wide) fibers cause pleural sarcomas and MMs regardless of their chemical composition. Studies using primates maintained for as long as 11.5 years after chronic exposures to short chrysotile failed to reveal premalignant lesions or tumors, also supporting the premise that short chrysotile asbestos fibers are not pathogenic (Stettler et al. 2008). Fiber diameter may also play a role in tumorigenesis as tangled nanomaterials resembling serpentine chrysotile in morphology do not cause tumors in rats after intraperitoneal administration to animals (Nagai et al. 2013). Moreover, thick fibers of >0.5 μm do not translocate to the human pleura from the lung (Lentz et al. 2003; Wylie and Korchevskiy 2023).

We used sized preparations of chrysotile asbestos and electron microscopy to show that long rodlike fibers selectively caused hyperplasia and squamous metaplasia, early changes in lung carcinogenesis, in tracheal organ cultures. Short fibers and cleavage fragments of minerals were inactive (Woodworth et al. 1983). In several of these studies, synthetic vitreous and other fiber types, including erionite, were used to demonstrate the importance of long needlelike shape in cell injury, inflammation, and proliferation. Similar to results in rodent assays exploring aneuploidy or polyploidy by crocidolite asbestos fibers, long fibers of a number of types induced parameters of cell proliferation. These included uptake of markers of proliferation in cells, synthesis of polyamines, that is, proteins that are increased in cells before cell division occurs, and increased expression of *fos* and *jun* protooncogenes that form the AP-1 transcription factor (Landesman and Mossman 1982; Marsh and Mossman 1988; Heintz et al. 1993; Janssen et al. 1994). Experiments using mineral fibers over a range of concentrations also demonstrate thresholds below which no increases in gene expression and markers of cell division are observed (Sesko and Mossman 1989; Heintz et al. 1993). These data and others argue against linear no-threshold models of disease (Cox 2018; Cox et al., 2023; Weill 2023).

Long fibers (>5–10 μm) that exceed the dimensions of cells cannot be engulfed, leading to "frustrated phagocytosis" whereas smaller fibers and fragments are taken up into phagosomes which then merge with lysosomes. Both fiber length and iron content of asbestos types are critical in generation of oxidants that cause genetic and epigenetic changes in cells during carcinogenesis (reviewed in Shukla et al. 2003; Mossman et al. 2013). Many studies have revealed the importance of iron chemistry as another mineral property linked to cancer causation as ferrous iron drives reactions of cells favoring production of oxidants. Both iron content and its surface availability are elevated in crocidolite and amosite asbestos as compared to chrysotile and other types (tremolite) of amphibole asbestos. Moreover, these fiber types induce intracellular mobilization of iron and activation of iron transport receptors in cells. Thus, a cell presented with high concentrations of long thin, iron-containing fibers has multiple pathways of oxidant generation that may overwhelm antioxidant defense mechanisms effective at lower concentrations of fibers. At high concentrations of fibers, it is likely that many of these events are initiated or perpetuated by a number of redox-sensitive protein signaling cascades that have been shown to be elicited by crocidolite asbestos fibers in rodent or human mesothelial cells (Janssen et al. 1995; Pache et al. 1998; Ramos-Nino et al. 2002).

Recently, a quantitative predictive model for the toxicity and disease potential of mineral fibers has been developed based on their complex features that govern their pathogenicity (Gualtieri et al. 2017; Gualtieri 2018; Mossman and Gualtieri 2020). Their physical, chemical, and morphologic features have been ranked in a hierarchy that includes: (1) morphometric properties, that is, mean fiber length, mean fiber diameter, crystal curvature, crystal habit, density, hydrophobic character, and specific surface area; (2) chemical parameters, that is, iron content, ferrous iron content, surface iron and its nuclearity, and other metal contents; (3) biodurability-related factors, that is, dissolution rate and iron release; and (4) surface activity-related parameters, that is, zeta potential, aggregation state, and cationic exchange capacity. A weighing scheme based on the scientific literature reporting *in vitro*, animal, human, and mineral studies has been linked to each parameter of the model for each mineral tested, giving rise to a Fiber Potential Toxicity Index (FPTI). Thus far, the FPTI has been calculated for several amphibole asbestos types, chrysotile asbestos from different sources, and the nonpathogenic mineral fibers, sepiolite and wollastonite. There is a distinct difference in scores in that all amphibole asbestos have FPTIs >2.5, chrysotile samples have scores from 2.0 to 2.3, and nonpathogenic mineral fibers <2.0. These results have been linked primarily to data showing that chrysotile and nonpathogenic dusts are cleared more rapidly and dissolve within lung tissue. This approach for ranking minerals is promising and being validated presently using other mineral fibers.

In summary, there are many properties other than morphology, length, and diameter that may drive the many molecular pathways linked to disease causation by mineral fibers. Although results of *in vitro* studies and rodent tumor models do not mimic the striking differences in fiber potency in the causation of human MMs (relative risks of crocidolite 371 fold > amosite 79 fold > chrysotile asbestos 1 fold) (Darnton 2023; Hodgson and Darnton, 2000; Garabrant and Pastula 2018), they have confirmed the importance of long (>5 µm) thin fibers, biodurability, and iron content/availability in a number of carcinogenic responses as well as in pulmonary fibrosis. It should be noted that the fiber length cut off of 5 µm for denoting a long pathogenic fiber in historical studies has been one of convenience in that fibers greater than this dimension could be measured by phase contrast light microscopy. However, some studies suggest that fibers >10 µm or more are those important in human pathology and lung cancers (Lippmann 2014; Roggli 2018). Both *in vivo* and *in vitro* studies have played complementary roles in establishing the importance of long fibers and dose–response effects in cellular and tissue responses. Moreover, they have documented thresholds of exposures below which adverse outcomes by asbestos fibers do not occur.

A Systems Biology Approach to Understanding Connections and Interactions Between Adverse Outcomes in Mineral Fiber-Induced Diseases

Data summarized in this chapter often support the adverse outcomes of lung cancers, MMs, and pulmonary fibrosis observed in human lungs after exposures to high concentrations of asbestos fibers in the past workplace. Experimental and epidemiologic data indicate a hierarchy of fiber potency that is especially indicated in MMs where confounding influences such as exposure to cigarette smoke, as is observed in human lung cancers, do not occur. The hierarchy of mineral fiber potency, erionite > crocidolite asbestos > amosite asbestos > Libby amphibole > chrysotile asbestos is suggested from studies in the literature (reviewed in Wylie and Korchevskiy), whereas convincing data showing the lack of pathogenicity of wollastonite, synthetic vitreous fibers, and PO fibers exist. In addition, animal and human data (Ilgren and Browne 1991; Cox 2018; Cox et al. 2023) suggest that thresholds exist below which MMs are not observed.

Figure 2.1 presents a plausible schema for integration of human, animal, and *in vitro* data on mechanisms for application to human risk assessment. It can be applied to the adverse outcomes of mineral fiber-induced cancers or pulmonary fibrosis as endpoints induced by pathogenic mineral fibers. It should be noted that both cancers and fibrosis are diseases attributed to chronic inflammation and increased cell proliferation. In addition, fibrosis, a disease of increased fibroblast proliferation and collagen production, has been linked to the development of lung cancers (reviewed in Mossman and Churg 1998) and MMs (Abayasiriwardana et al. 2019). Thus, molecular, cellular, and tissue responses may be interactive.

As shown in Figure 2.1 and described above, generation of reactive oxygen species (ROS) is a cellular response linked to priming and subsequent activation of the inflammasome in mesothelial cells, epithelial cells, and phagocytes. ROS generation also is key to a number of other molecular and cellular responses.

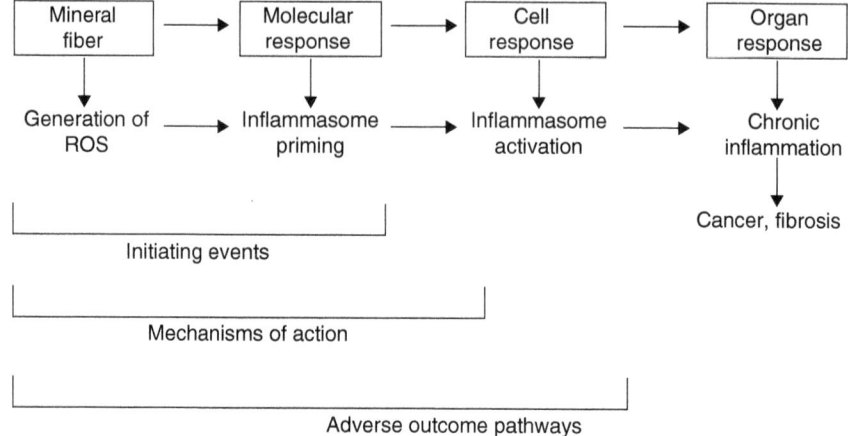

Figure 2.1 Integration of *in vitro* and *in vivo* responses to predict adverse outcomes.

At low concentrations of minerals, a number of antioxidant responses may curtail inflammasome activation. It is noteworthy that Cox (2018) has shown that each of several stages in inflammasome activation by mineral fibers in cells reflects thresholds that may be overwhelmed at high concentrations of fiber exposures resulting in chronic inflammation and disease. Chronic inflammation contributes to ROS generation in a positive feed-back loop and may be exacerbated because of the biopersistence of high concentrations of durable fibers in the lungs and pleura. These phenomena may explain the increased pathogenicity of erionite and certain amphibole asbestos fibers (crocidolite, amosite) in tumor development.

References

Abayasiriwardana, K.S., Wood, M.K., Prele, C.M., et al. (2019). Inhibition of collagen production delays malignant mesothelioma tumor growth in a murine model. *Biochemical and Biophysical Research Communications* 510(2): 198–204. https://doi.org/10.1016/j.bbrc.2019.01.057.

Adachi, S., Kawamura, K. and Takemoto, K. (2001). A trial on the quantitative risk assessment of man-made mineral fibers by the rat intraperitoneal administration assay using the JFM standard fibrous samples. *Industrial Health* 39(2): 168–174. https://doi.org/10.2486/indhealth.39.168.

Baris, I., Simonato, L., Artvinli, M., et al. (1987). Epidemiological and environmental evidence of the health effects of exposure to erionite fibres: a four-year study in the Cappadocian region of Turkey. *International Journal of Cancer* 39(1): 10–17. https://doi.org/10.1002/ijc.2910390104.

Bernstein D.M. (2022). The health effects of short fiber chrysotile and amphibole asbestos. *Critical Reviews in Toxicology* 52(2): 89–112. https://doi.org/10.980/10408444.2022.2056430.

Bernstein, D.M., Rogers, R., Sepulveda, R., et al. (2014). Evaluation of the deposition, translocation and pathological response of brake dust with and without added chrysotile in comparison to crocidolite asbestos following short-term inhalation: interim results. *Toxicology and Applied Pharmacology* 276(1): 28–46. https://doi.org/10.1016/j.taap.2014.01.016.

Bernstein, D.M., Rogers, R.A., Sepulveda, R., et al. (2015). Evaluation of the fate and pathological response in the lung and pleura of brake dust alone and in combination with added chrysotile compared to crocidolite asbestos following short-term inhalation exposure. *Toxicology and Applied Pharmacology* 283(1): 20–34. https://doi.org/10.1016/j.taap.2014.12.012.

Bernstein, D.M., Toth, B., Rogers, R.A., et al. (2018). Evaluation of the dose-response and fate in the lung and pleura of chrysotile-containing brake dust compared to chrysotile or crocidolite asbestos in a 28-day quantitative inhalation toxicology study. *Toxicology and Applied Pharmacology* 351: 74–92. https://doi.org/10.1016/j.taap.2018.04.033.

Bernstein, D.M., Toth, B., Rogers, R.A., et al. (2021). Final results from a 90-day quantitative inhalation study evaluating the dose-response and fate in the lung and pleura of chrysotile-containing brake dust compared to TiO_2, chrysotile, crocidolite or amosite asbestos: histopathological examination, confocal microscopy and collagen quantitation of the lung and pleural cavity. *Toxicology and Applied Pharmacology* 424: 115598. https://doi.org/10.1016/j.taap.2021.115598.

BéruBé, K.A., Quinlan, T.R., Moulton, G., et al. (1996). Comparative proliferative and histopathologic changes in rat lungs after inhalation of chrysotile or crocidolite asbestos. *Toxicology and Applied Pharmacology* 137(1): 67–74. https://doi.org/10.1006/taap.1996.0058.

Brand, K.G., Buoen, L.C., and Brand, I. (1976). Multiphasic incidence of foreign body-induced sarcomas. *Cancer Research* 36(10): 3681–3683.

Brass, D.M., Hoyle, G.W., Poovey, H.G., et al. (1999). Reduced tumor necrosis factor-alpha and transforming growth factor-beta1 expression in the lungs of inbred mice that fail to develop fibroproliferative lesions consequent to asbestos exposure. *American Journal Pathology* 154(3): 853–862. https://doi.org/10.1016/s0002-9440(10)65332-1.

Brody, A.R. and Overby, L.H. (1989). Incorporation of tritiated thymidine by epithelial and interstitial cells in bronchiolar-alveolar regions of asbestos-exposed rats. *American Journal of Pathology* 134(1): 133–140.

Campbell, W.J., Huggins, C.W., and Wylie, A.G. (1980). Chemical and physical characterization of amosite, chrysotile, crocidolite, and nonfibrous tremolite for oral ingestion studies by the National Institute of Environmental Health Sciences, U.S. Department of the Interior, U.S. Bureau of Mines. Report of Investigations 8452.

Chang, L.Y., Overby, L.H., Brody, A.R., et al. (1988). Progressive lung cell reactions and extracellular matrix production after a brief exposure to asbestos. *American Journal of Pathology* 131(1): 156–170.

Cox, L.A.T., Jr. (2018). Biological mechanisms of non-linear dose-response for respirable mineral fibers. *Toxicology and Applied Pharmacology* 361: 137–144. https://doi.org/10.1016/j.taap.2018.06.016.

Cox, L.A., Bogen, K.T., Conolly, R., et al. (2023). Mechanisms and shapes of causal exposure-response functions for asbestos in mesotheliomas and lung cancers. *Environmental Research* 230: 115607. https://doi.org/10.1016/j.envres.2023.115607.

Cullen, R.T., Searl, A., Buchanan, D., et al. (2000). Pathogenicity of a special-purpose glass microfiber (E glass) relative to another glass microfiber and amosite asbestos. *Inhalation Toxicology* 12(10): 959–977. https://doi.org/10.1080/08958370050138012.

Cyphert, J.M., Carlin, D.J., Nyska, A., et al. (2015). Comparative long-term toxicity of Libby amphibole and amosite asbestos in rats after single or multiple intratracheal exposures. *Journal of Toxicological and Environmental Health A* 78(3): 151–165. https://doi.org/10.1080/15287394.2014.947455.

Cyphert, J.M., McGee, M.A., Nyska, A., et al. (2016). Long-term toxicity of naturally occurring asbestos in male Fischer 344 rats. *Journal of Toxicological and Environmental Health A* 79(2): 49–60. https://doi.org/10.1080/15287394.2015.1099123.

Daniel, F.B. (1983). In vitro assessment of asbestos genotoxicity. *Environmental Health Perspectives* 53: 163–167. https://doi.org/10.1289%2Fehp.8353163.

Darnton L. (2023). Quantitative assessment of mesothelioma and lung cancer risk based on Phase Contrast Microscopy (PCM) estimates of fibre exposure: an update of 2000 asbestos cohort data. *Environmental Research* 230: 114753. https://doi.org/10.1016/j.envres.2022.114753.

Davis, J.M., Beckett, S.T., Bolton, R.E., et al. (1978). Mass and number of fibres in the pathogenesis of asbestos-related lung disease in rats. *British Journal of Cancer* 37(5): 673–688. https://doi.org/10.1038%2Fbjc.1978.105.

Denizeau, F., Marion, M., Chevalier, G., et al. (1985). Inability of chrysotile asbestos fibers to modulate the 2-acetylaminofluorene-induced UDS in primary cultures of rat hepatocytes. *Mutation Research* 155(1–2): 83–90. https://doi.org/10.1016/0165-1218(85)90029-1.

Donaldson, K., Murphy, F.A., Duffin, R., et al. (2010). Asbestos, carbon nanotubes and the pleural mesothelium: a review of the hypothesis regarding the role of long fibre retention in the parietal pleura, inflammation and mesothelioma. *Particle and Fibre Toxicology* 7: 5. https://doi.org/10.1186/1743-8977-7-5.

Dragon, J., Thompson, J., MacPherson, M., et al. (2015). Differential susceptibility of human pleural and peritoneal mesothelial cells to asbestos exposure. *Journal of Cell Biochemistry* 116(8): 1540–1552. https://doi.org/10.1002%2Fjcb.25095.

Drummond, G., Bevan, R., and Harrison, P. (2016). A comparison of the results from intra-pleural and intra-peritoneal studies with those from inhalation and intratracheal tests for the assessment of pulmonary responses to inhalable dusts and fibres. *Regulatory Toxicology and Pharmacology* 81: 89–105. https://doi.org/10.1016/j.yrtph.2016.07.019.

Eastman, A., Mossman, B.T., and Bresnick, E. (1983). Influence of asbestos on the uptake of benzo(a)pyrene and DNA alkylation in hamster tracheal epithelial cells. *Cancer Research* 43(3): 1251–1255.

Garabrant, D.H. and Pastula, S.T. (2018). A comparison of asbestos fiber potency and elongate mineral particle (EMP) potency for mesothelioma in humans. *Toxicology and Applied Pharmacology* 361: 127–136. https://doi.org/10.1016/j.taap.2018.07.003.

Giacobbe, C., Di Giuseppe, D., Zoboli, A., et al. (2021). Crystal structure determination of a lifelong biopersistent asbestos fibre using single-crystal synchrotron X-ray micro-diffraction. *IUCrJ* 8: 76–85. https://doi.org/10.1107/S2052252520015079.

Gualtieri, A.F. (2018). Towards a quantitative model to predict the toxicity/pathogenicity potential of mineral fibers. *Toxicology and Applied Pharmacology* 361: 89–98. https://doi.org/10.1016/j.taap.2018.05.012.

Gualtieri A.F. (2023). Journey to the centre of the lung. The perspective of a mineralogist on the carcinogenic effects of mineral fibres in the lungs. *Journal of Hazardous Materials.* 442: 130077. https://doi.org/10.1016/j.jhazmat.2022.130077.

Gualtieri, A.F., Mossman, B.T., and Roggli, V.L. (2017). Towards a general model for predicting the toxicity and of mineral fibres. In: *Mineral Fibres; Crystal Chemistry, Chemical-Physical Properties, Biological Interactions and Toxicity* (ed. A.F. Gualtieri), Vol. 18, 501–532. Great Britain and Ireland, European Mineralogical Union Notes in Mineralogy. http://dx.doi.org/10.1180/EMU-notes.18.15

Health Effects Institute-Asbestos Research (1991). *Asbestos in Public and Commercial Buildings: A Literature Review and Synthesis of Current Knowledge.* Cambridge, MA: Health Effects Institute.

Hei, T.K., Piao, C.Q., He, Z.Y., et al. (1992). Chrysotile fiber is a strong mutagen in mammalian cells. *Cancer Research* 52(22): 6305–6309.

Heintz, N.H., Janssen, Y.M., and Mossman, B.T. (1993). Persistent induction of *c-fos* and *c-jun* expression by asbestos. *Proceedings of the National Academy of Sciences of the United States of America* 90(8): 3299–3303. https://doi.org/10.1073%2Fpnas.90.8.3299.

Hesterberg, T.W., Miiller, W.C., Thevenaz, P., et al. (1995). Chronic inhalation studies of man-made vitreous fibres: Characterization of fibres in the exposure aerosol and lungs. *Annals of Occupational Hygiene* 39(5): 637–653. https://doi.org/10.1016/0003-4878(94)00091-E.

Hillegass, J.M., Miller, J.M., MacPherson, M.B., et al. (2013). Asbestos and erionite prime and activate the NLRP3 inflammasome that stimulates autocrine cytokine release in human mesothelial cells. *Particle and Fibre Toxicology* 10: 39. https://doi.org/10.1186/1743-8977-10-39.

Hodgson, J.T. and Darnton, A. (2000). The quantitative risks of mesothelioma and lung cancer in relation to asbestos exposure. *The Annals of Occupational Hygiene* 44(8): 565–601.

Ikegami, T., Tanaka, A., Taniguchi, M., et al. (2004). Chronic inhalation toxicity and carcinogenicity study on potassium octatitanate fibers (TISMO) in rats. *Inhalation Toxicology* 16(5): 291–310. https://doi.org/10.1080/08958370490428391.

Ilgren, E.B. and Browne, K. (1991). Asbestos-related mesothelioma: evidence for a threshold in animals and humans. *Regulatory Toxicology and Pharmacology* 13(2): 116–132. https://doi.org/10.1016/0273-2300(91)90017-p.

International Agency for Research on Cancer. (2002). *IARC Monograph on the Evaluation of Carcinogenic Risks to Humans: Man-Made Vitreous Fibres*, Vol. 81. Lyon, France: IARC Press. PMID: 2458547.

Janssen, Y.M., Heintz, N.H., Marsh, J.P., et al. (1994). Induction of *c-fos* and *c-jun* proto-oncogenes in target cells of the lung and pleura by carcinogenic fibers. *American Journal of Respiratory Cell and Molecular Biology* 11(5): 522–530. https://doi.org/10.1165/ajrcmb.11.5.7946382

Janssen, Y.M., Barchowsky, A., Treadwell, M., et al. (1995). Asbestos induces nuclear factor kappa B (NF-kappa B) DNA-binding activity and NF-kappa B-dependent gene expression in tracheal epithelial cells. *Proceedings of the National Academy of Sciences of the United States of America* 92(18): 8458–8462. https://doi.org/10.1073/pnas.92.18.8458.

Jaurand, M.C., Bignon, J., Sebastien, P., et al. (1977). Leaching of chrysotile asbestos in human lungs. Correlation with in vitro studies using rabbit alveolar macrophages. *Environmental Research* 14(2): 245–254. https://doi.org/10.1016/0013-9351(77)90036-6.

Jaurand, M.C., Gaudichet, A., Halpern, S., et al. (1984). In vitro biodegradation of chrysotile fibres by alveolar macrophages and mesothelial cells in culture: comparison with a pH effect. *British Journal of Industrial Medicine* 41(3): 389–395. https://doi.org/10.1136%2Foem.41.3.389.

Jensen, C.G. and Watson, M. (1999). Inhibition of cytokinesis by asbestos and synthetic fibres. *Cell Biology International* 23(12): 829–840. https://doi.org/10.1006/cbir.1999.0479.

Kelsey, K.T., Yano, E., Liber, H.L., et al. (1986). The in vitro genetic effects of fibrous erionite and crocidolite asbestos. *British Journal of Cancer* 54(1): 107–114. https://doi.org/10.1038%2Fbjc.1986.158.

Kodama, Y., Boreiko, C.J., Maness, S.C., et al. (1993). Cytotoxic and cytogenetic effects of asbestos on human bronchial epithelial cells in culture. *Carcinogenesis* 14(4): 691–697. https://doi.org/10.1093/carcin/14.4.691.

Landesman, J.M. and Mossman, B.T. (1982). Induction of ornithine decarboxylase in hamster tracheal epithelial cells exposed to asbestos and 12-O-tetradecanoylphorbol-13-acetate. *Cancer Research* 42(9): 3669–3675.

Larson, T.C., Williamson, L., and Antao, V.C. (2020). Follow-up of the Libby, Montana screening cohort: a 17-year mortality study. *Journal of Occupational and Environmental Medicine* 62(1): e1–e6. https://doi.org/10.1097/jom.0000000000001760.

Le, H.Q., Tomenson, J.A. Warheit, D.B., et al. (2018). A review and meta-analysis of occupational titanium dioxide exposure and lung cancer mortality. *Journal of Occupational and Environmental Medicine* 60(7): e356–e367. https://doi.org/10.1097/JOM.0000000000001314.

Lechner, J.F., Tokiwa, T., LaVeck, M., et al. (1985). Asbestos-associated chromosomal changes in human mesothelial cells. *Proceedings of the National Academy of Sciences of the United States of America* 82(11): 3884–3888. https://doi.org/10.1073/pnas.82.11.3884.

Lentz, T.J., Rice, C.H., Succop, P.A., et al. (2003). Pulmonary deposition modeling with airborne fiber exposure data: a study of workers manufacturing refractory ceramic fibers. *Applied Occupational and Environmental Hygiene* 18(4): 278–288. https://doi.org/10.1080/10473220301404.

Lippmann, M. (2014). Toxicological and epidemiological studies on effects of airborne fibers: coherence and public [corrected] health implications. *Critical Reviews in Toxicology* 44(8): 643–695. https://doi.org/10.3109/10408444.2014.928266.

MacCorkle, R.A., Slattery, S.D., Nash, D.R., et al. (2006). Intracellular protein binding to asbestos induced aneuploidy in human lung fibroblasts. *Cell Motility and the Cytoskeleton* 63(10): 646–657. https://doi.org/10.1002/cm.20151.

Marsh, J.P. and Mossman, B.T. (1988). Mechanisms of induction of ornithine decarboxylase activity in tracheal epithelial cells by asbestiform minerals. *Cancer Research* 48(3): 709–714.

McClellan, R.O., Miller, F.J., Hesterberg, T.W., et al. (1992). Approaches to evaluating the toxicity and carcinogenicity of man-made fibers: summary of a workshop held November 11–13, 1991, Durham, *North Carolina. Regulatory Toxicology and Pharmacology* 16(3): 321–364. https://doi.org/10.1016/0273-2300(92)90011-w.

McGavran, P.D., Moore, L.B. and Brody, A.R. (1990). Inhalation of chrysotile asbestos induces rapid cellular proliferation in small pulmonary vessels of mice and rats. *American Journal of Pathology* 136(3): 695–705.

Morrow, P.E. (1988). Possible mechanisms to explain dust overloading of the lungs. *Fundamental and Applied Toxicology* 10(3): 369–384. https://doi.org/10.1016/0272-0590(88)90284-9.

Mossman, B.T. (2018). Mechanistic in vitro studies: what they have told us about carcinogenic properties of elongated mineral particles (EMPs). *Toxicology and Applied Pharmacology* 361: 62–67. https://doi.org/10.1016/j.taap.2018.07.018.

Mossman, B.T. and Churg, A. (1998). Mechanisms in the pathogenesis of asbestosis and silicosis. *American Journal of Respiratory Critical Care Medicine* 157(5 Pt 1): 1666–1680. https://doi.org/10.1164/ajrccm.157.5.9707141.

Mossman, B.T. and Gualtieri A.F. (2020). Lung cancer: mechanisms of carcinogenesis by asbestos. In: *Occupational Cancers* (eds. S. Antilla and P. Boffetta). Switzerland: Springer Nature. https://doi.org/10.1007/978-3-030-30766-0-12.

Mossman, B.T., Eastman, A., Landesman, J.M., et al. (1983). Effects of crocidolite and chrysotile asbestos on cellular uptake and metabolism of benzo(a)pyrene in hamster tracheal epithelial cells. *Environmental Health Perspectives* 51: 331–335. https://doi.org/10.1289/ehp.8351331.

Mossman, B.T., Bignon, J., Corn, M., et al. (1990). Asbestos: scientific developments and implications for public policy. *Science* 247(4940): 294–301. https://doi.org/10.1126/science.2153315.

Mossman, B.T., Lippmann, M., Hesterberg, T.W., et al. (2011). Pulmonary endpoints (lung carcinomas and asbestosis) following inhalation exposure to asbestos. *Journal of Toxicology and Environmental Health B Crit Rev* 14(1–4): 76–121. https://doi.org/10.1080/10937404.2011.556047.

Mossman, B.T., Shukla, A., Heintz, N.H., et al. (2013). New insights into understanding the mechanisms, pathogenesis, and management of malignant mesotheliomas. *The American Journal of Pathology* 182(4): 1065–1077. https://doi.org/10.1016/j.ajpath.2012.12.028.

Muhle, H., Pott, F., Bellmann, B., et al. (1987). Inhalation and injection experiments in rats to test the carcinogenicity of MMMF. *Annals of Occupational Hygiene* 31(4B): 755–764. https://doi.org/10.1093/annhyg/31.4b.755.

Murphy, F.A., Poland, C.A., Duffin, R., et al. (2011). Length-dependent retention of carbon nanotubes in the pleural space of mice initiates sustained inflammation and progressive fibrosis on the parietal pleura. *American Journal of Pathology* 178(6): 2587–2600. http://dx.doi.org/10.1016/j.ajpath.2011.02.040.

Murphy, F.A., Schinwald, A., Poland, C.A., et al. (2012). The mechanism of pleural inflammation by long carbon nanotubes: interaction of long fibres with macrophages stimulates them to amplify pro-inflammatory responses in mesothelial cells. *Particle and Fibre Toxicology* 9: 8. https://doi.org/10.1186%2F1743-8977-9-8.

Nagai, H., Okazaki, Y., Chew, S.H., et al. (2013). Intraperitoneal administration of tangled multiwalled carbon nanotubes of 15 nm in diameter does not induce mesothelial carcinogenesis in rats. *Pathology International* 63(9): 457–462. https://doi.org/10.1111/pin.12093.

Oberdörster, G. (1995). Lung particle overload: implications for occupational exposures to particles. *Regulatory Toxicology and Pharmacology* 21(1): 123–135. https://doi.org/10.1006/rtph.1995.1017.

Pacaud, R., Brocard, E., Lalier, L., et al. (2014). The DNMT1/PCNA/UHRF1 disruption induces tumorigenesis characterized by similar genetic and epigenetic signatures. *Scientific Reports* 4: 4230. https://doi.org/10.1038/srep04230.

Pache, J.C., Janssen, Y.M., Walsh, E.S., et al. (1998). Increased epidermal growth factor-receptor protein in a human mesothelial cell line in response to long asbestos fibers. *The American Journal of Pathology* 152(2): 333–340.

Pott, F., Ziem, U., Reiffer, F.J., et al. (1987). Carcinogenicity studies on fibres, metal compounds, and some other dusts in rats. *Experimental Pathology* 32(3): 129–152. https://doi.org/10.1016/s0232-1513(87)80044-0.

Quinlan, T.R., Marsh, J.P., Janssen, Y.M., et al. (1994). Dose-responsive increases in pulmonary fibrosis after inhalation of asbestos. *American Journal of Respiratory and Critical Care Medicine* 150(1): 200–206. https://doi.org/10.1164/ajrccm.150.1.8025751.

Quinlan, T.R., BeruBe, K.A., Marsh, J.P., et al. (1995). Patterns of inflammation, cell proliferation, and related gene expression in lung after inhalation of chrysotile asbestos. *The American Journal of Pathology* 147(3): 728–739.

Ramos-Nino, M.E., Timblin, C.R., and Mossman, B.T. (2002). Mesothelial cell transformation requires increased AP-1 binding activity and ERK-dependent Fra-1 expression. *Cancer Research* 62(21): 6065–6069.

Reiss, B., Solomon, S., Tong, C., et al. (1982). Absence of mutagenic activity of three forms of asbestos in liver epithelial cells. *Environmental Research* 27(2): 389–397. https://doi.org/10.1016/0013-9351(82)90094-9.

Roggli, V.L. (2018). Measuring EMPs in the lung what can be measured in the lung: asbestiform minerals and cleavage fragments. *Toxicology and Applied Pharmacology* 361: 14–17. https://doi.org/10.1016/j.taap.2018.06.026.

Ross, M., Nolan, R.P., Langer, A.M., et al. (1993). Health effects of mineral dusts other than asbestos. In: *Health Effects of Mineral Dusts* (eds. G.D. Guthrie and B.T. Mossman), 361–407. Washington, DC: Mineralogical Society of America. Reviews in Mineralogy.

Sayan, M. and Mossman, B.T. (2016). The NLRP3 inflammasome in pathogenic particle and fibre-associated lung inflammation and diseases. *Particle and Fibre Toxicology* 13(1): 51. https://doi.org/10.1186/s12989-016-0162-4.

Schinwald, A., Murphy, F.A., Prina-Mello, A., et al. (2012). The threshold length for fiber-induced acute pleural inflammation: shedding light on the early events in asbestos-induced mesothelioma. *Toxicological Sciences* 128(2): 461–470. https://doi.org/10.1093/toxsci/kfs171.

Sesko, A.M. and Mossman, B.T. (1989). Sensitivity of hamster tracheal epithelial cells to asbestiform minerals modulated by serum and by transforming growth factor beta 1. *Cancer Research* 49(10): 2743–2749.

Shelby, M.D. (1988). The genetic toxicity of human carcinogens and its implications. *Mutation Research* 204(1): 3–15. https://doi.org/10.1016/0165-1218(88)90113-9.

Shukla, A., Gulumian, M., Hei, T.K., et al. (2003). Multiple roles of oxidants in the pathogenesis of asbestos-induced diseases. *Free Radical Biology and Medicine* 34(9): 1117–1129. https://doi.org/10.1016/s0891-5849(03)00060-1.

Shukla, A., Vacek, P. and Mossman, B.T. (2004). Dose-response relationships in expression of biomarkers of cell proliferation in in vitro assays and inhalation experiments. *Nonlinearity in Biology, Toxicology, Medicine* (2): 117–128. https://doi.org/10.1080/15401420490464420.

Sincock, A.M., Delhanty, J.D., and Casey, G. (1982). A comparison of the cytogenetic response to asbestos and glass fibre in Chinese hamster and human cell lines. Demonstration of growth inhibition in primary human fibroblasts. *Mutation Research* 101(3): 257–268. https://doi.org/10.1016/0165-1218(82)90157-4.

Smith, D.M., Ortiz, L.W., Archuleta, R.F., et al. (1987). Long-term health effects in hamsters and rats exposed chronically to man-made vitreous fibres. *Annals of Occupational Hygiene* 31(4B): 731–754. https://doi.org/10.1093/annhyg/31.4b.731.

Stanton, M.F., Layard, M., Tegeris, A., et al. (1981). Relation of particle dimension to carcinogenicity in amphibole asbestoses and other fibrous minerals. *Journal of the National Cancer Institute* 67(5): 965–975.

Stettler, L.E., Sharpnack, D.D., and Krieg, E.F. (2008). Chronic inhalation of short asbestos: lung fiber burdens and histopathology for monkeys maintained for 11.5 years after exposure. *Inhalation Toxicology* 20(1): 63–73. https://doi.org/10.1080/08958370701665566

Tatrai, E., Kovacikova, Z., Brozik, M., et al. (2004). Pulmonary toxicity of wollastonite in vivo and in vitro. *Journal of Applied Toxicology* 24(2): 147–154. https://doi.org/10.1002/jat.965.

Vincent, J.H., Johnston, A.M., Jones, A.D., et al. (1985). Kinetics of deposition and clearance of inhaled mineral dusts during chronic exposure. *British Journal of Industrial Medicine* 42(10): 707–715. https://doi.org/10.1136%2Foem.42.10.707.

Wagner, J.C. and Skidmore, J.W. (1965). Asbestos dust deposition and retention in rats. *Annals of the New York Academy of Science* 132(1): 77–86. https://doi.org/10.1111/j.1749-6632.1965.tb41091.x.

Wagner, J.C., Berry, G., Skidmore, J.W., et al. (1974). The effects of the inhalation of asbestos in rats. *British Journal of Cancer* 29(3): 252–269. https://doi.org/10.1038%2Fbjc.1974.65.

Wagner, J.C., Skidmore, J.W., Hill, R.J., et al. (1985). Erionite exposure and mesotheliomas in rats. *British Journal of Cancer* 51(5): 727–730. https://doi.org/10.1038/bjc.1985.108.

Warheit, D.B., Hartsky, M.A., McHugh, T.A., et al. (1994). Biopersistence of inhaled organic and inorganic fibers in the lungs of rats. *Environmental Health Perspectives* 102 Suppl 5: 151–157. https://doi.org/10.1289%2Fehp.94102s5151.

Weill D. (2023). Editorial. Proceedings of the Monticello II conference on elongate mineral particles (EMP). *Environmental Research* 230: 115776. https://www.sciencedirect.com/journal/environmental-research/vol/230/suppl/C.

Williams, G.M. (1979). Review of in vitro test systems using DNA damage and repair for screening of chemical carcinogens. *Journal of Association of Official Analytical Chemists* 62(4): 857–863.

Woodworth, C.D., Mossman, B.T., and Craighead, J.E. (1983). Induction of squamous metaplasia in organ cultures of hamster trachea by naturally occurring and synthetic fibers. *Cancer Research* 43(10): 4906–4912.

Wylie, A.G. and Korchevskiy, A.A. (2023). Dimensions of elongate mineral particles and cancer: a review. *Environmental Research* 230: 114688. https://doi.org/10.1016/j.envres.2022.114688.

Yamato, H., Oyabu, T., Ogami, A., et al. (2003). Pulmonary effects and clearance after long-term inhalation of potassium octatitanate whiskers in rats. *Inhalation Toxicology* 15(14): 1421–1434. https://doi.org/10.1080/08958370390248969.

3

Health Outcomes of Asbestos Exposure – A Pathology and Diagnostic Perspective

Bruce Case

Department of Pathology, Faculty of Medicine and Health Sciences, McGill University, Montreal, Canada

Introduction

Asbestos exposures can result in numerous changes in structure and function in humans, mainly in the thorax as a result of inhalation of fibers. These changes may or may not be defined as "disease," and when they are, they can be malignant or nonmalignant.

Since the body has limited ways to react to exogenous insult, most of the changes associated with asbestos exposure, including all of the commonly recognized asbestos-related diseases (ARD) are not exclusive to that exposure. Further, the term "asbestos" as used in general, mineralogical, and regulatory contexts is generic and changes over time, as new fibrous minerals which meet (or do not meet) changing definitions are added to the expanding list (Case et al. 2011). For our purposes in this chapter, we will consider any "fiber" having *either* asbestiform habit as defined mineralogically *or* chemical and physical attributes commonly described as *asbestos associated with health effects* as "asbestos." However, since subspecies of asbestos (and fiber categories with similarities to asbestos, such as nonasbestiform minerals having similar physical appearance and chemistry) differ in health effects, we'll come back to the differences of necessity on occasion. "Asbestos" is a term best used with quotation marks intact in the reader's head.

Of course, asbestiform fibers are not the only fibrous structures causing biological effects. There are differences in biological reactions and for some diseases, quantitative effects, of different types of asbestos. Similarly, other fibers or "elongated mineral particles" (see definitions elsewhere in this volume) vary in effect as well. With a few exceptions such as so-called "Libby Amphibole," this chapter will be

Health Risk Assessment for Asbestos and Other Fibrous Minerals, First Edition.
Edited by Andrey Korchevskiy, James Rasmuson, and Eric Rasmuson.
© 2024 John Wiley & Sons, Inc. Published 2024 by John Wiley & Sons, Inc.

limited to biological reactions to and health effects of the six regulated asbestos fibers, including chrysotile and the two commercial and three regulated noncommercial asbestiform amphiboles.[1] Where relevant, differences between effects of the latter may be mentioned, with the understanding that all of the regulated forms of asbestos can cause all of the usually listed malignant and nonmalignant asbestos-related diseases, with "considerable evidence" for potency differences for mesothelioma, and less for lung cancer (IARC 2012; Travis et al. 2015).

Nonmalignant Change in Structure or Function

Arguably, the first change in "structure" caused by exposure is the entry of fibers into the body, in that the fibers are themselves foreign in nature. To have foreign material within the human body is itself an "abnormality," albeit not always one that permanently alters any structure or function. Reactions to the presence of foreign material vary greatly, from rapid clearance to long-term consequences of uncertain pathogenesis ending in the development of neoplasms decades later.

What happens when asbestos fibers enter the body? We have three ways of approaching this; theoretical/modeled, real-time animal models, and human studies. Of these, human studies are the most appropriate but the least possible. Observing the actual inhalation of asbestos fibers by humans cannot be performed for obvious reasons, and what we know is based mostly on extrapolation from the other two approaches and from observations many years after first exposure. What "actually happens" after asbestos fibers enter the human body is not known directly, especially for carcinogenesis.

Early changes are related to interactions between asbestos fibers and cells, in the milieu in the alveolar spaces in which they arrive when inhaled. These vary by fiber type, and while many types of reaction have long been known, new discoveries are teaching us how fibers react differently.

As might be expected, of the six "classical" asbestos fibers, chrysotile reacts differently than the asbestiform amphiboles. Acid-leaching of magnesium from chrysotile structures takes place rapidly, and in animal experiments rapid clearance results. However, in humans under conditions of continuing exposure, the situation may be quite different. Recent work by Gualtieri and colleagues *in vitro* (Gualtieri et al. 2019) demonstrates that "chrysotile's fibrous structure induces cellular damage, mainly through physical interactions." Their toxicity model

---

1 That is, commercial amphiboles "amosite" and "crocidolite" and the asbestiform variants of the amphiboles tremolite, actinolite, and anthophyllite. This limitation is not intended to imply lack of toxicity of any other class of minerals, but to simplify this discussion.

indicates that "inhaled chrysotile fibers exert their toxicity in the alveolar space by physical and biochemical action. The fibers are soon leached by the intracellular acid environment into a product with residual toxicity, and the dissolution process liberates toxic metals in the intracellular and extracellular environment." Because chrysotile is less biopersistent than commercial and noncommercial amphiboles, these reactions occur early in exposure history but may not be sustained. Amphiboles on the other hand are more biopersistent and may exert their effects by other means for longer periods of time.

Nonmalignant Asbestos-Related Disease

Pleural

Asbestos Effusion

An "asbestos effusion" is the most common effect of asbestos exposure in humans during the first 10 years after first exposure to asbestos. Up to 2,000 ml fluid may accumulate in the pleural space, but asbestos effusions are usually small and often clinically undetected. Because such effusions can have other causes, including treatable infectious disease such as tuberculosis, these must be ruled out. Underlying malignancy, including mesothelioma but also pleural and pulmonary metastases from a wide variety of sites, must also be ruled out. An occupational history addressing asbestos exposure is needed, but diagnosis will depend upon pleural fluid cytology and especially imaging such as CT and PET scans; effusion must be confirmed by a "transient pleural change" at a minimum in serial chest films or by thoracentesis (Epler et al. 1982). Asbestos effusions should be carefully followed for at least 2 years to ensure there is no underlying malignancy, and longer if history and clinical findings merit. While asbestos effusions may be the first manifestation of any ARD, they are often, and in some series usually, associated with other lesions including pleural plaques, round atelectasis, diffuse pleural thickening, and parenchymal asbestosis (Porcel 2017).

Pleural Plaques and Localized Pleural Thickening (LPT)

The normal pleural space is so thin that it cannot be seen even on high-resolution CT unless there is pleural pathology. Pleural fibrosis caused by asbestos exposure takes several forms, and subtypes are still being identified. Mutsaers et al. (2004) hypothesized that a "key event" in pathogenesis of asbestos-related pleural fibrosis was transport of fibers in the lung interstitium to the subpleural space by means of lymphatic flow. They noted however that "it is impossible to predict whether or not asbestos exposure will cause pleural fibrosis, even in the presence of developing asbestosis in the lung parenchyma," leading to the suggestion that "the degree and pattern of fibrogenic disease is controlled by certain, as yet undefined, genes that control susceptibility" (Mutsaers et al. 2004). As with other forms of pleural fibrosis, subpleural fibroblasts in an extracellular matrix underlying the mesothelial cell layer are involved in the production of excess collagen and other components as a response to injury, with collagenous scar tissue development as the result. Mutsaers et al. suggest that "at least" two mechanistic pathways "are likely to be involved in asbestos-induced pleural fibrosis"; one involving cytokine and growth factor production (Mutsaers et al. 2004). The other, suggested mechanistically in a wide range of asbestos-related disease, is oxygen radical production known to be caused by interaction of cells with asbestos fibers (Case et al. 1986), through unknown mechanisms, possibly involving both structural iron in commercial amphiboles and surface iron for chrysotile asbestos.

Pleural plaques are circumscribed and discrete areas of hyaline or calcified fibrosis, which are localized on the parietal pleura of the lateral chest wall, the diaphragm, or the mediastinum (Gevenois et al. 1998). They are quite common in many general populations, found in 1–2% of males and less in females, and can be much higher in some autopsy series. Plaques are often missed (even when clearly present) on chest X-ray but can also be said to be present when they are not, including on CT scans. Calcification of plaques, which has been associated with amphibole exposures but not exclusively, enhances definitive identification. Estimates are that under 20% of plaques are calcified, with wide variation. Plaques are closely related to past asbestos exposures, but can be mimicked by other conditions, including normal anatomical variants (Alfudhili et al. 2016). Visceral pleural thickening can be confused with parietal plaques but can also be associated with past asbestos exposures. Scarring from old infectious and noninfectious chronic disease, including tuberculosis, fungal infections, and silicosis, can be misinterpreted as "plaques" radiologically. Pathological identification of plaques related to past exposure is distinct, as these usually smooth, elevated lesions have sharp borders and if not adherent to other structures are pathognomonic of past exposure. Past exposure to asbestos is sometimes included as a necessary part of the definition of a pleural plaque, but this is misleading. Past exposure can be unknown, and plaques identified radiologically or anatomically may lead to further questioning, which discovers the exposures. Alternatively, pseudo-plaques, especially if unilateral, should be assessed on their appearances rather than from exposure history.

True pleural plaques can occur at low dose, most commonly following amphibole asbestos exposures. Unlike asbestos effusions, they rarely appear until long (two decades or more) after first exposure.

Since they denote exposure, plaques are often associated with an increased risk of other asbestos-related lesions and diseases. There is a great deal of variation from study to study in the coexistence of such diseases, as might be expected.

A workforce or cohort with large cumulative exposures will inevitably have plaques in association with other ARD, including true parenchymal asbestosis (see below). On the other hand, general population studies or case series where populations may be exposed to asbestos in soil or air may document pleural plaques without other ARD (Churg and DePaoli 1988). As one example, farmers have a dusty occupation, and although "Fibrous mineral species are generally uncommon in soil" they have been reported as "important locally – for example, richterite and anthophyllite asbestos in Finland, tremolite in Cyprus, and erionite in Turkey and Nevada" (Schenker et al. 1998).

Although the issue of the presence of plaques as an "independent risk factor" for other ARD, including malignancy, has been studied, it is not possible to remove collinearity from any model that explores this. Further, there is no good biologic rationale to imply that plaques themselves, as opposed to the exposures, which caused them, could cause or contribute to the cause of other pleural or pulmonary diseases, with the exception of restrictive and painful syndromes seen in the circumstances of "LPT" as seen in exposures to Libby Amphibole (see below).

Pleural plaques are nearly always asymptomatic (BTS 2011), but some studies have indicated a marginal possible effect on pulmonary function. Bourbeau et al. (1990) studied a small group of 110 currently employed construction insulators. Just over half had pleural plaques without any other abnormality; 5.5% had diffuse pleural thickening. Routine chest X-rays were used for plaque assessment. Compared with those without plaques, "those with any pleural abnormality had a decrease in FEV_1 and FVC on average of 222 and 402 ml ($p<0.05$), and those with isolated pleural plaques, a decrease on average of 200 and 350 ml ($p<0.05$), after taking into account age, height, smoking status, and the presence of parenchymal abnormality as assessed by chest radiography and gallium uptake." In addition, the plaques were symptomatic, in the sense that shortness of breath complaints were "also significantly related to the width and extent of chest wall pleural thickening ($p<0.05$), independently of parenchymal disease."

A new syndrome involving similar and often radiologically indistinguishable lesions has recently been described where there may also be evidence of plaque-related pulmonary dysfunction. "Libby Amphibole Asbestos" is a combination of the three amphiboles tremolite, winchite, and richterite in association with vermiculite. This complex mineral grouping of amphiboles is seen in the mineral deposit at Zonolite Mountain near Libby, Montana, originally considered for asbestos ("tremolite") mining, but subsequently mined for vermiculite. The complex mixture of minerals included fibers with a wide range of morphologies from prismatic crystals to clearly asbestiform fibers (Meeker et al. 2003).

In the mid-1980s, epidemiological studies by McGill University and NIOSH investigators established a large risk of asbestos-related disease, including malignancies, in the Libby mining and milling workforce (McDonald et al. 1986, 2004;

Amandus and Wheeler 1987; Sullivan 2007), as well as in a group of workers exposed during extraction of vermiculite in mills to which the contaminated amphibole-contaminated vermiculite was shipped (Lockey et al. 1984). A huge body of research on exposure and disease in workers and neighborhood residents ensued, and ultimately US EPA published a risk assessment for "Libby Amphibole asbestos" (IRIS 2014) separate from that for "Asbestos" (IRIS 1988).

In the risk assessment for Libby Amphibole asbestos, a unique decision was made to not only assess carcinogenic risk, but also risk for noncarcinogenic health effects. This was in part because of, and almost entirely based on, the pleural effects observed, especially in a group of 280 vermiculite exfoliation plant workers in Marystown, Ohio. Libby Amphibole asbestos (LAA) exposure in these workers was estimated and cumulative exposure reconstructed estimate for each individual. Cumulative exposure ranged from 0.01 to 19.03 f/cc-yrs (mean = 2.48). The outcome measure for "nonmalignant respiratory disease" effect was LPT, corrected for smoking, age, etc. Over the time period of the studies from 1984 to 2008, the prevalence of pleural thickening increased from 2% to 28.7% (Rohs et al. 2007). The rationale – that LPT, which was not differentiated from pleural plaques, was not just an asymptomatic change in structure, but was a "disease" associated with "respiratory dysfunction" – was a nontraditional view based entirely on statistical differences in some studies in numerical values for pulmonary function tests. It was hard to determine "clinical significance," and others have criticized the IRIS decision. Maxim et al. (2015) critique most of the few studies that have suggested any pulmonary function effects from "plaques," and point out that, "With very few exceptions...investigators have concluded that most subjects with pleural plaques are nearly always asymptomatic...in the six cross-sectional studies of non-Libby populations that evaluated the association between pleural plaques and pulmonary function with adjustment for asbestos exposure results varied depending on the measure of lung function, with no consistent patterns of association with spirometric or symptom-based outcomes. Moreover, two studies did not distinguish LPT from (diffuse pleural thickening)." The authors also point out that in the Ohio Marystown studies, while those with LPT had lower predicted pulmonary function values compared with those with normal HRCT/CTs by 5.4% and 3.3% for FVC ($p < 0.05$) and FEV_1 ($p = 0.14$), respectively, this might not be judged clinically significant by most standards (Maxim et al. 2015).

On the one hand, to the degree that pulmonary dysfunction – or decrease in pulmonary testing values – is associated, it may well be due to "missed" other disease such as subclinical lung disease, etc. On the other hand, it is possible to look at the small risk the same way we look at the effects of lead on children's IQ scores. That is, while the changes in function as measured numerically may be small "on average," this in itself denotes a potentially serious problem in some part of the population at risk; for example, the upper quartile. Defining that population is another matter: both carcinogenic risk and noncarcinogenic risk for LAA have been defined on extremely small populations of workers, but many millions of individuals have had potential environmental exposure to LAA as a contaminant of in-place vermiculite.

With respect to the actual pathology of "LAA" plaques, or LPT, there may be a *biological* difference between these and what we have called "pleural plaques" in the past. There is in fact very little actual pathology material available for review in these cases, as they are nonfatal and do not require tissue diagnosis. However, symptoms in these patients, such as increased pain and atypical radiological appearances, differ from those with usual "pleural plaques."

CT scans have revealed that many of the LAA "plaques" show as "a thin layer of pleural thickening internal and parallel to the ribs, distinguished from subpleural fat." Pleural thickening is frequent, identified in 96 (48%) of 198 heavily exposed vermiculite miners. Of these 49 (51%) had only or predominantly circumscribed thickening, and 47 (49%) had only or predominantly lamellar pleural thickening (Miller et al. 2018). Similar findings were seen among a group exposed environmentally in childhood who then moved away from the area (Szeinuk et al. 2017).

Diffuse Pleural Thickening

Although as might be imagined by the name, diffuse pleural thickening is more severe in its clinical effects than LPT – or pleural plaques – it is not as well defined. Prevalence and incidence vary from era to era, job to job, and with method of and criteria for diagnosis. For the most part the definition is radiological and medicolegal – a necessary step in the determination of workman's compensation schemes, for example. Most definitions involve a specified thickness in the range of 1 mm to 1 cm, and in severe cases, the thickening can be even greater.

DPF affects the visceral pleural surface but is often adherent to the adjacent parietal pleural surface. The tissue consists of "pale gray diffuse thickening that blends at the edges with the more normal pleura" (ATS 2004). It can extend over the entire lung, or a lobe of lung, and may involve the costophrenic angles.

The degree to which it is accompanied by underlying asbestosis also varies related to exposure. Unlike the situation for pleural plaques, there is no doubt that diffuse pleural thickening, with or without underlying pulmonary disease, can be a cause of serious pulmonary dysfunction. It is, because of its effect on the lung and respiration, the physiological equivalent of restrictive parenchymal lung disease. It may occur with asbestosis, following asbestos effusions, or *de novo* without any accompanying disease. Radiologically, it can on occasion be confused with mesothelioma.

Pathogenetically, DPF may begin as immature granulation tissue and fibrin at the visceral pleural surface, with later scarring composed of mature collagen.

Scarring extends into the peripheral lung, giving a characteristic "crow's foot" radiological appearance.

Physiologically, the degree of pulmonary dysfunction varies with associated conditions such as asbestosis. In one study, constrictive (restrictive) respiratory dysfunction (as percent with Vital Capacity under 80%) was found in 91% of 96 asbestos-exposed patients, and obstructive respiratory dysfunction (FEV_1/FVC less than 70% predicted in 28%). Mixed respiratory dysfunction was found in 24%. Vital capacity deficit was correlated with degree of extension of DPT radiologically (Fujimoto et al. 2014).

Rounded Atelectasis

A consequence of pleuritis, this folding of the lung can have other causes, but is often associated with asbestos-induced pleural disease. The definition is radiological (chest X-ray or CT), often as "an incidental finding manifesting as a rounded, wedge-shaped or lentiform opacity" (Riley and Naidoo 2018). Historically, it has been confused clinically with lung cancer, and because of this, it is sometimes resected, although correct interpretation of imaging should avoid this. In images, rounded atelectasis is a peripheral mass of "rounded, wedge-shaped or lentiform morphology of variable size with the pathognomonic 'comet-tail' sign of converging bronchovascular markings" (Riley and Naidoo 2018).

Cases associated with asbestos exposure where resection was performed in error may show associated asbestos bodies histologically, in association with the typical folded and fibrotic visceral pleura with atelectasis. Pleural plaques and other forms of pleural thickening are often associated, and the lesion is most common in the lower lung zones, where it can be single or multiple. Pathophysiology is debated (ATS 2004).

Lung

Asbestosis

Asbestosis is diffuse interstitial fibrosis of the pulmonary parenchyma caused by inhalation of asbestos fibers. The lung has limited ways to respond to illness, and many other stimuli can lead to interstitial fibrosis. The first step in determining whether interstitial fibrosis is, in fact, caused by asbestos exposure and is therefore asbestosis is to define the terms that prove this to be true.

There are two ways to do this: using either a clinical or pathologic definition. Currently, in medicine, the clinical definition is most important, since evidently it would be counterproductive – indeed unethical – to insist upon removing tissue from an affected person to prove that they have the disease. Nevertheless, the underlying disorder is one of lung pathology, and therefore both "definitions" have merit.

under the heading of asbestosis. Their conclusion was that "bronchiolar wall fibrosis should not be referred to as asbestosis and (we prefer) the term asbestos airways disease for bronchiolar wall fibrosis associated with asbestos bodies" (Roggli et al. 2010).

The authors also directly tackled the question of the need to see asbestos and provided a quantitative approach using iron-stained routine sections, which had been developed previously by the lead author (Hammar and Abraham 2015).

As with the ATS radiological definition, these "new criteria" have been controversial, in part because they would tend to exclude (properly or improperly) cases previously diagnosed as "asbestosis." In the case of the new CAP pathologic guidelines, one criticism proffered was that the "new" pathologic criteria were insufficient in the authors' views "for recognizing asbestosis at its earliest stages; with statements focusing on the number of asbestos bodies needed in order to make a pathologic diagnosis of asbestosis" (Roggli and Pratt 1983). An assessment reporting an international consensus meeting largely agreed with the new definitions (Wolff et al. 2015), which was unsurprising given overlapping authorship: the consensus report states "The Roggli–Pratt modification of CAP-NIOSH system is recommended."

Given the movement toward more restrictive diagnosis of early asbestosis pathologically, and less restrictive definition clinically and radiologically, it is difficult to conclude what the minimum criteria for "asbestosis" actually are. It is also quite probably irrelevant clinically, since most cases of "asbestosis" in clinical practice have indices of both radiologic abnormality and (if pathology is even available) interstitial fibrosis with accumulations of asbestos bodies that far exceed the minimum. In an individual case, what is most important is the degree of pulmonary dysfunction caused by any abnormalities, and in the medicolegal situation, whether the exposures believed to have caused the abnormalities can be clearly defined and linked to the disease.

Malignant Diseases Attributable to Asbestos Exposure

General Comments

The 2012 IARC monograph on the carcinogenicity of arsenic, metals, fibers, and dusts, and specifically the section on "*Asbestos (Chrysotile, Amosite, Crocidolite, Tremolite, Actinolite, And Anthophyllite),*" is the most recent comprehensive description of carcinogenic hazard for "asbestos" (IARC 2012). Like all such compendia, it is not universally accepted. IARC's mission is to place materials and other matters (such as type of work) in a hierarchy of carcinogenic *hazard* rather than risk. For substances or groups of substances like "asbestos," which is an

Unfortunately, current medical science is moving in opposite directions with such definitions, with clinical definition becoming more loose and pathologic definition more restrictive.

The American Thoracic Society first evaluated criteria for asbestos diagnosis in 1986 and re-evaluated them in 2004 (ATS 1986, 2004). ATS's clinical definition logically requires a history of exposure to asbestos as one component – one cannot clinically "see" asbestos in the way one can under a microscope (see below). The amount of exposure necessary was "moderate to heavy asbestos exposure, typically, but not always, occupational and often protracted for many years." The second requirement is the presence of interstitial fibrosis, and this is generally confirmed by radiological criteria. These include reticular-linear diffuse opacities in the lower zones of the lung fields, usually using a radiologic grading scheme for parenchymal changes. Historically, such grading schemes have included the International Labour Organization system (ILO 2002), which classified "profusion scores" for small irregular opacities on a chest X-ray meeting certain minimum technical standards as 0/0, 0/1, 1/0, 1/1, 1/2, 2/1, and >2/2, based on expert reader's comparisons with a set of standard X-rays.

The clinical 1986 ATS definition included small irregular opacities of expert reading of radiologic profusion score of at least "1/1." In 2004 ATS noted that historically, "A critical distinction is made between films that are suggestive but not presumptively diagnostic (0/1) and those that are presumptively diagnostic but not unequivocal (1/0). This dividing point is generally taken to separate films that are considered to be 'positive' for asbestosis from those that are considered to be 'negative'" (ATS 2004). Since profusion itself is continuous, however, ATS went further in attempting to define the minimal necessary change as "A profusion of irregular opacities at the level of 1/0 is used as the boundary between normal and abnormal in the evaluation of the film, *although the measure of profusion is continuous and there is no clear demarcation between 0/1 and 1/0*" (emphasis added). Further, ATS relied heavily on a subgroup in a now more than 30-year-old small study (Kipen et al. 1987). That study, according to ATS (2004) found that "Among individuals with asbestosis confirmed by histopathologic findings, 15–20% had no radiographic evidence of parenchymal fibrosis...similar to the proportion of other interstitial lung diseases that present with normal chest films." In fact, the pre-1973 X-rays of 138 insulation workers that were evaluated had only 23 cases with moderate to severe histopathological fibrosis at radiological profusion less than 0/1 profusion score – only nine of whom had "Complete absence of X-ray changes including pleural abnormalities." This small number was a subgroup of 219 "with adequate X-rays," who were themselves a subgroup of "450 pathologically confirmed lung cancers." The purpose of the original study was to look at the question of whether asbestosis (as defined radiologically) was in fact "necessary" to determine attribution of lung cancer to exposure, a different issue altogether (see below).

The upshot was to make it "easier" to clinically (at 0/1 or even lower) determine that a person "has asbestosis"; a controversial decision that is the converse of the now more restrictive pathologic and anatomical diagnosis of the disease.

Determination of a "threshold" for the exposure needed to cause asbestosis in those so exposed is beyond the scope of this chapter, but generally, moderate to heavy exposure has been found necessary. Definitions of moderate to heavy are probabilistic rather than fixed but are quite consistent over time. For example, the Ontario Royal Commission of 1984, while accepting that susceptibility issues might allow for exceptions, concluded that, "On the basis of the available data, our best judgment as to the lifetime occupational exposure to asbestos at which the fibrotic process cannot advance to the point of clinical manifestation of asbestosis is in the range of 25 f/cc-yrs and below" (Dupré et al. 1984). Writing for the College of American Pathologists in their review 26 years later, Roggli et al. (2010) still agreed that "clinical asbestosis can be induced by cumulative asbestos exposure amounting to an estimated 25 f/ml-yrs," and that other previous studies have indicated a necessary dose "in the range of 25–100 f/ml-yrs."

Other elements useful in the diagnosis included pulmonary function abnormalities (principally but not exclusively restriction), and abnormal breath sounds (classically bibasilar rales) on auscultation, but the disease was ultimately diagnosed by radiological findings, history of exposure, and exclusion of other possible causes of the clinical picture.

The *pathologic* definition of asbestosis for ATS depended on past College of American Pathologist (CAP) and NIOSH determination of pathologic definition of asbestosis, especially using the 1982 criteria (Craighead et al. 1982). The 1982 minimum criteria necessary for a diagnosis of asbestosis were "discrete foci of fibrosis in the walls of respiratory bronchioles associated with accumulations of asbestos bodies" in histologic sections. "Accumulations" of asbestos bodies were not well defined. A grading scheme for histological assessment of asbestosis was provided but has not generally been followed in clinical pathology practice.

In 2010, the CAP revised this, realizing that similar minimum criteria for fibrosis can be seen with other dusts, and indeed in smokers (Roggli et al. 2010). The new CAP criteria specifically require the presence of fibrosis not only of the walls of the respiratory bronchioles but also alveolar septal fibrosis. This was a mor restrictive definition provided to avoid confusion of low-grade asbestosis wi "distinct and different from small airways mineral-dust disease, where multifo peribronchiolar fibrosis may be seen after the inhalation of a variety of min dusts, including asbestos, silicates, metal oxides, and coal (and also) seen monly in the lungs of tobacco smokers." They excluded bronchiolar wall fi in the *absence* of alveolar septal fibrosis in association with asbestos referred to by some previously as "early asbestosis," acknowledging th are arguments for and against the inclusion of bronchiolar wall fibro

Scarring extends into the peripheral lung, giving a characteristic "crow's foot" radiological appearance.

Physiologically, the degree of pulmonary dysfunction varies with associated conditions such as asbestosis. In one study, constrictive (restrictive) respiratory dysfunction (as percent with Vital Capacity under 80%) was found in 91% of 96 asbestos-exposed patients, and obstructive respiratory dysfunction (FEV_1/FVC less than 70% predicted in 28%). Mixed respiratory dysfunction was found in 24%. Vital capacity deficit was correlated with degree of extension of DPT radiologically (Fujimoto et al. 2014).

Rounded Atelectasis
A consequence of pleuritis, this folding of the lung can have other causes, but is often associated with asbestos-induced pleural disease. The definition is radiological (chest X-ray or CT), often as "an incidental finding manifesting as a rounded, wedge-shaped or lentiform opacity" (Riley and Naidoo 2018). Historically, it has been confused clinically with lung cancer, and because of this, it is sometimes resected, although correct interpretation of imaging should avoid this. In images, rounded atelectasis is a peripheral mass of "rounded, wedge-shaped or lentiform morphology of variable size with the pathognomonic 'comet-tail' sign of converging bronchovascular markings" (Riley and Naidoo 2018).

Cases associated with asbestos exposure where resection was performed in error may show associated asbestos bodies histologically, in association with the typical folded and fibrotic visceral pleura with atelectasis. Pleural plaques and other forms of pleural thickening are often associated, and the lesion is most common in the lower lung zones, where it can be single or multiple. Pathophysiology is debated (ATS 2004).

Lung

Asbestosis
Asbestosis is diffuse interstitial fibrosis of the pulmonary parenchyma caused by inhalation of asbestos fibers. The lung has limited ways to respond to illness, and many other stimuli can lead to interstitial fibrosis. The first step in determining whether interstitial fibrosis is, in fact, caused by asbestos exposure and is therefore asbestosis is to define the terms that prove this to be true.

There are two ways to do this: using either a clinical or pathologic definition. Currently, in medicine, the clinical definition is most important, since evidently it would be counterproductive – indeed unethical – to insist upon removing tissue from an affected person to prove that they have the disease. Nevertheless, the underlying disorder is one of lung pathology, and therefore both "definitions" have merit.

Unfortunately, current medical science is moving in opposite directions with such definitions, with clinical definition becoming more loose and pathologic definition more restrictive.

The American Thoracic Society first evaluated criteria for asbestos diagnosis in 1986 and re-evaluated them in 2004 (ATS 1986, 2004). ATS's clinical definition logically requires a history of exposure to asbestos as one component – one cannot clinically "see" asbestos in the way one can under a microscope (see below). The amount of exposure necessary was "moderate to heavy asbestos exposure, typically, but not always, occupational and often protracted for many years." The second requirement is the presence of interstitial fibrosis, and this is generally confirmed by radiological criteria. These include reticular-linear diffuse opacities in the lower zones of the lung fields, usually using a radiologic grading scheme for parenchymal changes. Historically, such grading schemes have included the International Labour Organization system (ILO 2002), which classified "profusion scores" for small irregular opacities on a chest X-ray meeting certain minimum technical standards as 0/0, 0/1, 1/0, 1/1, 1/2, 2/1, and >2/2, based on expert reader's comparisons with a set of standard X-rays.

The clinical 1986 ATS definition included small irregular opacities of expert reading of radiologic profusion score of at least "1/1." In 2004 ATS noted that historically, "A critical distinction is made between films that are suggestive but not presumptively diagnostic (0/1) and those that are presumptively diagnostic but not unequivocal (1/0). This dividing point is generally taken to separate films that are considered to be 'positive' for asbestosis from those that are considered to be 'negative'" (ATS 2004). Since profusion itself is continuous, however, ATS went further in attempting to define the minimal necessary change as "A profusion of irregular opacities at the level of 1/0 is used as the boundary between normal and abnormal in the evaluation of the film, *although the measure of profusion is continuous and there is no clear demarcation between 0/1 and 1/0*" (emphasis added). Further, ATS relied heavily on a subgroup in a now more than 30-year-old small study (Kipen et al. 1987). That study, according to ATS (2004) found that "Among individuals with asbestosis confirmed by histopathologic findings, 15–20% had no radiographic evidence of parenchymal fibrosis...similar to the proportion of other interstitial lung diseases that present with normal chest films." In fact, the pre-1973 X-rays of 138 insulation workers that were evaluated had only 23 cases with moderate to severe histopathological fibrosis at radiological profusion less than 0/1 profusion score – only nine of whom had "Complete absence of X-ray changes including pleural abnormalities." This small number was a subgroup of 219 "with adequate X-rays," who were themselves a subgroup of "450 pathologically confirmed lung cancers." The purpose of the original study was to look at the question of whether asbestosis (as defined radiologically) was in fact "necessary" to determine attribution of lung cancer to exposure, a different issue altogether (see below).

The upshot was to make it "easier" to clinically (at 0/1 or even lower) determine that a person "has asbestosis"; a controversial decision that is the converse of the now more restrictive pathologic and anatomical diagnosis of the disease.

Determination of a "threshold" for the exposure needed to cause asbestosis in those so exposed is beyond the scope of this chapter, but generally, moderate to heavy exposure has been found necessary. Definitions of moderate to heavy are probabilistic rather than fixed but are quite consistent over time. For example, the Ontario Royal Commission of 1984, while accepting that susceptibility issues might allow for exceptions, concluded that, "On the basis of the available data, our best judgment as to the lifetime occupational exposure to asbestos at which the fibrotic process cannot advance to the point of clinical manifestation of asbestosis is in the range of 25 f/cc-yrs and below" (Dupré et al. 1984). Writing for the College of American Pathologists in their review 26 years later, Roggli et al. (2010) still agreed that "clinical asbestosis can be induced by cumulative asbestos exposure amounting to an estimated 25 f/ml-yrs," and that other previous studies have indicated a necessary dose "in the range of 25–100 f/ml-yrs."

Other elements useful in the diagnosis included pulmonary function abnormalities (principally but not exclusively restriction), and abnormal breath sounds (classically bibasilar rales) on auscultation, but the disease was ultimately diagnosed by radiological findings, history of exposure, and exclusion of other possible causes of the clinical picture.

The *pathologic* definition of asbestosis for ATS depended on past College of American Pathologist (CAP) and NIOSH determination of pathologic definition of asbestosis, especially using the 1982 criteria (Craighead et al. 1982). The 1982 minimum criteria necessary for a diagnosis of asbestosis were "discrete foci of fibrosis in the walls of respiratory bronchioles associated with accumulations of asbestos bodies" in histologic sections. "Accumulations" of asbestos bodies were not well defined. A grading scheme for histological assessment of asbestosis was provided but has not generally been followed in clinical pathology practice.

In 2010, the CAP revised this, realizing that similar minimum criteria for fibrosis can be seen with other dusts, and indeed in smokers (Roggli et al. 2010). The new CAP criteria specifically require the presence of fibrosis not only of the walls of the respiratory bronchioles but also alveolar septal fibrosis. This was a more restrictive definition provided to avoid confusion of low-grade asbestosis with "distinct and different from small airways mineral-dust disease, where multifocal peribronchiolar fibrosis may be seen after the inhalation of a variety of mineral dusts, including asbestos, silicates, metal oxides, and coal (and also) seen commonly in the lungs of tobacco smokers." They excluded bronchiolar wall fibrosis in the *absence* of alveolar septal fibrosis in association with asbestos bodies, referred to by some previously as "early asbestosis," acknowledging that there are arguments for and against the inclusion of bronchiolar wall fibrosis alone under the heading of asbestosis. Their conclusion was that "bronchiolar wall fibrosis should not be referred to as asbestosis and (we prefer) the term asbestos airways disease for bronchiolar wall fibrosis associated with asbestos bodies" (Roggli et al. 2010).

The authors also directly tackled the question of the need to see asbestos and provided a quantitative approach using iron-stained routine sections, which had been developed previously by the lead author (Hammar and Abraham 2015).

As with the ATS radiological definition, these "new criteria" have been controversial, in part because they would tend to exclude (properly or improperly) cases previously diagnosed as "asbestosis." In the case of the new CAP pathologic guidelines, one criticism proffered was that the "new" pathologic criteria were insufficient in the authors' views "for recognizing asbestosis at its earliest stages; with statements focusing on the number of asbestos bodies needed in order to make a pathologic diagnosis of asbestosis" (Roggli and Pratt 1983). An assessment reporting an international consensus meeting largely agreed with the new definitions (Wolff et al. 2015), which was unsurprising given overlapping authorship: the consensus report states "The Roggli–Pratt modification of CAP-NIOSH system is recommended."

Given the movement toward more restrictive diagnosis of early asbestosis pathologically, and less restrictive definition clinically and radiologically, it is difficult to conclude what the minimum criteria for "asbestosis" actually are. It is also quite probably irrelevant clinically, since most cases of "asbestosis" in clinical practice have indices of both radiologic abnormality and (if pathology is even available) interstitial fibrosis with accumulations of asbestos bodies that far exceed the minimum. In an individual case, what is most important is the degree of pulmonary dysfunction caused by any abnormalities, and in the medicolegal situation, whether the exposures believed to have caused the abnormalities can be clearly defined and linked to the disease.

Malignant Diseases Attributable to Asbestos Exposure

General Comments

The 2012 IARC monograph on the carcinogenicity of arsenic, metals, fibers, and dusts, and specifically the section on *"Asbestos (Chrysotile, Amosite, Crocidolite, Tremolite, Actinolite, And Anthophyllite),"* is the most recent comprehensive description of carcinogenic hazard for "asbestos" (IARC 2012). Like all such compendia, it is not universally accepted. IARC's mission is to place materials and other matters (such as type of work) in a hierarchy of carcinogenic *hazard* rather than risk. For substances or groups of substances like "asbestos," which is an

IARC group 1 carcinogen ("Sufficient evidence of carcinogenicity to humans"), IARC and similar working groups consider "that a causal relationship has been established between exposure to the agent and human cancer...A statement that there is sufficient evidence is followed by a separate sentence that identifies the target organ(s) or tissue(s) where an increased risk of cancer was observed in humans. Identification of a specific target organ or tissue does not preclude the possibility that the agent may cause cancer at other sites" (IARC 2012). For asbestos, the identified target organs with sufficient evidence were pleura and other mesothelium (mesothelioma), lung, larynx, and ovary. Proportions of cases caused by asbestos exposure for each target organ are not specified; the hazard identification is a simple statement that for the organs specified, the material assessed has been proven to be *capable* of causing cancer in humans. Each such assessment is also made not by an agency or government, but by a group of individuals. The IARC assessment "...represents the views and expert opinions of an IARC Working Group on the Evaluation of Carcinogenic Risks to Humans, which met in Lyon, 17–24 March 2009" (IARC 2012). IARC, through its publication of the Monographs Series, is the best-known World Health Organization (WHO) group to assess carcinogenic hazard, but it is not the only one. From the standpoint of the WHO Classification of Tumours, known to pathologists as the "WHO Blue Books" for example, the most recent edition notes under "etiology" that, "The majority of mesotheliomas are caused by occupational exposure to asbestos, with the strength of the relationship being dependent on fiber type. Commercial amphibole asbestos types, such as amosite and crocidolite, are more carcinogenic than serpentine asbestos (chrysotile)" (Tsao et al. 2021; World Health Organization 2021).

The historical range of assessments of malignant tumors caused by asbestos is wide. No other material has been assessed in so many ways for so long, and for almost any malignancy published papers can be found with a wide range of opinion about asbestos exposure association or causation. In this discussion, only mesothelioma and asbestos-related lung cancer are included. There is an emphasis on the development of knowledge and on pathology and diagnosis; other chapters in this volume deal with epidemiological data and risk assessment. The latter have important roles to play, particularly in relating the diseases to different types of asbestos exposure, with different time relationships, and highly variable attributable risks.

The history of development of knowledge about the role of asbestos for these two diseases is relatively recent, with first observations in the first half of the twentieth century and consolidation of knowledge from 1955 through approximately 1970. The 1964 "New York Conference" hosted by Irving Selikoff and Jacob Churg and the publication of the proceedings as a separate volume of the Annals of the New York Academy of Sciences (Selikoff and Churg 1965) were turning points for acceptance of the relationship for asbestos-related lung cancer and for mesothelioma (Case 2008).

Asbestos-Related Lung Cancer

Until the early to mid-twentieth century, lung cancer was a fairly rare disease. Cigarette smoking changed this and continues to do so, with the combined histologic types of lung cancer rising rapidly in incidence and mortality until today, lung cancer is the leading cause of cancer death among both men and women in the United States, and the leading cause of cancer death among men and second among women worldwide (Torre et al. 2016). While attributable risks are highest with squamous cell and small cell subtypes, most lung cancers – perhaps in the range of 80% – are caused by cigarette smoking. Given the ascendance of that factor, it is not surprising that other factors – often *cofactors* – had only occasional mention. When lung cancer was first noted in cases with heavy asbestos exposure, it was usually with exposures so high they inevitably also caused asbestosis. In 1934, Wood and Gloyne reported two cases of lung cancer among 43 cases of asbestosis but did not in their paper claim a causal link to either exposure or fibrotic disease (Wood and Gloyne 1934; Enterline 1978, 1991; Enterline et al. 1978). In 1935, Gloyne specifically rejected a causal hypothesis in a report of an additional two cases found incidentally at autopsy in female asbestos plant workers (Gloyne 1935).

A single lung cancer case report by Lynch and Smith in 1935 again related the disease to "asbestosis" rather than asbestos exposure per se "by reason of chronic bronchial irritation...comparable with the current knowledge of the etiology of such tumors" (Lynch and Smith 1935; Enterline 1991). In Quebec, Paul Cartier, an asbestos mining and milling company physician, reported six lung cancer cases among 40 autopsy cases with asbestosis vs. seven among workers without asbestosis (denominator unknown). Cartier did not find "...any statistical relationship of a cause (sic) relationship" (Cartier 1955; Enterline 1991).

Most of the above studies were in fact not designed in ways that could properly assess causation issues at all; they were pathology collections or case reports. Only definitive epidemiological study, beginning with Doll in 1955 (Doll 1955) established asbestos once and for all as a cause of lung cancer, although even then asbestosis was often considered a necessary precondition.

This was a different era and while epidemiological studies were of importance most or all of the *cases* in those studies came not only histologically verified but having had individual autopsies – something which would be impossible today. Doll reviewed death registrations for 105 Coroner's autopsies from an asbestos textile plant performed between 1935 and 1952. Seventy-five had asbestosis, and of these 15 had lung cancer, while 3 of the remaining 30 without asbestosis had died with lung cancer – a smaller excess, but still an excess, without asbestosis. Separately, 113 male asbestos workers employed in certain areas of the plant for 20 or more years had lung cancer mortality risk ten times higher than that experienced by *age-matched members of the general population*. This population-based context of occupational cancer definitively established a relationship between asbestos *exposure* and lung cancer. Asbestosis, in retrospect, was a marker of exposure, and Doll never stated "asbestosis" itself to be a cause or precondition (Doll 1955).

The idea of asbestosis as a precursor, or precondition, or cause of asbestos-related lung cancer has died hard, and still has its adherents, but there is no real evidence that lung cancer causation by asbestos requires – or involves – asbestosis. That they are related at all is a result of the collinearity of asbestos exposure for the two: parallel dose–response relationships for both lung cancer and asbestosis. Asbestosis remains one of several good proxies for degree of asbestos exposure, the others being a detailed exposure history, when available, or a specified variable level of lung-retained fiber – which is almost never as available as in the past.

Smoking itself acts together with asbestos exposure, synergistically, to cause lung cancer. Lung cancer is caused by asbestos exposure in nonsmokers at exposures estimated by some at over 25 f/ml-yrs (Wolff et al. 2015), but historically far more asbestos-related lung cancers occur in smokers. The relationship between the two factors is more than additive, but probably less than multiplicative (Case 2006). Assigning the relative contribution of the two exposures is difficult in populations and very difficult in individual cases. The pathogenetic mechanisms are not known. Hypotheses involve the potentially carcinogenic compounds in cigarette smoke "traveling" into lung on asbestos fibers. Increased polycyclic aromatic hydrocarbon uptake from benzo(a)pyrene-coated asbestos fibers has been demonstrated *in vitro* (Eastman et al. 1983) and DNA strand breakage has been shown in rat respiratory tract epithelial cells after co-exposure to asbestos and cigarette smoke *in vivo* (Jung et al. 2000).

Lung cancer histological types have changed very little over the years. The principal categories of squamous cell, small cell, large cell, and adenocarcinoma were first proposed in 1924 and still account for most cases (Fraser et al. 1989). The most current fifth edition of the WHO classification (2) notes that almost all lung cancers are carcinomas, with "predominant histological types" adenocarcinoma, squamous cell carcinoma, small cell lung carcinoma, and large cell carcinoma. Changes are largely in subtypes (especially for adenocarcinoma) and related to molecular pathology testing and immunohistochemistry. These changes are important for treatment and for "personalized medicine," and additional molecular testing of tumor has become routine in major pathology centers.

All of the four major histological types of lung cancer can be caused by asbestos exposure, and despite many studies over the decades, there are no histologic types or subtypes that speak reliably to asbestos causation or lack of same. Rarer sarcomatoid and adenosquamous lung carcinomas have also been reported with and without asbestos exposure.

In some older studies, attempts were made to relate tumor site to asbestos causation with greater or lesser incidence in upper or lower lobes. Such studies were of no value for individual cases, and there are currently no accepted sites associated with exposure. Pathology may be most helpful in indication of asbestos causation in identifying the agent itself through the use of iron-stained sections of adjacent lung, where possible. However, there are at present no accepted values for "how many" asbestos bodies are necessary, and such testing is not a routine part of lung cancer assessment by pathologists. Lung-retained fiber analysis results may indicate exposure and even exposure levels, but methods are not standardized and very few laboratories perform this work today: most studies using this type of assessment are either based on medicolegal cases or are specific to national groups dedicated to asbestos research. Further, as asbestos-related lung cancer occurs in older age groups, fiber clearance may decrease previously suggested minimal lung "burdens," especially for chrysotile asbestos.

Given advances in molecular pathology and genetics there may, at some point, be value in using some such techniques in teasing out those lung cancers related to asbestos exposure. Nymark and colleagues in Finland have begun this work, and a constellation of genetic abnormalities has been identified, which parallel asbestos exposure and lung fiber content, but the changes are not yet adequately sensitive and specific for routine use (Nymark et al. 2013).

Mesothelioma – Accelerating Knowledge

Although malignant mesothelioma remains a rare disease, and in fact is decreasing in incidence in western countries, there has been an explosion in knowledge about the disease, which precludes a comprehensive or even up-to-date chapter section.

When Christopher J. Wagner first established the linkage between exposure to crocidolite asbestos and mesothelioma in the South African Cape mining area (Wagner et al. 1960), not all pathologists yet recognized it as a disease separate from lung cancer. A search of the PubMed database for "mesothelioma" finds the first listed case report in 1932 (Hashiba et al. 1932), although there are other cited historical cases. Hashiba et al. (1932) make clear that what was recognized was a "primary malignant tumor of the pleura." While pathologists recognized the entity, most preferred the term "endothelioma" of the pleura, and whatever it was called, it was so rare that there were less than 50 cases in the literature by one account. In 1901, Adler (Adler 1901) described a case and noted that nonetheless, even then "an extensive literature ha(d) gradually developed on the subject...(but) the older literature is practically useless." Adler himself held the belief, as did most, that there was no epithelial/epithelioid component in the malignancy, and that there had been insufficient attention to separating pleural metastases from primary tumors. Clarkson (1914), also calling the condition "primary endothelioma of the pleura," stated without reference that 2 of 10,829 autopsies at the "Pathological Institute of Munich" had been "primary endothelioma of the pleura," and summarized what he believed to be all 41 referenced cases to the date of publication in 1914. Although some have argued for some recognition as early as the eighteenth century, this is not well established. Not surprisingly, recognition of the disease itself began as incidence burst forth in the second half of the twentieth century as a result of asbestos exposures.

Wagner's own discovery was nearly missed, as was the case for many years to follow, with resulting underdiagnosis in many countries up to the present day. In South Africa, some cases were confused with tuberculosis. Sleggs, the general practitioner who co-authored Wagner's first paper, actually "found" the cases in and after 1952 during his work as Superintendent of a tuberculosis hospital, where he began to see cases of pleural tuberculosis that did not respond to therapy. He consulted the surgeon, Marchand, and cases were referred to Wagner. Wagner, whose father was a prominent geologist, had recently been appointed Asbestos Research Fellow at the South African Pneumoconiosis Research Institute, which he joined in 1952 (Hashiba et al. 1932). On 15 February 1956, he "examined the body of a black shower attendant and diagnosed a malignant mesothelioma on the macroscopic appearance postmortem. Because this was such a controversial diagnosis I called Professor Becker to come and see the postmortem and he agreed that this might be a case of diffuse MM – appropriate histological stains, to confirm the diagnosis were carried out" (Wagner's personal notes, per Case 2008).

The diagnosis was readily accepted, although the association with asbestos exposure was debated – within 10 years; however, it was well accepted per Rochdale's physician J.F. Knox in a memo dated 18 January 1967; by this time "mesothelioma" itself was well accepted as the disease rather than pleural endothelioma: "Although I have 'grown up with' the concept of mesothelioma as a new form of tumour found in association with exposure to asbestos dust in certain circumstances since 1958, I did not accept the hypothesis readily. From a position of frank disbelief, I gradually accepted a position of qualified acceptance...(but) It was not suggested – and has never been suggested – that mesothelioma of the pleura never occurred except with previous asbestos exposure."

Misdiagnosis as other diseases, especially tuberculosis, dogged the ability to establish the true incidence of mesothelioma and even pleural plaques, and probably still does today in less developed countries (Case 2016). Hillerdal (2000) pointed out that Baris in Turkey had found that the many "calcified pleural plaques in some villages had been dismissed by the local doctors as 'old tuberculosis' – just as many colleagues even today (elsewhere) tend to dismiss any apical pleural thickening as being due to this disease."

Mesothelioma is usually pleural, and most of the pathology literature has been developed around pleural neoplasms. Peritoneal cases and cases at other sites such as the tunica vaginalis have fundamental differences in pathology and possibly in pathogenesis. There is increasing evidence that peritoneal mesothelioma differs in molecular mechanisms, treatment response, and patterns of etiology, with the latter changing over time (Case 2016). Histological and immunohistochemical profiles for pleural and peritoneal tumors are similar, but Dragon et al. (2015) have suggested recently that there may be more genetic "events" potentially leading to pleural mesothelioma, which could explain a greater frequency of disease at that site and higher asbestos susceptibility.

The histologic classification of mesothelioma, like that of carcinoma of the lung, has changed very little over time. McCaughey (1958) first proposed a classification very similar to current classifications (Husain et al. 2009, 2013, 2018; Travis et al. 2015) at a time when many pathologists still considered the tumor an endothelioma or even a variant of carcinoma of the lung. McCaughey's classification was derived from a set of 13 cases, 11 diffuse and two localized, autopsied in Belfast in a single hospital pathology department. Immunohistochemistry did not yet exist, but differential diagnosis was aided by PAS and mucicarmine stains as well as routine H&E. Three diffuse pleural mesotheliomas were considered of "epithelial character," two of "mesenchymal type," and three "mixed," corresponding to our current epithelioid, sarcomatoid and biphasic (or mixed) categories. The other three diffuse cases were of "anaplastic type," and the detailed descriptions given correspond closely to what we today classify as the pleomorphic variant of epithelioid mesotheliomas. Of interest, all three of the latter were dead within weeks of the first symptoms, and recently, there have been calls to remove the pleomorphic variant from the epithelioid group due to its poor prognosis (Husain et al. 2009, 2013, 2018). The "localized" cases did not correspond well to the currently rare "localized mesothelioma." One was "a large bilobed tumor" measuring 23 cm in greatest dimension and weighing 3.7 kg, attached to the antero-medial aspect of the left lower lobe of lung by a short vascular one cm thick pedicle, but with focal metastases elsewhere. Grossly, the principal difficulty faced for making a diagnosis was differentiation from lung cancer; of the 11 diffuse cases "in at least four... the manner in which the tumor infiltrated the lung substance did not permit the exclusion of a pulmonary neoplasm." Histology, however, was distinctive: McCaughey reviewed 305 lung cancers and found none that duplicated the "microscopic characteristics" of the mesotheliomas. Overall, for diffuse mesothelioma, what was notable was "the completeness with which they encase the lung and the diversity of their microscopic structure which may mimic that of either a carcinoma or a sarcoma. A papillary structure is usual in those tumours with epithelial characteristics. Differentiation from metastatic growths is often exceedingly difficult. The presence of malignant elements of both epithelial and mesenchymal form in some of these growths is of diagnostic importance."

Mesothelioma diagnosis remained limited to gross description and microscopic pathology similar to that used by McCaughey until near the end of the twentieth century. Immunohistochemistry use in a mesothelioma case was first reported if differential diagnosis by Wang and colleagues at McGill University in 1979 (Wang et al. 1979), but this was limited to *negative* immunostaining for carcinoembryonic antigen (CEA)-like material found in nine cases of diffuse mesothelioma and three "localized," while absent from 14 lung cancers (12 of which were adenocarcinomas or bronchoalveolar carcinoma). Transmission and scanning electron microscopy and conventional light microscopy were used as the gold standard for comparisons (Wang et al. 1979). This then was not a positive identification of mesothelioma but a diagnosis of exclusion by a positive immunohistochemical result for CEA-like material in lung cancers.

As immunohistochemistry came into further use, additional tests were applied with mixed results, but immunohistochemistry remained unreliable for identifying mesothelial tumor cells themselves. As of 1992, one group summarized findings as "Once a malignant diagnosis is arrived at by careful pathological examination, the tumor is classified as mesothelioma if mesothelial cells are identified as the constituent cells of the neoplasm. Mesothelial cells are recognized by (1) their main ultrastructural (that is, electron microscopic) features: slender and elongated microvilli, abundant intermediate filaments, and lacking secretary granules; and (2) their characteristic immunocytochemical reactivity: positivity for cytokeratin, EMA, and vimentin, and negativity for carcinoembryonic antigen (CEA), B72-3, Leu-M1, and other gland-cell markers" (Bedrossian et al. 1992). None of the "positive" stains mentioned were specific, however, especially cytokeratins, which were generally positive in both mesothelioma and carcinomas. It was not until 1996 that the first (and still one of the best) immunostaining procedures were published that positively identified mesothelial origin in the tumor cells of mesothelioma. Calretinin was expressed, mainly in cytoplasm, in "a specific and reproducible manner in tumor cells of epithelial-type mesotheliomas and in the epithelial component of the mixed type" in 23 pleural mesotheliomas (Gotzos et al. 1996). Tumor cells of the sarcomatoid subtype or sarcomatoid portions of the "mixed type" were negative, as were those in four adenocarcinomas primary in lung. Calretinin, as applied today, has a different appearance; both cytoplasmic and nuclear staining are necessary for a positive test. By the turn of the century, many of the immunohistochemical tests now considered most reliable for determining mesothelial cell origin in tumors, such as Wilm's Tumur Gene (WT-1) and cytokeratin 5/6 had first come into practice: "from the practical point-of-view, calretinin and cytokeratin 5/6 appear to be the most sensitive and specific positive markers for distinguishing between malignant epithelial mesotheliomas and adenocarcinomas metastatic to the serosal membranes. Among the antibodies, which are considered to be negative markers for mesothelioma, MOC-31, CEA, and BG-8 seem to be the best diagnostic discriminators. Other markers, such as B72.3, Leu-M1 (CD15), Ber-EP4, and thrombomodulin can be used as secondary markers if the results obtained with the previously mentioned markers are equivocal" (Ordóñez 1999). With the exception of the added TTF-1 for carcinomas and D2-40 for mesotheliomas, these remain baseline tests used in practice, with at least two positive and two negative markers as well as pankeratin applied in most cases (Husain et al. 2009, 2013, 2018; Case 2016). Specialist readers should note, however, that the application of immunohistochemical tests has a constantly changing landscape, and the "best" tests at the time of this writing are likely to be supplanted by better ones at a later date.

This has marked a sea-change in diagnostic habit over the last 20 years, with some consequences. Transmission or scanning electron microscopy, formerly thought a gold standard for mesothelial cell identification but really most useful in well-differentiated and well-preserved epithelioid tumors, was used in *none* of a series of over 748 British mesotheliomas reported by this author (Case 2016). Instead, of recorded immunohistochemical stains specific for mesothelial cell origin, calretinin (95%) and CK 5/6 or CK5 alone (84%) were by far the most common. Calretinin and CK 5/6 or CK 5 alone were also most sensitive and positive in 92% of cases. Ninety percent of cases had at least one immunohistochemical marker for possible lung carcinoma applied, with BER-Ep4 and TTF-1 the most frequent at 68% and CEA at 58%.

It is important to realize several things about this "diagnostic revolution" in the pathology of mesothelioma, however.

First, immunohistochemistry remains most useful for the epithelioid variant and the epithelioid portion of biphasic or mixed tumors. Sarcomatoid tumors are typically positive for keratin stains, unlike true sarcomas, but not for the positive mesothelial markers.

Second, application of immunostains is not a panacea; the majority of cases could be diagnosed on microscopy and clinical grounds alone, and historically *have been*. In fact, most of what we know about mesothelioma and almost all of the mesothelioma cases in the epidemiological series upon which our knowledge of asbestos etiology is based relied on conventional light microscopy using hematoxylin and eosin (H&E) stains. The final US Armed Forces Institute of Pathology monograph on these tumors noted that "The fundamental diagnosis of mesothelioma depends on routine hematoxylin and eosin (H&E) stains, and in most cases the diagnosis of mesothelioma is histologically obvious in routinely stained sections" (Churg et al. 2006).

Nevertheless, the standard of care currently is to use "at least two mesothelial and two carcinoma markers with greater than 80% sensitivity and specificity for the diagnosis of mesothelioma when all clinical, radiologic, and histologic features are concordant" (Arif and Husain 2015).

Third, specific diagnostic dilemmas remain, such as the distinction between (benign) atypical mesothelial hyperplasia and epithelioid mesothelioma, especially in small biopsies: even expert panels (US-Canadian Mesothelioma Reference Panel and the French panel Group MesoPath) "had a 22% to 47% disagreement on whether the process was benign or malignant in cases circulated to the entire panel" (Husain 2014). Most suggested markers have proved either insufficiently sensitive or (more frequently) specific, again leaving careful study of clinical features, especially imaging, together with routine microscopy as important.

Distinguishing sarcomatoid mesotheliomas, especially the relatively acellular desmoplastic variant, from sarcomatoid metastatic sarcomatoid lung carcinoma presents another diagnostic dilemma. GATA binding protein 3 (GATA3) is the most recent candidate. GATA3 stains in sarcomatoid/desmoplastic malignant mesotheliomas showed strong diffuse staining in 100% of 19 cases, while only two of 13 sarcomatoid carcinomas of the lung stained at all, and then staining was "weak and patchy" (Berg and Churg 2017). Since GATA3 stains other metastatic tumors such as breast carcinoma, care must be taken with a positive stain. Indeed, GATA3 is a recommended marker for possible metastatic breast origin in cytology cell blocks from pleural effusions, as are tumor origin in pleural metastases from TTF-1 for lung adenocarcinoma, p40 for squamous cell lung cancer, etc. (Porcel 2018).

Inevitably, molecular pathology techniques will become more routine and useful. Mesothelioma is increasingly identified as a disease of loss of tumor suppressor genes, and the tests being applied and developed relate to this.

One of the most common genetic alterations in mesothelioma is the homozygous deletion of the 9p21 locus within a gene cluster that includes cyclin-dependent kinase inhibitor 2A (CDKN2A). p16/CDKN2A deletions have been reported in up to 80% of primary pleural mesotheliomas, but more in sarcomatoid mesothelioma – close to 100% – than in epithelioid tumors (Husain et al. 2013, 2018). A positive test thus has a prognostic implication, and biphasic tumors, which are positive are likely to be those with a greater sarcomatoid element and poor relative prognosis.

Loss of BAP1 protein expression evaluated by immunohistochemistry is also frequent in malignant mesothelioma. However, not all mesothelioma cases are positive; "specificity for MPM is high, but their sensitivity is not completely satisfactory, since some MPM cases do not have aberrations in these markers" (Churg et al. 2018). p16/CDKN2A deletions and BAP1 loss are also not seen in atypical mesothelial hyperplasia, allowing their differentiation from malignancy.

Overall, despite increased reliance on immunohistochemistry and increasingly on molecular testing in specialized centers, "The quintessential role of the pathologist as master of *morphology-based* diagnosis is as relevant today as ever," and "Histopathologic assessment of tumor tissue remains a rapid, cost-effective, and multifaceted tool for the pathologist's armamentarium" (Chapel et al. 2019).

None of the mesothelioma pathology discussion above mentions "asbestos," and that is important. As asbestos use decreases, mesothelioma incidence and mortality are decreasing, and as a result, while more cases are exposure-related than not, spontaneous unrelated cases are increasing as a proportion of the total. This is complicated by the importance of time from first exposure as a causative factor, since increased life expectancy leads at the same time to *higher* incidence in the oldest age groups. Certainly, a history of asbestos exposure – or the absence of such history – has no place in the actual diagnosis of mesothelioma; any more than smoking history can determine whether or not a neoplasm is or is not lung cancer. Mesothelioma diagnosis is "based on clinical, radiologic, and, ultimately, pathologic features, and the issue of asbestos exposure is irrelevant" (Husain et al. 2013). This does not mean that, once diagnosed, pathologists have no role in determining attributability of a given case to past exposures. Indeed, in many jurisdictions, mesothelioma and asbestos-related lung cancers are reportable diseases. In Quebec, for example, cases must be reported to the Workman's Compensation Board (Commission des normes, de l'équité, de la santé et de la sécurité du travail) shortly after first diagnosis, although this is usually done by treating physicians.

References

Adler, I. (1901). Remarks on primary endothelioma of lung and pleura, with demonstrations. *The Journal of Medical Research* 6(1): 175–186.3.

Alfudhili, K.M., Lynch, D.A., Laurent, F., et al. (2016). Focal pleural thickening mimicking pleural plaques on chest computed tomography: tips and tricks. *British Journal of Radiology* 89(1057). https://doi.org/10.1259/bjr.20150792.

Amandus, H.E. and Wheeler, R. (1987). The morbidity and mortality of vermiculite miners and millers exposed to tremolite-actinolite: part II. Mortality. *American Journal of Industrial Medicine* 11(1): 15–26. https://doi.org/10.1002/ajim.4700110103.

American Thoracic Society (ATS). (1986). Medical Section of the American Lung Association: the diagnosis of nonmalignant diseases related to asbestos. *The American Review of Respiratory Disease* 134(2): 363–368. https://doi.org/10.1164/arrd.1986.134.2.363.

American Thoracic Society (ATS). (2004). Diagnosis and initial management of nonmalignant diseases related to asbestos. Official statement of the American

Thoracic Society was adopted by the ATS Board of Directors. *American Journal of Respiratory and Critical Care Medicine* 170: 691–715. https://doi.org/10.1164/rccm.200310-1436ST.

Arif, Q. and Husain, A.N. (2015). Malignant mesothelioma diagnosis. *Archives of Pathology and Laboratory Medicine* 139(8): 978–980. https://doi.org/10.5858/arpa.2013-0381-ra.

Bedrossian, C.W.M., Bonsib, S., and Moran, C. (1992). Differential diagnosis between mesothelioma and adenocarcinoma: a multimodal approach based on ultrastructure and immunocytochemistry. *Seminars in Diagnostic Pathology* 9(2): 124–140.

Berg, K.B. and Churg, A. (2017). GATA3 Immunohistochemistry for distinguishing sarcomatoid and desmoplastic mesothelioma from sarcomatoid carcinoma of the lung. *The American Journal of Surgical Pathology* 41(9): 1221–1225. https://doi.org/10.1097/pas.0000000000000825.

Bourbeau, J., Ernst, P., Chrome, J., et al. (1990). The relationship between respiratory impairment and asbestos-related pleural abnormality in an active work force. *The American Review of Respiratory Disease* 142(4): 837–842. https://doi.org/10.1164/ajrccm/142.4.837.

British Thoracic Society (BTS). (2011). *Pleural Plaques: Information for Healthcare Professionals*. London: BTS.

Cartier, P. (1955). Some clinical observations of asbestosis in mine and mill workers. *American Medical Association Archives of Industrial Health* 11(3): 204–207. PMID: 14349406.

Case, B.W. (2006). Asbestos, smoking, and lung cancer: interaction and attribution. *Occupational and Environmental Medicine* 63(8): 507–508. https://doi.org/10.1136/oem.2006.027631.

Case, B.W. (2008). From cotton-stone to the New York Conference. In: *Asbestos and Its Diseases* (eds. J. Craighead and A.R. Gibbs), 3–22. New York, USA: Oxford University Press.

Case, B.W. (2016). Pathology analysis for mesothelioma study in the United Kingdom: current practice and historical development. *Journal of Toxicology and Environmental Health. Part B Critical Reviews* 19(5–6): 201–212. https://doi.org/10.1080/10937404.2016.1195320.

Case, B.W., Ip, M.P., Padilla, M., et al. (1986), Asbestos effects on superoxide production. An in vitro study of hamster alveolar macrophages. *Environmental Research* 39(2): 299–306. https://doi.org/10.1016/s0013-9351(86)80056-1.

Case, B.W., Abraham, J.L., Meeker, G., et al. (2011). Applying definitions of "asbestos" to environmental and "low-dose" exposure levels and health effects, particularly malignant mesothelioma. *Journal of Toxicology and Environmental Health* 14(1–4): 3–39. https://doi.org/10.1080/10937404.2011.556045.

Chapel, D.B., Churg, A., Santoni-Rugiu, E., et al. (2019). Molecular pathways and diagnosis in malignant mesothelioma: a review of the 14th International

Conference of the International Mesothelioma Interest Group. *Lung Cancer* 127: 69–75. https://doi.org/10.1016/j.lungcan.2018.11.032.

Churg, A. and DePaoli, L. (1988). Environmental pleural plaques in residents of a Quebec chrysotile mining town. *Chest* 94(1): 58–60. https://doi.org/10.1378/chest.94.1.58.

Churg, A., Cagle, P.T., and Roggli, V.L. (eds). (2006). *Tumours of the Serosal Membranes*, Fourth Series, Fascicle 3. Washington DC: Armed Forces Institute of Pathology, Atlas of Tumour Pathology, 147.

Churg, A., Nabeshima, K., Ali, G., et al. (2018). Highlights of the 14th international mesothelioma interest group meeting: pathologic separation of benign from malignant mesothelial proliferations and histologic/molecular analysis of malignant mesothelioma subtypes. *Lung Cancer* 124: 95–101. https://doi.org/10.1016/j.lungcan.2018.07.041.

Clarkson, F.A. (1914). Primary endothelioma of the pleura. *Canadian Medical Association Journal* 4(3): 192–196. PMCID: PMC406585; PMID: 20310472.

Craighead, J.E., Abraham, J.L., Churg, A., et al. (1982). The pathology of asbestos-associated diseases of the lungs and pleural cavities: diagnostic criteria and proposed grading schema. Report of the Pneumoconiosis Committee of the College of American Pathologists and the National Institute for Occupational Safety and Health. *Archives of Pathology and Laboratory Medicine* 106(11): 544–596. PMID: 6897166.

Doll, R. (1955). Mortality from lung cancer in asbestos workers. *British Journal of Industrial Medicine* 12(2): 81–86. https://doi.org/10.1136/oem.12.2.81.

Dragon, J., Thompson, J., MacPherson, M., et al. (2015). Differential susceptibility of human pleural and peritoneal mesothelial cells to asbestos exposure. *Journal of Cellular Biochemistry* 116(8): 1540–1552. https://doi.org/10.1002/jcb.25095.

Dupré, J.S., J. Fraser Mustard, Robert J. Uffen, et al. (1984). Report of the Royal Commission On Matters Of Health And Safety Arising from the use of asbestos in Ontario. Ontario Ministry of the Attorney General, Queen's Printer for Ontario. (3 volumes) Toronto. https://archive.org/details/royalcommissions?query=asbestos (accessed 25 January 2024).

Eastman, A., Mossman, B.T., and Bresnick, E. (1983). Influence of asbestos on the uptake of benzo(a)pyrene and DNA alkylation in hamster tracheal epithelial cells. *Cancer Research* 43(3): 1251–1255.

Enterline, P.E. (1978). Asbestos and cancer: the International lag. *The American Review of Respiratory Disease* 118(6): 975–978.

Enterline P.E. (1991). Changing attitudes and opinions regarding asbestos and cancer 1934-1965. *American Journal of industrial Medicine* 20(5): 685–700. https://doi.org/10.1002/ajim.4700200511. Accessed January 25th 2024

Enterline, P.E., Sussman, N., and Marsh, G.M. (1978). *Asbestos and Cancer: The First Thirty Years*. Pittsburgh: Philip E. Enterline.

Epler, G.R., McLoud, T.C., and Gaensler, E.A. (1982). Prevalence and incidence of benign asbestos pleural effusion in a working population. *Journal of the American Medical Association* 247(5): 617–622. PMID: 7054563.

Fraser, R.G., Paré, P., Paré, P., et al. (1989). *Diagnosis of Diseases of the Chest*. 3rd Edition, Vol. 2. Philadelphia: WB Saunders.

Fujimoto, N., Kato, K., Usami, I., et al. (2014). Asbestos-related diffuse pleural thickening. *Respiration* 88(4): 277–284. https://doi.org/10.1159/000364948.

Gevenois, P.A., de Maertelaer, V., Madani, A., et al. (1998). Asbestosis, pleural plaques and diffuse pleural thickening: three distinct benign responses to asbestos exposure. *European Respiratory Journal* 11(5): 1021–1027. https://doi.org/10.1183/09031936.98.11051021.

Gloyne, S.R. (1935). Two cases of squamous carcinoma of the lung occurring in asbestosis. *Tubercle* 17(1): 5–10. https://doi.org/10.1016/S0041-3879(35)80795-2.

Gotzos, V., Vogt, P., and Celio, M.R. (1996). The calcium binding protein calretinin is a selective marker for malignant pleural mesotheliomas of the epithelial type. *Pathology – Research and Practice* 192(2): 137–147. https://doi.org/10.1016/S0344-0338(96)80208-1.

Gualtieri, A.F., Lusvardi, G., Pedone, A., et al. (2019). Structure model and toxicity of the product of biodissolution of chrysotile asbestos in the lungs. *Chemical Research Toxicology* 32(10): 2063-2077. https://doi.org/10.1021/acs.chemrestox.9b00220.

Hammar, S.P. and Abraham, J.L. (2015). Commentary on pathologic diagnosis of asbestosis and critique of the 2010 Asbestosis Committee of the College of American Pathologists (CAP) and Pulmonary Pathology Society's (PPS) update on the diagnostic criteria for pathologic asbestosis. *American Journal of Industrial Medicine* 58(10): 1034–1039. https://doi.org/10.1002/ajim.22512.

Hashiba, G.K., Cowan, A.B., and Nixon, C.E. (1932). Mesothelioma of the pleura: report of case. *California and West Medicine* 37(6): 385–387.

Hillerdal, G. (2000). Environmental dangers: asbestos and tuberculosis. *Respiration* 67(2): 134. http://dx.doi.org/10.1159/000029498.

Husain, A.N. (2014). Mesothelial proliferations: useful marker is not the same as a diagnostic one. *American Journal of Clinical Pathology* 141(2): 152–153. https://doi.org/10.1309/ajcpoi1desu8mthz.

Husain, A.N., Colby, T.V., Ordonez, N.G., et al. (2009). Guidelines for pathologic diagnosis of malignant mesothelioma: a consensus statement from the International Mesothelioma Interest Group. *Archives of Pathology and Laboratory Medicine* 133(8): 1317–1331. https://doi.org/10.5858/133.8.1317.

Husain, A.N., Colby, T., Ordonez, N., et al. (2013). Guidelines for pathologic diagnosis of malignant mesothelioma: 2012 update of the consensus statement from the International Mesothelioma Interest Group. *Archives of Pathology and Laboratory Medicine* 137(5): 647–667. https://doi.org/10.5858/arpa.2012-0214-oa.

Husain, A.N., Colby, T.V., Ordóñez, N.G., et al. (2018). Guidelines for pathologic diagnosis of malignant mesothelioma. 2017 update of the consensus statement from the International Mesothelioma Interest Group. *Archives of Pathology and Laboratory Medicine* 142(1): 89–108. https://doi.org/10.5858/arpa.2017-0124-ra.

IARC. (2012). Working Group on the Evaluation of Carcinogenic Risks to Humans. Arsenic, Metals, Fibres, and Dusts: A Review of Human Carcinogens. *IARC Monogr Eval Carcinog Risks Hum*. 100C. *International Agency for Research on Cancer, Lyon, France*.

Integrated Risk Information System (IRIS). (1988). US Environmental Protection Agency Chemical Assessment. National Center for Environmental Assessment. Asbestos CASRN 1332-21-4. https://cfpub.epa.gov/ncea/iris2/chemicalLanding.cfm?substance_nmbr=371. Accessed October 23rd, 2019.

Integrated Risk Information System (IRIS). (2014). US Environmental Protection Agency Chemical Assessment. National Center for Environmental Assessment. Libby Amphibole asbestos. https://cfpub.epa.gov/ncea/iris2/chemicalLanding.cfm?substance_nmbr=1026. Accessed October 23rd, 2019.

International Labour Organization (ILO) (2002). *ILO Guidelines for the Use of the ILO International Classification of Radiographs of Pneumoconioses, 2000 edition*, Occupational Safety and Health Series, No. 22 (rev. 2000). Geneva: International Labour Office.

Jung, M., Davis, W.P., Taatjes, D.J., et al. (2000). Asbestos and cigarette smoke cause increased DNA strand breaks and necrosis in bronchiolar epithelial cells in vivo. *Free Radical Biology and Medicine* 28(8): 1295–1299. https://doi.org/10.1016/s0891-5849(00)00211-2.

Kipen, H.M., Lilis, R., Suzuki, Y., et al. (1987). Pulmonary fibrosis in asbestos insulation workers with lung cancer: a radiological and histopathological evaluation. *British Journal of Occupational and Environmental Medicine* 44(2): 96–100.

Lockey, J.E., Brooks, S.M., Jarabek, A.M., et al. (1984). Pulmonary changes after exposure to vermiculite contaminated with fibrous tremolite. *The American Review of Respiratory Disease* 129(6): 952–958. https://doi.org/10.1164/arrd.1984.129.6.952. Accessed January 25th 2024

Lynch, K.M. and Smith, W.A. (1935). Pulmonary asbestosis III: carcinoma of the lung in asbestos-silicosis. *American Journal of Cancer* 24(1): 56–61. https://doi.org/10.1158/ajc.1935.56.

Maxim, L.D., Niebo, R., and Utell M.J. (2015). Are pleural plaques an appropriate endpoint for risk analyses? *Inhalation Toxicology* 27(7): 321–334. https://doi.org/10.3109/08958378.2015.1051640.

McCaughey, W.T.E. (1958). Primary tumours of the pleura. *The Journal of Pathology and Bacteriology* 76(2): 517–529. https://doi.org/10.1002/path.1700760222.

McDonald, J.C., McDonald, A.D., Armstrong, B., et al. (1986). Cohort study of mortality of vermiculite miners exposed to tremolite. *British Journal of Industrial Medicine* 43(7): 436–444. https://doi.org/10.1136/oem.43.7.436.

McDonald, J.C., Harris, J., and Armstrong, B. (2004). Mortality in a cohort of vermiculite miners exposed to fibrous amphibole in Libby, Montana, *Occupational and Environmental Medicine* 61(4): 363–366. https://doi.org/10.1136/oem.2003.008649.

Meeker, G.P., Bern, A.M., Brownfield, I.K., et al. (2003). The composition and morphology of amphiboles from the Rainy Creek complex, near Libby, Montana. *American Mineralogist* 88 (11–12 PART 2): 1955–1969.

Miller, A., Szeinuk, J., Noonan, C.W., et al. (2018). Libby amphibole disease: pulmonary function and CT abnormalities in vermiculite miners. *Journal of Occupational and Environmental Medicine* 60(2): 167–173. https://doi.org/10.1097/jom.0000000000001178.

Mutsaers, S.E., Prele, C.M., Brody, A.R., et al. (2004). Pathogenesis of pleural fibrosis. *Respirology* 9(4): 428–440. https://doi.org/10.1111/j.1440-1843.2004.00633.x.

Nymark, P., Aavikko, M., Mäkilä, J., et al. (2013). Accumulation of genomic alterations in 2p16, 9q33.1 and 19p13 in lung tumours of asbestos-exposed patients. *Molecular Oncology* 7(1): 29–40. https://doi.org/10.1016/j.molonc.2012.07.006.

Ordóñez, N.G. (1999). The immunohistochemical diagnosis of epithelial mesothelioma. *Human Pathology* 30(3): 313–323. https://doi.org/10.1016/s0046-8177(99)90011-4.

Porcel, J.M. (2017). Persistent benign pleural effusion. *Revista Clinica Espanola* 217(6): 336–341. https://doi.org/10.1016/j.rce.2017.03.008.

Porcel, J.M. (2018). Biomarkers in the diagnosis of pleural diseases: a 2018 update. *Therapeutic Advances in Respiratory Disease* 12: 1753466618808660. https://doi.org/10.1177/1753466618808660.

Riley, J.Y. and Naidoo, P. (2018). Imaging assessment of rounded atelectasis: a pictorial essay. *Journal of Medical Imaging and Radiation Oncology* 62(2): 211–216. https://doi.org/10.1111/1754-9485.12710.

Roggli, V.L. and Pratt, P.C. (1983). Numbers of asbestos bodies on iron-stained tissue sections in relation to asbestos body counts in lung tissue digests. *Human Pathology* 14(4): 355–361. https://doi.org/10.1016/s0046-8177(83)80122-1.

Roggli, V.L., Gibbs, A.R., Attanoos, R., et al. (2010). Pathology of asbestosis – an update of the diagnostic criteria: Report of the asbestosis committee of the College of American Pathologists and Pulmonary Pathology Society. *Archives of Pathology and Laboratory Medicine* 134(3): 462–480. https://doi.org/10.5858/134.3.462.

Rohs, A.M., Lockey, J.E., Dunning, K.K., et al. (2007). Low-level fiber-induced radiographic changes caused by Libby vermiculite: a 25-year follow-up study. *American Journal of Respiratory and Critical Care Medicine* 177(6): 630–637. https://doi.org/10.1164/rccm.200706-841oc.

Schenker, M.B., Christiani, D., Cormier Y., et al. (1998). American Thoracic Society respiratory health hazards in agriculture. Official Conference Report of the American Thoracic Society. *American Journal of Respiratory Critical Care Medicine* 158: S1–S76.

Selikoff, I.J. and Churg, J., Eds. (1965). *Biological Effects of Asbestos*. New York: Annals of the New York Academy of Science, 132.

Sullivan, P.A. (2007). Vermiculite, respiratory disease, and asbestos exposure in Libby, Montana: update of a cohort mortality study. *Environmental Health Perspectives* 115(4): 579–585. https://doi.org/10.1289/ehp.9481.

Szeinuk, J., Noonan, C.W., Henschke, C.I., et al. (2017). Pulmonary abnormalities as a result of exposure to Libby amphibole during childhood and adolescence – the Pre-Adult Latency Study (PALS). *American Journal of Industrial Medicine* 60(1): 20–34. https://doi.org/10.1002/ajim.22674.

Torre, L.A., Siegel, R.L., and Jemal, A. (2016). Lung cancer statistics. *Advances in Experimental Medicine and Biology* 893: 1–19. https://doi.org/10.1007/978-3-319-24223-1_1.

Travis, W.D., Brambilla, E., Burke, A.P., et al. (2015). WHO Classification of Tumours of the Lung, Pleura, Thymus and Heart. *WHO Classification of Tumours*, 4th Edition, Volume 7. ISBN-13 978-92-832-2436-5.

Tsao, M.S., Galateau-Salle, F., Nicholson, A.G., et al. (2021). Tumours of the pleura and pericardium: introduction: pleural tumours. In: *WHO Classification of Tumours Editorial Board. Thoracic Tumours*. Lyon (France): International Agency for Research on Cancer; 2021 [cited 2023 August 4]. (*WHO Classification of Tumours Series*, 5th Edition, volume 5). https://tumourclassification.iarc.who.int/chapters/35.

Wagner, J.C., Sleggs, C.A., and Marchand, P. (1960). Diffuse pleural mesothelioma and asbestos exposure in the North Western Cape Province. *British Journal of Industrial Medicine* 17(4): 260–271. https://doi.org/10.1136/oem.17.4.260.

Wang, N.S., Huang, S.N., and Gold, P. (1979). Absence of carcinoembryonic antigen-like material in mesothelioma: an immunohistochemical differentiation from other lung cancers. *Cancer* 44(3): 937–943. https://doi.org/10.1002/1097-0142(197909)44:3%3C937::aid-cncr2820440322%3E3.0.co;2-k.

Wolff, H., Vehmas, T., Oksa, P., et al. (2015). Asbestos, asbestosis, and cancer, the Helsinki criteria for diagnosis and attribution 2014: recommendations. *Scandinavian Journal of Work, Environment and Health* 41(1): 5–15. https://doi.org/10.5271/sjweh.3462.

Wood, W.B. and Gloyne, S.R. (1934). Pulmonary asbestosis: a review of one hundred cases. *Lancet* 224 (S808): 1383–1385. https://doi.org/10.1016/S0140-6736(00)43332-5.

World Health Organization. (2021). *WHO Classification of Tumours Editorial Board. Thoracic Tumours*. Lyon (France): International Agency for Research on Cancer (WHO Classification of Tumours Series, 5th Edition, Volume 5). https://publications.iarc.fr/595.

Part II

Exposure Assessment

4

Principles of Exposure Assessment for Elongate Mineral Particles (EMPs)

Eric Rasmuson, James Rasmuson, and Andrey Korchevskiy

Chemistry & Industrial Hygiene, Inc., Lakewood, CO, USA

As described in Science and Decisions, Advancing Risk Assessment by the National Research Council (NRC) of the National Academy of Sciences (NAS) (NRC 2009), the National Institute of Occupational Safety and Health (NIOSH) Practice in Occupational Risk Assessment (NIOSH 2020a), and the United States Environmental Protection Agency (US EPA) Science Policy Council Handbook on Risk Characterization (US EPA 2000), exposure assessment is an essential component of the risk assessment process (see Figure 4.1).

Principles, methodologies, standards, tools, and analytical methodologies for conducting exposure assessments have been exhaustively covered in publications by the exposure sciences, industrial hygiene, environmental, and occupation health communities (AIHA 2009, 2015; US EPA 2019; NIOSH 2020b; Hinds and Zhu 2022; AIHA 2024; and US EPA 2024a,b,c,d). Within these references, there is specificity to Elongate Mineral Particle (EMP) exposure assessment methods, but situations will dictate the use of targeted publications for a given exposure scenario. Specifically focusing on EMPs, a plethora of publications have been authored on the evaluation of exposures in a multitude of settings, for a wide variety of conditions, and for many distinct similar exposure groups to characterize current, prospective, and retrospective exposures. Topically covering the nature and extent of these publications is beyond the scope of this chapter. The reader should refer to the available literature for detailed analyses and explanations of exposure-related information associated with targeted products, distinct processes, naturally occurring EMPs, and for different environments.

Health Risk Assessment for Asbestos and Other Fibrous Minerals, First Edition.
Edited by Andrey Korchevskiy, James Rasmuson, and Eric Rasmuson.
© 2024 John Wiley & Sons, Inc. Published 2024 by John Wiley & Sons, Inc.

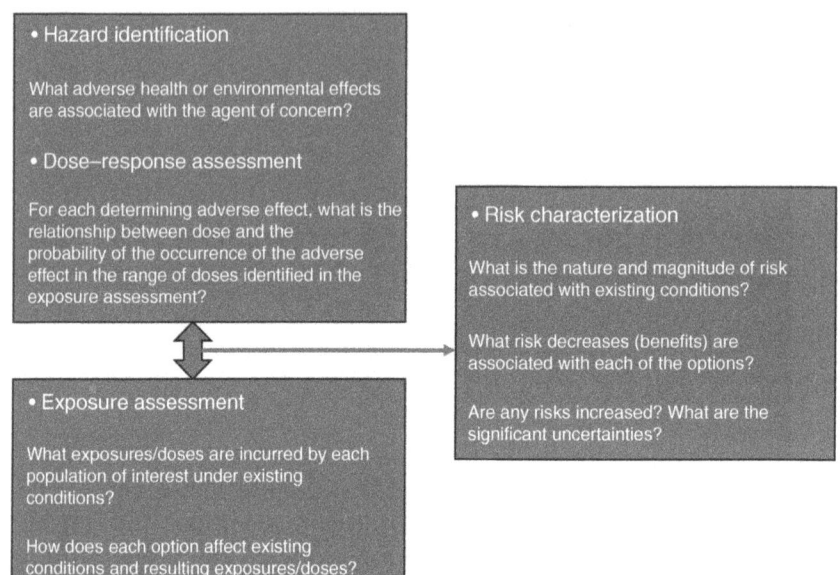

Figure 4.1 NRC/NAS risk assessment process (the Risk Assessment flowchart of a framework for risk-based decision-making).

This chapter highlights several concepts that are important to consider when evaluating exposures that can ultimately be used to evaluate average lifetime or cumulative lifetime EMP exposures. Familiarity and mastery of the concepts of frequency, duration, proximity, and intensity of a given exposure scenario are necessary to ultimately evaluate human health risks related to an exposure scenario.

In terms of human-health risk evaluations related to specific exposures, it is not scientifically sound to conclude that a determined short-term or eight-hour TWA exposure concentration is sufficient, alone, to determine the nature and extent of chronic health risks. It is widely accepted in epidemiology, occupational health, and risk sciences that average lifetime or cumulative lifetime exposure information is necessary to evaluate the risk of chronic disease endpoints, such as lung cancer and mesothelioma. According to NIOSH (NIOSH 2020a):

> Ideally, the choice of metric is determined from adequate information on which metrics (if any) best predict risk. In lieu of this information, the characteristics of the adverse effect might imply the most appropriate choice. For example, exposure indices used to examine acute toxicity effects are typically based on short-term or instantaneous intensity (e.g. peak

airborne concentration), whereas cumulative dose (i.e. the time integral of exposure intensity) is generally preferred for chronic effects in which biologic damage appears proportional to the delivered dose quantity (e.g. silica and chronic silicosis or ionizing radiation and cancer) [Checkoway and Rice 1992 Rappaport 1991].)

General Principles and Methods

EMP exposure assessments can be current, prospective, or retrospective. Current exposure assessments estimate exposure at the time of the study and might provide short-term or long-term exposure estimations in specific current conditions. Prospective exposures are based on a prediction of EMP exposures that could happen in the future. Retrospective exposure assessments are intended for a reconstruction of past EMP exposure levels for some persons, workplaces, or populations. These concepts have been covered in referenced publications above and example calculations have been provided in Chapter 13 of this book, "Risk Characterization for Occupational and Environmental Exposure to Asbestos: Case Studies."

Gathering Information

When performing exposure assessments for EMPs, gathering data regarding the nature and source of potential exposures is critical in the determination of the average lifetime or cumulative lifetime exposure estimates. For current and future exposure scenarios, basic chemico-physical properties of the products of concern and analytical results of air sampling during product use or during representative working or living conditions are the ideal approach to evaluate EMP exposures. For the determination of EMP exposures in products, processes, or on premises that have occurred in the past, the following topics should be considered:

1) General—product information, exposure assessment data associated with the products, and regulatory compliance.
   - Corporation—years of operation
   - Supplier information
     - Details of raw materials used—short or long fiber, fiber-type, etc.
   - EMP-containing products manufactured, supplied, and/or distributed
     - Timeframes products were manufactured, supplied, and/or distributed
     - Manufacturing formulas and instructions
     - Product specifications

- Material Safety Data Sheets/Safety Data Sheets
        - Documentation describing the scope and nature of the use of the products
        - Representative product catalogues
        - Representative product marketing materials
        - Common product use name brands for the product
        - Installation instructions
        - Maintenance and repair instructions
   - Historical exposure-related data
        - Corporate and/or OSHA EMP—monitoring data
        - Literature regarding similar exposure groups using similar products
        - Modeled EMP exposures
        - Records regarding exposure monitoring data: other than EMPs
   - Description of product distribution chain
        - Geographical distribution area
        - Distribution volume
        - Distribution timeframe
2) Corporate knowledge
   - Miscellaneous documentation of product history and use (invoices, receipts, etc.)
   - Written procedures for product installation, maintenance, and repair
   - Health and safety programs during timeframe of product use
3) Are there current or past sites that can be visited to better conceptualize the use of the product?

Evaluating the Quality of Data

As short-term and eight-hour TWA exposure concentrations are used as input variables to determine the average lifetime or cumulative lifetime exposure concentrations used in risk calculations, it is of the utmost importance to characterize the quality of the data used in the exposure assessment.

Evaluating the weight of the evidence in order to make credible decisions based on a full set of data is in the interest of all scientists as well as the public. However, methods on the application and ultimately on the inclusion or exclusion of publications in the scientific decision-making process on the basis of quality or biases is an issue of current debate. According to Berman and Case (2012):

> ...It is generally better to include as much of the available data as possible in these analyses while formally addressing uncertainty as part of the analysis itself... rather than to sequentially exclude studies based on one type of limitation or another. Throwing out data without clearly proving some type of bias is never a good idea because it will limit both the power to test

various hypotheses and the confidence that can be placed in any findings that are derived from the resulting, truncated data set.

On the other hand, Montibeller and von Winterfeldt (2015) noted that biases significantly impact study outcomes or the assessment of the outcomes by expert and policy risk analysts, leading to the conclusion that publications with significant biases should be excluded in the risk assessment process:

> Behavioral decision research has demonstrated that judgments and decisions of ordinary people and experts are subject to numerous biases. Decision and risk analysis were designed to improve judgments and decisions and to overcome many of these biases. However, when eliciting model components and parameters from decisionmakers or experts, analysts often face the very biases they are trying to help overcome. When these inputs are biased they can seriously reduce the quality of the model and resulting analysis. Some of these biases are due to faulty cognitive processes; some are due to motivations for preferred analysis outcomes.

Regardless, for a given set of referenced publications under consideration for inclusion or exclusion in summary exposure or risk estimates, it is useful to evaluate and document the quality assessment criteria used and potential biases observed in the analysis.

The evaluation of the quality of data and their use in hazard and risk assessment as a systematic approach are described by Klimisch et al. (1997). Definitions are proposed for reliability, relevance, and adequacy of data.

The following definitions are proposed here to be used in hazard and risk assessment processes, but are also useful in the evaluation of exposure assessment data sets:

Reliability—Evaluating the inherent quality of a test report or publication relating to preferably standardized methodology and the way that the experimental procedure and results are described to give evidence of the clarity and plausibility of the findings.

Relevance—Covering the extent to which data and/or tests are appropriate for a particular hazard identification or risk characterization.

Adequacy—Defining the usefulness of data for risk assessment purposes. When there is more than one set of data for each effect, the greatest weight is attached to the most reliable and relevant.

Klimisch et al. developed categories/codes of reliability:

Code 1   Reliable without restriction
Code 2   Reliable with restrictions

Code 3 Not reliable
Code 4 Not assignable

Note Code 3: "Not reliable includes studies or data in which test systems were used which are not relevant in relation to the exposure and carried out in an unacceptable method and the documentation is not sufficient for an assessment and which is not convincing for an expert judgment."

Measurement Techniques

EMP exposure assessments may include some of the following components:

1) Measurements of airborne EMP concentrations,
2) Measurements of EMP content in other environmental media, like soil or dust, or in agents present in the environment (like EMP content in industrial talc),
3) Modeling of EMP concentrations based on physical processes, like dust dispersion in the air, or EMP release from various products,
4) Measurements of EMP fibers/bodies in biological media (so called "burden"), for examples, in lung tissue, lymph nodes, or ovaries.

It should be noted that for EMPs, as for other respiratory agents, there is a significant difference between exposure and dose. Exposure is a metric of the impact of specific agents. One of the components of EMP exposures is the concentration of EMPs in the air, measured in fibers or structures per cubic centimeter (f/cc) (or, historically, in million particles per cubic foot, mppcf). The EMP dose, however, is the number of EMPs that are inhaled and reach the target levels of respiratory system. The dose can be measured in terms of the number of fibers deposited and retained in the pulmonary area of the respiratory system. For EMPs, exposure concentrations are more straightforward and better characterized than the dose, and the cumulative exposure concentrations (duration × concentration) are widely used and accepted as a surrogate for dose.

In the United States, the primary method to evaluate airborne asbestos concentrations is the NIOSH 7400 phase-contrast optical microscopy method (NIOSH 2019).

NIOSH 7400 requires that sampling be performed utilizing open-face cassettes with 0.45–1.2 μm mixed cellulose ester membrane filters measuring 25 mm. Pumps with flow rate of 0.5–16 l/min are recommended. Electrically grounding the cassette cowl is highly recommended where possible during area sampling, especially under conditions of low relative humidity.

For optimal counting, the sampling flow rate, Q (l/min), and time, t (min), are expected to be adjusted to produce a fiber density, E, of 100–1,300 fibers/mm^2

(3.85×10^4 to 5×10^5 fibers per 25-mm filter with effective collection area, A_c, of 385 mm²). These variables are related to the concentration of fibers in the air, L (fibers/cc), as can be seen in the following formula:

$$t = \frac{A_c \times E}{Q \times L \times 1{,}000}$$

NIOSH 7400 indicates that the purpose of adjusting sampling times is to obtain optimum fiber loading on the filter. The collection efficiency does not appear to be a function of flow rate in the range of 0.5–16 l/min for asbestos. However, counting efficiency is a function of filter loading, with lower loadings typically resulting in higher proportional concentrations. In the absence of significant amounts of nonasbestos dust, a sampling rate of 1–4 l/min for eight hour has been determined to be appropriate in atmospheres containing approximately 0.1 fiber/cc. Dusty atmospheres require smaller sample volumes ($\leq$400 L) to obtain countable samples.

According to NIOSH 7400, it is important not to overload the filter with background dust during the sampling. If $\geq$50% of the filter surface is covered with particles, the filter may be too overloaded to count and will bias the measured fiber concentration. It should be noted that OSHA regulations specify a minimum sampling volume of 48 l for an excursion measurement, and a maximum sampling rate of 2.5 l/min for all personal asbestos sampling.

For sample preparation, the acetone clearance procedure or the DMF/acetic acid clearance procedure are used. In either procedure, the filter material is placed on a glass microscope slide and is then made transparent (clarified). Then, either triacetin or Euparal are placed on the clarified filter and a cover slip is placed on top.

The prepared phase-contrast test slide is placed under the phase objective. The slide contains seven blocks of grooves (approximately 20 grooves per block) in descending order of visibility. For asbestos counting, it is intended that some blocks of lines are completely visible and one or more are completely invisible when centered in the graticule area (blocks in between may be partially visible).

The following counting rules are used:

a) Count any fiber longer than 5 µm that lies entirely within the graticule area.
   i) Count only fibers longer than 5 µm.
   ii) Measure the length of curved fibers along the curve. Count only fibers with a length-to-width ratio equal to or greater than 3 : 1.
b) For fibers that cross the boundary of the graticule field:
   i) Count as 1/2 fiber any fiber with only one end lying within the graticule area, provided that the fiber meets the criteria of rule "a" above.
   ii) Do not count any fiber that crosses the graticule boundary more than once.
   iii) Reject and do not count all other fibers.

c) Count bundles of fibers as one fiber unless individual fibers can be identified by observing both ends of a fiber.
d) Count enough graticule fields to yield 100 fibers. Count a minimum of 20 fields. Stop at 100 graticule fields regardless of count.

After counting, fiber density on the filter, E (fibers/mm^2), is calculated by dividing the mean fiber count per graticule field, F/n_f, minus the mean field blank count per graticule field, B/n_b, by the graticule field area, A_f:

$$E = \frac{\left(\dfrac{F}{n_f}\right) - \left(\dfrac{B}{n_b}\right)}{A_f}, \text{fibers/mm}^2$$

Following NIOSH 7400, A_f is equal to 0.00785 mm^2 for a graticule with a projected diameter of 100 µm. The method recommends that fiber counts above 1,300 fibers/mm^2 and fiber counts from samples with >50% of filter area covered with particulate shall be reported as "uncountable" or "probably biased." Fiber counts outside the 100–1,300 fiber/mm^2 range shall be reported as having "greater than optimal variability" and as being "probably biased."

The concentration, (C) in fibers/cc, of fibers in the air volume sampled, (V) in liters, using the effective collection area of the filter, A_c (nominally 385 mm^2 for a 25-mm filter) is calculated as:

$$C = \frac{E \times A_c}{V \times 1{,}000}$$

NIOSH 7400 indicates that asbestos fibers thinner 0.05 µm for amphibole asbestos and 0.15 µm for chrysotile asbestos will not be detected. These numbers are based on published research and are lower than the previously suggested width limit of 0.25 µm for the PCM method. It should be noted that for all types of elongate mineral particles longer than 5 µm with aspect ratio greater than 3 : 1, about 54% are thinner than 0.25 µm, 33% are thinner than 0.15 µm, and only 6.3% are thinner than 0.05 µm (based on the database information, described in Wylie et al. 2022). This shows that historical data based on PCM measurements most probably allowed for most of the elongate particles to be counted.

The NIOSH 7400 method is associated with significant inter- and intra-laboratory uncertainty that is difficult to evaluate. The overall relative standard deviations (S_r) for the method were reported at the level of 0.10–0.12. However, based on a theoretical Poisson distribution of fiber counts,

$$S_r = \frac{1}{\sqrt{N}}$$

The NIOSH 7400 method stated that the actual S_r found in a number of studies is greater than the theoretical number.

The main limitation of the NIOSH 7400 method is that it cannot determine the mineral type of fibers, and therefore the reported concentrations are not necessarily asbestos (or other toxicologically relevant asbestiform fibers). The NIOSH 7402 method allows to supplement the NIOSH 7400 method with transmission electron microscopy (TEM) tools for an estimation of the fraction of counted fibers that are expected to be "true asbestos" (NIOSH 1994a). Higher magnification is utilized (10,000×) to determine fiber dimensions and countability under the acceptance criteria. Energy-dispersive X-ray (EDX) spectra and selected area electron diffraction (SAED) are obtained for the fibers of interest to determine mineralogical fiber type.

According to the method, all particles with diameter greater than 0.25 µm that meet the definition of a fiber (aspect ratio 3 : 1, longer than 5 µm) are counted. However, it should be noted that criteria of particles visible by PCM have not been updated in the NIOSH 7402 method, as it was done for the NIOSH 7400. Therefore, the NIOSH 7402 method would artificially exclude from counting some fraction of thin fibers that might be visible by PCM and would also be potentially toxicologically relevant (as for asbestiform amphibole fibers thinner than 0.25 µm).

The relative standard deviation (S_r) of the TEM method has been shown to be in the order of 0.275 (in an evaluation of mixed amosite and wollastonite fibers). The estimate of the asbestos fraction, however, had a precision of $S_r = 0.11$. When this fraction was applied to the PCM count, the overall precision of the combined analysis was 0.20.

Other important methods used for asbestos exposure assessment are:

1) PLM (Polarized-Light Microscopy)—bulk sample analysis, primary asbestos type identification (identified as % of sample) (NIOSH 9002:1994b).
2) SEM (Scanning Electron Microscopy)—another type of electronic microscopy most often used for bulk samples but that can be applied to air samples (ISO 14966:2019a). The method is known to miss very thin fibers, as compared to TEM.
3) XRD—analysis of mineral crystal structure (US EPA 1993).
4) Chatfield/NYS NOB—ashing and acid digestion of samples to release fibers bound in an organic matrix (US EPA 1993).
5) Indirect sample preparation—allows for cassettes overloaded by environmental contaminants (dust, diesel fumes etc.) to be analyzed (ISO 13794:2019b).

Real-time instruments are gaining traction in current exposure assessment approaches worldwide, though no definite breakthroughs have yet been achieved. For example, Stopford et al. (2013) proposed a method for real-time detection of airborne asbestos by light scattering from magnetically re-aligned fibers.

This methodology is based on the fact that asbestos, uniquely among fibrous materials, has a magnetic susceptibility that leads to a magnetic torque when the material is in the presence of a magnetic field. A similar effect was studied by Timbrell (1975). Later Ulanowski and Kaye suggested that the alignment of crocidolite and chrysotile fibers in a magnetic field was due to the anisotropy of paramagnetic susceptibility within the fibers (Ulanowski and Kaye 1999). Ulanowski and Kaye mentioned that though "very few respirable fiber types can be confused with asbestos on the basis of their magnetic properties," crystalline fibers like fibrous carbon, or glassy fibers with significant amounts of iron can be mistaken for asbestos. It is unclear how this method would react on erionite, balangeroite, or any variety of nonasbestiform EMPs.

Near-infrared instruments (NIR) have also been used for real-time detection of asbestos in air (Zholobenko et al. 2021). While it was claimed that this methodology is efficient for in situ spectroscopic identification of the six types of asbestos, it clearly would not distinguish between asbestiform and nonasbestiform varieties (like amosite and grunerite).

Also, it is unclear if real-time instruments would be able to determine the concentrations of asbestos fibers, besides the simple fact of their presence or absence.

Asbestos loading on a surface is usually assessed by ASTM method D5755. Samples are taken by vacuuming dust and debris from a specified surface area (usually $100\,cm^2$) into a standard open-face cassette, through a nozzle with a 45° angle opening. The cassette should be attached to a pump pulling approx. 2 lpm creating a velocity of approx. 100 cm/s. According to Millette et al. (1990), concentrations greater than $1,000\,s/cm^2$ would be considered elevated, and concentrations greater than 100,000 would be "in the range found when an abatement project barrier had been breached." Newman (2004) agrees, indicating that asbestos concentrations greater than $100,000\,s/cm^2$ "should be cause for concern" and that concentrations above $10,000\,s/cm^2$ would generally be considered greater than background levels.

Comparison of the Results of Different Analytical Methodologies

The relationship between various methods for asbestos exposure assessment is complex. Verma and Clark (1995) compared transmission electron microscope (TEM) and phase contrast microscope (PCM) fiber counts from the same filter. Industrial hygiene samples from operations within a chrysotile mine, crusher, mill and tailings site, a brake manufacturing industry, and a taping products manufacturer were used. A total of 10,318 individual fibers were analyzed from 65 filter samples. The aim of the study was to derive the relationships between TEM counts and PCM counts that could be used in the extrapolation of risk estimates from occupational exposure to low-level nonoccupational exposure to asbestos. The results show that ratios of TEM (PCME) to PCM counts varied between 1.4 and 3.2. The asbestos concentrations for samples prepared by indirect methods should be treated with caution, as suggested by Eypert-Blaison et al. (2010) who argued against the equivalency of the data for indirect and direct sample preparation. Remarkably, when Hwang and Wang (1983) also compared direct and indirect methods using a TEM, for a set of 25 samples, fiber concentrations measured using the indirect method were, on average, 15.5 times higher than those measured using the direct method.

Proximity to the Emission Source

Proximity to the emission source also is a key variable in determining the level of exposure.

Figure 4.2 demonstrates the typical pattern of the decrease of asbestos concentration from emission source with distance (the data from Donovan et al. 2011).

The data determined by Donovan can be approximated by a nonlinear regression:

$$\text{Bystander Factor} = 85 \times 10^{-0.048\,\text{Distance}},$$

where Distance is the distance from emission source in ft. This is only an approximation for many indoor scenarios. However, confined areas, large open areas,

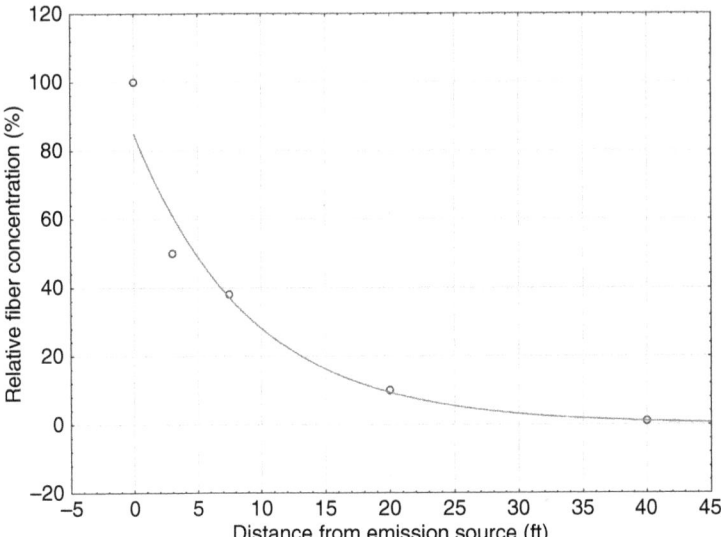

Figure 4.2 "Bystander factor" for the asbestos exposure.

varying ventilation rates, room geometries, and equipment layout, can vary this factor significantly. More accurate calculation of bystander factors can be performed for indoor scenarios as described in Chapter 6 of this book, "Asbestos Exposure Modeling Using Advanced Tools."

Adjusting Results for Censored Data

A special procedure is needed for exposure assessment if censored datapoints (values "less than" limit of detection) appear among the measured asbestos concentrations. For example, some of the PCM concentrations can be reported as "<0.1 f/cc." While different methods can be proposed to determine statistical parameters of the datasets with mixed regular and censored datapoints (as in Hewett and Ganser 2007), the US EPA noted that for asbestos risk assessment, the statistical methods replacing censored datapoints with some hypothetical values can artificially increase risk estimation to unrealistic levels. The US EPA (2021) recommended replacing censored values with 0 or removing them from the calculations.

Correlation of EMP Exposures and Lung Burden Analysis

Rasmuson et al. (2014) performed a detailed evaluation of the correlation and linearity of industrial hygiene retrospective exposure assessment (REA) for cumulative EMP exposure with EMP lung burden analysis (LBA). During the study, a panel of four experienced industrial hygiene raters independently estimated the cumulative EMP exposure for 363 cases with limited exposure details in which EMP LBA had been independently determined. LBA for EMP bodies was performed by a pathologist by both light microscopy and scanning electron microscopy (SEM) and, for free EMP fibers, by SEM. Precision, reliability, correlation, and linearity were evaluated via intraclass correlation, regression analysis, and analysis of covariance. Linear relationships between REA and LBA were found when adjustment was made for EMP fiber-type exposure differences. A significant correlation between REA and LBA was found with amphibole EMP lung burden and mixed fiber types, but not with chrysotile. The intraclass correlation coefficients (ICC) for the precision of the industrial hygiene rater cumulative EMP exposure estimates and the precision of repeated laboratory analysis were found to be in the good to excellent range. It was suggested that both REA and pathology assessment are reliable and complementary predictive methods to characterize EMP exposures. Correlation analysis between the two methods effectively validates both REA methodology and LBA procedures within the determined precision, particularly for cumulative amphibole EMP exposures since chrysotile fibers, for the most part, are not retained in the lung for an extended period of time.

References

[AIHA] American Industrial Hygiene Association. (2024). Risk Assessment Tools. https://www.aiha.org/public-resources/consumer-resources/apps-and-tools-resource-center/aiha-risk-assessment-tools. Accessed January 30, 2024.

[AIHA] American Industrial Hygiene Association, Exposure Assessment Strategies Committee. (2015). *A Strategy for Assessing and Managing Occupational Exposures*, 4th Edition. AIHA Press.

[AIHA] American Industrial Hygiene Association, Exposure Assessment Strategies Committee, Modeling Subcommittee. (2009). *Mathematical Models for Estimating Occupational Exposure to Chemicals*, 2nd Edition. AIHA Press.

Berman, D.W. and Case, B.W. (2012). Overreliance on a single study: there is no real evidence that applying quality criteria to exposure in asbestos epidemiology affects the estimated risk. *The Annals of Occupational Hygiene* 56 (8): 869–878. https://doi.org/10.1093/annhyg/mes027.

Checkoway, H. and Rice, C.H. (1992). Time-weighted averages, peaks, and other indices of exposure in occupational epidemiology. *American Journal of Industrial Medicine* 21 (1): 25–33. https://doi.org/10.1002/ajim.4700210106.

Donovan, E.P., Donovan, B.L., Sahmel, J., et al. (2011). Evaluation of bystander exposures to asbestos in occupational settings: a review of the literature and application of a simple eddy diffusion model. *Critical Reviews in Toxicology* 41 (1): 52–74. https://doi.org/10.3109/10408444.2010.506639.

Eypert-Blaison, C., Veissiere, S., Rastoix, O., and Kauffer, E. (2010). Comparison of direct and indirect methods of measuring airborne chrysotile fibre concentration. *The Annals of Occupational Hygiene* 54 (1): 55–67. https://doi.org/10.1093/annhyg/mep066.

Hewett, P. and Ganser, G.H. (2007). A comparison of several methods for analyzing censored data. *The Annals of Occupational Hygiene* 51 (7): 611–632. https://doi.org/10.1093/annhyg/mem045.

Hinds, W.C. and Zhu, Y. (2022). *Aerosol Technology: Properties, Behavior, and Measurement of Airborne Particles*, 3rd Edition. Wiley.

Hwang, C.-Y. and Wang, Z.M. (1983). Comparison of methods of assessing asbestos fiber concentrations. *Archives of Environmental Health: An International Journal* 38 (1): 5–10. https://doi.org/10.1080/00039896.1983.10543972.

ISO. (2019a). Ambient air – Determination of numerical concentrations of inorganic fibrous particles – Scanning electron microscopy method. ISO 14966:2019. ISO 14966:2019 - Ambient air — Determination of numerical concentration of inorganic fibrous particles — Scanning electron microscopy method.

ISO. (2019b). Ambient air – Determination of asbestos fibers – Indirect-transfer transmission electron microscopy method. ISO 13794:2019. ISO 13794:2019 – Ambient air — Determination of asbestos fibres — Indirect-transfer transmission electron microscopy method.

Klimisch, H.J., Andreae, M., and Tillmann, U. (1997). A systematic approach for evaluating the quality of experimental toxicological and ecotoxicological data. *Regulatory Toxicology and Pharmacology* 25 (1): 1–5. https://doi.org/10.1006/rtph.1996.1076.

Millette, J.R., Kremer, T., and Wheeles, R.K. (1990). Settled dust analysis used in assessment of buildings containing asbestos. *Microscope* 38: 215–220.

Montibeller, G. and von Winterfeldt, D. (2015). Cognitive and motivational biases in decision and risk analysis. *Risk Analysis* 35 (7): 1230–1251. https://doi.org/10.1111/risa.12360.

[NIOSH] National Institute of Occupational Safety and Health. (1994a). Asbestos by TEM. 7402. 7402 (cdc.gov).

[NIOSH] National Institute of Occupational Safety and Health. (1994b). Asbestos (bulk) by PLM. 9002. C:TEMP900280.PDF (cdc.gov).

[NIOSH] National Institute of Occupational Safety and Health. (2019). Asbestos and Other Fibers by PCM. 7400, v.3. NMAM METHOD 7400 (cdc.gov).

[NIOSH] National Institute of Occupational Safety and Health. (2020a). Current intelligence bulletin 69: NIOSH practices in occupational risk assessment. By Daniels, R.D., Gilbert, S.J., Kuppusamy, S.P., Kuempel, E.D., Park, R.M., Pandalai, S.P., Smith, R.J., Wheeler, M.W., Whittaker, C., Schulte, P.A. Cincinnati, OH: U.S. Department of Health and Human Services, Centers for Disease Control and Prevention, National Institute for Occupational Safety and Health. DHHS (NIOSH) Publication No. 2020-106, (Revised 03/2020) https://doi.org/10.26616/NIOSHPUB2020106revised032020.

[NIOSH] National Institute of Occupational Safety and Health. (2020b). Andrews, R. and O'Connor, P.F. *NIOSH Manual of Analytical Methods (NMAM)*, 5th Edition. Washington, DC. https://www.cdc.gov/niosh/nmam/pdf/NMAM_5thEd_EBook-508-final.pdf. Accessed January 30th, 2024.

[NRC] National Research Council. (2009). *Science and Decisions: Advancing Risk Assessment*. Washington, DC: The National Academies Press. https://doi.org/10.17226/12209

Newman, D.E. (2004). On the Issue of Microvac Sampling. In: EPA World Trade Center Expert Technical Review Panel. Comments, May 3, 2004.

Rappaport, S.M. (1991). Assessment of long-term exposures to toxic substances in air. *Annals of Occupational Hygiene* 35 (1): 61–121. https://doi.org/10.1093/annhyg/35.1.61.

Rasmuson, J.O., Roggli, V.L., Boelter, F.W., et al. (2014). Cumulative Retrospective Exposure Assessment (REA) as a predictor of amphibole asbestos lung burden: validation procedures and results for industrial hygiene and pathology estimates. *Inhalation Toxicology* 26 (1): 1–13. https://doi.org/10.3109/08958378.2013.845273.

Stopford, C., Kaye, P.H., Greenaway, R.S., et al. (2013). Real-time detection of airborne asbestos by light scattering from magnetically re-aligned fibers. *Optics Express* 21 (9): 11356–11367. https://doi.org/10.1364/oe.21.011356.

Timbrell, V. (1975). Alignment of respirable asbestos fibres by magnetic fields. *Annals of Occupational Hygiene* 18 (4): 299–311. https://doi.org/10.1093/annhyg/18.4.299.

Ulanowski, Z. and Kaye, P.H. (1999). Magnetic anisotropy of asbestos fibers. *Journal of Applied Physics* 85 (8): 4104–4109. https://doi.org/10.1063/1.370318.

US EPA. (1993). Method for the Determination of Asbestos in Bulk Building Materials. EPA/600/R-93/116 (1993). Accessed January 30, 2024. https://www.nist.gov/system/files/documents/nvlap/EPA-600-R-93-116.pdf.

US EPA. (2000). Science Policy Council Handbook, Risk Characterization, Office of Science Policy, Office of Research and Development, EPA 100-B-00-002, December. Accessed January 30, 2024. https://www.epa.gov/sites/default/files/2015-10/documents/osp_risk_characterization_handbook_2000.pdf.

US EPA. (2019). Guidelines for Human Exposure Assessment. Washington, D.C.: Risk Assessment Forum. EPA/100/B-19/001, October. Accessed January 30, 2024. https://www.epa.gov/sites/default/files/2020-01/documents/guidelines_for_human_exposure_assessment_final2019.pdf.

US EPA. (2021). Framework for Investigating Asbestos-Contaminated Comprehensive Environmental Response, Compensation and Liability Act Sites. Asbestos Committee of the Technical Review Workgroup of the Office of Land and Emergency Management. OLEM No. 9200.0-90. Accessed January 30, 2024. https://semspub.epa.gov/work/HQ/100002942.pdf.

US EPA. (2024a). Exposure Assessment Tools by Tiers and Types – Deterministic and Probabilistic Assessments. https://www.epa.gov/expobox/exposure-assessment-tools-tiers-and-types-deterministic-and-probabilistic-assessments. Accessed January 30, 2024.

US EPA. (2024b). Exposure Assessment Tools by Approaches – Exposure Reconstruction (Biomonitoring and Reverse Dosimetry). www.epa.gov/expobox/exposure-assessment-tools-approaches-exposure-reconstruction-biomonitoring-and-reverse. Accessed January 30, 2024.

US EPA. (2024c). Exposure Assessment Tools by Approaches – Exposure Reconstruction (Biomonitoring and Reverse Dosimetry). www.epa.gov/expobox/

exposure-assessment-tools-approaches-exposure-reconstruction-biomonitoring-and-reverse. Accessed January 30, 2024.

US EPA. (2024d). EPA's Exposure Factors Handbook (EFH). www.epa.gov/expobox/about-exposure-factors-handbook. Accessed January 30, 2024.

Verma, D.K. and Clark, N.E. (1995). Relationship between phase contrast microscopy and transmission electron microscopy results of samples from occupational exposure to airborne chrysotile asbestos. *American Industrial Hygiene Association Journal* 56 (9): 866–873. https://doi.org/10.1080/15428119591016494.

Wylie, A.G., Korchevskiy, A.A., Van Orden, D.R., and Chatfield, E.J. (2022). Discriminant analysis of asbestiform and non-asbestiform amphibole particles and its implications for toxicological studies. *Computational Toxicology* 23: 100233. https://doi.org/10.1016/j.comtox.2022.100233.

Zholobenko, V., Rutten, R., Zholobenko, A., and Holmes, A. (2021). In situ spectroscopic identification of the six types of asbestos. *Journal of Hazardous Materials* 403: 123951. https://doi.org/10.1016/j.jhazmat.2020.123951.

5

Asbestos Exposure Measurements: Principles of Current and Historical Data Interpretation

Garry Burdett

Research and Measurement Scientist (Retired), UK Health and Safety Executive

Aim and Background

The aim of this chapter is to discuss the historical developments in asbestos exposure measurement and their effect on quantitative epidemiological studies included in meta-analyses. The type, quantity, quality, and the extent of the airborne exposure measurements available in different epidemiological studies are fundamental to the adequate estimation of exposure-response gradients. Epidemiological asbestos studies quantify the worker exposure in units of f/ml years, where the f/ml refer to the estimated concentration of fibers per milliliter of air for a particular process in terms of optical phase contrast microscopy (PCM) counts of "defined or regulatory" fibers averaged over a working year. The current method for measuring workplace exposure is based on personal sampling onto a membrane filter (MF) placed near the breathing zone, which is then analyzed by PCM for visible fibers >5 μm long (MF-PCM).

In many individual worker cohort studies, higher airborne concentrations of asbestos were prevalent before the current regulatory unit of exposure based on PCM fiber counting was developed. This meant the measurements were made with different metrics based on different sampling and analysis methods, so approximate conversions and broad estimates of exposure have to be made. The detail of what was measured is often rolled up with other variables and expressed as the overall uncertainty of the dose-response gradient in the mathematical modeling.

More recent meta-analyses have often used a panel of experts, or even a systematic appraisal, to assess the limitations of individual cohort studies and

---

Health Risk Assessment for Asbestos and Other Fibrous Minerals, First Edition.
Edited by Andrey Korchevskiy, James Rasmuson, and Eric Rasmuson.
© 2024 John Wiley & Sons, Inc. Published 2024 by John Wiley & Sons, Inc.

have tended to focus on the quality and quantity of the exposure monitoring. These assessments are then used to decide whether they merit inclusion in the meta-analysis and/or what weight they should be given. As dose-response estimates from different cohorts may vary by 2–3 orders of magnitude, their inclusion or noninclusion in the meta-analyses can result in considerable differences in the estimate of risk. For example, a recent meta-analysis (Darnton 2023) found that lung cancer rates in chrysotile-only cohorts vary by up to 2 orders of magnitude and mesothelioma rates for cohorts using different asbestos types vary by up to 2.5 orders of magnitude.

The purpose of this chapter is to set out how exposure measurements were made (e.g. from static sampling of total mass/number of particulates to analytical electron microscopy analysis of fiber types and sizes) and how the current metric of PCM f/ml for air concentration has evolved and changed over the last 50 years. Chapter 10 considers what air sampling and analysis methods were used in various countries, their shortcomings, and how well (and when) they were converted to the MF-PCM "standard" unit of exposure used in the meta-analyses.

Causes of Asbestos-Related Lung Disease and Their Relationship to Exposure Assessment

The ability of dust to reach the lung (i.e. respirable) is of fundamental importance for exposure assessment. The current definition of respirable (ISO 7708:1995) is in the form of a probability function of the likelihood that a particle (i.e. of an aerodynamic diameter below 10 μm) can penetrate to the nonciliated region of the lung. As the aerodynamic diameter of a fiber is relatively independent of its length, this also allows elongated particles and fibers (many tens of micrometers in length) to penetrate and deposit in the lung. Some of the longer fibers tend to deposit in the upper airways (bronchiole region of the lung) due to interception at the bifurcations, but they can also penetrate and deposit on the surface (epithelium) of the air-exchange (alveolar) regions of the lung.

The lung clearance mechanisms are poorly adapted to deal with elongate particles and fibers greater than around 15–20 μm in length and their continued presence (i.e. bio persistence) is sufficient to both initiate and promote tumors and a fibrotic response. Merewether (1938) and others argued that if the fibers were longer than the human macrophage cells (~18 μm), this resulted in incomplete "phagocytosis," which caused scarring and damage to the lung associated with fibrosis and asbestosis. Early animal experiments (Vorwald et al. 1951) supported this view and by the late 1950s, the UK asbestos industry when establishing the Asbestosis Research Council (ARC) concluded that fibers >5 μm long were responsible for asbestosis and were the most relevant metric for monitoring and controlling worker exposure to asbestosis (Walton 1982). At the same time, a fiber was defined as a particle with an aspect ratio (length:width) of >3 : 1.

Asbestos-related lung cancers are often associated with the upper airways of the lung, but other causes of lung cancer (e.g. tobacco smoke and radon) initially made the association with asbestos exposure more difficult to diagnose and identify until epidemiological studies of asbestos mining, milling, and manufacturing cohorts were established (Doll 1950). This means that asbestos cohorts are assessed for their excess lung cancer deaths as compared to a representative unexposed population.

Until the latter part of the twentieth century, mesothelioma was a rare cancer of the lung lining, a cavity between the chest wall (pleura) and the abdomen (peritoneum). The cavity is bounded on the lung side by the visceral lining and on the outside by the parietal lining. It was initially associated with the crocidolite-producing mines in the North-Western Cape region of South Africa (Wagner et al. 1960), but has been subsequently reported in many asbestos worker cohorts. Mesothelioma is widely considered a nonthreshold carcinogen whose occurrence is primarily related to asbestos (and some other types of bio-durable fibers). The relatively low idiopathic occurrence of mesothelioma means that any occurrence in an asbestos cohort is assumed to be due to asbestos exposure.

Many series of animal experiments based on implanting, injecting, and breathing specially prepared fibers of different sizes, have been conducted to assess what fiber dimensions are associated with asbestos-related cancers. An opinion from a recent review of the many studies on the role of fiber length (Barlow et al. 2017a) considered that exposure to fibers longer than 10 µm and perhaps 20 µm are required to significantly increase the risk of developing asbestos-related diseases in humans. However, the mechanisms relating fiber dimensions to mesothelioma continues to be a focus of research for nanomaterials with fibers blocking the stoma clearance to the lymphatic system in the parietal pleura suggested as the possible cause of mesothelioma (Kane 1991). As individual stomata are up to about 6 µm in width and have bifurcations, the translocation and clearance of >5 µm long fibers can become blocked. This was observed to take place in animal injection experiments, using specially prepared and sized samples of carbon nanotubes and nickel nanowires (Murphy et al. 2011). Therefore, the same length criteria (>5 µm long) adopted by the UK asbestos industry over 70 years ago remains a relevant index of exposure.

The information on how the dimensions of particles affect their probability of causing disease (or potency) has been reflected in the development of air monitoring methods and analysis. However, as expressed by the Pott hypothesis (Pott 1978), there is likely to be a gradation of potency with fiber dimensions, rather than a single sharp cut-off. This would make sense in that there are variations in both the dimensions of human (and animal) lungs and the biologically relevant clearance mechanism, with age and between individuals. Pott also demonstrated that the type of fiber (i.e. mineral and synthetic/man-made) also had different potencies with fiber of low bio durability, causing little or no tumor formation in animal injection experiments. The bio durability of fibers in the lung has been recognized as an important modifier of their potency and is now used as the basis for the carcinogenic classification of man-made vitreous fibers in the EU (EU JRC 1999). Chrysotile has been shown to have low bio durability in animal inhalation experiments (i.e. a half-life of a few days) and can be rapidly cleared from the lung (Bernstein et al. 2013). In contrast, crocidolite has an estimated half-life of ~9 years (Berry et al. 2009).

While a simple mechanistic role of fiber dimensions with disease is attractive for measurement science: the cellular and molecular mechanisms of carcinogenesis that cause the initiation and promotion of cancers remain uncertain and widely debated, after many decades of research.

Although not reviewed here, excess deaths due to other types of cancer have also been studied in some asbestos cohorts and the same cumulative air exposure data is used as a measure of dose for risk estimation.

Exposure Measurement

Employment records and job histories usually provide epidemiologists with good evidence of the duration of exposure in a particular area of the mine, mill, or factory. It is the estimation of the intensity of exposure using the relevant metric for the disease risk, which is by far the more difficult task. Measurements, were usually carried out infrequently, and mostly representing short-term snapshots of the air concentration, taken to check how well a control system was working, or later, for regulatory compliance. How representative these short periods of measurement were of the average exposure over a year, is usually very limited or unknown. As many of the quantitative asbestos cohort studies were set up from the mid-1960s once the association with mesothelioma and lung cancer had become established, they often included asbestos workers who were first exposed several decades previously (i.e. retrospective studies). This means that most of their exposure took place before the current index of measurement (>5 µm long fibers) was established. As the methods for air sampling and analysis and the units of measurement have evolved over time (and continue to do so), this complicates both the interpretation of past exposure measurements and the assessment of the risk associated with current exposure measurements.

A short summary of the historic exposure measurement methods that have been used for asbestos is given below.

Historic Methods of Asbestos Exposure Measurement

Occupational hygiene-related measurements of dust exposure began at the very start of the twentieth century in South African gold mines and its use spread to other mining industries where work-related diseases such as silicosis were occurring (Walton 1982). In 1919, records show that over 35,000 gravimetric samples were analyzed for underground mining activities in South Africa, including the asbestos mines and mills (Walton and Vincent 1998). High rates of lung disease were also associated with the manufacture of asbestos products (e.g. producing mattresses stuffed with asbestos fiber for thermal insulation of steam engines) and the air sampling carried out in mines was later extended into downstream manufacturing.

Gravimetric Methods

Early measurements were gravimetric assessments of total dust using various filtration media. Hand pumps were used to sample air through a filter medium (e.g. moss, cotton wool, sugar etc.) often packed in a glass tube. The dust sampled was recovered using ignition or dissolving in water, as appropriate. From 1923 onward, paper thimble-type collectors became popular and with the development of electrically powered air pumps could sample over several hours. The gravimetric sampling of total dust continues to be the main exposure/control index for asbestos mining and manufacturing in China and Russia (Schonfeld et al. 2017).

The use of total dust monitoring to reduce the levels of industrial disease was not found to be particularly effective and the importance of the smaller respirable particles was identified when lung tissue was digested and examined (McCrae 1913). This observation meant that the microscopical evaluation of smaller (respirable) particles, often using short-term sampling, increasingly became the preferred measurement method from the 1920s (Walton and Vincent 1998), but gravimetric dust surveys using sugar tubes were carried out into the 1930s in some US asbestos factories.

Evidence of improved correlations between the finer particles and lung diseases continued to accumulate from the microscopical analysis, and in 1952, the British Medical Research Council (BMRC 1952) published a probability curve for the penetration of "respirable" particles into the lung. This prompted the development of samplers, such as elutriators, to separate the respirable fraction of the airborne dust. Although respirable mass measurements were widely adopted for assessing airborne dust exposure in other sectors of industry, the widespread use and relevance of particle and fiber number counts, meant that respirable mass monitoring was rarely used in the asbestos industry. Further work by the BMRC (Timbrell 1965) also showed that for asbestos fibers, a 10 μm aerodynamic diameter was broadly equivalent to a maximum microscopic width of ~3 μm.

Over time, this value was adopted for PCM fiber counting, to differentiate between respirable and nonrespirable fibers during the analysis, instead of using a respirable size selector inlet for the sampling.

Impaction Sampling and Microscopic Particle Counting

The use of inertial impaction to collect dust samples for microscopic evaluation was first developed in South Africa (Kotze 1919). The konimeter used a spring-loaded piston to rapidly accelerate 5 ml of air through a small 5 mm diameter nozzle directly onto (at a right angle to the air and 5–6 mm away) the surface of a glass slide coated with petroleum jelly. The deposit was then analyzed microscopically (e.g. ×150 darkfield microscopy) to give particle number concentrations. The ease of use and portability of this hand-held instrument contributed to its widespread and extensive use in many western countries and became the standard in several countries (e.g. South Africa and Australia). The basic design was further developed by several instrument manufacturers over the following three decades in South Africa (Verma et al. 1987) and elsewhere (e.g. the Ziess Konimeter built in Germany and the Owens jet counter in the United Kingdom) (Walton 1982). These included a rotating glass disc allowing the operator to collect up to 30 samples; the incorporation of a microscope to allow immediate on-site counting; and the use of wet blotting paper to supersaturate the air, which removed the need for the petroleum jelly coating.

Konimeters were widely used in asbestos mines and mills but as in other mining sectors, did not discriminate by particle shape, so measurements were expressed as all visible particles per cubic centimeter (p/cm^3). The konimeter was a hand-held instrument so the operator could hold it close to the breathing zone of a worker (personal sample), or at a selected position in a processing area (static sample). The sampling time was essentially instantaneous, so several samples were often collected throughout the day to estimate the shift average. As different versions of the instrument tended to be favored in different countries, and even between different mining sectors in the same country, some variations in instrument performance will occur. The analysis of the particle count is likely to give a number of biases and temporal variations may be due to operator variables, as well as changes in plant operation. Also, the conversions from konimeter to the MF-PCM method will be both facility and area specific, as the percentage of the host rock particles is changed during mining and milling operations and other types of particles may be added during the manufacture of asbestos products.

Impinger Sampling and Microscopic Particle Counting

Attempts to compare the existing sampling methods were made by the US Bureau of Mines in 1922, in collaboration with some universities. The review favored the development of an impinger sampling and particle counting method (Greenberg and Smith 1922; Greenberg and Bloomfield 1932). An impinger differs from impaction in that the air is accelerated onto an impaction plate submerged in a liquid. The impinger developed (i.e. the Greenburg-Smith Impinger) accelerated the sampled air through a glass tube of decreasing diameter to a 2.3 mm diameter jet positioned 5 mm above the base of a flat glass plate (or a flat-bottomed tube) containing a liquid (e.g. water or water and isopropyl alcohol). After sampling, a 1 ml aliquot of the liquid was pipetted into a glass sedimentation cell and left to allow the particles to settle onto the flat bottom of the cell. The number of visible particles on the cell bottom was counted as seen using light microscopy at either ×100 or ×150 magnification. Results were reported in terms of million particles per cubic foot (mppcf) based on the volume of air drawn through the apparatus and the fractional volume of the liquid transferred into the sedimentation cell.

The method had poor collection efficiency for the respirable-sized particles that can penetrate deep into the lung. The velocity of the jet and collection in water also resulted in the break-up and division of fiber bundles and matrices that were present in the air. However, delays in counting (recommended within 36 hours) resulted in agglomeration of the fibers and particles. The Greenburg-Smith Impinger was only suitable for static sampling requiring a flow rate of 1 cubic foot per minute (cfpm) using an electrically driven suction device (weighing 45 lbs). The development of a smaller "midget impinger" operating at one-tenth of the flow rate (Littlefield et al. 1937) and the introduction of a hand-cranked pump by a second operator and the later micro-impinger, allowed the possibility for personal samples to be taken.

Impinger sampling and particle counting were widely used for asbestos dust surveys in North America and was the basis for the first US asbestos dust standard, until the late 1960s, when the MF-PCM method was adopted. There were however many variables and difficulties associated with the Impinger method, ranging through the relative short sampling times (10–30 minutes), the practical issues of using a liquid collection method, to the variability in procedures and precision for the manual counting. While alternative filter sampling methods were investigated (Brown 1944), the most fundamental limitations were that the method was shown to have poor collection efficiency for <1 μm particles (Silverman et al. 1950) and counted all particles without discrimination so did not report a fiber count.

Thermal Precipitator (TP) Sampling and Microscopic Particle Counting

The development of the TP (Green and Watson 1935) produced the first sampling technology with a high collection efficiency for respirable particles. Air was drawn into the instrument between two coverslips placed parallel to a heated wire and particles deposited in two strips due to thermophoresis pushing particles toward the colder surface. The low sampling rates (1 ml/minute) meant that it did not break up the asbestos fibers and matrices, and the longer sampling times offered a better estimate of the average shift exposure. The low flow rate was achieved by a needle valve draining water from a liquid reservoir, which meant that the TP was predominantly used for static sampling. But the use of a backpack or a longer connecting tube to the reservoir made personal sampling feasible provided the worker movements were over a limited area (e.g. spinning and twisting in asbestos textile manufacture). This method was used mainly in the UK asbestos manufacturing industry.

The introduction of the TP was also associated with the greater awareness that asbestos fibers were the probable cause of fibrosis (asbestosis) and that fiber concentration based on microscopic examination of the deposit on the coverslips was likely to be a more appropriated index of exposure. While thermophoresis is very efficient for particles <5 µm, the design of the standard TP (STP) was eventually recognized by the asbestos industry (Holmes 1965) to have limited efficiency for sampling fibers >10 µm long.

Further modifications (MTP) to the STP were made in the 1950s (Kitto and Beadle 1952) to produce a better sampling geometry followed by a larger instrument with a size-selective elutriator capable of whole shift sampling onto a glass slide (Hamilton 1956). The resulting long-running thermal precipitator (LRTP) was used for static sampling in the British coal mines and asbestos industry in the 1960s (Walton and Vincent 1998), although it appears that the horizontal elutriator was often removed for asbestos sampling, to reduce the loss of long fibers (Holmes 1965). The thermal gradient around and distance from the heated wire produced a size gradient of the particles deposited, so the microscopical counting methods (usually ×150 darkfield) had to take account of the inhomogeneous nature of the deposit.

Direct comparisons with the STP vs. MF-PCM method showed that the routine TP counts were about three-quarters of the MF-PCM counts with losses of long fibers in the former (if/when they occurred), most likely due to the inlet slit (500 µm width) becoming blocked with loose agglomerates of fibers (Burdett 1998). The long experience of particle and fiber counting with the TP in the United Kingdom was used to inform the development of the UK MF-PCM method for asbestos fiber concentration (Holmes 1965).

Direct Reading Instruments for Particle and Fiber Counting

The manual effort required for microscopic analysis of air samples led to the use of automated light-scattering instruments for "real-time" monitoring and control of dust in the asbestos industry. The Royco instrument (Royco Corporation, California) developed for clean-room monitoring of particulates in air, was widely investigated and used in the UK asbestos industry (Addingley 1965) at the same time as the MF-PCM method was being developed. The real-time continuous monitoring of airborne dust concentrations in the production industry was widely seen as the best way to control exposures in manufacturing processes and give supporting data on the exposure of the workers provided by periodical MF-PCM personal sampling. Although these instruments were bulky and expensive and could only count particles, they could be placed at representative locations to check that controls were effective and that the required dust concentrations were not exceeded over a continuous period. Many attempts to produce a fiber monitor have been made over the years, with various manufacturers bringing models onto the market. These new technologies have yet to supplant the MF-PCM method and remain rarely used, but the continued reliance on human fiber counting remains a significant source of bias and uncertainty.

Early Sampling Strategies
The bulkiness and operational limitations of the available early sampling technology meant that area (static) samples were usually taken, with the sampling position chosen to represent the general levels in the area of a particular process, for example milling, spinning, and weaving. While measurement scientists were aware that personal (breathing zone) exposures were likely to be higher, the use of backpacks for the TP and/or long rubber piping for the TP or impinger was rarely practicable enough to be deployed widely (Burdett 1998; Cherrie 2003). This meant that the "far-field" static samples collected would underestimate the actual exposure to the worker. The amount of underestimation would depend on the type of process, the level of ventilation, the number of workers in the area, the size of the work area, the time into the shift when the sample was collected, and the length of the sampling period, etc. In an area of relatively well-mixed air with many closely spaced, workers carrying out the same activity, near-field static samples may approach the personal exposure but a factor of up to ×10 may exist in other indoor circumstances (Cherrie 1999).

While the hand-held konimeter could be used to take a near-field or personal sample, the general practice in industry often ignored such potential, preferring to use well-established static sampling points to assess that the plant conditions were stable and at or below a "datum" level by: visible dust levels (Merewether and Price 1930), gravimetric methods (Schonfeld et al. 2017), or other newly developed particle and fiber counting techniques (Merewether 1938). Static sampling continued in many factories after hygiene limits were established to control the costs of installing and running extraction and filtration equipment and the supply of warm make-up air.

The amount of dust in the air and the design of the early sampling instrument meant that the length of sampling times varied from near instantaneous to around an hour. Awareness of the importance of whole-shift sampling to measure the time-weighted average (TWA) exposure, lead to the development of the LRTP and to the current MF-PCM monitoring method (WHO 1997).

Development of the Current Analytical Methods for Fiber Counting

As with any analytical method, there is often a substantial lag due to method development, standardization, and accreditation before the measurement method and the index used can be considered as stable and routine. Similarly, the development of a hygiene standard/occupational exposure and its adoption by national and international regulatory authorities, will also take time, to enable differences between industry, research, and national methods to be reduced and reconciled. Arguably, for asbestos, this is still something of a work in progress.

Membrane Filter Sampling and Phase Contrast Microscopy Fiber Counting (MF-PCM)

The development and wider availability in the 1950s of cellulose-type MFs (Paulus et al. 1957) and personal battery-powered sampling pumps (Sherwood and Greenhalgh 1960) made the change from mostly (static) area sampling to personal sampling of airborne particulates both possible and practical to do. MF-PCM fiber counting of asbestos was first described in 1959 and was initially developed in the early 1960s by the asbestos industry's Asbestosis Research Council (ARC), as described by Walton (1982). In the United States, a parallel development of an MF method was being undertaken to replace the liquid impinger methods (Lynch and Ayer 1968). This led the P&CAM 239 method, which was superseded by the NIOSH 7400 method in 1979, now in its third revision (NIOSH 2019).

The adoption of MF-PCM methods further established an exposure index based on the fiber concentration. The work undertaken in the United Kingdom (BOHS 1968) and in the United States by the American Industrial Hygiene Association (AIHA) and the American Conference of Governmental Hygienists (ACGIH) (Barlow et al. 2017b) in setting a workplace hygiene standard, helped establish the MF-PCM method as an "index of exposure" for asbestos epidemiology. However, a standard method for MF-PCM analysis did not come rapidly and several significant changes were made over the next 20+ years to further improve and standardize the method.

During the initial development period, the new MF-PCM method started to be used alongside previously used methods, to enable comparisons between the methods and conversion factors to be estimated. Undoubtedly, during the period, the experience of the persons involved, and the equipment and procedures used, would have led to significantly lower conversion ratios compared with the current MF-PCM standard. While some authors have attempted to estimate the effect of the various improvements on early fiber counts (Rickards 1994), the magnitude of the difference depends on the date that the conversion measurements were carried out, and which standard or procedure was used for the MF-PCM counting. Therefore, the date the conversion of the earlier measurement methods to an

MF-PCM count was made is a further large source of error, which has yet to be properly addressed in many of the quantitative cohort studies.

Alongside the technical issues associated with the assessment methods, the biases and systematic errors involved in the conversion factors, the selective nature of areas and activities sampled (and more importantly not sampled) plus the length of sampling period and plant conditions at the time, will produce substantial biases and errors in the exposure assessment of the different asbestos worker cohorts.

Membrane Filter Sampling and Electron Microscopy (EM) Analysis

Although MF-PCM has been used as an "index of asbestos exposure" for over 50 years, it has a number of intrinsic instrumental limitations: for example many of the asbestos fibers in air are either too thin to be visible or too short to be counted and the PCM cannot identify or discriminate between fiber types. While this was well understood and while state-of-the-art ×2000 oil immersion light microscopy was used in some early UK investigations in the 1930's (Burdett 1998), it was not until transmission electron microscopy (TEM) was available that researchers were able to assess the thinner fibers. Some limited access to TEM was available at the time the MF-PCM method was being established and the much higher resolution and magnifications meant that it was possible to measure the full fiber size distribution and/or to assess the performance of the PCM for $>5\,\mu m$ long fiber counting. Due to the TEM image quality on the phosphor screen and the range of magnifications available, it was possible to size (in discrete size ranges) and determine mineral type (chrysotile of amphibole using electron diffraction) on the screen in real time. High-quality photographic images could also be recorded, for more accurate off-line size distribution analysis. Due to the size range of the individual fibers (fibrils) and the complexity of the asbestos structures present in air, it was used as a research tool to further assess workplace exposures from the 1970s.

The availability of digital-based measurement technology from the 1980s and the ability to capture increasingly high-quality digital images from the 1990s, also made more rapid individual fiber sizing and identification possible. The availability of "analytical" TEMs from the early 1950s, fitted with wavelength dispersive X-ray analyzer (microprobe) and in the late 1960s with energy dispersive X-ray analyzer (EDXA), made this the instrument of choice for more detailed examination of asbestos exposure in a wide range of environments. An analytical TEM-EDXA can in combination with its high resolution and image contrast and its electron diffraction capability (to measure crystal lattice spacings) identify individual fibers and determine whether they are one of the six types of regulated asbestos. One other advantage was that portions of the same MF can be used for both PCM and TEM analysis (Burdett and Rood 1983), and this has been particularly used for studies of the North and South Carolina chrysotile textile cohorts.

Scanning electron microscopy (SEM) became more readily and cheaply available than the TEM during the 1970s and offered routine counting of approximately the same size of fibers as seen by the PCM. While higher resolution and magnifications were available, longer viewing and integration times were needed to overcome the signal:noise ratio visibility issues for thin fibers. Also, the speed and sensitivity of the EDXA system on conventional tungsten filament SEMs meant that it was usually only sufficient to discriminate between asbestos and nonasbestos fibers that would be visible by PCM (>0.2 µm width). Therefore SEM-EDXA analysis has also been used for the discrimination between asbestos and nonasbestos fibers (AIA 1992; ISO 14496:2002; VDI 1991) in downstream uses of the asbestos products (e.g. installing/removing and ambient/indoor air) when other fibers are likely to be present.

One disadvantage of the SEM is that it requires the sample to be collected on a polycarbonate filter, which is not suitable for PCM analysis. This means duplicate sampling is necessary to comply with the MF-PCM hygiene limit, which does not allow fiber discrimination. However, the sample preparation of these polycarbonate filters for SEM analysis is both quicker and cheaper than for TEM, and SEM has been used by several European countries (e.g. Austria, Germany, and Holland). The development of the high illumination intensity Field Emission Gun (FEG-SEM) in the 1990s has improved the SEM resolution and signal:noise ratio, so that it is practicable to observe the full fiber size distribution. Other improvements in the secondary electron and EDX sensors have made its performance similar to that of the TEM but at a much higher cost than the SEMs used for routine monitoring of asbestos concentration.

The development, acceptance, and standardization of EM analysis also has had a considerable lag time, further exacerbated by its higher cost and limited availability. The first published EM standard was (VDI 1991) developed in Germany along with the asbestos industry (AIA 1992). The International Standards Organization (ISO) set up working groups for a MF-PCM standard in 1980 and this has developed a number of asbestos methods under the long-term chairmanship of the working group by Eric Chatfield. The first ISO TEM standard was published in 1995 (ISO 10312:1995) followed by an indirect TEM method (ISO 13794:1999) and an SEM method (ISO 14496:2002). The ISO standards were initially developed for ambient air and have since been adopted for indoor air measurement. All ISO standards are scheduled for a 5-year review (last updated 2019). ISO standards do not override nationally adopted standards for occupational hygiene compliance. With the improvements in EM analysis, it is now possible to revisit the question of what proportion of the asbestos fiber size distribution should be evaluated (Yamani et al. 2012). However, in many western countries the largest group regularly handling asbestos products are maintenance workers, and their asbestos awareness and behavioral mitigation of exposure, are arguably far more important than refining the index of exposure, should you be able to take a representative personal sample.

Limitations of Current Indices of Exposure Assessment

The review of historic methods shows how the measured exposure has evolved from total airborne dust to respirable dust, to fiber counts, as defined and visible by MF-PCM. It was well understood that EM was necessary to evaluate fibers below the limit of visibility of PCM. Exposure measurement is still in the process of evolving to better define the disease-specific fiber dimensions. Additionally, the measurement of the exposure (i.e. respirable fibers that can penetrate the nonciliated regions of the lung) is not a measure of the fibers that deposit in the lung (dose). Most of the inhaled particles (ISO 13138:2012) and fibers (Nielsen and Koponen 2018) are exhaled and not retained in the lung. When deposition occurs, the lung surfactant layer will start to divide the fiber bundles longitudinally and break down fiber matrices into their component fibers and smaller particles. The macrophage clearance mechanism will try to engulf and remove short fibers from the lung via the lymphatic system and will break up the longer fibers laterally, to produce more short fibers, which can then be removed. Long biopersistent fibers may be coated with an iron-rich deposit to form asbestos bodies.

The MF-PCM fiber count has only ever been seen as an "index of exposure" based on >5 μm long fibers that are visible. This is due to workplace exposures containing a range of complex structures other than single fibers, and counting rules have been developed to help standardize what analysts report. For example, the ISO TEM methods counts fibers >0.5 μm long and >5 μm long, as well as classifying "structures" as: fibers, bundles, dispersed clusters, compact clusters, dispersed matrices, and compact matrices, with up to nine fibers being counted in a single structure. While these TEM methods give greater insight into the exposure, it can also give rise to confusion, as sample handling and preparation can divide and break up the bundles, clusters, and matrices. "Direct" preparation methods, as used for PCM and ISO 14496 and ISO 10312, are essentially trying to preserve the airborne structures for analysis; however, "indirect" preparation methods (e.g. ISO 13794) can have a large effect on the number and size distribution of the particulates collected, so the results will be very different and can give rise to confusion, as it is now a different "Index of exposure." However, due to the lack of historical exposure-response data for indirect preparation method and different size indices, the data collected is primarily research data and can rarely be used to estimate disease rates from current exposures, without the use of conversion factors to PCM-MF fiber counts.

As epidemiological studies and meta-analysis are being applied to a much greater range of diseases in asbestos cohort studies (e.g. gastro-intestinal cancers and heart disease etc.), using the same MF-PCM exposure measurements for lung penetration may not be an appropriate index. This further underlined that the MF-PCM fiber count has only ever been seen as an approximate "index of exposure." This suggests that the variations in disease rates between cohorts are also likely to be due to the limitation of the index (i.e. it cannot see some thin fibers) (Wylie and Korchevskiy 2023), as well as other imprecisions from the limited number and types of exposure measurements made. For example, the higher refractive indices of amosite and crocidolite due to their much higher iron content will increase the visibility of the finer fibers in the PCM compared to chrysotile (Kenny et al. 1987). However, as so few samples were taken in amosite and crocidolite cohorts where epidemiology is available, it is difficult to calculate the extent of the bias.

Variability of the MF-PCM Index Over Time

Airborne asbestos fibers have a range of dimensions with lengths from <1 to several hundred micrometers (µm), and widths from 0.01 to >10 µm. The MF-PCM index of exposure used for epidemiology and risk assessments, only counts visible fibers >5 µm long and <3 µm width. A fiber is defined as a particle with an aspect ratio of >3 : 1. Due to the intrinsic limitations of visible light, the PCM optics will struggle to reveal fibers below about 0.2 µm in width. The actual limit of visibility will depend on the way the microscope is set up, the quality of the illumination and the optics, the visual acuity of the operator, the refractive index (RI) difference between the fiber and the filter mount, the length of the fiber and various other subjective and environmental influences.

The changes to the MF-PCM method over time can be grouped into five categories:

1) Sampling method,
2) Sample preparation method,
3) Microscope equipment and set up,
4) Fiber definitions, and
5) Counting procedures and performance.

Sampling Method

Key issues for sampling are whether personal or area (static) samples were collected. The inventors of the personal sampling pump compared 39 paired week-long personal and area samples and found that on average the personal sample exposures were five times higher. Comparison from across industries (nonasbestos) suggests that for full shift samples the personal exposure is unlikely to exceed the static exposure by more than an order of magnitude (Cherrie 1999). However, for simulations involving short-term samples, even in a small unventilated work area ($12.7\,m^3$), some asbestos release activities can far exceed an order of magnitude difference between personal and static samples (ASEA 2016). However, one of the earliest comparisons of static and personal sampling, in asbestos (crocidolite) textile manufacturer in 1938, found little difference between personal and static TP samples collected in conditions of good ventilation (Burdett 1998). Therefore, it is not possible to apply a general conversion factor even when personal sampling is used, because depending on which shoulder/lapel the filter is placed, the results may give a factor of two difference.

Large fluctuations in concentrations can occur during the day, so full-shift sampling is important, but this is only possible in relatively low dust concentrations. To avoid filter over-loading when sampling high airborne concentrations, multiple short-term samples are required, and a TWA calculated. Flow rate will also determine the particle loading on the filter and the size of particles that can reach the filter. To help avoid tampering or damage to the filter, an entry cowl was added to the open-face sampling head. While this was primarily used to prevent damage to the filter, it also acted as a vertical elutriator and limited the sampling of larger particles with aerodynamic diameters of >20 µm at the flow rate of 1 l/minute used for personal sampling. As with any sampling operation, accurate measurement of the flow rate against a calibrated method is important and significant errors can occur from the use of inline or uncalibrated rotameters. Also often overlooked, is the effectiveness of the seal between the filter and the filter holder.

The procedure used for loading and unloading the filter is also important. Loss of the filter deposit can occur due to over-sampling, careless handling, the past use of fixative sprays, and the use of highly static plastic containers to store the sampled filters. Many studies mention that some MFs were too overloaded with particles to count, and their exclusion will result in an underestimate of the airborne concentration.

Sample Preparation

The choice of filter type and pore size will have an influence. MFs, made from cellulose nitrate and/or a mixture with cellulose acetate, were and remain the normal filter formulations used for asbestos sampling. These have a sponge-like texture where interconnecting pores or a framework gives a nominal pore size. The size of the pores will determine the filtration efficiency as well as how far particles penetrate into the ~160-µm-thick filter. Therefore, there will be significant differences between using a 0.8 µm pore size filter instead of an 8 µm pore size filter, as was used in some early measurements in the South African mines and mills.

The filter material has an RI of around 1.51 (slightly below that of chrysotile) and appears white in air, due to the air-filled pores. To render it transparent for counting either liquids of a similar RI must be introduced to fill the pores, or the pores collapsed to remove the air. In the US (Paulus et al. 1957), the liquid of choice was a mixture of dimethylpthalate (n_D 1.515) and diethyl oxalate (n_D 1.410) – so overall RI would depend on the mix, plus the amount of dissolved filter material. This method became codified as the NIOSH P&CAM 239 method (Edwards and Lynch 1968; NIOSH 1979). In the United Kingdom, the use of glycerol triacetate "triacetin" (n_D 1.43) was specified (DEP 1970; ARC 1971), which gave the cleared filter a RI of about 1.47–1.48. These pore-filling semisoluble methods produced samples that were stable for only a short period, as liquid and particles would migrate toward the edges of the slide. In addition, some of the particles and fibers collected would be buried inside the filter, requiring the microscopist to rack the focal plane down into the filter for each field of view, to find all the fibers.

The exposure of filters to acetone vapor was found to collapse the filter pore structure. This had some advantages in that it produced a more permanent mount, helped to align the fibers horizontally to the filter/slide and distributed them in a much narrower horizontal plane, close to the surface. These were introduced in various national and international methods in the decade from late 1979. A mount with the coverslip was still needed and a thermoset resin (e.g. Euparal) (Le Guen and Galvin 1981) or triacetin was used (AIA RTM1 1979). It was reported that the filter collapsing methods gave higher mean values than the methods which cleared the filters by filling the pore spaces (e.g. P&CAM 239 mount) by almost 2 : 1 (Gonzalez-Fernandez and Martini 1986). A further comparison of chrysotile asbestos cement industry samples found even larger differences, when comparing the two methods (Gonzalez-Fernandez et al. 1987). The final RI of the mount is important for the visibility of chrysotile, which has RIs of around 1.55. Use of liquids to clear the filter with RIs close to the MF material (RI = 1.51) will decrease the visibility of chrysotile fibers and underestimate the fiber concentration.

Microscope Equipment and Set-Up

The type and quality of the optics, the power of the light source, and how well set up the microscope is will have a large effect on the fiber count. Bright field, oil-immersion, phase contrast, and dark field microscopy have all been used for fiber counting with magnifications from ×100 to ×2000. The importance of using PCM for counting chrysotile fibers has already been noted and early PCM counting comparisons showed fibers were particularly likely to be undercounted by inexperienced counters (e.g. a factor of 3, Beckett and Attfield 1974). Until a standard test slide was introduced (Le Guen et al. 1984) to aid microscopists to set up the microscope to a uniform level of visibility, it was difficult to know by how much each count was influenced by the microscope equipment and set-up. As the width/thickness distribution of the sampled asbestos fibers is both above and below the limit of visibility, a microscope capable of seeing 0.2 µm width fibers, will give a higher fiber count (e.g. a factor of 2–3) than one only seeing 0.4 µm width fibers. The lower limit of visibility is particularly important when analyzing samples of chrysotile and crocidolite due to their predominantly lower median width of their fiber size distributions. The metamorphic conditions of formation at individual mines also determine the fiber width distribution.

The intensity of the light source and whether Kohler illumination was used are also important for the visibility of fibers. The use of high-intensity halogen illumination sources gave further improvements in fiber visibility, compared to the older tungsten bulbs. Over the last decade, high-intensity LED illumination systems are replacing the halogen bulbs. This and other improvements in microscope performance will usually result in further increase in the visibility of thin fibers, despite the use of the test slide to set up and limit the performance.

Given the higher concentrations being measured during the early use of the MF method for asbestos analysis, the use of full-field counting was particularly responsible for the undercounting of fibers (Beckett and Attfield 1974). They reported that using an eyepiece graticule (i.e. Porton graticule), fiber counts were increased by a factor of about 1.5 for amosite and 2.5 for chrysotile, as compared with full-field counting. Results reported from a study comparing three different types of microscope eyepiece graticules (Porton: Walton-Beckett: Patterson) that had been widely used for asbestos fiber counting, gave counts in the approximate ratios of 1.6 : 1.2 : 1, respectively (Gonzalez-Fernandez and Martini 1986). The adoption of a specialized microscope graticule for asbestos with a ~100 µm diameter circular field of view (Walton and Beckett 1977) has helped to standardize counts.

Fiber Definition

For MF-PCM fiber counting, the basic definition of a fiber (>5 µm long, <3 µm wide with an aspect ratio >3 : 1) has remained relatively unchanged over time and jurisdiction, except for minor changes to the upper width (i.e. ≤3 µm) and upper length limit (<100 µm).

The aspect ratio was introduced to discriminate between "fibers" and "particles" as counted previously in some of the historical measurements of exposure. A separate count of particles and fibers had been in use in the TP counting in the United Kingdom and the >3 : 1 ratio was adopted from this previous experience. The >3 : 1 aspect ratio, has proven to be highly controversial as being too low for fibers of an asbestiform "habit" and includes many acicular particles and elongate cleavage and rock fragments, which are common in other minerals and rock formations. It was soon recognized that more consistent counting performance can be obtained with a larger minimum aspect ratio (>5 : 1 and >10 : 1) but this change has not gained widespread acceptance to date.

While fiber dimensions for counting have been relatively unchanged, the way fiber bundles, split fibers, and complex structures were counted varied until more rigorous counting rules were adopted. More recently, the practice of whether fibers attached to nonrespirable particles were counted has been an important change in some sectors of the asbestos manufacturing industry. Previously, samples attached to particles >3 µm width, were not counted and only in 2006 was the counting of all fibers in the sample, obligatory in the EU. This was particularly important when the asbestos was combined with other materials in the manufacturing process. This included most manufacturing uses of asbestos (e.g. asbestos cement, insulating boards, friction materials, reinforced plastics, etc.) except textile manufacturing.

It was common in textile manufacturing to also use cotton fibers and the use of subjective discrimination was commonplace in early MF-PCM counts as observed by Walton (1982), who stated: "In practice up to the present day, some laboratories in the ARC member companies and elsewhere attempt such discrimination while others do not." Discrimination between asbestos and nonasbestos fibers due to other factors was widely used and a source of much debate. The application of these unofficial industry-based discrimination rules was a source of considerable biases between different parts of the asbestos manufacturing industry. While never appearing in the regulatory or consensus definitions for counting, they will have produced considerable differences in the early estimates of fiber exposures between cohorts.

Counting Procedures and Performance

Although some subjective assessment of dimensions is inevitable with human counters, there were large differences between the laboratories that performed the MF counts. This became apparent when standard methods started to be produced and inter-laboratory counting exercises took place (Beckett and Attfield 1974). The difference in counts between different industry laboratories on the same slides was sometimes very large, and what started as casual inter-laboratory comparisons between government and asbestos manufacturing industry laboratories, developed and spread to become national and international proficiency testing schemes (e.g. the UK HSE's Regular Inter-laboratory Counting Exchange [RICE] and the asbestos industry Asbestos Fibre Regular Informal Counting Arrangement [AFRICA] and the US Proficiency Analytical Testing [PAT] schemes). Participation in such proficiency testing schemes soon became a regulatory requirement and the fiber counting performance of participating laboratories is assessed several times a year. Further requirements and regulations for full laboratory accreditation, which looked at quality assurance and control procedures and their implementation from sample collection to reporting, were also widely introduced. These schemes, originally run as independent national schemes (e.g. the US PAT and the United Kingdom Accreditation Service [UKAS]), now have to comply with one or more global accreditation schemes (e.g. the International Laboratory Accreditation Co-operation [ILAC] and the International Accreditation Forum [IAF]).

Variations between counters (intra-laboratory) and between laboratories (inter-laboratory) have been studied and increasingly documented and controlled over time. The published MF-PCM methods have also been updated to include more detail and codify how the method is applied, with the aim to further reduce intra- and inter-laboratory differences. However, the precision of a fiber count is limited by the underlying statistical distribution. It is assumed that the sampling and counting are based on their being a random distribution of fibers on the filter. This means that the best statistical precision that can be achieved is limited by the assumption of random (Poisson) distribution of fibers, but there will always be a number of additional subjective errors and biases due to the use of people to carry out the analysis. In a well-controlled intra-laboratory environment, this subjective element may be as low as a standard deviation of ~20% (0.2) of the mean count, superimposed on the Poisson variation (Ogden 1982). The current NIOSH 7400 method gives details on the expected precision of fiber counts within the loading range of 100–1300 fibers/mm^2 filter area; for instance, a count of 10 fibers will have a 95% confidence interval of between 4.2 and 20 fibers being found in a recount (NIOSH 2019).

Effect of Changes to the MF-PCM Counts Over Time

Many of the improvements and refinements to the MF-PCM method over the last 50 years have resulted in an increase in the fiber count. The cumulative effects of the changes and improvements to the five categories reviewed above can have a large effect on the fiber concentrations reported. For example, a small cumulative increase of 10% (×1.1) on average in each of four or five categories over this period would imply that the overall MF-PCM fiber count would increase by around ×1.46 to ×1.6, while a 50% (×1.5) increase in all five categories would give an increase of ×7.6 in the count. Rickards estimated overall increase of ×32 from the early fiber counts made in the United Kingdom (Rickards 1994). (Note: this is the same as assuming a factor of 2 improvement for each of the five categories.) Therefore, considerable biases depending on the version of the MF-PCM method in use will be present in the exposure assessment for each individual cohort. This is particularly important in that many conversions of the previous historic measurement indices to MF-PCM counts were made before the mid-1980s, when improved intra- and inter-laboratory standardization and the checking of the microscope performance against a test slide became routine. Any adjustments will be study/cohort specific and where PCM vs. TEM comparisons have been made, with good practice, the maximum variation is likely to be <×3. The experience from the North and South Carolina chrysotile textile plants showed that ~×1.5 was possible.

Conclusion

Retrospective epidemiological studies have many issues of exposure measurement to resolve. Only one type of historical air monitoring method based on thermal precipitation is directly comparable to the current index of exposure. All the other historical methods express results in total mass of particulates or the number of particulates. At best, these measurements may help determine the relative dust exposure concentrations in various areas of the mine, mill, or manufacturing plant at a certain point in time. These relationships also depend on where the sampling took place, for how long and how well sampling personnel followed a protocol or were creatures of habit. Meta-analyses require the units of cumulative exposure to be expressed in terms of (f/ml) × years for relevance to current measurements and for extrapolation to be used in risk assessment and the setting hygiene limits. Therefore, there is a high reliance on the actual exposure measurements made for the individual cohorts and when and how they were converted to the current MF-PCM f/ml index of exposure.

It is likely that only prospective studies set up since the 1980s will be able to reduce the high level of uncertainty from air concentration measurements. The most important conclusion is that when the relationships between fiber type and size and asbestos-related disease is still only poorly, or at best partially understood, the taking and storing of MF samples with full sampling and subject details is essential. Even now, routine mounting of the whole filter for PCM analysis means that any further chance of future reanalysis of fiber types and dimensions is thrown away. It is simple to cut a 25 mm diameter (or larger filter) in half, and analyze one half and store the other half for future analysis. This will allow the future deployment of artificial intelligence and the EM analysis of thinner fibers that are not visible by PCM but are increasingly implicated with mesothelioma and lung cancer (Wylie and Korchevskiy 2023). This is the one simple action that will improve the precision of prospective individual epidemiological studies and meta-analyses, giving greater confidence and precision that future hygiene limits for the various types of asbestos-related diseases are fit-for-purpose.

While most Western manufacturing has ceased (some chrysotile manufacturing and new installation of chrysotile products continues in the United States [US EPA 2020]), much of the original manufacturing equipment and process was relocated to less-developed and less-regulated countries. This industry, along with the continuation of mining and manufacturing in Russia and China, means that there are still many opportunities to document worker personal exposures using both PCM and EM methods, for processes and conditions of control for which many retrospective studies had no similar measurements. If the use of asbestos and its mining and manufacturing cannot be prohibited, it should at least be possible for international and national agencies, manufacturers, and the vast asbestos litigation industry, to help instigate better monitoring and controls. This will have the mutual benefit of both better-protecting workers and increasing the understanding of risk from both past and current use of asbestos.

Acknowledgements

The Late Henty Walton for his 1982 review and Jean Prentice.

References

Addingley, C.G. (1965). Dust measurement and monitoring in the asbestos industry. *Annals of the New York Academy of Sciences* 132(1): 298–305. https://doi.org/10.1111/j.1749-6632.1965.tb41110.x.

AIA RTM1. (1979). *Reference Method for the Determination of Asbestos Fibre Concentrations at Workplaces by Light Microscopy (Membrane Filter Method)*. London, England: Asbestos International Association.

AIA, RTM2. (1992). *Method for the Determination of Airborne Asbestos Fibres and Other Inorganic Fibres by Scanning Electron Microscopy*. London, England: Asbestos International Association.

ARC. (1971). *Technical Note 1. Measurement of Airborne Asbestos Dust by the Membrane Filter Method*. Rochdale, England: Asbestosis Research Council.

ASEA. (2016). Measurement of asbestos fibre release during removal works in a variety of diy scenarios. Report by Monash University for the Australian Asbestos Safety and Eradication Agency. https://www.asbestossafety.gov.au/research-publications/measurement-asbestos-fibre-release-during-removal-works-variety-diy-scenarios (accessed 08 September 2023).

Barlow, C.A., Grespin, M., and Best, E.A. (2017a). Asbestos fiber length and its relation to disease risk. *Inhalation Toxicology* 29(12–14): 541–554. https://doi.org/10.1080/08958378.2018.1435756.

Barlow, C.A., Sahmel, J., Paustenbach, D.J., et al. (2017b). History of knowledge and evolution of occupational health and regulatory aspects of asbestos exposure science: 1900–1975. *Critical Reviews in Toxicology* 47(4): 286–316. https://doi.org/10.1080/10408444.2016.1258391.

Beckett, S.T. and Attfield, M.D. (1974). Inter-laboratory comparison of the counting of asbestos fibres sampled on membrane filters. *The Annals of Occupational Hygiene* 17(2): 85–96. https://doi.org/10.1093/annhyg/17.2.85.

Bernstein, D., Dunnigan, J., Hesterberg, T., et al. (2013). Health risk of chrysotile revisited. *Critical Reviews in Toxicology* 42(2): 154–183. https://doi.org/10.3109/10408444.2012.756454.

Berry, G., Pooley, F., Gibbs, A., et al. (2009). Lung fiber burden in the Nottingham gas mask cohort. *Inhalation Toxicology* 21(2): 168–172. https://doi.org/10.1080/08958370802291304.

British Medical Research Council (BMRC). (1952). British Medical Research Council report. London, England.

British Occupational Hygiene Society (BOHS). (1968). Hygiene standards for chrysotile asbestos dust. *The Annals of Occupational Hygiene* 11(2): 47–69. https://doi.org/10.1093/annhyg/11.2.47.

Brown, C.E. (1944). Filter paper method for obtaining dust concentration results comparable to the impinger results. Report of Investigations 3788. US Bureau of Mines.

Burdett, G. (1998). A comparison of historic asbestos measurements using a thermal precipitator with the membrane filter-phase contrast microscopy method. *The Annals of Occupational Hygiene* 42(1): 21–31. https://doi.org/10.1016/s0003-4878(97)00048-3.

Burdett, G.J. and Rood, A.P. (1983). Membrane-filter, direct-transfer technique for the analysis of asbestos fibers or other inorganic particles by transmission electron microscopy. *Environmental Science & Technology* 17(11): 643–648. https://doi.org/10.1021/es00117a004.

Cherrie, J.W. (1999). The effect of room size and general ventilation on the relationship between near and far-field concentrations. *Applied Occupational and Environmental Hygiene* 14(8): 539–546. https://doi.org/10.1080/104732299302530.

Cherrie, J.W. (2003). The beginning of the science underpinning occupational hygiene. *Annals of Occupational Hygiene* 47(3): 179–185. https://doi.org/10.1093/annhyg/meg030.

Darnton, L. (2023). Quantitative assessment of mesothelioma and lung cancer risk based on phase contrast microscopy (PCM) estimates of fibre exposure: an update of 2000 asbestos cohort data. *Environmental Research* 230: 114753. https://doi.org/10.1016/J.ENVRES.2022.114753.

Department of Employment and Productivity (DEP). (1970). Standards for asbestos dust concentrations for use with the Asbestos Regulations. 1969 Technical Data, Note 13.

Doll, R. (1950). Mortality from lung cancer in asbestos workers. *British Journal of Industrial Medicine* 12: 81–86. https://doi.org/10.1136/bmj.2.4682.739.

Edwards, G.H. and Lynch, J.R. (1968). The method used by the US Public Health Service for enumeration of asbestos dust on membrane filters. *The Annals of Occupational Hygiene* 11(1): 1–6. https://doi.org/10.1093/annhyg/11.1.1.

European Commission Joint Research Centre (EU JRC). (1999). *Method for the Determination of the Hazardous Properties for Human Health of Man Made Mineral Fibres (MMMF) (EUR 18748)*. European Commission Joint Research Centre. Edited by Bernstein, D. and Riego Sintes, J.

Gonzalez-Fernandez, E. and Martini, F.R. (1986). Comparison of the NIOSH and AIA methods for evaluating asbestos fibre: effects of asbestos type, mounting medium, graticule type and counting rules. *The Annals of Occupational Hygiene* 30(4): 397–410. https://doi.org/10.1093/annhyg/30.4.397.

Gonzalez-Fernandez, E., de La Osa, P.D., and Martin, F.R. (1987). Comparison of the AIA and NIOSH methods on asbestos fibre measurements in the workplace. *The Annals of Occupational Hygiene* 31(3): 363–373. https://doi.org/10.1093/annhyg/31.3.363.

Green, H.L. and Watson H.H. (1935). Physical method for the estimation of dust hazard in industry. Special Report 199. MRC. London: HMSO. https://wellcomecollection.org/works/zty27qzk.

Greenberg, L. and Bloomfield, J.J. (1932). Dust sampling apparatus used by the US public health service. *Public Health Reports* 47: 654–675. https://doi.org/10.2307/4580381.

Greenberg, L. and Smith, G.W. (1922). A new instrument for sampling aerial dust. US Bureau of Mines Report of Investigation 2392.

Hamilton, R.J. (1956). A portable instrument for respirable dust sampling. *Journal of Scientific Instruments* 33(10): 395–399. https://doi.org/10.1088/0950-7671/33/10/310.

Holmes, S. (1965). Developments in dust sampling and counting techniques in the asbestos industry. *Annals of the New York Academy of Sciences* 132(1): 288–297. https://doi.org/10.1111/j.1749-6632.1965.tb41109.x.

ISO 10312. (1995). *Ambient Air – Determination of Asbestos Fibres – Direct Transfer Transmission Electron Microscopy Method*. Geneva, Switzerland: International Standards Organisation.

ISO 13138. (2012). *Air Quality – Sampling Conventions for Airborne Particle Deposition in the Human Respiratory System*. Geneva, Switzerland: International Standards Organisation.

ISO 13794. (1999). *Ambient Air – Determination of Asbestos Fibres – Indirect-Transfer Transmission Electron Microscopy Method*. Geneva, Switzerland: International Standards Organisation.

ISO 14496. (2002). *Ambient Air – Determination of Numerical Concentration of Inorganic Fibrous Particles – Scanning Electron Microscopy Method*. Geneva, Switzerland: International Standards Organisation.

ISO 7708. (1995). *Air Quality – Particle Size Fraction Definitions for Health Related Sampling*. Geneva, Switzerland: International Standards Organisation.

Kane, A.B. (1991). Fiber dimensions and mesothelioma: a reappraisal of the Stanton hypothesis. In: *Mechanisms in Fibre Carcinogenesis* (eds. R.C. Brown, J.A. Hoskins, and N.F. Johnson), NATO ASI Series, 223. Boston, MA: Springer. https://doi.org/10.1007/978-1-4684-1363-2_14.

Kenny, L.C., Rood, A.P., and Blight, B.J.N. (1987). A direct measurement of the visibility of amosite asbestos fibres by phase contrast optical microscope. *The Annals of Occupational Hygiene* 33: 261–264. https://doi.org/10.1093/annhyg/31.2.261.

Kitto, P.H. and Beadle, D.G. (1952). A modified form of thermal precipitator. *Journal of the Chemical, Metallurgical and Mining Society of South Africa* 52: 284–287.

Kotze, R. (1919). Final report of the Miners Phthisis Prevention Committee. Miners' Phthisis Commission of Enquiry. South Africa.

Le Guen, J.M.M. and Galvin, S. (1981). Clearing and mounting techniques for the evaluation of asbestos fibres by the membrane filter method. *The Annals of Occupational Hygiene* 24(3): 273–280. https://doi.org/10.1093/annhyg/24.3.273.

Le Guen, J.M.M., Ogden, T.L., Shenton-Taylor, T., et al. (1984). THE HSE/NPL phase contrast test slide. *The Annals of Occupational Hygiene* 28(2): 237–247. https://doi.org/10.1093/annhyg/28.2.237.

Littlefield, J.B., Fiecht, F.L., and Schrenk, H.H. (1937). The Bureau of Mines midget impinger for dust sampling. Report of Investigation 3360. Washington, DC: US Bureau of Mines. http://catalog.hathitrust.org/Record/005982337.

Lynch, J.R. and Ayer, H.E. (1968). Measurement of asbestos exposure. *Journal of Occupational Medicine* 10(1): 21–24. PMID: 5635968.

McCrae J. (1913). *The Ash of Silicotic Lungs*. Johannesburg: The South African Institute for Medical Research.

Merewether, E.R.A. (1938). Dust counts on asbestos flyer spinning process under normal conditions without exhaust: consideration of the so-called "Datum" level. LAB 62/18, Document 17. London: Public Records Office.

Merewether, E.R.A. and Price, C.W. (1930). Report on effects of asbestos dust on the lungs and dust suppression in the asbestos industry. Part II. Processes giving rise to dust and methods for its suppression in the asbestos industry. Part II. London: HMSO.

Murphy, F.A., Poland, C.A., Duffin, R., et al. (2011). Length-dependent retention of carbon nanotubes in the pleural space of mice initiates sustained inflammation and progressive fibrosis on the parietal pleura. *American Journal of Pathology* 178(6): 2587–2600. https://doi.org/10.1016/j.ajpath.2011.02.040.

Nielsen, G.D. and Koponen, I.K. (2018). Insulation fiber deposition in the airways of men and rats. A review of experimental and computational studies. *Regulatory Toxicology and Pharmacology* 94: 252–270. https://doi.org/10.1016/j.yrtph.2018.01.021.

NIOSH. (1979). *USPHS/NIOSH Membrane Filter Method for Evaluating Airborne Asbestos Fibers* (eds. N.A. Leidel, S.G. Bayer, R.D. Zumwalde, and K.A. Busch). Cincinnati, OH: US Department of Health, Education, and Welfare, Center for Disease Control, National Institute for Occupational Safety and Health, DHEW (NIOSH) Publication No. 79-127.

NIOSH. (2019). Asbestos and other fibers by PCM: method 7400. *NIOSH Manual of Analytical Methods (Fifth Edition)* 2(3): 1–40. https://www.cdc.gov/niosh/nmam/pdf/7400.pdf.

Ogden T. (1982). The reproducibility of asbestos fibre counts. Research Paper No. 18. Sheffield, England: Health and Safety Laboratory, Health and Safety Executive. ISBN 0 7176 0101 3.

Paulus, H.J., Talvitie, N.A., Fraser, D.A., et al. (1957). Use of membrane filters in air sampling. *American Industrial Hygiene Association Quarterly* 18(3): 267–273. https://doi.org/10.1080/00968205709343505.

Pott, F. (1978). Some aspects on the dosimetry of the carcinogenic potency of asbestos and other fibrous dusts. *Staub-Reinhalt Luft* 38: 486–490.

Rickards, A.L. (1994). Levels of workplace exposure. *The Annals of Occupational Hygiene* 38(4): 469–475. https://doi.org/10.1093/annhyg/38.4.469.

Schonfeld, S.J., Kovalevskiy, E.V., Feletto, E., et al. (2017). Temporal trends in airborne dust concentrations at a large chrysotile mine and its asbestos-enrichment factories in the Russian federation during 1951–2001. *Annals of Work Exposures and Health* 61(7): 797–808. https://doi.org/10.1093/annweh/wxx051.

Sherwood, R.J. and Greenhalgh, D.M.S. (1960). A personal air sampler. *Annals of Work Exposure and Health* 2(2): 127–132. https://doi.org/10.1093/annhyg/2.2.127.

Silverman, L., First, M.W., Reichenbach, G.S. Jr., et al. (1950). Final progress report for the US Atomic Energy Commission. NYO-1527. February 1, 1950. https://digital.library.unt.edu/ark:/67531/metadc1022125/m2/1/high_res_d/4437141.pdf.

Timbrell V. (1965). Human exposure to asbestos: dust controls and standards. The inhalation of fibrous dusts. *The Annals of the New York Academy of Sciences* 132(1): 255–273. https://doi.org/10.1111/j.1749-6632.1965.tb41107.x.

United States Environmental Protection Agency (US EPA). (2020). Risk evaluation for asbestos Part I: chrysotile asbestos EPA Document # EPA-740-R1-8012, December 2020. https://www.epa.gov/sites/production/files/2020-12/documents/1_risk_evaluation_for_asbestos_part_1_chrysotile_asbestos.pdf.

VDI, 3492-1. (1991). Measurement of inorganic fibrous particles in ambient air; scanning electron microscopy method. *Verlag des Vereins Deutscher Ingenieure*.

Verma, D.K., Sebestyen, A., and Muir, D.C.F. (1987). An evaluation of konimeter performance-II. A field comparison of konimeters. *The Annals of Occupational Hygiene* 31(4A): 451–461. https://doi.org/10.1093/annhyg/31.4A.451.

Vorwald, A.J., Durkan, T.M., and Pratt, P.C. (1951). Experimental studies of asbestosis. *A.M.A. Archives of Industrial Hygiene and Occupational Medicine* 3(1): 1–43. https://doi.org/10.5694/j.1326-5377.1951.tb88509.x.

Wagner, J.C., Sleggs, C.A., and Marchand, P. (1960). Diffuse pleural mesothelioma and asbestos exposure in the Northwestern Cape Province. *British Journal of Industrial Medicine* 17(4): 260–271. https://doi.org/10.1136/oem.17.4.260.

Walton, W.H. (1982). The nature, hazards, and assessment of occupational exposure to airborne asbestos dust: a review. *The Annals of Occupational Hygiene* 25: 115–247 https://doi.org/10.1093/annhyg/25.2.117.

Walton, W.H. and Beckett, S.T. (1977). A microscope eyepiece graticule for the evaluation of fibrous dusts. *The Annals of Occupational Hygiene* 20(1): 19–23. https://doi.org/10.1093/annhyg/20.1.19.

Walton, W.H. and Vincent, J.H. (1998). Aerosol instrumentation in occupational hygiene: an historical perspective. *Aerosol Science and Technology* 28(5): 417–438. https://doi.org/10.1080/02786829808965535.

World Health Organization (WHO). (1997) *Determination of Airborne Fibre Number Concentrations A Recommended Method, by Phase-Contrast Optical Microscopy (Membrane Filter Method)*. Geneva: World Health Organisation. ISBN: 9241544961.

Wylie, A.G. and Korchevskiy, A.A. (2023). Dimensions of elongate mineral particles and cancer: a review. *Environmental Research* 230: 114688. https://doi.org/10.1016/J.ENVRES.2022.114688.

Yamani, M., Boulanger, G., Catelinois, E., et al. (2012). Revision of French occupational exposure limits of asbestos and recommendation of measurement method: can the dimensional characteristics of the asbestos fibers (long, thin, short) be taken into account? *Critical Reviews in Environmental Science and Technology* 42: 1441–1484. https://doi.org/10.1080/10643389.2011.556895.

6

Asbestos Exposure Modeling Using Advanced Tools Including Computational Fluid Dynamics (CFD)

Daniel Hall, James Rasmuson, and Cassidy Strode

Chemistry & Industrial Hygiene, Inc., Lakewood, CO, USA

Introduction

Modeling is a term that is utilized frequently to represent something about which we do not have much information, something of which we do not have a tangible real-world version, or something that does not currently exist. A model often intends to predict a future situation or recreate the past. For instance, when a model walks out on the catwalk, he or she is showing the world what could be the future of fashion. The fashion designers create many fashions, some are out on the fringe, and some are more conservative and in line with the general populace, but, nonetheless, these fashions represent the range of tastes. Why, possibly, are they called models? Because they represent what "you" can look like in the future. True, some of these fashions will never make it to market, but many versions encompass the future fashion trends.

In a similar manner, meteorologists develop models to track hurricanes. The National Hurricane Center (NHC) lists 28 different "forecast models" on their website that are able to represent the potential hurricane track, intensity, and/or wind radii. The models provide a broad range of hurricane paths, each developed by emphasizing different variables and assumptions that all suggest different landfall targets along a thousand miles of shoreline. Some of the modeled paths are much closer to the true path; however, none of the models follow the actual path. If models can be so inaccurate, and some just outright wrong, and the use of models continues with increasing frequency, why are models still used and accepted? Because models give us a glimpse into the future of what may be true, and the results help us make decisions or give us comfort about the future.

Health Risk Assessment for Asbestos and Other Fibrous Minerals, First Edition.
Edited by Andrey Korchevskiy, James Rasmuson, and Eric Rasmuson.
© 2024 John Wiley & Sons, Inc. Published 2024 by John Wiley & Sons, Inc.

Why are these hurricane models inaccurate? Because the actual number of variables is too great and complex (and likely unknown) in order to capture and understand every nuance of weather phenomenon. In fact, some meteorologists believe we have approached a point where the prediction cannot become any more accurate because of the large number of unknown variables. But what might it take to improve these models? More data and the academic and experimental knowledge to improve an existing model.

In the occupational health profession, when an air sample is collected on filter media, it represents the airborne concentration of one specific event, at a specific time, under specific conditions that occurred in the past. And because the variables of a future situation will never be the same as the conditions under which the samples were collected, it is very unlikely that one sample result will represent the results of a future situation. Even philosopher, Heraclitus, was paraphrased by Plato as saying, "... that all things pass and nothing stays," and comparing existing things to the flow of a river, he says, "you could not step twice in the same river." Yet even knowing that a future situation will never be exactly the same as our prediction, we continue to depend on our understanding of current situations to predict and make decisions on future or past unknown events. So, we do not part from modeling, and instead, with the continual advent of advancing technologies, we find ourselves applying these technologies in increasing complexity to include more and more parameters to enhance the predictive power of modeling, continually validating models using real-world results.

The occupational health and safety professional applies mathematical or computer models to many differing applications, such as, retrospectively quantifying a person's exposure to a workplace contaminant, optimizing capture efficiency of local exhaust ventilation, defining community exposure, determining the effect of various working conditions, and many other applications. The American Industrial Hygiene Association's (AIHA) Exposure Assessment and Strategies Committee (EASC) developed Mathematical Models for Estimating Exposures to Chemicals (Keil et al. 2009) publication to compile useful models that benefit the occupational health practitioner. It includes models ranging from the simple screening level well-mixed box model to the complex computational fluid dynamics (CFD) model that requires involved training and expertise to set up and proficiently implement. When the need to understand contaminants outside the indoor environment is essential, other models, not covered in this publication, are useful, such as the United States Environmental Protection Agency's (US EPA) validated American Meteorological Society/Environmental Protection Agency Regulatory Model (AERMOD) or the US EPAs California Puff (CALPUFF) atmospheric dispersion models.

CFD is a fully customizable and useful modeling tool, based on Navier Stokes equations that govern fluid flow, which can be applied to an infinite number of situations. The accuracy of the airflow predictions utilizing CFD has been demonstrated for many years in such applications as the design of airplanes, wind turbines, and the space shuttle. The applicability of CFD has been illustrated and suggested for use in exposure modeling by the EASC and has been demonstrated and evaluated for complex exposure modeling of toxic substances related to potential terrorist attacks, hazardous pollutant transport in outdoor industrial facilities, and prediction of cooling tower drift (Meroney 2004; Kisa et al. 2006). Studies have shown that CFD is a valid technology for the prediction of downwind contaminant dispersion in industrial facilities (Tang et al. 2006). As well, Hall et al. have generated CFD models and performed validation studies that have shown good correlation and accuracy between CFD and real-world indoor and outdoor exposure scenarios, especially for scenarios with some known movement of air indoors, or where there is an adequate quantity and quality of meteorological data outdoors (Hall et al. 2007, 2011, 2012, 2013, 2016). In this section, the focus is on the use of CFD to determine a worker's exposure in conjunction with other accepted models.

Validation and Application of CFD Air Dispersion Modeling

In order to gauge the reliability of CFD results, ongoing "validation" is being performed where modeled results are compared to experimentally derived, measured results. Because CFD has the capability to quantify the distribution and magnitude of air contaminant dispersion throughout a space, the exposure assessment and epidemiology communities have begun to apply the predictive power of CFD to estimate short-term and long-term worker contaminant exposures to increase the effectiveness of worker health protection. In 2009, the AIHA specifically emphasized that...

> ... These tools [CFD] may be useful when... determining how contaminant concentration may vary throughout an existing workplace when other models provide uncertainty intervals that are unacceptably large... CFD will not provide a quick and easy solution, but it can provide a detailed analysis of air movement and contaminant transport within a work environment (Keil et al. 2009, pp. 137–152).

The experimental validation of CFD models for ventilation and air quality studies has been performed and published by various authors. For example, Yang et al. (2001) demonstrated the applicability of this type of model to predict the contaminant concentrations of emissions from fossil-fuel-powered resurfacing equipment in ice skating rinks. The authors stated that the numerical results of

CFD modeling agreed reasonably with the corresponding experimental data for both steady-state and transient conditions. Hanna et al. (2006) applied five different CFD models to evaluate tracer gas dispersion in the urban setting of downtown Manhattan and found the models not only predictive for tracer gas observations, but also provided insights into the "hold up" of tracer material in recirculating zones behind buildings or in blocked regions with very low velocities. Hanna et al. (2006) characterized the results as an outcome very important for emergency response. Li and Guo (2006) applied CFD and CALPUFF models to simulate odor dispersion from a sow-farrowing farm. The authors emphasized important advantages of the CFD model as compared to CALPUFF, stating, "The CFD model employed computational time intervals ranging from seconds to hours; thus, it has the potential to characterize instantaneous odor concentrations downwind." Gousseau et al. (2011) validated a high-resolution CFD simulation model for air pollution in downtown Montreal, using two different wind tunnels; good agreement was found between numerical simulation and wind-tunnel measurements. Scargiali et al. (2005) simulated the dispersion of heavy gas clouds over a large topographically complex area using a general-purpose CFD code and investigated the influence of different parameters (meteorological conditions, terrain complexity, and others) on the model output, and the results were found useful for emergency evacuation planning. In 2009, Hanna et al. (2009) used CFD to model chlorine gas release from a railcar in a large urban area, demonstrating that buildings and topography greatly impact the contaminant flow and dispersion. Blocken and Gualtieri (2012) analyzed the performance of CFD modeling in environmental studies, including air pollution dispersion testing in urban areas (e.g. natural ventilation of the Amsterdam Arena football stadium). Meroney (2006) compared the applicability of CFD modeling along with the US EPA-approved ISCST3 and SACTI for prediction of the drift of small water droplets from cooling towers (as one of the examples of air dispersion of potentially hazardous substances) and found CFD results closer to the observed values than either the ISCST-3 or other tested models. Meroney et al. (2016) also reviewed various CFD guidelines for dispersion modeling and gave examples of success stories, including the Atomic Energy Society of Japan (AESJ) guidelines that were made to utilize CFD models for dispersion calculations for nuclear safety evaluation of radiation dose. Per Meroney et al. (2016), CFD modeling effectively substituted wind tunnel experiments for nuclear safety evaluation in Japan. Kisa et al. (2004) tested CFD modeling for predicting pollutant concentrations in the vicinity of damaged apparatus or pipeline, and emphasized that "…CFD modeling is more appropriate than the simpler models (Gauss model, box models, K-theory models etc.)…as it is able to solve the balance equation for any possible geometric situation." The authors found CFD with significantly better agreement with the experimental data of chlorine gas release than the ALOHA model.

Hall et al. (2016) validated a CFD model of an area toxic gas release in a complex building and mountainous terrain, utilizing a tracer gas study. The regression between log-transformed CFD modeling results for 18 locations (LOGCFD) and measured tracer gas levels (LOGEXP) demonstrated that the Pearson correlation coefficient between measured and modeled levels was 0.85, with 72% of the variability between data points explained by the model. If merely concentrations greater than 0.5 ppb had been included in the analysis, the correlation coefficient between measured and modeled was 0.97, with 93% of the variation in the data explained (Hall et al. 2016). Figure 6.1 shows the agreement between the modeled and measured results. It was demonstrated that CFD had better agreement with the experimental data ($r = 0.9$), while AERMOD results had a lower level of correlation ($r = 0.6$). Approximately 35 days of sampling was performed at 19 sampling locations to compare the modeled results. The CFD agreement with respirable dust measurements was good, given the uncertainty associated with estimating dust emission rates from 136 emission sources over an area of 90 mi^2.

The USEPA team from Research Triangle Park, North Carolina reported their results of CFD simulation for short-range atmospheric dispersion over open-field and within array of buildings and concluded, "CFD methods can well simulate the atmosphere-like boundary layer and plume dispersion over an open field... Broader use of CFD modeling should be expected..." (Tang et al. 2006).

After comparing results of various mathematical models for occupational exposure assessment, Bennett et al. (2000) stated:

> ...CFD was found to be a powerful tool for understanding worker exposure and contaminant dispersion. Significantly, it provided a means for evaluating exposure model accuracy, and of conducting numerical experiments to explore the impact of factors such as flow rate, source location, and receptor location on model performance.

Overview of CFD General Methodology

CFD is a computationally intensive method of solving fluid flow equations for mass, momentum, and energy within a three-dimensional domain divided into thousands to millions of discrete elements to predict the characteristics of flow and contaminant dispersion within that domain, which has only recently become widely accessible due to the increase processing capabilities of desktop computers. There are many providers of CFD code including open-source options that require more time and expertise to navigate through simulations.

Developing a CFD simulation requires specific software training to set up, process, and critique the validity of the results and a computer capable of processing

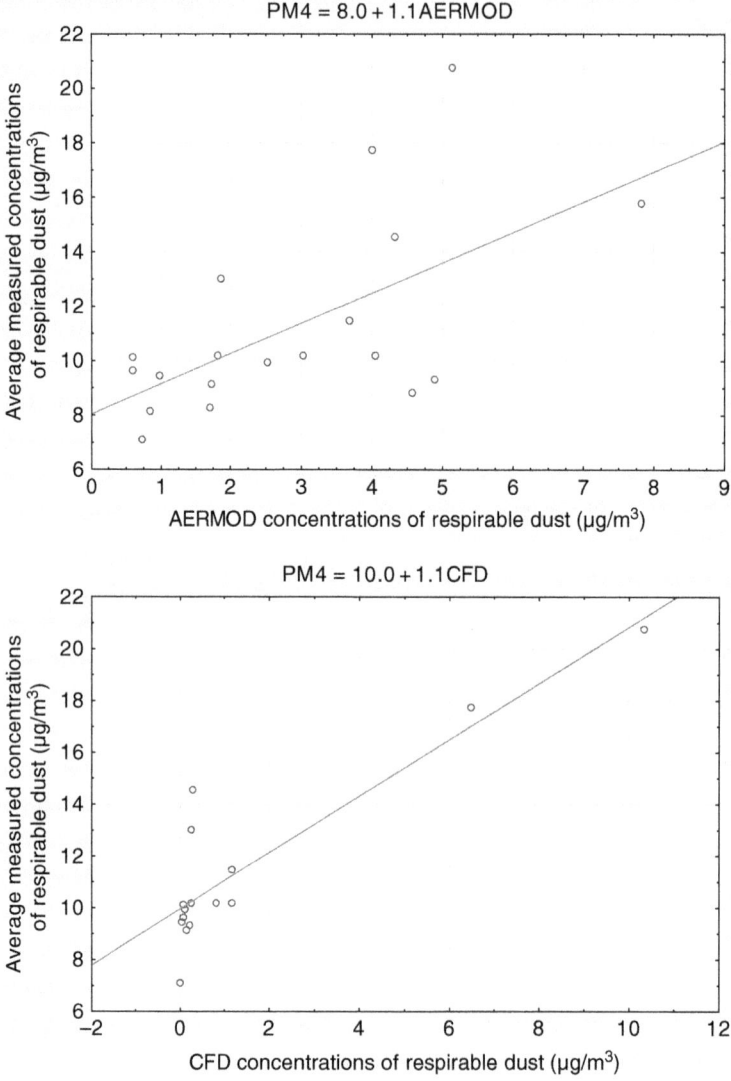

Figure 6.1 Nine-week average results of dust modeling, multiple estimated process source terms, complex terrain with variable wind conditions. Top chart – correlation between USEPA AERMOD and field measured results ($r = 0.6$, $p < 0.05$); Bottom chart – correlation between CFD and field measured results ($r = 0.9$, $p < 0.05$).

the simulation. Basic training can be accomplished in several weeks and, as technology and code complexity advances, training can extend throughout one's career by way of practical experience or advanced degrees, all of which improve knowledge and proficiency of CFD simulation. In this section, we will employ CFD simulation as a method for quantifying a worker's or workers' exposure(s) to asbestos under different situations, which can be applied to many other workplace contaminant exposures.

All CFD simulations are comprised of the following tasks: geometry creation, meshing, parameter set-up, solving, and post-processing and will be discussed in detail in the following section.

CFD Simulation Set-Up

Geometry Creation and Set-Up

Geometry creation consists of designing the fluid domain, which is composed of physical "wall" boundaries and boundary conditions at inlet and outlet openings through which the fluid will flow. In occupational exposure simulations, the primary fluid is air, with the balance being comprised of gaseous, particulate, or fibrous components that are transported along with the fluid and additionally dispersed by diffusion, gravity and, when necessary, chemical reaction. The three-dimensional indoor geometry/domain representing a real-world facility, similar to what is shown in Figures 6.2 and 6.3, is often created in computer-aided design or other three-dimensional design software and imported into the CFD software.

Figure 6.2 Real-world GoogleEarth® building image. *Source:* Google LLC.

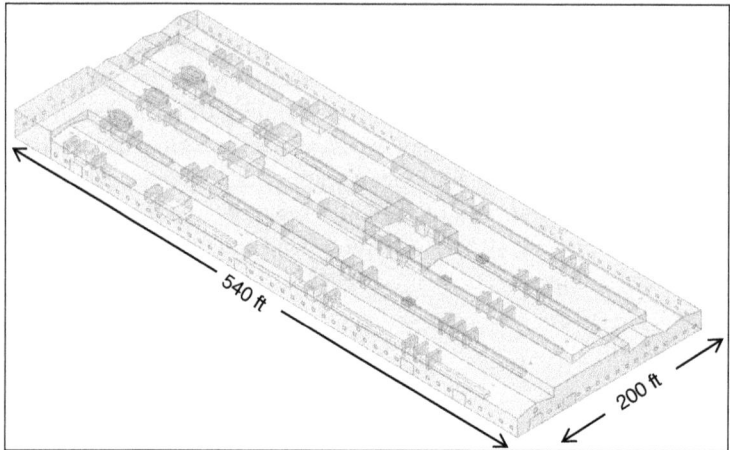

Figure 6.3 Representative computational indoor geometric domain.

The geometry is simplified by eliminating usually smaller unnecessary elements that will not significantly affect the fluid flow or contaminant dispersion.

Mesh Creation

The process of meshing consists of dividing the computational geometric domain into small discrete volumes called elements between which the change in mass, momentum, and energy equations will be calculated. Domains may consist of thousands or tens of millions of subdivided mesh elements as demonstrated in Figure 6.4. Mesh refinement combined with appropriate simplification provides a balance between simulation precision and computer processing time.

Parameter Set-Up

Parameter set-up consists of describing the physics involved in the fluid flow. The following list provides an example of some physics that may be included:

- Domain fluid (air) properties: temperature, pressure, density, and humidity
- Inlet velocities, turbulence, and temperatures
- Outlet pressure differential
- Interior fan velocities or pressure differentials
- Contaminant source (see example in Figure 6.5)
- Gas or particle
- Source flow rate
- Contaminant injection velocity
- Chemical reaction (if applicable)
- Surface conditions, roughness, particle adherence, or reflection
- Wall slip conditions

Overview of CFD General Methodology | 161

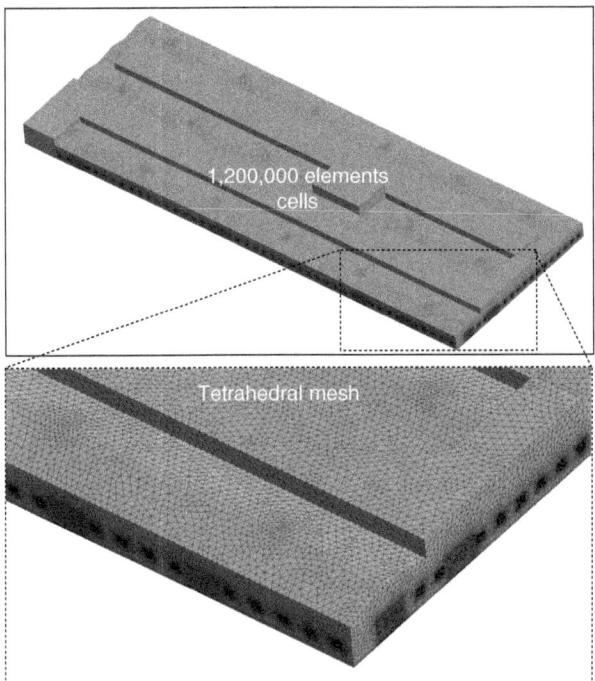

Figure 6.4 Building indoor surface mesh (top); zoomed in view of mesh (bottom).

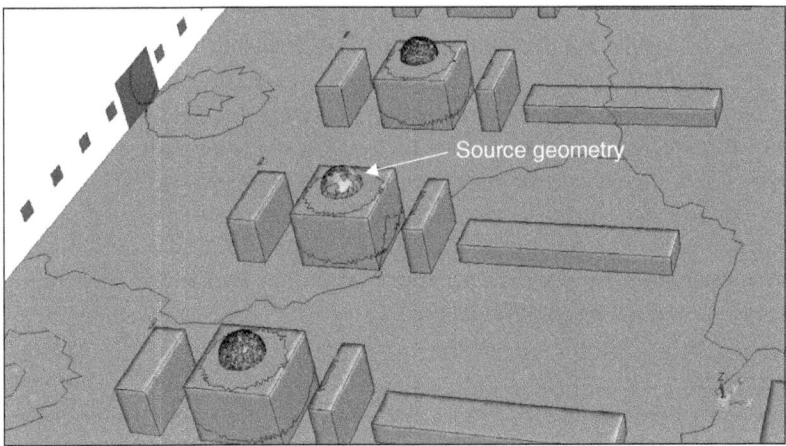

Figure 6.5 Contaminant source geometry.

Computational Solve

Solving is synonymous with running the simulation. This is the crux of the entire simulation process and is the step that takes the most computational resources. The number of computer processing cores and memory required to run the simulation is determined based on the size of the geometry, complexity of the physics, constraint of cost, and the project deadline. Processing can be conducted on a powerful multicore desktop computer or on nearly unlimited cloud computing resources, accessing hundreds or even thousands of processors to reduce the simulation solver run time from days or weeks to minutes or hours. The software performs calculations for mass, momentum and energy, and any additional degrees of freedom, for each cell within the mesh, each calculation of which is defined as an "iteration." The software then calculates the imbalance or error of all conserved variables across the entire domain and parameters are adjusted accordingly. Consecutive iterations are performed until the error for mass, momentum, and energy is reduced below a user-specified level, often 1×10^{-3} or less. This is equivalent to 0.001, and no simulation can ever achieve 0% error perfect solution; therefore, some value greater than zero, such as 0.001, is selected as the value to trigger that solution has been completed. As the combined error consistently moves toward zero, this is referred to as "convergence," whereas if the error is moving away from zero, increasing error, this is referred to as "divergence" and indicates there is a problem with the simulation. If this occurs, the simulation engineer must identify which aspect or parameter(s) may be causing the divergence. Often divergence is caused by minutely small, poorly meshed artifacts at specific locations within the geometry, and therefore, the mesh must be corrected. Once the convergence criteria are met (e.g. 1×10^{-3}) the simulation is complete and a data file(s) containing the results is exported.

Either steady-state or transient simulations can be conducted based on the requirement of the results and required modeling of turbulent effects. If all conditions of the simulation are constant and turbulent effects are low, a steady-state simulation is often the most appropriate simulation to conduct. However, if results are required as time changes or if there are significant turbulent effects or highspeed flows, then a transient simulation would be necessary. For instance, if the time it takes for a room to reach a specific concentration is needed, then a transient simulation is necessary. Steady-state simulations require much less time to solve than transient simulations because a transient simulation must achieve convergence for each time step simulated.

Post-processing

Post-processing produces descriptive pictures, diagrams, tables of data, or charts that represent the outcome of the simulation to either confirm or contradict initial hypotheses and help arrive at specific conclusions. Figure 6.6 demonstrates the

Overview of CFD General Methodology | 163

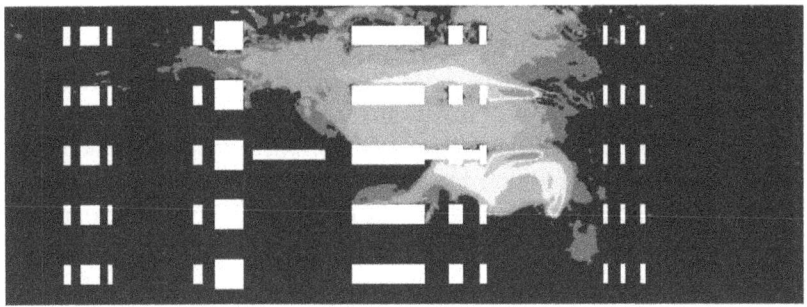

Figure 6.6 Post-processing showing contaminant concentration contours within the building.

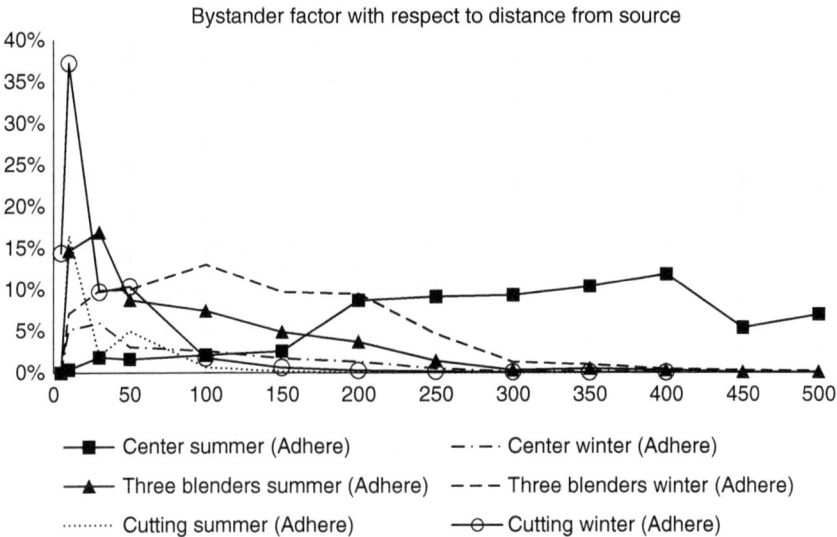

Figure 6.7 Bystander factors with respect to distance.

concentration gradient throughout a building and the chart in Figure 6.7 demonstrates how the bystander factor decays with respect to distance from the source.

Complementary Modeling Software Tools

In addition to CFD, other modeling tools, such as CALPUFF and AERMOD, can be used to provide a method of general validation of the CFD results. These two models are only designed for use in outdoor environments, whereas CFD is customizable for indoor or outdoor or a combination of the two.

California Puff (CALPUFF) atmospheric dispersion model and American Meteorological Society/Environmental Protection Agency Regulatory Model (AERMOD) atmospheric dispersion model describe the dispersion of airborne particles or gaseous contaminants based on various meteorological conditions and wind in a three-dimensional spatial manner. These models are accepted models for air pollution permitting and are used for tracking long-range dispersion of estimating pollutant emissions up to hundreds of kilometers (km) from a source(s), and pollutant concentration for nearby receptors can be determined by decreasing the grid size.

Other Software Tools

In addition to the modeling software, meteorological and statistical analysis tools such as Lakes Environmental WRPlot® and Microsoft Excel, respectively, are extensively used to interpret large meteorological datasets and compile and analyze large CFD output datasets into reasonably concise descriptive statistics.

Indoor and Outdoor Modeling Examples

To demonstrate the process of using CFD for modeling of indoor and outdoor contaminant dispersion, two examples will be described below, with methodologies and results presented in detail.

First Example – Indoor CFD Modeling

In the first example, the concept of modifying a primary worker's exposure by a factor, referred to as bystander factor, was the desired model outcome. A bystander factor is a function of both ventilation and distance from a source has been reported and quantified in the literature (Cherrie 1999; Semple et al. 2003). Bystander exposures from nearby primary workers working with asbestos-containing materials, for example, can be increased by nearby or distant sources of fiber release caused by other workers in the workplace (Semple et al. 2003). Such exposure would be expected to result in relatively low-level exposures compared with the primary worker's exposure, depending on the level of ventilation, surrounding geometry, and other factors (Donovan et al. 2011). The bystander exposure factor is defined as the ratio of a bystander exposure to a primary worker's or workers' exposure(s) who is working with, for example, a material that contains asbestos. Estimates of exposure ranges for primary workers can be found in the literature, internal or external databases, and from simulation studies.

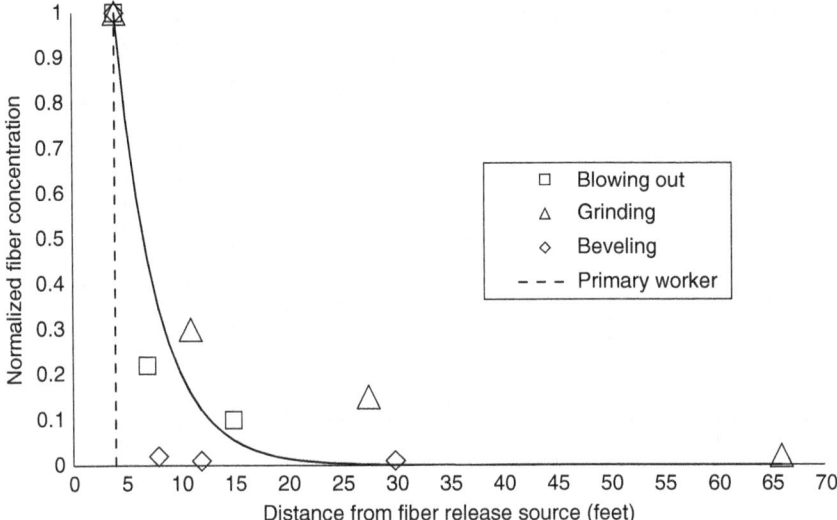

Figure 6.8 Concentration drop-off with respect to distance for brake-related operations.

For indoor bystander factors, either modeling or estimation from the literature can be utilized to predict bystander factors.

The rapid drop-off in asbestos concentration with distance from brake-related operations, for example, is shown in Figure 6.8, as a function of distance from the primary worker performing brake work.

The rule-of-thumb diagram shown in Figure 6.9 represents bystander factors for varying distances from the primary worker based on the combination of a mathematical model and available published data for indoor activities (Donovan et al. 2011). This is preferable to relying solely on one source of information to evaluate potential bystander exposure factors.

For CFD models, reasonable estimates for parameters are applied to achieve the most accurate results possible. Where data were not readily available, worst-case or exaggerated assumptions were made based on available data, knowledge of modeling, and exposure assessment expertise in order to likely overestimate the predicted breathing zone concentrations.

In this first example, five CFD simulations were performed to determine the final bystander exposure factors. One preliminary outdoor CFD simulation was performed in order to determine the effect of the wind on the ventilation characteristics within the building. Four indoor simulations were conducted, two under low and two under high building ventilation characteristics for each source activity, pipe insulation removal, and reheat furnace brick removal operations. Table 6.1 lists the simulations that were performed.

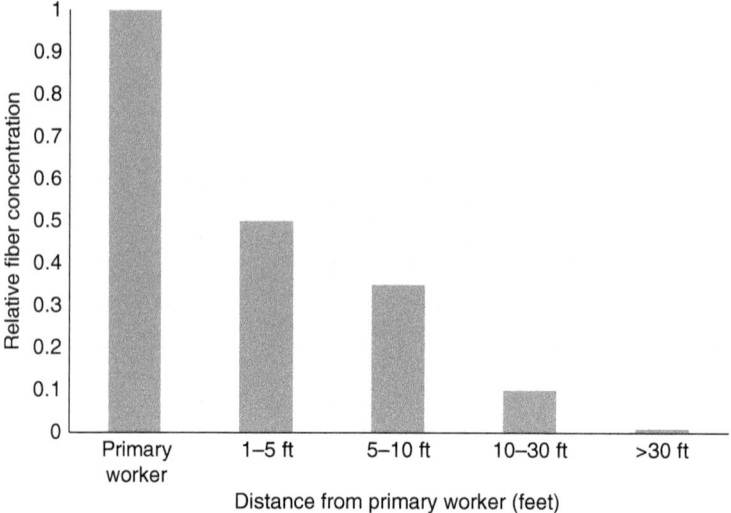

Figure 6.9 Rule-of-thumb bystander factors.

Table 6.1 CFD simulations performed.

Simulation	Activity simulated	Ventilation
1	External wind effect on internal air movement	Preliminary outdoor wind to determine whether wind speed had an effect on indoor ventilation
2	Reheat furnace brick removal	Low indoor ventilation
3	Pipe insulation removal	Low indoor ventilation
4	Reheat furnace brick removal	High indoor ventilation
5	Pipe insulation removal	High indoor ventilation

Preliminary Outdoor CFD Wind Simulation – Effect on Indoor Ventilation

Figure 6.10 shows the physical size of the CFD model domain (3,934 × 2,936 × 656 ft, East-West × North-South × Height) used in the preliminary outdoor simulation. The outlines of the building were obtained from GoogleEarth Image of the hot strip steel Mill.

Hourly daytime wind speeds from a nearby airport were used to calculate an average wind speed. This data contained all three shifts (0:00–24:00 midnight)

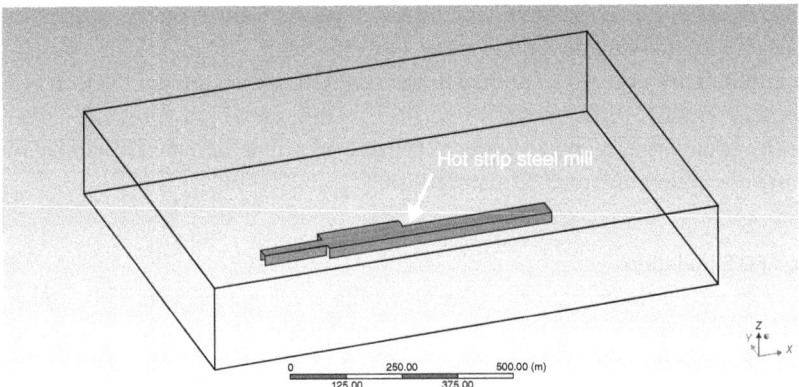

Figure 6.10 Preliminary CFD modeling domain.

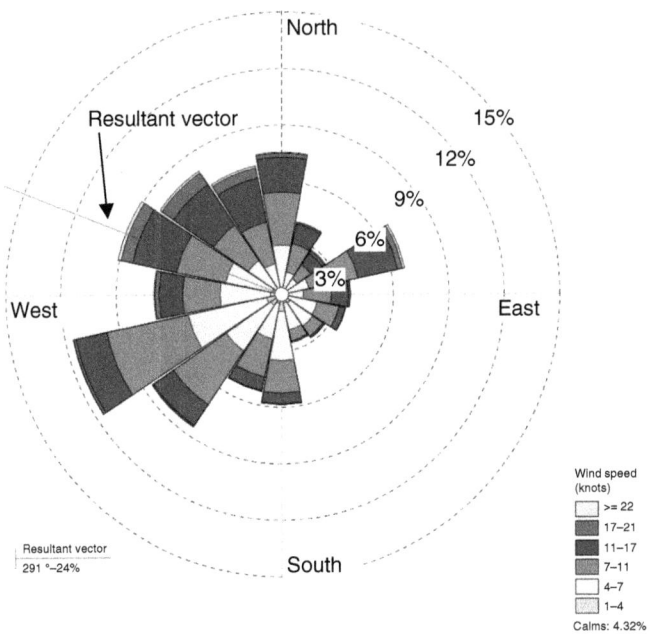

Figure 6.11 16-Petal (24-hours, 1960–1964) wind rose developed from airport weather station data.

which represented three worker shifts, and hourly data between 1960 and 1964 was selected. Figure 6.11 shows a 16-petal wind rose, generated by more than 35,000 meteorological data records, and indicates the frequency and magnitude for each wind direction for the 4 years between 1 January 1961 and

31 December 1964. The resultant wind direction vector, located on the wind rose, indicates the mean wind direction for the 4-year dataset.

The preliminary outdoor CFD model was run with the mean wind velocity of 8.0 knots (9.2 mph) 291 degrees from north. The differential pressures calculated across the openings at each end of the hot strip steel mill building were applied to the four subsequent indoor CFD simulations.

Indoor CFD Simulations

In order to model the bystander exposure within the hot strip steel mill, a new three-dimensional modeling domain of only the building interior was generated to perform the four indoor CFD simulations listed in Table 6.1, and the differential pressures determined in the outdoor CFD simulation were applied to the building openings. Figure 6.12 shows the indoor modeling domain and the objects within the mill.

The size, shape, and location of the objects situated within the domain were estimated based on a general understanding of the hot strip mill process. The domain included: finishing mill rollers, a strip steel conveyor, four reheat furnaces, gravity ventilators, windows, and building openings.

Mill Ventilation

In addition to the differential pressures applied to the building openings, there were three rows of gravity ventilators, which ran the length of the roof, two rows of large ventilators near the edge, and one row of small ventilators along the center. Gravity ventilators provide natural ventilation for mill buildings, and the general building geometries and ventilation arrangements, for this

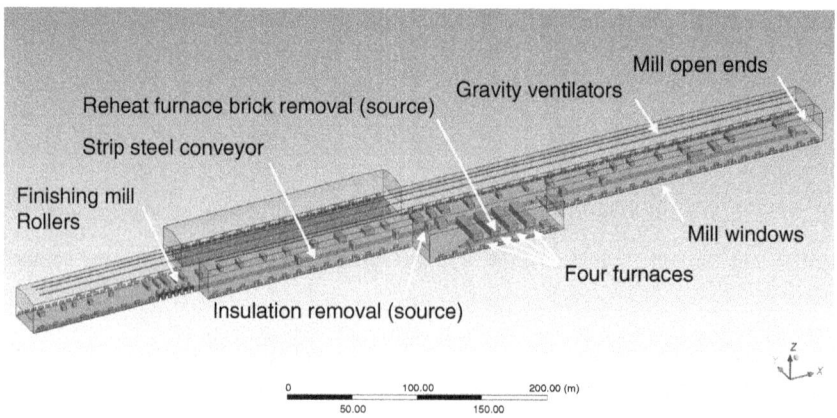

Figure 6.12 Hot strip steel mill indoor CFD modeling domain (2,215 × 340 × 80 ft).

example, were gathered from the 1965 Steel Mill Ventilation manual (American Iron and Steel Institute 1965, p. 17). Based upon Patty's text (Witheridge 1948), two quantities of ventilation air, 4 and 30 air changes per hour (ACH), were selected based on the minimum and maximum requirements. The 4 and 30 ACH ventilation air was distributed equally among the three gravity ventilators and air was allowed to enter the domain through the open windows and building ends.

Other Model Parameters

The temperature of the strip of rolled steel was chosen based on the steel exiting the reheat furnace "cherry red." One steel industry temperature color chart represents "cherry red" as likely around 1,450 °F (West Yorkshire Steel 2017). This temperature was applied to the surface of the strip steel conveyor shown in Figure 6.12.

Particle adherence was applied to all surfaces in this CFD model, implying that a fiber adheres to a surface and is not re-entrained into the model when it contacts a surface.

Dr. Mort Corn (1965, 1966, personal communication), the former director of OSHA, has demonstrated that when respirable fibers do make contact with surfaces, electrostatic forces tend to tightly bind the microscopic respirable fibers, making re-entrainment essentially impossible. Asbestos fibers are electrostatically charged, and the high charge-to-mass ratio and the geometry of the fiber cause the fiber to bind very strongly to a surface on which the fiber has landed or impacted. On the other hand, re-entrainment can occur in buildings or from surfaces where there is visible larger-particle asbestos-containing dust and/or debris.

In a book on sampling and analysis of settled asbestos dust, Millette and Hays (1994) also discuss these principles:

> The electrostatic forces between particles and other particles and between particles and surfaces are proportional to surface area, and the effect of the force is proportional to the mass of the particles. Smaller particles are influenced by these attractive forces more strongly than larger particles. Individual small particles can be strongly attached to a surface by electrostatic forces…It was observed that small particles (below 2 microns) require local air flow velocities near 100 miles per hour to overcome surface attraction forces…A single layer of very small particles, uniform in size, will be hard to dislodge from a surface.

The small particles that Millette and Hays (1994) refer to were essentially spherical particles. In the case of respirable asbestos fibers, the binding forces would be expected to be even greater due to the greater proportion of surface-to-volume contact.

Source Descriptions

The reheat furnace brick removal and pipe insulation removal sources present in each model were run independently to determine the contribution of each to the final bystander factor result. The arrows in Figure 6.12 point to the reheat furnace brick removal source and the insulation removal source. The magnitude of each source generation rate (kg/s) was chosen to approximate the actual generation rate for asbestos fibers based on previous experience with asbestos fiber CFD simulation. Utilizing the exact generation rate is not critical to the results because the bystander factor is calculated by determining the proportion of the bystander's breathing zone concentration with respect to the primary worker's breathing zone concentration, a unit-less percentage.

CFD has the capability to introduce as many customized fiber sizes into the simulation as needed. However, in this example, adding multiple fiber sizes would have unnecessarily increased the complexity and time required to run the simulation without adding significant benefits to the results because a fiber exits the domain or adheres to a surface within a few minutes after introduction into the domain. The settling distance that a fiber would experience during that time would be insignificant between various micron-sized fibers. In this simulation, an arbitrary single fiber size of 20 μm in length and 0.2 μm in diameter was selected, with a specific gravity of 3.05, similar to that of Actinolite.

Reheat Furnace Brick Removal Source

The four reheat furnaces were placed at the beginning of the strip steel conveyor (Figure 6.13). The furnace dimensions were selected to be $98 \times 26 \times 20$ ft ($30 \times 8 \times 6$ m) with a brick removal opening of 13.1×6.6 ft (4×2 m). A flow rate was applied to the furnace openings to represent two pedestal fans that were used

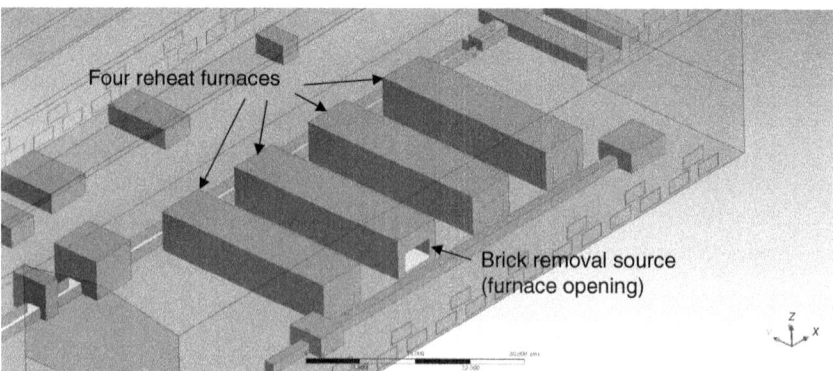

Figure 6.13 Reheat furnace removal source.

to ventilate the furnace during the brick removal. Flowrates from current-day industrial pedestal fan specification sheets were averaged to achieve an average flowrate of 8,656 cubic feet per minute (CFM) (4.1 cubic meters per second [m^3/s]). Assuming two fans of this type and a 13.1×6.6-foot furnace opening, the air velocity exiting the furnace would be 201 feet per minute (1.02 meters per second [m/s]).

Pipe Insulation Removal Source

Another source was included in the model to represent pipe insulation removal (Figure 6.14). The pipe was located near the strip steel conveyor with the centerline positioned 5.25 ft (1.6 m) above the floor at the approximate height of the worker's breathing zone. The source area was estimated to be a 3.3 square foot (0.34 m^2) area with a source inlet velocity of 49 feet per minute (0.25 m/s) of air.

Tables 6.2 and 6.3 display the source parameters and the physical parameters for the CFD model, respectively.

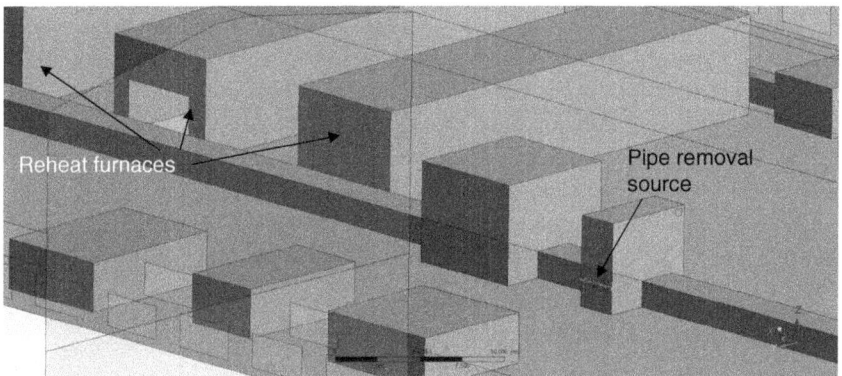

Figure 6.14 Pipe insulation removal source.

Table 6.2 Source parameters.

CFD source description	Source area (m^2)	CFD source inlet velocity (m/s)	Source temp (°C)	Source generation rate (kg/s)
Pipe insulation removal	0.31	0.25	Ambient	5.0×10^{-7}
Reheat furnace brick removal	8.0	1.02	66	5.0×10^{-7}

Table 6.3 Summary of CFD model parameters.

Parameter	Description
Hot strip steel mill domain size, $L \times W \times H$	$2{,}215 \times 340 \times 82$ ft ($675 \times 104 \times 25$ m)
Fiber size	0.2 μm diameter × 20 μm length
Fiber density (specific gravity)	3.05
Number of elements	4.3 million

CFD Results

Figures 6.15 and 6.16 show the turbulent nature of the vertical and horizontal distribution of airflow, along a plane intersecting the furnace where brick removal activities were modeled. The length of the arrows indicates the relative magnitude of the velocity.

The percent bystander concentration was determined by calculating the average concentration within bands between two, three-dimensional concentric circles centered on the source at various distances from the center location of the fiber release. The widths of these bands were approximately equivalent to the bands indicated in the Rule of Thumb percent bystander percent asbestos exposure illustrated earlier. This average bystander concentration was divided by the calculated primary worker breathing zone concentration to achieve the percent bystander concentration (bystander factor). Bystander factors were generated with respect to a bystander located at ground level. Additional 20-foot-wide zones were created within concentric annular volumes expanding farther distances from the Rule-of-Thumb distance distances.

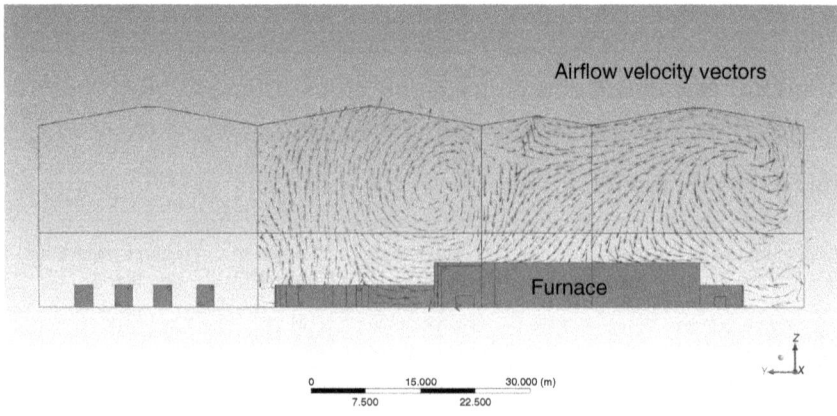

Figure 6.15 Vertical airflow distribution.

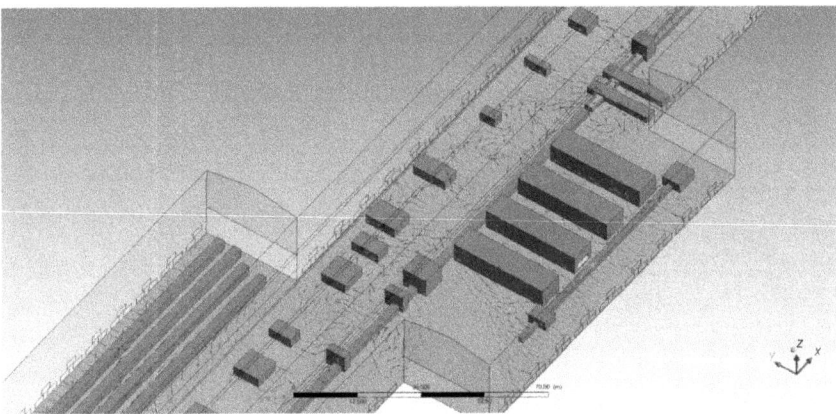

Figure 6.16 Horizontal breathing zone airflow distribution.

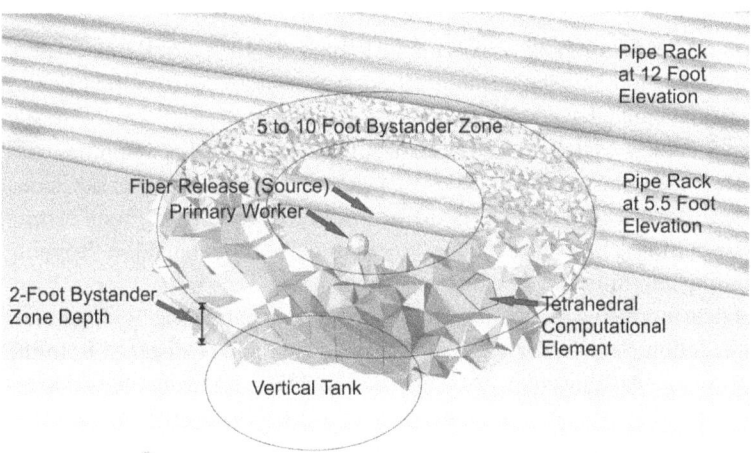

Figure 6.17 CFD 5–10-foot bystander zone.

Figure 6.17 provides an example of the three-dimensional -5 to 10-foot bystander zone volume used to calculate bystander factors. The model generates concentrations within each of the computational elements shown (e.g. three-dimensional tetrahedral shapes) and an average bystander exposure was calculated for that zone volume. This volume average calculation is conducted for each of the bystander zones illustrated in the Rule of Thumb mentioned earlier, 2.3–5, 5–10, 10–30, 30–50 ft, etc.

The result of the reheat furnace brick removal is illustrated in Figure 6.18. The various shades of grey, contour zones (better depicted via color graphics)

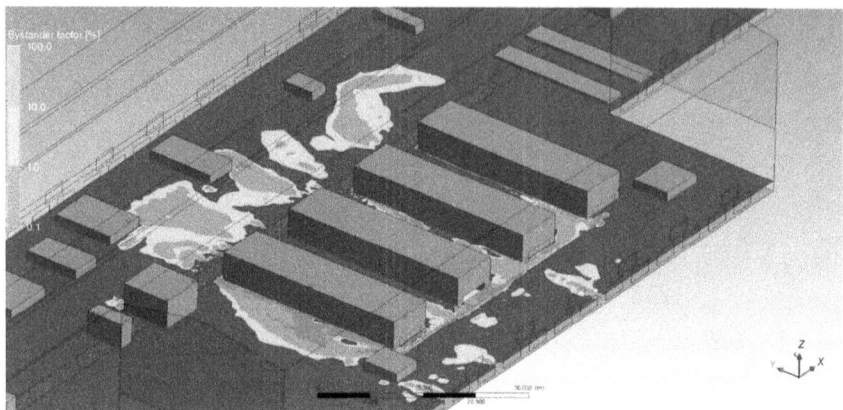

Figure 6.18 Reheat furnace brick removal bystander factor (%) for 4 ACH.

indicate bystander factor ranges of 0.1–1%, 1–10%, 10–100%, and greater than 100%. These zones indicate the bystander's exposure with respect to the primary worker's CFD-calculated average breathing zone concentration.

Figure 6.19 depicts bystander factor distributions for 4 ACH. The plume contour colors represent the bystander exposure factor percentage as earlier described in Figure 6.21. Figure 6.22 represents the upward mobility of the plume exiting the furnace opening. The plume makes its way under a room partition dropping from the ceiling and fully exits the building to the left.

Table 6.4 lists the average bystander factors for the reheat furnace brick removal and pipe insulation removal for 4 and 30 ACH as a function of distance from the primary worker's breathing zone. Also shown in Table 6.4 are the published Rule-of-Thumb bystander exposure factors as a function of distance. The bystander factors displayed in Figure 6.20 for the 10–30, 30–50, 50–70, 70–90, and 90–110 foot, etc., CFD calculated bystander factors are shown in order from top to bottom respectively with the legend, except the Rule-of-Thumb values which are not shown in this manner.

Second Example – Outdoor CFD, AERMOD, and CALPUFF Models

In addition to the first example, indoors, CFD was used in an outdoor setting, in combination with other modeling software tools, as a method for estimating a distant bystander's exposure. The purposed of this CFD simulation was to determine the exposure of an individual who worked at a property adjacent to an asbestos cement pipe manufacturing plant (plant) that conducted outdoor airborne asbestos-producing activities.

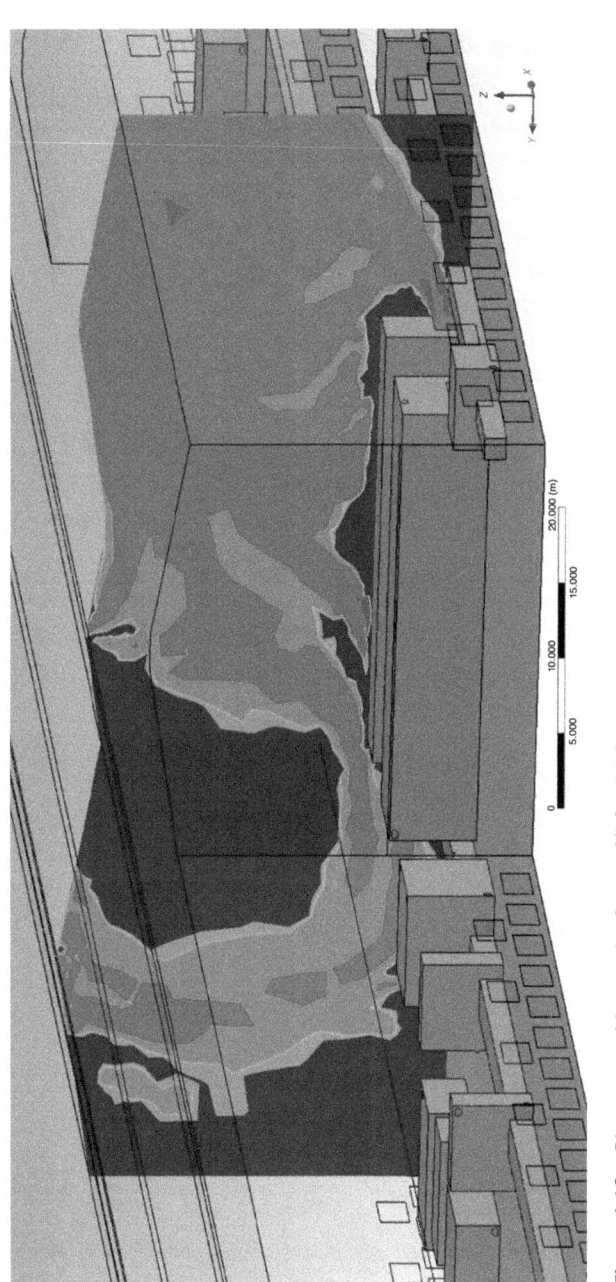

Figure 6.19 Pipe removal bystander factor (%) for 4 ACH.

Table 6.4 CFD modeled bystander factors.

Distance from source centroid (ft)	Rule of thumb bystander factor (%)	Reheat furnace brick removal		Pipe insulation removal	
		4 ACH (%)	30 ACH (%)	4 ACH (%)	30 ACH (%)
Primary worker (1.3–2.3)	100	100	100	100	100
2.3–5	50	95.8	105.6	0.8	10.5
5–10	35	104.6	82.8	0.0	0.9
10–30	10	10.4	0.5	0.0	0.0
30–50	1	0.5	0.0	0.0	0.0
50–70	—	0.02	0.0	0.0	0.0
70–90	—	0.5	0.0	0.0	0.0
90–110	—	0.6	0.0	0.0	0.0
110–150	—	0.0	0.0	0.0	0.0
150–300	—	0.02	0.0	0.00006	0.0
300–1,000	—	0.0	0.0	0.0	0.0

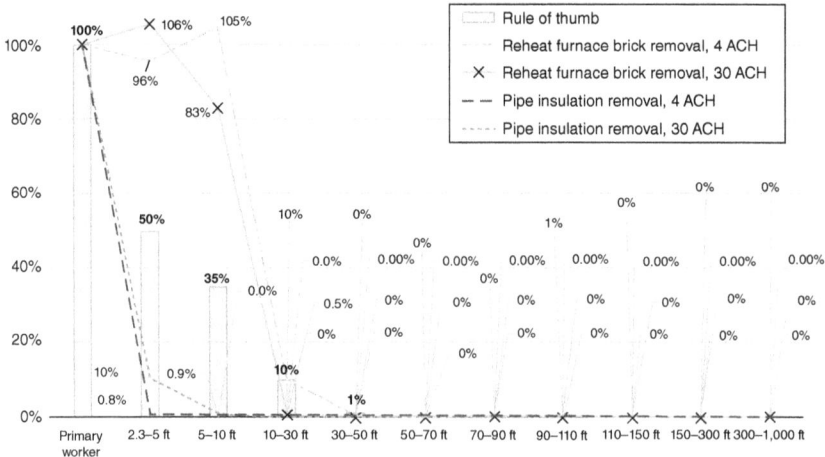

Figure 6.20 Rule-of-thumb and CFD bystander factors (bystander with respect to primary worker).

Model Geometry

Figure 6.21 shows the physical size of the CFD model domain (6,496 × 5,479 × 1,707 m, East-West × North-South × Height), plant building geometries, and buildings at the adjacent receptor property (receptor) that were included in the outdoor domain. The terrain data were downloaded from a publicly available USGS DEM and visually compared to the GoogleEarth™ three-dimensional terrain. Building ceiling heights were estimated for the five buildings based on visual estimation from Google Earth three-dimensional satellite images.

Receptor Descriptions

The area covering the northeast section of the receptor was selected as the receptor site where concentration data was extracted from the model because it is likely the location where the bystander parked his car when arriving at work and spent time loading his work truck prior to leaving the site.

Figure 6.22 shows a close-up view of the plant building layout. Additionally, Figures 6.23 and 6.24 show a perspective and overhead December 1993 Google Earth image of the plant and the receptor location.

Source Descriptions

Figure 6.25 shows the nine airborne asbestos-producing sources (five buildings, two baghouse groups, one crusher, and one shipping area) included in the model.

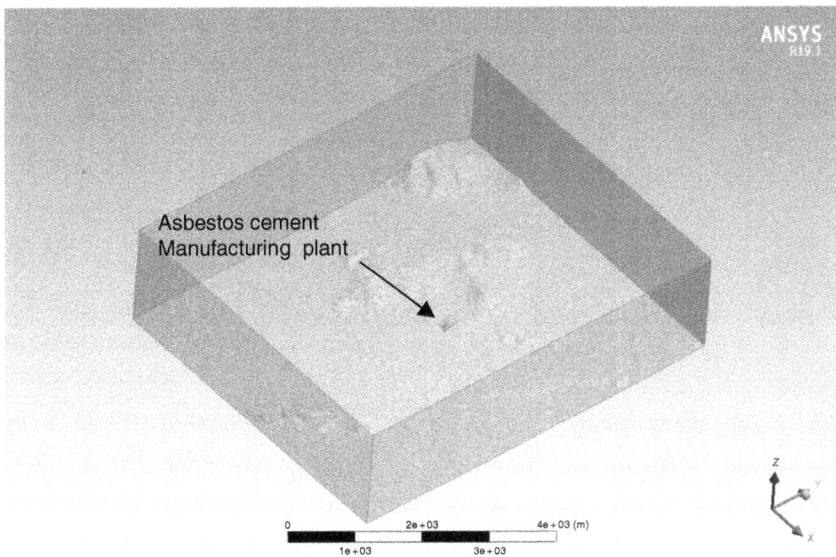

Figure 6.21 CFD modeling domain (6,496 × 5,479 × 1,707 m).

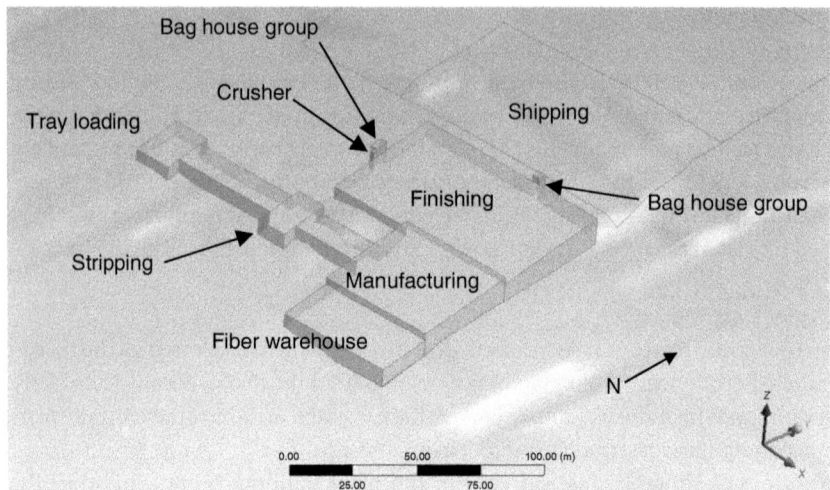

Figure 6.22 CFD model of asbestos cement manufacturing plant buildings.

Figure 6.23 Plant and receptor property location.

The 10 baghouses were grouped into manufacturing and finishing baghouses. The manufacturing baghouses were grouped into one source and located to the west of the finishing building. The finishing baghouses were grouped into one source and located to the north of the finishing building. These sources were selected based on their potential contribution to airborne fiber concentrations at

Indoor and Outdoor Modeling Examples | 179

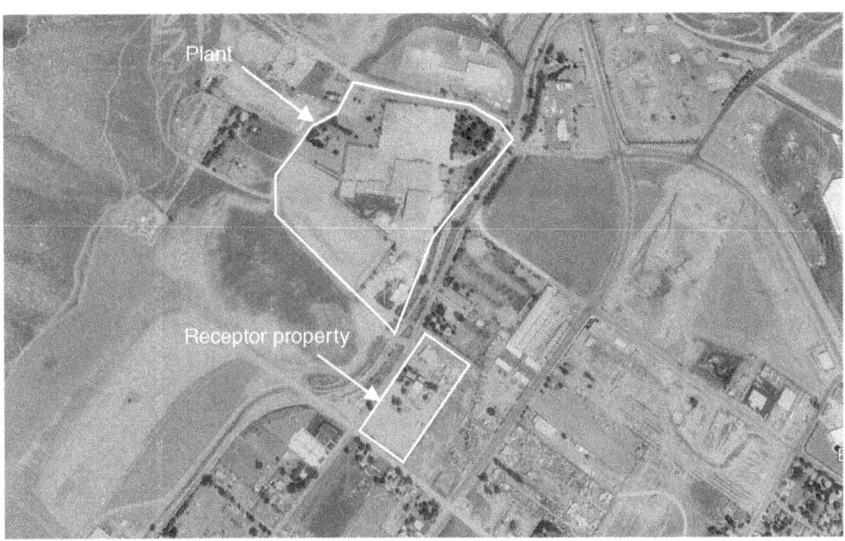

Figure 6.24 Plant and receptor property.

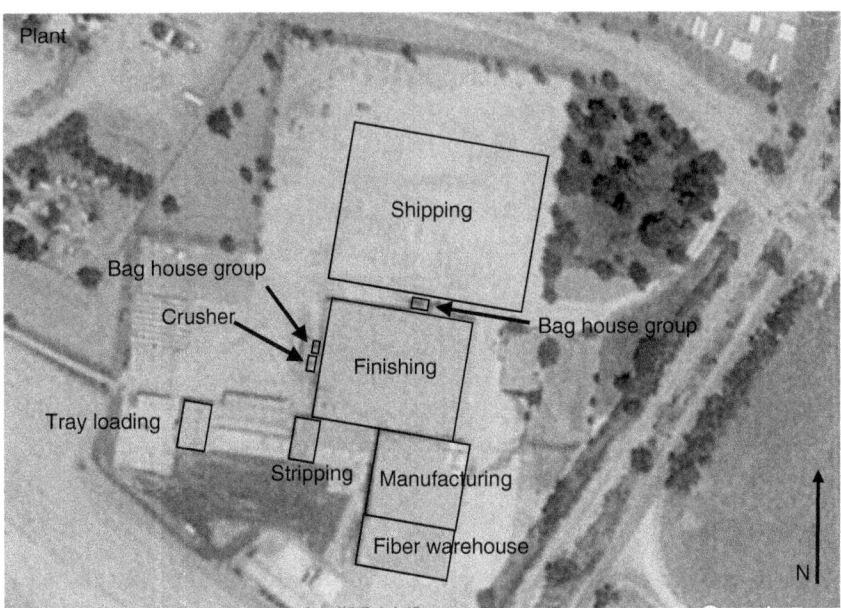

Figure 6.25 Plant source locations.

the receptor. Table 6.7, at the end of this Source Descriptions section, shows the magnitude of each of the source generation rates, kilograms per second (kg/s), and other associated source parameters.

The physical characteristics of the fibers included in this model are identified in Table 6.9. The fibers were assigned the characteristic which, if intercepted by the ground or wall, they would adhere (trap) to the surface for the duration of the simulation and not be re-entrained into the fluid. As in the indoor CFD model example shown earlier, because the average wind velocity would carry a particle through the domain in less than 7 minutes, the magnitude of the flow through the domain far outweighed the settling velocity of a falling micron-sized fiber; therefore, it was appropriate to introduce a single sized asbestos fiber source into the model.

Fugitive Plant Emission – Manufacturing, Finishing, Fiber Warehouse, Tray Loading, and Stripping Station

The plant had five main production buildings: (1) manufacturing, (2) finishing, (3) fiber warehousing, (4) tray loading, and (5) stripping. Likely exaggerated average asbestos concentrations were determined within each of the five building volumes.

Because definitive building ventilation estimates were not known, a worst-case assumption was made that fugitive emissions emanated from roofs and some exterior vertical building surfaces (walls) at a rate of six ACH. Ideally, fugitive emissions would best be represented by area samples collected within each of the buildings; however, because an adequate number of area sample results were not available, the average of personal samples for each area was used to characterize building interior air concentrations. Because air concentrations of contaminants decrease both horizontally and vertically away from sources, the use of personal air samples greatly exaggerated the actual indoor air concentrations within the various production spaces. The average value for area and personal samples collected between 1986 and 1992 for the five buildings are summarized in Table 6.5.

With a reasonable assumed air exchange rate of 6 ACH in combination with the building average concentrations, separate generation rates were calculated for each of the buildings. The calculated volumetric flow rates, the horizontal and vertical surface areas, and the average air concentrations of the three building spaces were used to calculate the generation rates shown in Table 6.5. These generation rates were not adjusted to reflect the assumed average percent activity for each location and the "worst-case" operational shift was assumed to be 24 hours per day, 7 days per week.

These five fugitive emission building sources represented 45.7% of the total emissions assumed in the model. That generation rate was equally distributed across the entire surface area for each building fugitive emission source.

Table 6.5 Building source generation rates.

Building source ID	Ceiling heights (m)	Number of area and/or personal samples (utilized to calculate source terms) (f/cc)	Average interior building area and personal samples (utilized to calculate source terms) (f/cc)	Calculated, worst-case, fiber generation rate (f/s)	Corresponding mass generation rate (kg/s)
Manufacturing	7	138	0.085	1.3×10^7	6.8×10^{-9}
Finishing	10	184	0.088	2.2×10^7	1.1×10^{-8}
Fiber Warehouse	7	9	0.074	4.1×10^6	2.2×10^{-9}
Stripping	7	6	0.018	2.5×10^5	1.3×10^{-10}
Tray loading	7	20	0.012	2.1×10^5	1.1×10^{-10}

Baghouse Source Emission Rates

Baghouses were used, in the manufacturing and finishing buildings, to collect and contain airborne asbestos fibers and dust at the generation sources within the plant. This dust was delivered through ducts to the baghouses where the air was filtered and exhausted outside the building. Because there was no definitive information on whether this exhaust air was delivered back into or to the exterior of the plant, as a worst-case assumption, it was assumed that all air that was exhausted through baghouses was exhausted to the outside of the plant. Ten baghouse locations were listed on a dust collector inventory sheet and five were dedicated to the manufacturing building and five to the finishing building. Two of the five dedicated to the manufacturing building collected dust for the silica and cement silos, which were not expected to be asbestos-producing sources, so they were not included in the baghouse source. There were two locations marked on plant layout maps indicating baghouses located on the north and west sides of the finishing building. The exhaust flow rates for the three manufacturing baghouses were combined into one flow rate and assigned to the west baghouse location and the flow rates for the five finishing baghouses were combined into one flow rate and assigned to the north baghouse location shown in Figure 6.25.

The average of two samples collected from baghouse exhausts, 0.16 and 0.0037 f/cc, from two plant locations, were used to estimate the average concentration exhausting from each modeled baghouse. By including the formerly listed higher concentration, collected from another plant, the exhausted fiber concentration represented a "worst-case," compared with the actual measured emission concentrations collected at the site being studied. Using the individual and combined exhaust flow rates in CFM, an estimated exhaust flow rate of 2.0 m/s, and the resulting average concentration, a source generation rate was determined for each baghouse location listed in Table 6.6 and input into the model.

It was assumed that the baghouses operated 24 hours per day, 7 days per week. The mass generation rate produced by the baghouses represented approximately 10.7% of the total emissions assumed in the model.

Because the manufacturing and finishing buildings were likely negatively pressured due to the volumetric flow exhausted from the baghouses, and the air concentration within the building would be exhausted through those same baghouses, the fugitive emission sources for the manufacturing and finishing

Table 6.6 Baghouse generation and flow rates.

Baghouse ID	Fiber generation rate (fibers/s)	Mass generation rate (kg/s)
Manufacturing baghouse	1.52×10^6	7.91×10^{-10}
Finishing baghouse	7.76×10^6	4.04×10^{-9}

buildings were likely exaggerated by assuming all fibers escaped from the building prior to being filtered in the baghouse. Essentially, the potential emissions from the building spaces were accounted for twice, once as a fugitive emission from within the building, and again relative to the baghouse emissions.

Pipe Storage and Shipping Yard Source Emission Rate

Based on air sampling reports of personnel involved in forklift hauling and loading operations in the area to the north of plant, a generation rate for this potential source was determined using an approximate volume surrounding a moving forklift, an average airborne asbestos concentration within that volume, and an average annual daytime wind velocity. The volume surrounding the forklift was estimated as a 2.5×1.5-m footprint and a 4-m height, which would put the breathing zone of the forklift operator in the vertical center of that volume, where the sample was collected. Ten measured air samples (f/cc) from the plant's shipping and forklift operators were averaged for 1986, 1987, and 1990, and an average breathing zone concentration of 0.021 f/cc was calculated. The annual daytime average wind velocity (2.81 m/s) in the direction of the receptor, was used to determine the flow rate through the aforementioned forklift volume. In the same manner, the tractor-trailer volume was estimated with a 23×2.5-m tractor-trailer footprint and an eight-meter estimated height, which was combined with the shipping yard source.

As a worst case, it was assumed that three forklifts and one tractor-trailer were moving around the shipping and pipe storage yard continually. The volumetric flow rate and the average concentration for the three forklifts and one tractor-trailer were used to calculate the generation rate, which was 6.04×10^6 fibers/s or 3.14×10^{-9} kg/s. This represented 6.9% of the total emissions assumed in the model. That generation rate was equally distributed across the entire surface area ($6{,}500 \, m^2$) for the pipe storage and shipping yard source.

Crusher Source Emission Rate

Based on air sampling reports of personnel involved in outdoor crusher operations, a generation rate for this potential source was determined with the approximate volume of the crusher, a calculation of the approximate airborne asbestos concentration within that volume, and average annual daytime wind velocity. The crusher volume was estimated as a 3×6-m crusher building footprint and a seven-meter building height. Twenty-one measured air samples (f/cc) for the crusher operator collected between 1989 and 1992 were averaged, and an average breathing zone concentration 0.151 f/cc was calculated. That concentration was assumed to be the average concentration where waste pipe was delivered to be crushed below ground. The annual daytime average wind velocity (2.81 m/s) was used to determine the flow rate through the estimated crusher building volume.

The volumetric flow rate from the crusher building and the average concentration inside of the crusher exterior building were used to calculate the generation rate, which was 3.19×10^7 fibers/s or 1.66×10^{-8} kg/s. This represented 36.7% of the total emissions assumed in the model.

The crusher building was enclosed on three sides with a roof. Therefore, the assumption of an air exchange rate based on wind velocity through the relatively enclosed space represents an exaggerated or "worst-case" assumption.

Table 6.7 displays the compilation of the parameters associated with the generation rates for the CFD model.

Meteorology

Weather research forecasting (WRF) model meteorological data were used. WRF data were developed in a collaborative partnership primarily between the National Center for Atmospheric Research (NCAR) and the National Oceanic and Atmospheric Administration (NOAA) beginning in the 1990s and has provided three-dimensional hourly meteorological data between 2006 and present. WRF uses objective analysis of global weather reports and satellite data, combining energy and momentum equations to account for the atmospheric conditions for the indicated time periods. In order to represent the bystander's 1989–1993 time working most closely at the receptor property, the earliest available (2006)

Table 6.7 Summary of source descriptions and magnitudes.

	Source magnitude	Raw source generation rate[a]	Source inlet area	Percent of total generation ate
	(fibers/cc)	(kg/s)	(m^2)	(%)
Manufacturing (fugitive)	0.085	6.80×10^{-9}	2,500	15.0%
Finishing (fugitive)	0.088	1.15×10^{-8}	5,848	25.4%
Fiber warehouse (fugitive)	0.074	2.15×10^{-9}	1,300	4.8%
Stripping station (fugitive)	0.018	1.31×10^{-10}	325	0.3%
Tray loading (fugitive)	0.012	1.08×10^{-10}	400	0.2%
Shipping yard	0.021	3.14×10^{-9}	6,500	6.9%
Crusher operator (outside)	0.151	1.66×10^{-8}	18.0	36.7%
Manufacturing baghouses (exhaust outside plant)	0.082	7.91×10^{-10}	2.5	1.7%
Finishing baghouses (exhaust outside plant)	0.082	4.04×10^{-9}	12.8	8.9%

[a] based on a $1.92\mathrm{e}^{-15}$ kg/fiber respirable C&IH selected fiber.

set of 50×50-km data was used. Figure 6.26 shows a 36-petal wind rose representing the 2006 WRF data. Each wind rose petal indicates the frequency and magnitude from each wind direction.

The CFD simulations were run for the average wind velocities for the sectors of WRF data that intersected the receptor property from all of the various plant sources.

Figure 6.27 shows the radially lined template, overlaid on the Google Earth image, used to select six wind directions (330°, 340°, 350°, 0°, 10°, and 20° from north) to be modeled based on the likelihood of receptor interception. In order to determine whether the range of wind directions chosen accounts for all of the airborne fibers at each receptor, it is important to make sure that the range of wind directions is bounded by wind directions where the airborne asbestos concentration is zero or near zero for each receptor.

Table 6.8 shows the selected wind directions to be simulated in CFD, their frequency of occurrence, and average wind velocities. The wind directions in these tables are labeled in degrees from north, where north equals zero degrees, east equals ninety degrees, and so forth. In order to account for the potential variation

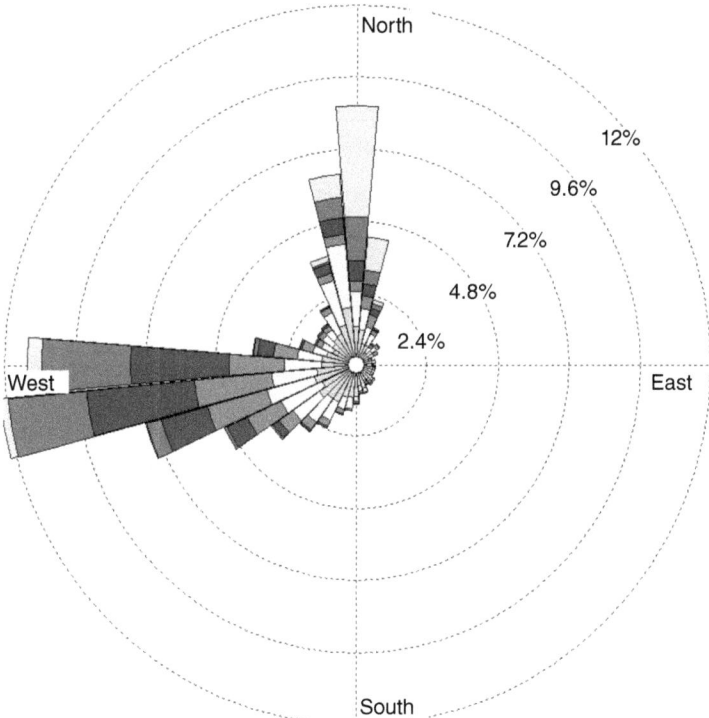

Figure 6.26 Wind rose for 2006 WRF plant data.

Figure 6.27 Radial-lined template superimposed on receptor property boundaries.

Table 6.8 Wind direction and velocity analysis.

		2006 WRF	
Sim. ID	Wind direction (degrees from North)	Wind direction frequency (%)	Average wind speed (m/s)
1	330	2.2	2.30
2	340	3.8	3.51
3	350	6.4	5.06
4	0	8.6	8.56
5	10	4.3	6.78
6	20	2.3	4.40

in the wind frequencies measured for different times of day, a weighted average was calculated for each wind direction frequency. Additionally, daytime meteorological data were used to most closely represent the Bystander's work hours. Table 6.9 lists the primary parameters included in this model.

CFD Results

Figure 6.28 shows the dispersion and concentration dilution of the CFD results for the 330-° wind direction simulation with a 2.82 m/s average wind speed. Below the 330-° wind direction graphic are thumbnail images of all the CFD simulations performed. The average concentration was calculated for the breathing zone height with the CFD result, which is shown in Table 6.10.

Table 6.9 Summary of primary CFD model parameters.

Parameter	Description
Domain size (N to S × E to W × Height)	6,496 × 5,479 × 1,707 m
Fiber size	0.2 μm diameter × 20 μm length
Fiber density (specific gravity)	3.05 SG
Number of elements	14.1 million elements

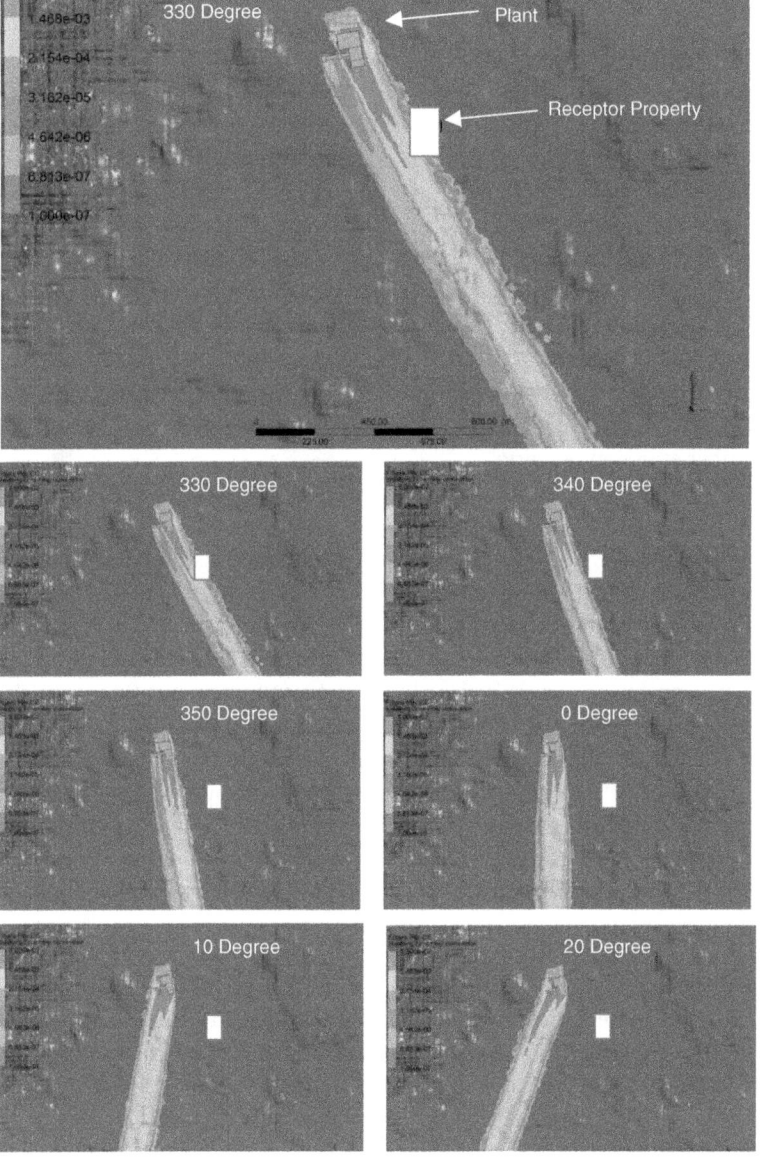

Figure 6.28 CFD simulation result for various wind directions.

Table 6.10 CFD result at receptor.

Annual average concentration (f/cc)
0.000127 (1.27×10^{-4})

Because the emission source outputs from CFD were in kilograms per cubic meter (kg/m^3), a factor of $1.92\,\mathrm{e}^{-15}$ kg/fiber was used to convert to fibers per cubic centimeter as mentioned previously.

EPA Outdoor Dispersion Models

To reinforce the validity of the steady-state CFD model, two US EPA time-dependent models, CALPUFF and AERMOD, were employed that include annual hourly average wind speeds, temperatures and conditions. While CFD has better spatial resolution and potentially superior predictive ability for small-scale models (less than a few kilometers), the increased ability to perform time-dependent and variable weather condition-analyses is a strength of the US EPA-validated models.

CALPUFF is a United States Environmental Protection Agency (US EPA) model accepted for air pollution permitting. Primarily, it is used for tracking long-range dispersion of pollutant emissions up to hundreds of kilometers from the source; however, in this study, the grid resolution was improved in order to improve the model's application to the smaller distances (under a few kilometers) present in this plant fiber emission scenario.

CALPUFF software Version 7 (2015) was used because it contained the most up-to-date features. The latest version of Lakes Environmental CALPUFF View v8.6.1 was used as the CALPUFF graphical user interface software.

In previous CALPUFF simulations where receptors were near the source, several simulations were generated with different size domains to determine the sensitivity of the model with respect to smaller grid sizes (higher resolution) for this smaller domain. It was determined that a 10×10-kilometer (km) domain with a 0.2 km grid resolution produced reasonably representative results, and grid resolutions below 0.2 km did not significantly improve the model accuracy. However, in this CALPUFF model, a grid resolution of 0.1 km was chosen because computer processing time (30 hours for a fast computer) was not identified as a limitation. Table 6.11 shows the model parameters that were run. Primarily, default CALPUFF parameters were used during the set-up of the model; nondefault parameters are discussed in this write-up.

Geophysical Set-Up

Terrain and land use parameters were entered into the model using the CALPUFF Geophysical Processor. The Shuttle Radar Topography Mission (SRTM1) 30-m Version 3 terrain data from National Aeronautics and Space

Table 6.11 CALPUFF model parameters.

Set-up description	MET grid size (km)	Computational grid size (km)	MET and computational grid spacing (km)	Sampling grid size (km)	Receptor spacing (km)
2006	10×10	10×10	0.1	10×10	0.1

Administration (NASA) were used, and the United States Geological Survey's (USGS) Contiguous United States (CONUS), 30-m resolution 1992 National Land Cover Data (NLCD92) land use data were also used.

CALMET Set-Up

CALMET is the meteorological portion of the software that handles the inputs related to meteorological conditions.

Weather research and forecasting (WRF) model's three-dimensional meteorological data were used for a 50×50-km domain surrounding the plant with a 1.0×1.0 km resolution for 2006. The year 2006 was selected based on it being the earliest available year of data nearest the 1989–1993 time of interest. For CALPUFF simulations where grid spacing is less than that WRF data resolution (1.0 km), the software applies interpolation. The model was run for the entire 2006 year using 3,600-second time steps.

CALPUFF Processor

Asbestos was introduced into the model according to the mass flow rates indicated in Table 6.12. Due to the extremely small size of respirable asbestos fibers, it is also appropriate to model them as a PM0.56 particle because the mean aerodynamic diameter would be similar to respirable asbestos fibers. It is also important to note that air currents overpower settling velocities for the critical distances, and respirable fibers tend to stick on surfaces. Theory suggests that emission sources near the ground intercept and adhere to the ground in a substantial manner. However, in this instance, the settling rate was generally not a significant factor for the distances involved. Moreover, the CALPUFF model was performed without particles sticking to the ground, as discussed below, and in contrast to the CFD model.

The shapes of the five building area sources were drawn according to the roof outline with an effective height approximately equivalent to the vertical distance from the ground to the top of the roof.

Each of the five "area" (building surfaces) sources, crusher sources, and shipping yard sources were assigned an emission rate in grams per meter squared second ($g/m^2 \cdot s$) and each of the two "point sources" (baghouses) in grams per

Table 6.12 Source descriptions and emission rates.

Source description	Source area (m²)	Emission rate (g/(m²·s))	Emission rate (g/s)
Manufacturing (fugitive)	2,500	2.72×10^{-9}	N/A
Finishing (fugitive)	5,848	1.97×10^{-9}	N/A
Fiber warehouse (fugitive)	1,300	1.66×10^{-9}	N/A
Stripping station (fugitive)	325	4.03×10^{-10}	N/A
Tray loading (fugitive)	400	2.69×10^{-10}	N/A
Shipping yard (outside)	6,500	4.84×10^{-10}	N/A
Crusher operator (outside)	18.0	9.23×10^{-7}	N/A
Manufacturing baghouses (exhaust)	Point	N/A	7.91×10^{-7}
Finishing baghouses (exhaust)	Point	N/A	4.04×10^{-6}

second (g/s). Figure 6.29 shows the source shapes of the building sources and the locations of the baghouse sources.

The TERRAD radius of influence of terrain features was selected to be 5.0 km for the annual average data due to the approximate distances between hilltops in the area.

Modeling concentration only and not "concentration and deposition" was chosen as the method of calculation, a worst-case consideration.

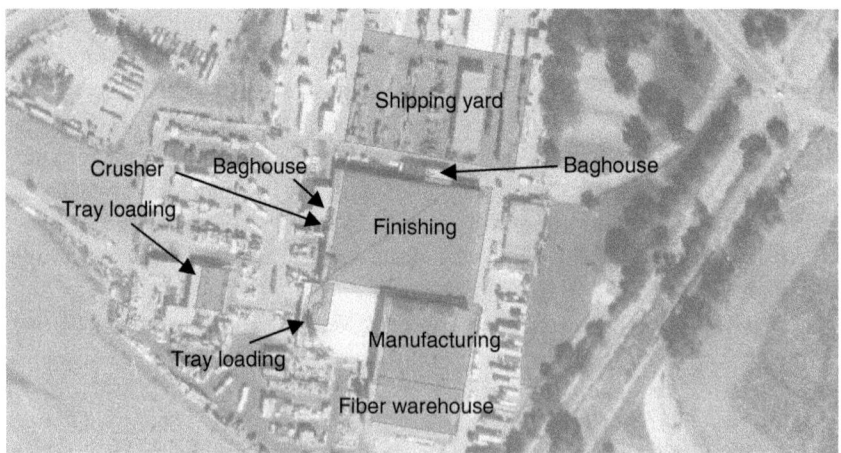

Figure 6.29 CALPUFF model sources. *Source:* Google LLC.

CALPUFF Results

Because the emission sources from CALPUFF were entered into the model as mass generation rates, the 1.92×10^{-15} kg/fiber conversion factor was used to convert mass generation rate to fibers per cubic centimeter. The predicted receptor concentrations are independent of the assumed dimensions and specific gravity of a respirable fiber because the relationship between f/cc and mass in the initial source term calculations is the same as that used to convert mass per cubic meter back into f/cc at the receptor locations. Figure 6.30 shows an example of the contours for the annual average asbestos concentrations in mass concentration of fibers per cubic centimeter (f/cc) with the receptor labeled. The data for the receptor was converted to fibers per cubic centimeter using the conversion factor and the results are listed in Table 6.13.

AERMOD Model

In addition to the CFD and CALPUFF models, an American Meteorological Society/Environmental Protection Agency Regulatory Model (AERMOD) atmospheric dispersion model was utilized to determine the modeled concentration at each of the receptor locations. AERMOD is a United States Environmental Protection Agency (US EPA) accepted model for air pollution permitting. Similar to CALPUFF, it is primarily used for tracking mid-range dispersion of pollutant emissions up to around 50 km from the source. For this application, the grid resolution was decreased in order to apply it to the smaller distances for this model. The Lakes Environmental AERMOD View v8.2.0 was used as the AERMOD graphical user interface software. Primarily, default AERMOD parameters were used during the set-up of the model, nondefault parameters are described in this write-up.

Geophysical Set-Up

The terrain and land use parameters entered into the model using the AERMOD Geophysical Processor were the same as those entered into the CALPUFF model above.

Meteorology Set-Up

AERMOD requires single station surface and upper air meteorological data to model hourly wind flow patterns throughout the year.

One year (2006) of hourly meteorological data was provided by Lakes Environmental and was centered on a single point at the plant interpolated from the 2006 WRF data described earlier in this modeling section of the report.

The upper air data was generated using the "Upper Air Estimator" in the American Meteorological Society/Environmental Protection Agency Regulatory Meteorology (AERMET) processor within the Lakes Environmental AERMET View software. (See Table 6.14.)

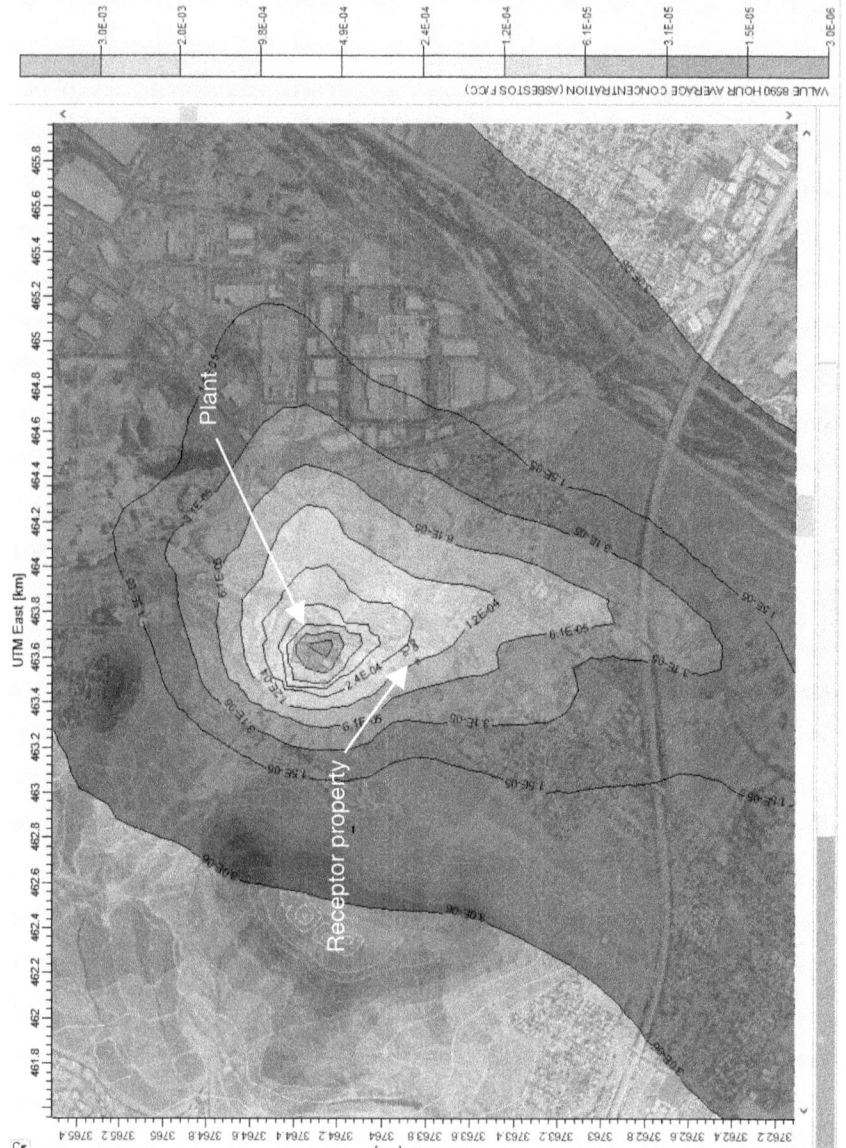

Figure 6.30 Annual average CALPUFF asbestos concentration (f/cc).

Table 6.13 CALPUFF results at receptor.

Annual average concentration (f/cc)
0.000204 (2.04×10^{-4})

Table 6.14 AERMOD model.

Set-up description	Computational domain size (km)	MET and computational grid spacing (km)	Receptor spacing (km)
2006	10×10	0.1	0.1

AERMOD Set-Up

Asbestos was introduced into the model according to the mass flow rates indicated in Table 6.15. Each of the five "area" (building surfaces), the crusher, and the shipping yard sources were assigned an emission rate in grams per meter squared second (g/m$^2 \cdot$s) and each of the two "point sources" (baghouses) in grams per second (g/s). The source data were exported from the CALPUFF simulation discussed earlier.

The AERMOD simulation was run using the source values from Table 6.15 and the final results are shown in Table 6.16.

Table 6.15 Source descriptions and emission rates.

Source description	Source area (m²)	Emission rate (grams/(m²·s))	Emission rate (g/s)
Manufacturing (fugitive)	2,500	2.72×10^{-9}	N/A
Finishing (fugitive)	5,848	1.97×10^{-9}	N/A
Fiber warehouse (fugitive)	1,300	1.66×10^{-9}	N/A
Stripping station (fugitive)	325	4.03×10^{-10}	N/A
Tray loading (fugitive)	400	2.69×10^{-10}	N/A
Shipping yard	6,500	4.84×10^{-10}	N/A
Crusher operator (outside)	18.0	9.23×10^{-7}	N/A
Manufacturing baghouses (exhaust outside plant)	Point	N/A	7.91×10^{-7}
Finishing baghouses (exhaust outside plant)	Point	N/A	4.04×10^{-6}

Table 6.16 AERMOD results at adjacent property receptor.

Receptor	Annual average concentration (f/cc)
2006 WRF Data	0.001 (1.0×10^{-3})

AERMOD Results

Figure 6.31 shows an example of the contours for the annual average asbestos concentrations in f/cc with the receptor labeled.

Table 6.16 shows the result for each set of wind data for the receptor in the model. The data for the receptor was displayed in fibers per cubic centimeter (f/cc).

Comparison of CFD, CALPUFF, and AERMOD Results

Table 6.17 shows a list of the annual average asbestos concentrations (f/cc) for the receptor for CFD, CALPUFF, and AERMOD.

Discussion and Conclusions

These two examples have shown that CFD modeling has proved to be a useful tool that provides relevant exposure results where none exist. If modeling had not been employed, the worker exposure estimates would have been based on a professional assessment of exposures from similar site conditions, collected data, and/or historical literature research, which is often limited to a discrete point in time. Consequently, it is still important to apply this professional judgment to interpreting whether the modeled results make sense or vary significantly from professional experience. For instance, from the case studies provided above: It is important to ask oneself, what was the approximate background concentration of asbestos in the United States, and do the modeled results generally make sense with such considerations and other common-sense considerations, or do the CALPUFF, AERMOD, and CFD models compare reasonably well?

Although modeling can be a useful method to fill in the gaps in characterizing a worker's exposure, one must spend time contemplating several key aspects of CFD model set-up to ensure that the results are as accurate as possible, such as:

- Calculating an accurate contaminant source generation rate
- Selecting appropriate geometric boundaries
- Selecting deliverables before modeling begins
- Using appropriate averaging techniques
- Deciding upon steady state vs. transient simulation
- Choosing a bystander factor vs. direct exposure

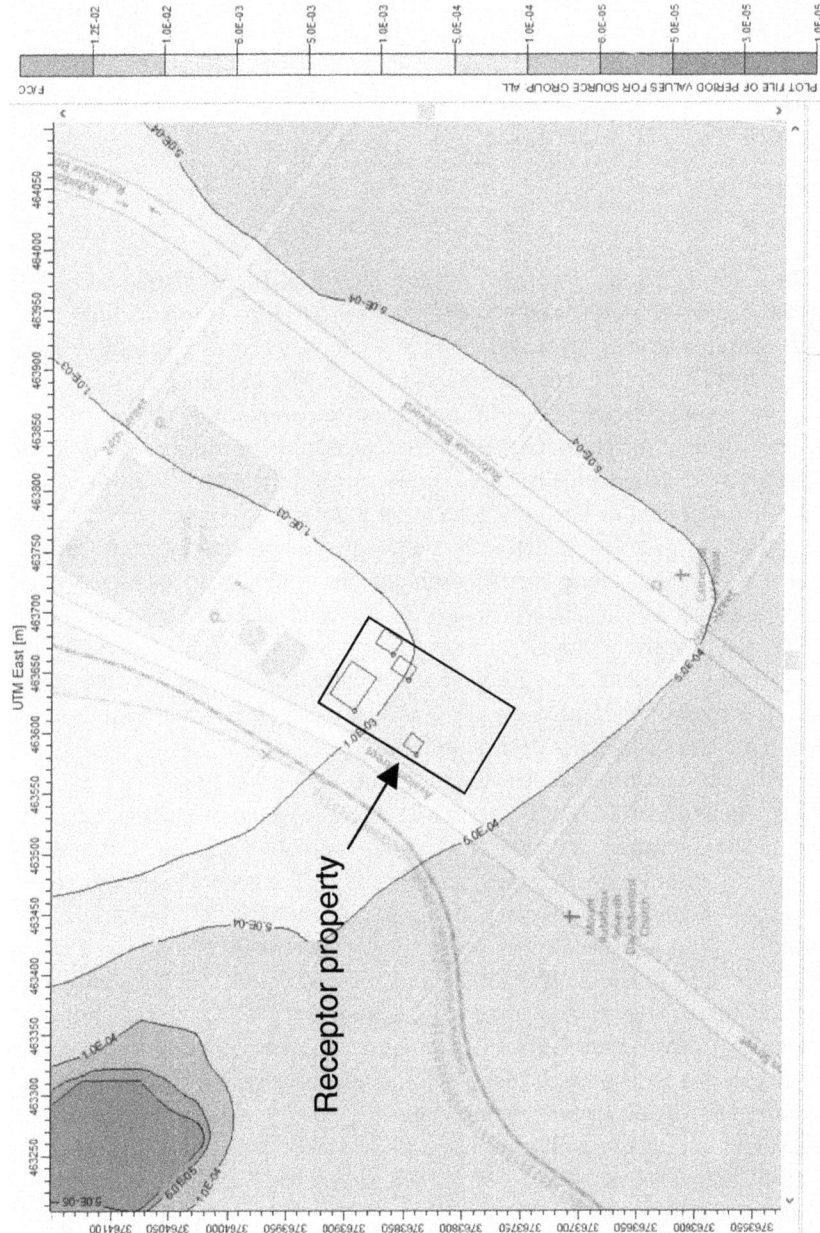

Figure 6.31 Annual average AERMOD asbestos concentration (f/cc).

Table 6.17 Annual average asbestos concentration for CFD, CALPUFF, and AERMOD.

Model	Annual average concentration (f/cc)
CFD	0.000127 (1.27×10^{-4})
CALPUFF	0.000204 (2.04×10^{-4})
AERMOD	0.001 (1.0×10^{-3})

In the following paragraphs, these key aspects will be discussed in order.

Calculating an accurate source generation rate requires an understanding of the process that created the release and a general understanding of the air movement introducing that contaminant into the workplace. The importance of this step cannot be overstated, and a significant amount of time and energy should be spent investigating the characteristics of the source. Determining the quantity of a material that is generated, dispensed, or used over a timeframe is the key component in developing an accurate generation rate. A time-dependent mass balance can be the best way to determine whether the assumptions made for the source generation rate are correct; for example, the number of 55-gallon drums of material used per day, the quantity of material removed per hour, etc. Additionally, the momentum of the contaminant material entering the model is defined by the quantity and intensity of the airflow and is usually described by the volumetric flow rate, for instance an exhaust stack that is carrying gaseous compounds or an injection nozzle injecting a solid or liquid into a model, etc.

An indoor geometry should be selected if the exposure were indoors, and the building supply and exhaust ventilation were the only factors impacting the distribution of the contaminant. However, if indoor ventilation is influenced by exterior wind, for instance, in the case of a leaky building or open windows or doors, then a combination of an indoor and outdoor geometry should be used. If the exposure was entirely outside, then the geometry may consist of a cubical domain with the terrain and buildings as the ground, the four vertical walls assigned specific wind velocity entering or leaving the simulation and the top sky-boundary allowing air to pass in or out of the simulation. It is also important to create a geometry that is large enough to capture the effects of ventilation or wind that may impact the distribution of contaminants.

Deliverables needed to answer a question should be defined before beginning the simulation. This may consist of numerical results, graphics, charts, and or videos. Review the hypothesis and ensure that the outputs will provide the appropriate information.

Applying area or volume averaging versus single-point data extraction is important because it dampens the effect of spatial variability, whereas a datum extracted from several adjacent points, within a model, could vary significantly. When selecting the volume to be averaged, expert judgment should be applied combining an understanding of the process, environment, ventilation, etc.

A steady state simulation is the fastest CFD simulation to run, and represents conditions at time of infinity. However, if a process is not consistent, a transient simulation may be required to demonstrate the concentration at a specific point in time with respect to a process.

Determining that a bystander factor is the needed parameter of interest may preclude one having to model the actual exposure concentration to describe a worker's exposure, and merely relative concentrations will suffice. The exact magnitude of the source generation rate is not critical to the determination of the bystander worker's exposure because it will be compared to primary work exposures that have already been historically established, thus reducing the need to acquire accurate source characteristics.

Throughout the modeling process, it is important to continue to refer to the original purpose of the modeling, so that, along the way, one is still on target for the intentions of the modeling.

At the end of modeling, when comparing the results from different models and along with expert judgment, a range of results does not necessarily indicate a problem, and is essentially reasonable, because it indicates that, even though the various models use different algorithms and methods to arrive at results, the results are generally in the ballpark of one another. Likewise, to reinforce this point, even real-world measurements collected over multiple iterations of the same experiment will have different results. And at the same time, one must be aware that models are merely a representation of reality and never exact because, as the statistician George E. P. Box has been attributed as saying, "All models are wrong, but some are useful" (Box 1979).

References

American Iron and Steel Institute, New York. (1965). *Steel Mill Ventilation*. Committee on Industrial Hygiene.

Bennett, J.S., Feigley, C.E., Khan, J., and Hosni, M.H. (2000). Comparison of mathematical models for exposure assessment with computational fluid dynamics. *Applied Occupational and Environmental Hygiene* 15(1): 131–144. https://doi.org/10.1080/104732200301953.

Blocken, B. and Gualtieri, C. (2012). Ten iterative steps for model development and evaluation applied to computational fluid dynamics for environmental fluid mechanics. *Environmental Modelling and Software* 33: 1–22. https://doi.org/10.1016/j.envsoft.2012.02.001.

Box, G.E.P. (1979). Robustness in the strategy of scientific model building. In: *Robustness in Statistics*, (eds. R.L. Launer and G.N. Wilkinson), 201–236. New York: Academic Press. https://doi.org/10.1016/C2013-0-11050-1.

Cherrie, J.W. (1999). The effect of room size and general ventilation on the relationship between near and far-field concentrations. *Applied Occupational and Environmental Hygiene* 14(8): 539–546. https://doi.org/10.1080/104732299302530.

Corn, M. (1966). The adhesion of particles. In: *Aerosol Science* (ed. C.N. Davies), 359–392. New York: Academic Press.

Corn, M. and Stein, F. (1965). Re-entrainment of particles from a plane surface. *American Industrial Hygiene Association Journal* 26(4): 325–336. https://doi.org/10.1080/00028896509342739.

Donovan, E.P., Donovan, B.L., Sahmel, J. et al. (2011). Evaluation of bystander exposures to asbestos in occupational settings: a review of the literature and application of a simple eddy diffusion model. *Critical Reviews in Toxicology* 41(1): 52–74. https://doi.org/10.3109/10408444.2010.506639.

Gousseau, P., Blocken, B., Stathopoulos, T., et al. (2011). CFD simulation of near-field pollutant dispersion on a high-resolution grid: a case study by LES and RANS for a building group in downtown Montreal. *Atmospheric Environment* 45(2): 428–438. http://dx.doi.org/10.1016/j.atmosenv.2010.09.065.

Hall, D.R., Birkner, J., Strode, R.D., and Rasmuson, J.O. (2007). Utility of computational fluid dynamics (CFD) for determining flow and re-entrainment of indoor airborne respiratory-ranged sized fibrous and non-fibrous particles. *Presented at the American Industrial Hygiene Conference* (Philadelphia, Pennsylvania) June, 2007.

Hall, D., Strode, C., Rasmuson, E., and Rasmuson, J. (2011). Comparison of a two-zone (near field-far field) exposure model with computational fluid dynamics (CFD) and spatial concentration distributions measured in a simulation chamber to estimate breathing zone concentrations and bystander exposure factors. *Presented at the American Industrial Hygiene Conference*, Portland, Oregon (May, 2011), 120–121.

Hall, D., Strode, C., Rasmuson, E., and Rasmuson, J. (2012). Using computational fluid dynamics (CFD) to assist in the determination of sample placement with respect to topographic features for airborne outdoor dust exposure assessment. *Presented to the American Industrial Hygiene Conference*, Indianapolis, Indiana (June, 2012). Paper Number SR-101.-4.

Hall, D., Strode, C., Rasmuson, J., et al. (2013). Computational fluid dynamics validation utilizing a tracer gas study related to a mine mill area toxic gas release for emergency response planning. *Manuscript accepted for presentation at NAFEMS World Congress 2013* (Salzburg, Austria), June, 2013.

Hall, D., Strode, C., Rasmuson, J., et al. (2016). Computational fluid dynamics validation utilizing a tracer gas study related to a mine mill area toxic gas release for emergency response planning. *NAFEMS International Journal of CFD Case Studies* 11: 31–46. https://doi.org/10.59972/7tlrkx6h.

Hanna, S., Brown, M.J., Camelli, F., et al. (2006). Detailed simulation of atmospheric flow and dispersion in downtown Manhattan: an application of five computational fluid dynamics models. *Bulletin of the American Meteorological Society* 1713–1726. https://doi.org/10.1175/BAMS-87-12-1713.

Hanna, S.R., Hansen, O.R., Ichard, M., and Strimaitis, D. (2009). CFD model simulation of dispersion from chlorine railcar releases in industrial and urban areas. *Atmospheric Environment* 43(2): 262–270. https://doi.org/10.1016/j.atmosenv.2008.09.081.

Keil, C.B., Simmons, C.E., and Anthony, T.R. (2009). *Mathematical Models for Estimating Occupational Exposures to Chemicals*, 2nd Edition. A Publication from the American Industrial Hygiene Association.

Kisa, M., Jelemensky, L., Mierka, O., and Stopka, J. (2004). Comparison of CFD modeling with small-scale field experiments of chlorine dispersion. *Chemical Papers* 58(6): 429–434. https://www.researchgate.net/publication/288436021_Comparison_of_CFD_modelling_with_small-scale_field_experiments_of_chlorine_dispersion.

Kisa, M., et al. (2006). Comparison of CFD Modelling with Small-Scale Field Experiments of Chlorine Dispersion. *Chemical Papers*.

Li, Y. and Guo, H. (2006). Comparison of odor dispersion predictions between CFD and CALPUFF models. *Transactions of the ASABE (American Society of Agricultural and Biological Engineers)* 49(6): 1915–1925. http://dx.doi.org/10.13031/2013.21120.

Meroney, R.N. (2004). *CFD Prediction of Cooling Tower Drift*. Wind Engineering and Fluids Laboratory, Civil Engineering Department, Colorado State University. https://www.engr.colostate.edu/~meroney/PapersPDF/CEP04-05-1.pdf.

Meroney, R.N. (2006). CFD prediction of cooling tower drift. *Journal of Wind Engineering and Industrial Aerodynamics* 94(6): 463–490. http://dx.doi.org/10.1016/j.jweia.2006.01.015.

Meroney, R., Ohba, R., Leitl, B., et al. (2016). Review of CFD guidelines for Dispersion modeling. *Fluids* 1(2). https://doi.org/10.3390/fluids1020014.

Millette, J.R. and Hays, S.M. (1994). Resuspension of settled dust. In: *Settled Asbestos Dust, Sampling and Analysis* (eds. S.M. Hays and J.R. Millette), 59. Lewis Publishers (CRC Press). https://doi.org/10.1201/9780203739808.

National Hurricane Center. n.d. NHC Track and Intensity Models (noaa.gov). https://www.nhc.noaa.gov/modelsummary.shtml June 11, 2019.

National Weather Service. (2017). Tropical Storm Irma Strom Track. https://www.weather.gov/chs/tropicalstormirma-Sept2017. Accessed October 31st 2023.

Scargiali, F., Di Rienzo, E., Ciofalo, M., Grisafi, F., Brucato, A. (2005). Heavy gas dispersion modelling over a topographically complex mesoscale: a CFD based approach. *Process Safety and Environmental Protection* 83(3): 242–256. https://doi.org/10.1205/psep.04073.

Semple, S., Proud, L.A., and Cherrie, J.W. (2003). Use of Monte Carlo simulation analysis to investigate uncertainty in exposure modeling. *Scandinavian Journal of Environmental Health* 29(5): 347–353. https://doi.org/10.5271/sjweh.741.

Tang, W., et al. (2006). Application of CFD simulations for short-range atmospheric dispersion over open fields and within arrays of buildings. *National Research Council at National Exposure Research Laboratory and the USEPA Presentation at the AMS 174th Joint Conference on the Applications of Air Pollution Meteorology with the A&WMA*.

West Yorkshire Steel. (2017). Steel hardening & forging temperatures colour chart. https://www.westyorkssteel.com/technical-information/steel-heat-treatment/hardening-temperatures/. Accessed March 2017.

Witheridge, W.N. (1948). Ventilation. In: *Industrial Hygiene and Industrial Toxicology* (ed. F.A. Patty), 281. New York: Interscience Publishers, Inc.

Yang, C., Demokritou, P., Chen Q. et al. (2001). Experimental validation of a computational fluid dynamics model for IAQ applications in ice rink arenas. *Indoor Air* 11(2): 120–126. https://doi.org/10.1034/j.1600-0668.2001.110206.x.

Part III

Dose-Response Assessment

7

Asbestos Dose–Response Assessment: The Peto Model and Its Application in the US EPA and Berman and Crump Studies

Andrey Korchevskiy

Chemistry & Industrial Hygiene, Inc., Lakewood, CO, USA

Rationale and Meaning of the Peto Model

Time- and age-dependent patterns of mesothelioma and lung cancer mortality bear significant implications for asbestos health risk assessment. Understanding of risk levels for individuals differentiated by age, time since first exposure, and age at first exposure is necessary for a proper interpretation of epidemiological data, for determination of risks associated with a specific level of exposure, but also for an effective projection of cancer mortality (a valuable tool for public policy implication of the risk assessment procedure).

White et al. (2014) emphasized that, "age…is used in virtually all studies of cancer epidemiology and is one of the most studied risk factors for cancer." In this relation, White quoted Robert Browning's verse: "Grow old along with me! The best is yet to be, the last of life, for which the first was made." Cancer risks are known to grow along with a person, demonstrating increased incidence of most cancers with age, with a swift drop of rates observed for numerous cancer sites after the age of 90 (Harding et al. 2012).

Various models were proposed to explain the dynamics of cancer incidence and mortality. In 1954, Armitage and Doll proposed an age-distribution model of cancer based on a multistage theory of carcinogenesis. The authors referred to the landmark observations of Fisher and Hollomon (1951) and Nordling (1953), according to which the cancer death rate increased proportionally with the sixth power of age. Armitage and Doll demonstrated that the cancer rate in the short time interval $(t, t + dt)$ can be seen as being proportional to the probability that a critical last change happens in a sequence of r changes following a specific order.

Health Risk Assessment for Asbestos and Other Fibrous Minerals, First Edition.
Edited by Andrey Korchevskiy, James Rasmuson, and Eric Rasmuson.
© 2024 John Wiley & Sons, Inc. Published 2024 by John Wiley & Sons, Inc.

Armitage and Doll found the following formula for the probability of cancer:

$$\frac{p_1 p_2 \ldots p_i \ldots p_r t^{r-1}}{(r-1)!} dt \qquad (7.1)$$

where p_i is a probability of ith "change" and r is the number of critical changes in cells (Armitage and Doll, 1954).

Armitage and Doll considered these changes to be mutations, but theoretically, other types of cell changes (like irreversible epigenetic damage to genetic material) can be a part of the cancer development stages modeled by Eq. (7.1).

The factorial $(r-1)!$ in the denominator of Eq. (7.1) is included to emphasize that only a specific sequence of r cell changes is producing the specific type of cancer. Actually, $(r-1)$ changes considered in Eq. (7.1) can happen in any possible order with the total of $(r-1)!$ combinations, but only one order of them is supposed to produce cancer. This is why the probability of cancer in Eq. (7.1) is divided by $(r-1)!$ (that means $1 \times 2 \times \ldots \times (r-1)$).

Armitage and Doll hypothesized that an external carcinogenic agent is affecting a single stage of carcinogenesis. For example, in this case, p_r would be proportional to the dose of a carcinogen affecting the body, but in other probabilities, p_i would be independent of the dose.

Therefore, the overall probability of cancer in the time period from t to dt would be linearly proportional to the dose and the age t in the power of $(r-1)$, but not to the higher power of dose if the carcinogenic agent affected several stages of carcinogenesis.

Equation (7.1) provides an estimation of the "instant" probability of cancer at a very short interval of dt. The actual mortality or incidence rate from cancer for a specific age group can be calculated as a sum, or integral, of the probability at each short interval of time (mathematically, it means integrating the Eq. (7.1)). Therefore, the observed incidence rate would follow the equation with varying exponents:

$$\text{Death (incidence) rate} \propto t^r \qquad (7.2)$$

where r, as previously, is hypothesized as the number of "stages," or mutations, necessary to produce the cancer.

It should be noted that the time processes with the probabilities of events being proportional to the certain power of time are usually modeled based on the so-called "Weibull random process," widely used in survival theory and theory of systems' reliability (Billings et al. 2022).

Armitage and Doll determined the exponential power of the age-dependency models for various types of cancer. In male patients, the parameter r for Eq. (7.1) was estimated as 6.26 for esophagus cancer, 5.91 for stomach, 5.18 for colon, 5.62

for rectum, and 5.76 for pancreas. In females, the coefficient was 5.27 for stomach, 4.97 for colon, 5.03 for rectum, and 6.48 for pancreas.

Cook et al. (1969) successfully tested the Armitage–Doll model for successive 5-year age-specific incidence rates for 31 types of cancer in 11 populations and demonstrated that

$$I = bt^k \qquad (7.3)$$

where I is the incidence at age t, and b and k are constants. In this formula, obviously, k is the analogy of the parameter r of the original Armitage–Doll model.

Cook, Doll, and Fellingham, however, hypothesized that the risk of cancer might be determined not by the age of the subject, but by the "prevalence of carcinogenic agents and the length of time the subject had been exposed to them." In this case, Cook, Doll, and Fellingham proposed a modification to Eq. (7.3):

$$I = b(t-w)^k \qquad (7.4)$$

where t is the time since first exposure, and w is the "lag" of time, or the minimal period needed for the sequence of the cell changes to occur.

Differences in the estimated value of r or k, (the power of the effective exposure time in the equation), were examined by Cook, Doll, and Fellingham between (i) types of cancer and (ii) populations. The results were not wholly consistent but suggested that the value of k might be a biological constant characteristic of the tissue in which the cancer is produced.

In 1982, Peto et al. (1982) published a paper describing the age and time-pattern of mesothelioma and lung cancer among North American insulators. Mortality among 17,800 members of the International Association of Heat and Frost Insulators and Asbestos Workers had been monitored between 1967 and 1979. This study, earlier described by Selikoff et al. (1979), revealed a substantial excess of nonmalignant respiratory disease, an approximately fourfold excess of lung cancer, and elevated incidence of mesothelioma in the cohort. Specifically, 87 pleural and 149 peritoneal tumors were observed up to age 80.

The authors concluded that the mesothelioma mortality in the cohort followed well the age-related function discovered by Armitage and Doll (Eqs. 7.1 and 7.3). The model fit was especially efficient for the subcohort with the time of first exposure between 1922 and 1946.

We reconstructed the modeling performed by Peto based on the data from Peto, Seidman, and Selikoff. The best fit for the model (Eq. 7.3) is with the parameters estimated as $b = 6.31 \times 10^{-8}$ and $k = 3.14$, with correlation between the log transformed observed and predicted values of $R = 0.98$ and $R^2 = 0.96$, $F = 119.5$, $p < 0.00011$.

The model fit is illustrated in Figure 7.1.

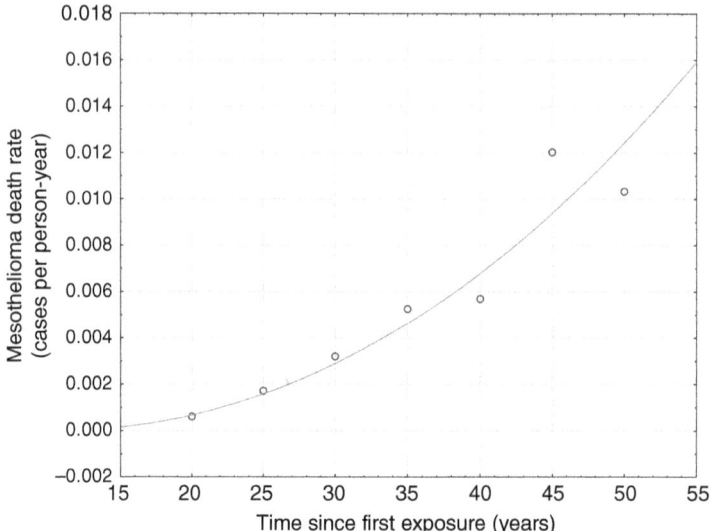

Figure 7.1 Mesothelioma death rate for American insulators modeling fit. The red line – fitting curve ($Y = 6.31*10^{-8}X^{3.14}$), where Y – mesothelioma death rate per person-year, X – time since first exposure (years) ($R = 0.98$, $R^2 = 0.96$, $F = 119.5$, $p < 0.00011$).

Peto et al. (1982) also hypothesized that the risk of mesothelioma is independent of the age of first exposure. Figure 7.2a and b recreates the relationship of cumulative mesothelioma mortality in the study with the age and years since first exposure for different onset ages (assuming there are no other causes of death in the cohort) based on the data from Peto et al.

It should be noted that if other causes of death are taken into account, the onset age of exposure would certainly affect the lifetime mesothelioma risk. The earlier the exposure happens, the larger is the remaining size of the surviving population for more cases of mesothelioma to occur.

At the same time, the incidence rate of mesothelioma at every age group fully depends on the exposure duration, time since first exposure, and exposure concentration, and is independent of the onset age, as Peto et al. suggested. In any case, the cumulative (but not lifetime) mesothelioma risk at a certain age also depends on the time since first exposure, but not the age of exposure onset.

Peto, Seidman, and Selikoff also quoted several other sources confirming the proposed model for mesothelioma. In particular, it was demonstrated that the model can be fitted to various other data sets on mesothelioma mortality in asbestos-exposed workers:

Newhouse and Berry (1976) – factory workers (mixed fiber type exposure):

$$k = 3.2, b = 4.95 \times 10^{-8}$$

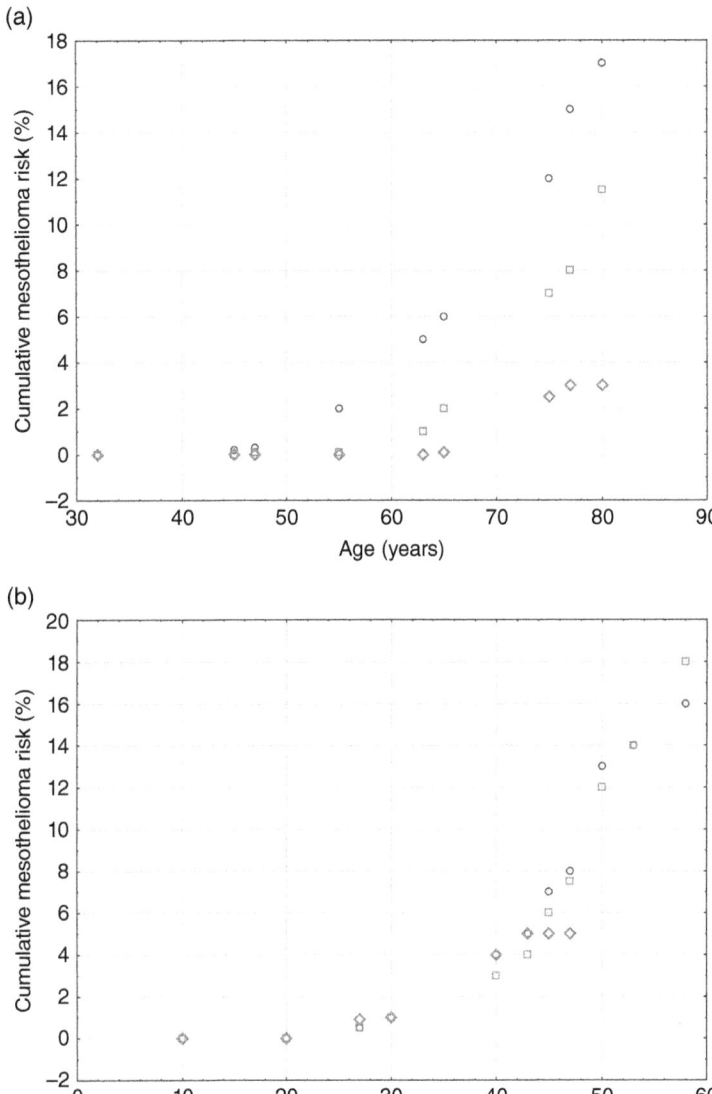

Figure 7.2 Mesothelioma risk among American insulators (the circles – first exposure at the age of 15–24; the squares – first exposure at the age of 25–34; the rhombus – first exposure at the age of more than 35 years). (a) Cumulative mesothelioma risk depending on age and (b) cumulative mesothelioma risk depending on time since first exposure.

Peto (1980) – textile workers (chrysotile exposure, with around 5% amphibole asbestos contamination):

$$k = 3.2, b = 2.94 \times 10^{-8}$$

Hobbs et al. (1980) – Australian miners (crocidolite exposure):

$$k = 3.2, b = 5.15 \times 10^{-8}$$

Seidman et al. (1979) – US factory (amosite exposure):

$$k = 3.2, b = 4.91 \times 10^{-8}$$

It should be noted that the exponent of the model is apparently not related to asbestos exposure level, but rather the coefficient b, in general, should reflect the level of exposure and the potency.

Peto et al. also noted that incidence and mortality rates are similar for mesothelioma because the interval from diagnosis to death is usually short. The time taken for the tumor to grow to a clinically detectable size may, however, be substantial, which could account for the anomalously low mortality 10–15 years after first exposure. The model $b(t-10)^2$, a quadratic time-dependence model with a lag of 10 years, fit the various cohorts better than the model $bt^{3.2}$.

Our own retesting of the data from Peto, Seidman, and Selikoff confirmed that the model $b(t-10)^2$ provides a better fitting to the data than the original model $bt^{3.14}$. For the model with the lag, $R = 0.987$, $R^2 = 0.974$, $F = 190$, $p < 0.000036$, vs. reported above ($R = 0.98$, $R^2 = 0.96$, $F = 119.5$, $p < 0.00011$ for the model with no lag).

The shape of the dose–response for asbestos and bronchial carcinoma (lung cancer) is significantly different than for mesothelioma. As Peto and coauthors emphasized, the suggestion that the excess relative risk for lung cancer (RR − 1) is roughly proportional to cumulative asbestos dose (with or without lag) has been widely accepted as a useful approximation for practical purposes (Peto 1978; Acheson and Gardner 1979). Apparently, Julian Peto (Wittenmore 1977) was the first to propose the relative risk model for lung cancer.

Peto, Seidman, and Selikoff suggested that asbestos "acts immediately and cumulatively to increase the rate of the final stage in bronchial carcinogenesis," which seemed to them to be "biologically implausible but perhaps not impossible."

The relative risk age-dependent model for lung cancer is apparently a deviation from the original Armitage–Doll model. If relative risk, and not absolute risk of lung cancer, depends on the time since first exposure, in this case, at least one of the transitional probabilities p_i in Eq. (7.1) should be a function of t (though Armitage and Doll suggested all of them should be time-independent).

The general form of the Peto model for mesothelioma can be written as follows:

$$\text{EIR}_M = b(t-w)^k \tag{7.5}$$

where b is the coefficient depending on mineral type of fibers and exposure intensity in f/cc, k is an exponent, w is the lag, usually assumed to be 10 years, and EIR_M is the excess rate of mesothelioma.

For lung cancer, the following model is used:

$$\text{RR} = \left(\text{EIR}_L + \text{BR}_L\right)/\text{BL}_L = 1 + c \times \text{CE}_w \qquad (7.6)$$

where c is the coefficient, depending on mineral type of fibers, w is the lag, usually assumed to be 10 years, EIR_L is the excess rate of lung cancer, BR_L is the baseline rate of lung cancer, RR is the relative risk of lung cancer from asbestos exposure, and CE_w is the cumulative exposure in f/cc-years, the last w years not being counted.

Both Equations (7.5) and (7.6) will be referred to as the "Peto model" in this book to emphasize the input of Julian Peto to quantitative assessment of the asbestos risk of respiratory cancers. It should be noted that the term "Peto model" is usually used for Eqs. (7.5), and (7.6) sometimes is referred to as "common knowledge."

The Peto model implies that two to three critical changes in cells (mutations) are needed to produce mesothelioma. For lung cancer, the required mutation pattern can be more complex because Eq. (7.9) also includes the term for baseline, or background incidence of lung cancer (because relative risk is a ratio of absolute risk to baseline risk). The interaction of asbestos with smoking is especially important because the majority of "background" cancer mortality in the population is smoking-related. Regarding this interaction, see Chapter 12.

It should be noted that in a general sense, the multistage model for carcinogenesis, and in particular the Armitage and Doll formulas, were criticized as just a theoretical approach not fully supported by observations. Freedman and Navidi (1989) concluded that "the multistage model provided a family of curves that often fit cancer incidence data but may not capture the underlying biological reality."

Even Armitage and Doll themselves suggested that their multistage model of carcinogenesis might need corrections based on various ideas of initiation and proliferation in regard to the development of cancer. In 1957, Armitage and Doll proposed a two-stage carcinogenic model to potentially simplify their initial multistage mode based on a proposed different biological basis. The authors stated, however, that there is no direct proof that the two-stage hypothesis is true vs. the multistage theory of cancer.

Moolgavkar et al. (2009) developed a so-called "two-stage clonal expansion" (TSCE) model reflecting the two-stage process with cells initiated via a Poisson process, undergoing clonal expansion and malignant conversion via a birth-death-mutation process:

$$h(t) = \frac{v}{\alpha} pq \frac{e^{-qt} - e^{-pt}}{qe^{-pt} - pe^{-qt}} \qquad (7.7)$$

where t – age, $h(t)$ – cancer death rate at the age t, q, p, v, and α – parameters that are related to the probability of mutation and proliferation of malignant cells. Moolgavkar et al. (2009) applied the two-stage model of age-dependency to the mesothelioma information in the National Institute of Health (NIH) Surveillance, Epidemiology, and End Results (SEER) Program database on mesothelioma. The following parameters were determined for pleural mesothelioma:

$$p = -0.12, q = 0.000014, \frac{v}{\alpha} = 0.00028$$

Remarkably, Moolgavkar et al. also fitted the Armitage–Doll model to the SEER data for pleural mesothelioma and found the following parameters:

$$k = 5.16 \left(95\% \text{ CI } 4.95, 5.34\right) \text{ and}$$

$$b = 3.36 \times 10^{-15} \left(95\% \text{ CI } 1.41 \times 10^{-15}, 7.99 \times 10^{-15}\right) \left(\text{no lag was included}\right).$$

The Armitage–Doll model and the TSCE model apparently fit the SEER data with a similar level of precision (though Moolgavkar preferred the TSCE model).

We attempted to fit the model (Eq. 7.10) to the original Peto, Seidman, and Selikoff data on mesothelioma mortality in insulators.

The following fit model was found:

$$\text{Death rate} = -5.3 \times 10^{-6} \left(\frac{e^{-0.00042t} - e^{0.15t}}{0.00042 e^{0.15t} + 0.15 e^{-0.00042t}} \right) \tag{7.8}$$

(correlation between log-transformed observed and predicted values $R = 0.956$, $R^2 = 0.914$, $p = 0.00078$), where t is the time from first exposure.

The model fit is demonstrated in Figure 7.3.

We also compared the Peto model with 10 years lag and the TSCE model for their performance in reconstructing the mesothelioma mortality rate in American insulators. The results are provided in Table 7.1.

Generally, the Peto model looks slightly preferable or similar in reconstructing the age-resolved mesothelioma rates.

Also, it is not necessarily clear how exposure level would affect the parameters of the TSCE model, while there is an understanding of this for the Peto model (the exposure affects the linear coefficient, but not the exponent power). The parameters p, q, v, and α of the TSCE model depend on the exposure scenario, but more studies are required to determine their relationship with exposure, duration, intensity, and exposure onset age. For the TSCE model for asbestos, it is actually impossible to determine a specific dose–response relationship.

In general, the Peto model is a universal, simple, and convenient tool for estimation of mesothelioma and lung cancer rates, depending on age and time since first exposure to asbestos. The Peto model for mesothelioma was created as a

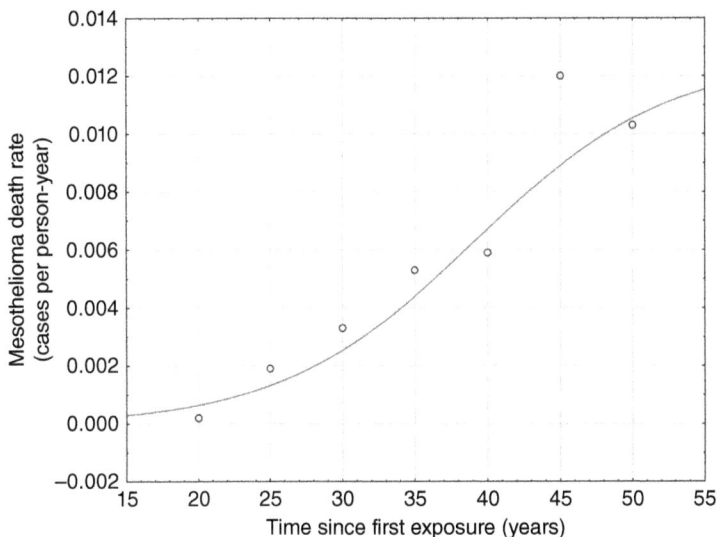

Figure 7.3 Fitting of the TSCE model to the mesothelioma mortality of American insulators. The red line – fitting curve ($Y = -5.3 \times 10^{-6} \frac{e^{-0.00042X} - e^{0.15X}}{0.00042 e^{0.15X} + 0.15 e^{-0.00042X}}$) ($R = 0.956$, $R^2 = 0.914$, $p = 0.00078$) (X – time since first exposure, Y – mesothelioma death rate in person-years).

Table 7.1 Comparison between Peto and TSCE models for the American insulators' mesothelioma mortality data.

	Peto model	TSCE model	Explanation of the parameter
Pearson correlation (R) between logged variables	0.987	0.985	Greater for better fit
AIC (Akaike Information Criteria)	−86	−83	Smaller for better fit
Maximum likelihood estimation of the variance	2.62×10^{-6}	2.99×10^{-6}	Smaller for better fit

development of the Armitage and Doll multistage model of carcinogenesis. For lung cancer, the Peto model seems to be a deviation of the Armitage and Doll theory (because not absolute but rather relative risk is modeled), but this can be evidence of different underlying biological mechanisms between mesothelioma and lung cancer development. The Peto model fits well with mesothelioma and lung cancer incidence data and is a direct mathematical framework for evaluation of the impact of various aspects of the exposure scenario, such as fiber type, fiber dimension, fiber habit, and exposure intensity, which is linearly related with the linear coefficients of the dose–response formulas.

The Peto model was utilized for characterization of mesothelioma and lung cancer mortality in various published studies. For example, Pira et al. (2007) stated that "the incidence of mesothelioma rises as function of the third or fourth power of time since first asbestos exposure," characterizing a cohort of 1,966 subjects employed by an Italian asbestos (mainly textile) company in the period from 1946 to 1984, followed to 2004. Reid et al. (2014) confirmed that the risk of malignant mesothelioma increases with the third or fourth power of time since first exposure to crocidolite; however, it appears to grow at a slower power after 45 years following the first exposure (apparently with no lag). Loomis et al. (2019) applied the Peto model with the lag 10 years and $k = 2$ to mesothelioma mortality of 5,397 asbestos textile manufacturing workers in North Carolina, USA. Lash et al. (1997) determined linear coefficients relating cumulative exposure to asbestos to relative risk of lung cancer in 15 cohorts with various lags and found that industry (mining and milling vs. asbestos cement vs. asbestos textile) is a source of heterogeneity of the coefficients.

Utilization of the Peto Model by the US EPA

Based on the Peto model and available epidemiological data, the US EPA in 1986 published its methodology for lung cancer and mesothelioma health-risk assessment. (The document was authored by W. Nicholson.)

In the case of lung cancer, the model was proposed in the form of:

$$I_L(t) = I_L(a, y, t, d, f) = I_E(a, y)\left[1 + K_L \times f \times d(t - 10)\right] \quad (7.9)$$

where $I_L(t) = I_L(a,y,t,d,f)$ is the lung cancer incidence observed or projected in a population of age, a, observed in calendar period, y, at t years from onset of an asbestos exposure, $I_E(a,y)$ is the age and calendar year lung cancer incidence expected in the absence of asbestos exposure, f is the intensity of asbestos exposure to PCM fibers longer than 5 μm (f/cc), $d(t-10)$ is the duration of exposure up to 10 years from the moment y, and K_L is a proportionality constant that is a measure of the carcinogenic potency of the asbestos exposure.

Equation (7.9) can be interpreted differently, depending on the time since first exposure, t, as related to the duration of exposure d.

It is easy to demonstrate that

$$\begin{aligned} I_L(t) = I_L(a,y,t,d,f) &= I_E(a,y)\left[1 + K_L \times f \times d\right] && \text{for}: t > 10 + d, \\ &= I_E(a,y)\left[1 + K_L \times f \times (t - 10)\right] && \text{for}: 10 + d > t > 10 \quad (7.10) \\ &= I_E(a,y) && \text{for}: 10 > t \end{aligned}$$

The model for mesothelioma was proposed as Eq. (7.11) in the form of:

$$I_M(t) = I_M(t,d,f) = K_M \times f \times \left[(t-10)^3 - (t-10-d)^3\right] \quad \text{for}: t > 10+d$$
$$= K_M \times f \times (t-10)^3 \quad \text{for}: 10+d > t > 10$$
$$= 0 \quad \text{for}: 10 > t$$

(7.11)

where $I_M(t) = I_M(t,d,f)$ is the mesothelioma incidence at t years from onset of exposure to asbestos for duration above, d at a concentration f of PCM fibers longer than $5\,\mu m$ (f/cc), and K_M is carcinogenic potency that depends on fiber type and dimensionality.

The derivation of Eq. (7.11) is as follows: The mesothelioma incidence rate at the moment t is assumed to be equal to the sum of contributions from every moment u: $0 < u < t-10$. It means that the variable u covers the entire range of time from 0 to $t-10$. Based on the Peto model, this contribution is found as $E(u) \times (t-u-10)^2$ where $E(u)$ is the exposure at the moment u. For each $u \le d$ (where d is the duration of exposure), we assume that $E(u) = f$, and the exposure is 0 if $u > d$.

In this case, the incidence of mesothelioma at t year after the onset of exposure can be estimated as:

$$I_M(t) = 3 \times K_M \times \int_0^{t-10} E(u)(t-u-10)^2 \, du \quad (7.12)$$

(the integral in this case means the summation of all contributions for each moment u since the first exposure). The arbitrary coefficient 3 is put in Eq. (7.3) just for convenience of calculations (as in Berman and Crump 2008a).

If $t < 10$, the integral will be equal to 0, because no exposure during the last 10 years contributes to mesothelioma incidence (this is the meaning of 10 years of lag).

If $10 < t < d+10$, E(u) in the integral is equal to f for every u. In this case,

$$3 \times K_M \times \int_0^{t-10} E(u)(t-u-10)^2 \, du = -K_M \times f \times (t-u-10)^3 \Big|_0^{t-10}$$
$$= K_M \times f \times (t-10)^3$$

(7.13)

Interestingly, in Eq. (7.13), the duration d is not a part of the formula. However, if $d+10 < t$, we would have

$$3 \times K_M \times \int_0^{t-10} E(u)(t-u-10)^2 \, du$$
$$= 3 \times K_M \times \int_0^d E(u)(t-u-10)^2 \, du + 3 \times K_M \times \int_d^{t-10} E(u)(t-u-10)^2 \, du$$
$$= 3 \times K_M \times \int_0^d f(t-u-10)^2 \, du + 3 \times K_M \times \int_d^{t-10} 0(t-u-10)^2 \, du$$
$$= K_M \times f \times \left[(t-10)^3 - (t-10-d)^3\right]$$

(7.14)

It should be noted that lifetime risk of mesothelioma and lung cancer is calculated based on Eqs. (7.10) and (7.11) in the following way:

$$\text{Risk} = \sum_{t=10}^{S} \left(I_M(t) + I_L(t)\right) l_{A+t} \qquad (7.15)$$

where S is the time since first exposure until the maximum age considered, A is the age of first exposure, and l_{A+t} is the number of the population survivors to the age of $A + t$. The maximum age of 100 years is usually considered (with 85 years and higher being the last life table age group taken into consideration).

The summation in Equation (7.15) is theoretically assumed to be performed by a calendar year. However, the lifetables with the l_{A+t} values are usually incremented by 5- or 10-year intervals. In this case, the incidences should be taken for specific incremented time intervals. For example, 5-year incidence rate of cancer from Eqs. (7.10) and (7.11) should correspond to 5-year lifetables. The annual incidence $I_M(t)$ in this case is being multiplied by the step of the life table (5 or 10 years). Similarly, yearly incidence rate $I_L(t)$ can be recalculated to 5 years lifetable step if instead of yearly lung cancer baseline incidence rate in Eq. (7.11), 5 years baseline incidence of lung cancer $I_L(a,y)$ would be used.

The US EPA (US EPA/Nicholson 1986) evaluated lung cancer potency coefficients for the proposed model based on the information available for their study. For chrysotile cohorts, a geometric mean of K_L was found at the level of 0.098×10^{-2} (95% CI 0.028×10^{-2} to 0.34×10^{-2}) for mining and milling operations, $0.023*10^{-2}$ (95% CI 0.010×10^{-2} to 0.51×10^{-2}) for friction products manufacturing, and 2.0×10^{-2} (95% CI 0.96×10^{-2} to 4.2×10^{-2}) for textile production. For amosite insulation products cohorts, the geometric mean for K_L was estimated at 4.3×10^{-2} (95% CI 0.84×10^{-2} to 7.4×10^{-2}). No pure crocidolite cohorts were apparently involved in the study.

For evaluation of mesothelioma potency coefficients, only four epidemiological studies were considered (US EPA/Nicholson 1986). For insulation workers, K_M was estimated at 1.5×10^{-8}, for textile workers, 1.0×10^{-8}, for amosite factory workers, 3.2×10^{-8}, and for cement factory workers, 1.2×10^{-8}. No pure chrysotile cohort data was available for the US EPA study.

Based on these results, the US EPA recommended using uniform potency coefficients across the different fiber types and dimensional distributions, in particular, $K_L = 1.0 \times 10^{-2}$ and $K_M = 1.0 \times 10^{-8}$. In fact, Nicholson came to believe that the mesothelioma potency factor of amosite and chrysotile were identical, but that crocidolite had a potency of about five times higher.

From those averaged coefficients, utilizing the lifetables and smoking data of US males and females from the 1977s, the Inhalation Unit Risk (IUR) was calculated as 0.23 per 1 f/cc per 1,000,000 population as an excess lifetime cancer risk increase during the lifetime of populations exposed uniformly and constantly during their life span. This estimate was published in 1988 and is currently used in the US Environmental Protection Agency Integrated Risk Information System (IRIS) (IRIS 1988).

Based on this approach, asbestos-related lifetime cancer risks for a 24-hour per day and 7-day per week lifetime exposure can be calculated as follows:

$$\text{RISK} = 0.23 \times f \times 1{,}000{,}000 \tag{7.16}$$

Where RISK is the excess number of individuals expected to develop cancer per each population of 1,000,000, over their lifetimes; 0.23 is the IUR (Inhalation Unit Risk) for asbestos; and f is an assumed constant continuous lifetime exposure to asbestos based on PCM criteria. It should be noted that the estimation of IUR at the level of 0.23 can be considered as some kind of an "average" value between the various mineral types of asbestos. Better estimations of the parameters for specific mineral types are given in Chapter 12 of this book, as well as in other publications (for example, in Korchevskiy et al. 2020).

In 2008, the US EPA (US EPA 2008) proposed that their dose–response model (IRIS) be expanded to take into account two important modifying factors: duration and starting (onset) age of exposure, rather than average lifetime exposure. Instead of utilizing the constant IUR equal to 0.23, the US EPA recommended that the following value be applied:

$$\text{UR}_{a,\,\text{RT}} = k_1 \left[1 - \exp\left(-k_2 \times D\right)\right] \tag{7.17}$$

where $\text{UR}_{a,D}$ = Unit risk for a continuous exposure beginning at age, a, of onset, and extending for a duration of D years,

$$
\begin{aligned}
k_1 &= b_1 + b_2 \times \exp(-a/b_3), \quad \text{and} \\
k_2 &= b_4 + b_5 \times \exp(-a/b_6), \quad \text{where} \\
b_1 &= -0.0176401, \qquad b_2 = 0.2492567 \\
b_3 &= 24.7806941, \qquad b_4 = 0.0415839 \\
b_5 &= 0.0039973, \qquad b_6 = -18.2212632
\end{aligned}
\tag{7.18}
$$

This approximation, as the US EPA stated, is completely based on the previously quoted US EPA publication from 1986 and is intended to reproduce the age-specific calculations performed in that study. The derivation of Equations (7.16) and (7.17) was empirical, with exponential functions fitted to the numerical data reported by Nicholson (US EPA/Nicholson 1986).

It is noteworthy that the original IRIS Equation for asbestos risk is linear with an average lifetime cumulative dose (IRIS 1988). However, with the introduction of US EPA's correction coefficient, the formula is linear with exposure intensity (if all other parameters of exposure are fixed), but not always with the cumulative dose.

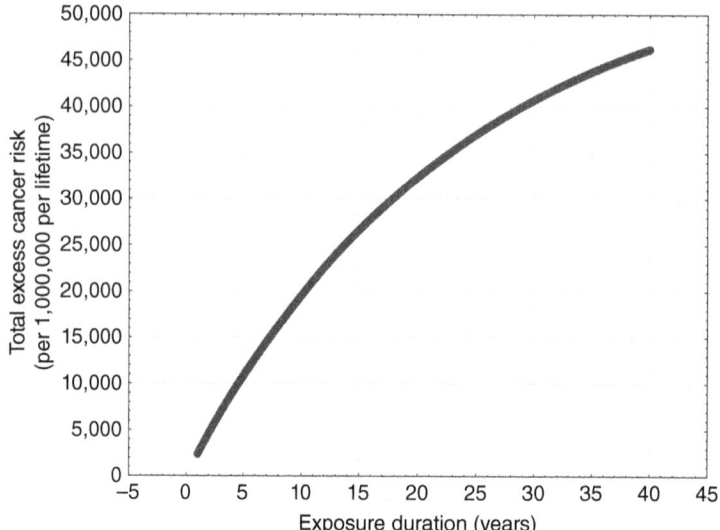

Figure 7.4 The lifetime cancer risk by theoretical formula from the US EPA IRIS as a function of exposure duration (exposure onset age 30 years, exposure intensity 1 f/cc) (calculated up to the age group of 85–100).

Figure 7.4 illustrates the theoretical relationship between risk level and exposure duration, assuming a constant exposure onset age of 30 years and an exposure intensity of 1 f/cc.

Similarly, the total lifetime cancer risk as a function of exposure onset age is demonstrated in Figure 7.5 (exposure duration of 5 years and exposure intensity of 1 f/cc are assumed).

Later, the US EPA used the Peto model in various toxicological evaluations.

In particular, the Peto model was used to evaluate the mesothelioma mortality for Libby, Montana workers; the estimation of $k = 2$ for the exponent was found as the best fit for the data (that corresponds to the third power in Eq. (7.12), in agreement to the Peto model) (US EPA 2014). Lags of 10 and 15 years were tested. The additional term for asbestos fiber clearance $\exp(-\lambda \times t)$ was also introduced as a variation of Peto's model. With the decay coefficient $\lambda = 0.15$, the exponent of $k = 5.4$ was fitting to the data, and with $\lambda = 0.068$, k was found to be equal to 3.9. Obviously, introduction of the decay term would significantly alter the Armitage–Doll basis of Peto's model. In general, the proposed dependence of the exponent k on the clearance coefficient λ is highly questionable. For example, Korchevskiy and Wylie (2022) demonstrated that the coefficient λ for chrysotile can be estimated as 6.45 (95% CI 5.35, 8.01). In this case, the exponent k for Peto's model would increase even higher than for amphiboles, but it was apparently never observed in chrysotile cohorts.

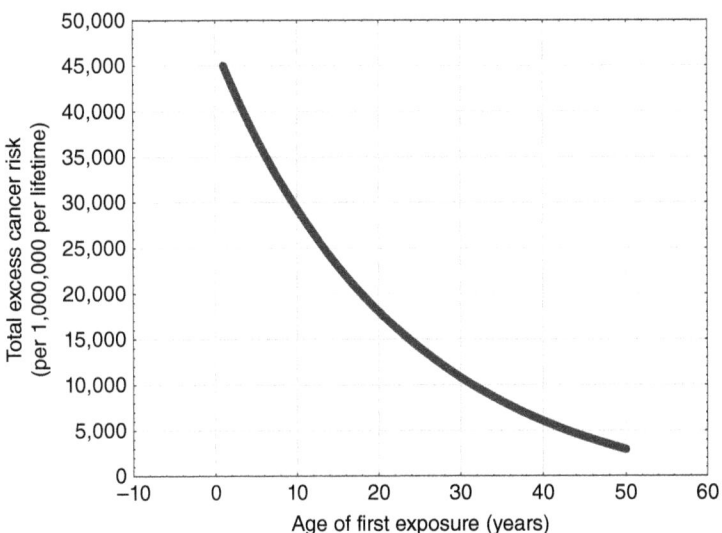

Figure 7.5 Lifetime cancer risk by theoretical formula from the US EPA IRIS as a function of exposure onset age (exposure duration 5 years, exposure intensity 1 f/cc) (calculated up to the age group of 85–100).

The US EPA also utilized the Peto mesothelioma model (without decay coefficient and with $k = 2$) for chrysotile risk evaluation (US EPA 2020).

For lung cancer, the US EPA used a modified version of Peto model in the following formulation:

$$RR = \alpha \left(1 + CE \times K_L\right) \qquad (7.19)$$

where

RR = Relative risk of lung cancer
CE = Cumulative exposure to asbestos (f/cc-yrs), equals the product of exposure concentration (f/cc) and the duration of exposure (years), lagged for 10 years
K_L = Lung cancer potency factor (f/cc-yrs)$^{-1}$.
α = the ratio of baseline (unexposed) risk in the study population compared to the reference population. It is emphasized in the document that if the reference population is well matched to the study population, α is usually assumed to be constant = 1 and is not treated as a fitting parameter. If the general population is used as the reference population, then α may be different from 1 and is treated as a fitting parameter.

It should be noted that the US EPA used various values of the parameter α for several of the considered cohorts. In particular, the value of $\alpha = 1.35$ was proposed for the South Carolina asbestos textile cohort, $\alpha = 1.18$ for North Carolina asbestos textile cohort, and $\alpha = 1.15$ for Quebec mining and milling cohort. It should be seen that higher values of α assumed for the lung cancer model (Eq. 7.19) would correspond to the lower levels of lung cancer potency factor K_L and vice versa. It makes it reasonable to utilize the estimations of $\alpha = 1$ and corresponding coefficients K_L for risk assessment purposes.

Berman and Crump Meta-analysis Based on Peto Model

In 2003 and 2008, Berman and Crump published a comprehensive meta-analysis based on updated data of asbestos-related disease epidemiology (Berman and Crump 2003, 2008a, 2008b).

To update the potency factors from the US EPA studies, Berman and Crump used generalized models derived from the original Peto model.

In particular, for mesothelioma, Berman and Crump utilized the following equation:

$$I_M(t) = 3 \times K_M \times \int_0^{t-10} E^K(u)(t-u-10)^z \, du \tag{7.20}$$

where the notation is as in Eq. (7.12), and K and z are the parameters that can be found by the maximum-likelihood estimation method. The parameter K allows testing for nonlinearity, with $K = 1$ suggesting a linear model by exposure intensity, $K<1$ and $K>1$ introducing respectively supralinear and sublinear dose–response relationship. The parameter $z = 2$ corresponded with the original assumptions from Peto's model (as adapted by the US EPA).

Parameter z was tested on three epidemiological cohorts: Wittenoom crocidolite miners (fitted value $z = 1.83$, for test of $z = 2$, $P = 0.29$, not statistically significant), Quebec miners and millers at Thetford Mines, the city of Asbestos, and factory workers ($z = 2.12$, $P = 0.79$, not statistically significant), and South Carolina asbestos textile workers ($z = 1.21$, $P = 0.56$, not statistically significant). The statistical tests mean that the value of $z = 2$ in the Peto model appears to be not fully invariant between cohorts.

The test for $K = 1$ was statistically significant at Wittenoom and Thetford mines, but not significant for the city of Asbestos, Quebec factory workers, and South Carolina. This means that Berman and Crump did not reject the nonlinearity of mesothelioma risk by exposure intensity at least for some of the cohorts. Berman and Crump suggested, however, that the superlinearity in several of the tested cohorts could be a result of the bias of exposure estimation, as was mathematically demonstrated, for example, in Crump (2005). Virtually all European and Chinese case-control studies generally have apparent superlinearity except when lung fiber burden is considered to be proportional to the dose rather than industrial hygiene or other estimation of exposure (Rödelsperger et al. 2001; Gilham et al. 2016). Errors in exposure reconstruction tend to flatten exposure-response relationships, and therefore tend to make the relationships appear non-linear, just as a shuffling of exposure values in a dose–response relationship would flatten the dose–response.

For lung cancer relative risk, Berman and Crump used the equation identical to Eq. (7.18), with parameter α being tested by maximum likelihood estimation (MLE). Berman and Crump also found that the hypothesis that $\alpha = 1$ in the lung cancer model was rejected for several cohorts, signaling above background lung cancer rates in the cohorts higher than the assumed average general population level in the corresponding age-group.

For the original assumptions of Peto's model ($K = 1$, $z = 2$, and $\alpha = 1$), Berman and Crump calculated potency coefficients (K_L and K_M) related to different environments (mineral types of fibers and specific cohorts). A coefficient of $\alpha = 1$ means that baseline cancer risks were identical to the general population, likely implying similar smoking rates. Selected potency coefficients from the Berman and Crump study are listed in Tables 7.2 and 7.3. It should be noted that for lung cancer coefficients K_L the value of $\alpha = 1$ is assumed. In addition, the values of the coefficients α found from the MLE and included in the table.

The potency factors found by Berman and Crump demonstrated a significant difference between cancer potency for various fiber mineral types. For example, for chrysotile mining and milling, the mesothelioma potency factors K_M were found in the interval from 0.0065 to 0.029×10^{-8}. To the contrary, the potency for crocidolite was found in the range from 11 to 14×10^{-8}. The potency factor for amosite was between 2.6 and 5.7×10^{-8}. For lung cancer, the difference in potency between chrysotile and amphiboles was less pronounced, but still significant.

Lenters (2011) criticized the meta-analysis performed by Berman and Crump (2008b) arguing that the difference between chrysotile and amphibole potency factors for lung cancer could be significantly affected, and partly explained, by insufficient quality of exposure assessment in some of the studies involved. Berman and Case (2012) evaluated the approach of Lenters and found it to be not fully valid, suggesting that the ratio between K_L levels derived from "best studies" and "all studies" in the Berman and Crump meta-analysis would not exceed a factor of 2.5, and is equal to 1, if the South Carolina textile cohort were excluded from consideration. Berman and Case called the approach of Lenters the "overreliance on a single study" (meaning that South Carolina data actually drives Lenters' conclusions. At the same time, the "outlier" quality of the South Carolina textile may have been caused by significant amphibole exposure by the workers. Though the

Table 7.2 Lung cancer dose–response coefficients (K_L).

Fiber type	Operation	Cohort	K_L*100	90% Confidence interval	The maximum likelihood estimation of the parameter α
Chrysotile	Mining and milling	Quebec mines and mills	0.029	0.019–0.051	1.15
		Italian mine and mill	0.051	0–0.57	0.937
	Friction products	Connecticut plant	0	0–0.61	1.49
	Cement manufacture	New Orleans plants	0.25	0–0.70	1.14
	Textile	South Carolina plant	1.8	1.1–3.7	1.35
		South Carolina plant (different cohort)	1	0.44–2.5	1.07
Crocidolite	Mining and milling	Wittenoom	1.1	0.75–5.3	2.81
Amosite	Insulation manufacture	Patterson, NJ factory	2.4	1.8–7.6	3.32
		Tyler, Texas factory	0.28	0–2.2	2.48
Libby amphiboles	Vermiculite mines and mills	Libby, Montana	0.26	0–1.3	1.50
		Libby, Montana (different cohort)	0.36	0.03–3.6	1.91

South Carolina asbestos textile plant was widely considered a "chrysotile only" production, it has been reported that crocidolite yarn in the amount of 2,000 lbs. was imported by this location annually and amosite was acquired here for experimental purposes in the late 1950s (McDonald 1998). However, Dement and Loomis (2023) deny that this was the case.

Additionally, the asbestos textile plant in Charleston, South Carolina was situated in close proximity to a Navy shipyard that employed at some point up to 29,000 workers. This shipyard was one of the largest bases in the United States, apparently employing up to 40% of the total population of the city. It was demonstrated that many of the workers at this shipyard were exposed to asbestos and it

Table 7.3 Mesothelioma dose–response coefficients (K_M).

Fiber type	Operation	Cohort	K_M*10^8	90% Confidence interval
Chrysotile	Mining and milling	Asbestos, Quebec	0.012	0.0065–0.021
		Thetford mines	0.021	0.014–0.029
	Friction products	Connecticut plant	0	0–0.12
	Cement manufacture	New Orleans plants	0.2	0.065–0.45
	Textile	South Carolina plant	0.15	0.047–0.33
		South Carolina plant (different cohort)	0.088	0.0093–0.32
Crocidolite	Mining and milling	Wittenoom	12	11–14
Amosite	Insulation manufacture	Patterson, NJ factory	3.9	2.6–5.7

affected the lung cancer mortality rate in the whole county, in particular, during the period from 1950 to 1969. During this period, in the county where the shipyard was situated, the lung cancer incidence was 66.5 cases per 100,000 white males vs. 37.8 cases in the state (age adjusted). In 1975, the US EPA found amphibole asbestos in process water from the North Charleston asbestos textile plant (US EPA 1976).

In the asbestos lung burden of workers from the South Carolina textile cohort, 6.8% of the fibers in lungs was found to be tremolite, vs. 1.62% in controls, perhaps indicating more than the usual tremolite concentrations in long-fibered chrysotile (Green et al. 1997). Amosite/crocidolite and anthophyllite were also found in the lungs of the South Carolina textile workers. The fiber concentration of amosite/crocidolite exceeded 1 million fibers per g of dry lungs in 28.5% of the workers sampled vs. 12.9% in controls in Charleston (Green et al. 1997; Roggli 2013), indicating substantial lung cancer and mesothelioma risk in these workers. Both chrysotile and tremolite lung burdens were positively associated with the grade of pulmonary fibrosis in workers; however, the concentration of tremolite in the lung was a better predictor of pulmonary fibrosis than the chrysotile concentration (Green et al. 1997).

Berman and Crump published several studies attempting to estimate the potency of various size categories of asbestos fibers.

Berman et al. (1995) developed a model of a potency for asbestos structure to induce lung tumors and mesothelioma in rats following inhalation and determined that zero lung cancer potency was attributed to structures shorter than

5 μm. The authors proposed a so-called "optimum exposure index" assigning relative potency of:

- 0.0017 for structures <0.3 μm in width and between 5 and 40 μm in length;
- 0.853 for structures <0.3 μm in width and more than 40 μm in length; and
- 0.145 for structures ≥5 μm in width and more than 40 μm in length (p = 0.24).

The authors expressed some skepticism for this model, which assigned relatively high potency to the very thick asbestos structures (suggesting that this could be an artifact caused by limitations in the data). Alternatively, another index was proposed for the relative potency of various asbestos fibers in producing lung tumors:

- 0.0024 for structures <0.4 μm in width and between 5 and 40 μm in length and
- 0.9976 for structures <0.4 μm in width and more than 40 μm in length (p = 0.09).

Remarkably, both models revealed low statistical significance, with $p > 0.05$. The very limited number of mesotheliomas in the experiment has not allowed the authors to develop a model for mesothelioma potency. The hypothesis that mesothelioma incidence is proportional to lung tumor incidence was rejected by Berman et al. for the combination of amphibole and chrysotile fibers, but not separately for each of the categories. The amphiboles were found to be more potent than chrysotile in producing lung tumors.

In 2001, Berman and Crump (Berman and Crump 2001) proposed a methodology to incorporate dimensional metrics into the risk assessment for asbestos fibers. In particular, for risk assessment purposes, they recommended the following metric with weights assigned to the fibers of different sizes:

$$C_{opt} = 0.003 C_S + 0.997 C_L \qquad (7.21)$$

where: C_S is the concentration of asbestos structures between 5 and 10 μm in length that are also thinner than 0.5 μm; and C_L is the concentration of asbestos structures longer than 10 μm that are also thinner than 0.5 μm.

In 2008, Berman and Crump published a landmark study (2008b) where they explained the variability between the potency factors in different human cohorts as in their previous publication (2008a) by fiber type and dimensionality factors. Berman and Crump not only differentiated the cohorts by mineral types and dimensional characteristics of fibers, but also estimated the fraction of amphibole asbestos in various cohorts with mixed exposure. In particular, Berman and Crump assumed that the amphibole fraction for Quebec mines and mills exposure was 0–4%, for the Connecticut friction product plant was 0–2%, for the South Carolina textile plant was 0–2%, for the Rochdale factory was 2.5–15%, for the Wittenoom mine was 95–100%, etc. It should be noted that the fraction of amphiboles in the exposure assumed by Berman and Crump, is approximate and appears not fully comparable between studies.

Based on the available dimensional data, Berman and Crump utilized the following major metrics:

- The "thin" metric ($L > 10\,\mu m$, $W < 0.4\,\mu m$);
- The "thinnest" metric ($L > 10\,\mu m$, $W < 0.2\,\mu m$)
- The "all-widths" metric (fibers and bundles, $L > 10\,\mu m$, $W < 3\,\mu m$),
- The PCME fibers ($L > 5\,\mu m$, $W > 0.20\,\mu m$).

It should be noted that the definition of PCME fibers in this study was apparently not fully adequate (see Chapter 44). Based on the visibility of fibers by optical microscopy, it was estimated previously that the cutpoint for width in the PCME category should be about $0.15\,\mu m$ for chrysotile and as low as $0.06\,\mu m$ for amphiboles compared with $0.25\,\mu m$ for both fiber types as it was usually considered by NIOSH (2019). This means that the estimation of potency factors for the PCME category in the Berman and Crump study might be overly conservative (because the true PCME exposure was likely higher than that indicated by a cutpoint of $0.25\,\mu m$.). It should also be noted that Berman and Crump used $0.2\,\mu m$ for their estimation of the PCME fraction in exposure for different cohorts.

To determine potency factors for asbestos fibers in different size categories, Berman and Crump used the following model to fit their updated potency factors for various mineral types of asbestos:

$$K_j = K_A^* \times \left(f_{Lj} + \text{rps} \times f_{Sj}\right) \times \left[f_{Aj} + \text{rpc} \times \left(1 - f_{Aj}\right)\right] / f_{\text{PCME}j} \qquad (7.22)$$

where K_j is the observed potency factor in the epidemiological cohort j, K_A^* is the corresponding potency of long amphibole fibers in a specific width category, f_{Lj} is a fraction of long amphibole fibers for the cohort exposure (the fraction was conditionally assumed to be equal for chrysotile and amphiboles), f_{Sj} is a fraction of short amphibole fibers for the cohort, f_{Aj} is a fraction of amphibole fibers in the cohort, $f_{\text{PCME}j}$ is a fraction of PCME fibers for the cohort, rps is a relative potency of short to long fibers, rpc is the relative potency of pure chrysotile to amphibole fibers. Equation (7.22) was used separately for lung cancer and mesothelioma.

The meaning of Eq. (7.22) is in the determination of several components that predict the total number of mesothelioma cases in each cohort. This number is assumed to be composed of the cases caused by long and short amphiboles of a specific width group, and long and short chrysotile of the same width group. Different width groups are not combined together, but Berman and Crump attempted to model the overall potency of fibers separately for each age group, checking how well each of the groups would predict the total potency. Instead of using different potency factors for amphiboles and chrysotile, and short and long fibers, Berman and Crump utilized potency ratios of amphiboles vs. chrysotile and long vs. short fibers for Eq. (7.22).

Berman and Crump utilized various estimations of the amphibole fraction in asbestos exposure f_{Aj} for each cohort, for example 0–4% for Quebec mines and mills, 2.5–15% for the Rochdale textile plant, and 95–100% for Wittenoom, Australia. To estimate size distribution parameters (f_{Lj} and f_{Sj}) for the cohorts, Berman and Crump paired each of the cohorts with asbestos size distribution data available from several publications: Dement and Harris (1979); Gibbs and Hwang (1980); Hwang and Gibbs (1981); and Dement et al. (2007). Actually, for the 16 cohorts taken into account in this study (corresponding to the cohorts in Tables 7.2 and 7.3 of this chapter), Berman and Crump used nine size distributions of asbestos fibers, often not specifically corresponding to a certain cohort, but assessed as fitting the best, according to the information available.

Fitting the equation to the epidemiological data, Berman and Crump determined the potency coefficients for the asbestos structures belonging to those two groups. These are listed in Table 7.4.

It should be noted that Berman and Crump found the results for the "thinnest" category of fibers (with the length longer than 10 μm and width less than 0.2 μm) to be not fully statistically consistent in their study; in particular, for this category, the relative mesothelioma potency of longer ($L > 10$ μm) to shorter ($5 < L < 10$ μm) amphibole fibers was not statistically significantly different from 1, that Berman

Table 7.4 Dose–response coefficients for the selected metrics.

Category	Mesothelioma (K_M*10^8, 95% CI)		Lung cancer (K_L*10^8, 95% CI)	
	Amphiboles	Chrysotile	Amphiboles	Chrysotile
The "thin" fibers: Width <0.4 μm, length >10 μm	30.8 (16.5, 61.5)	0 (0, 0.34)	7.7 (1.6, 26.6)	0.49 (0.092, 1.4)
The "thinnest" fibers: Width <0.2 μm, length >10 μm	32.0 (0, 89.9)	0 (0, 0.27)	24.5 (7.6, 66.3)	0.38 (0, 1.3)
The "all width": Width <3 μm, length >10 μm	13.8 (3.5, 26.3)	0 (0, 0.14)	2.7 (0.56, 9.9)	0.29 (0.083, 0.73)
The PCME Width >0.2 μm, length >5 μm	8.5 (3.5, 19)	0.009 (0, 0.16)	1.4 (0.23, 5.9)	0.20 (0, 0.55)

and Crump considered to be in contradiction to the toxicological mechanisms of asbestos carcinogenesis.

Berman and Crump suggested that central tendency potency factors for amphibole and chrysotile fibers depend on the size category. In particular, the authors found that mesothelioma potency between amphibole and chrysotile statistically differs for all considered size categories. Also, lung cancer potencies for chrysotile and amphiboles were statistically significantly different for the "long, thin" category of fibers. It is worth mentioning that, assuming a certain fraction of amphiboles in different cohorts, Berman and Crump demonstrated that mesothelioma potency of long chrysotile fibers was not statistically different from zero for all metrics: long, thin; long, thinnest; long, all width; and so-called "PCME."

Wylie and Korchevskiy (2023) reevaluated the estimations performed by Berman and Crump (2008b) for cancer potency factors in fibers with various size characteristics. The full continuous distribution of dimensional data was utilized as described in Wylie et al. (2022). The full database allowed for better characterization of the size distribution for mineral types of fibers analyzed in the Berman and Crump study. It was demonstrated that the data used by Berman and Crump apparently underappreciated the highly significant input of very thin fibers into the average mesothelioma potency for each category.

In particular, Berman and Crump used discrete categorization of the fiber sizes, being unable to calculate fractions of fiber with certain length and width. Some of the size categories used by Berman and Crump actually overlapped (like all width and "thin" fibers).

Instead, Korchevskiy and Wylie (2022) were able to utilize a virtually continuous distribution of fiber size categories. For example, it can be demonstrated that potency factors K_M for non-serpentine mineral types of fibers can be approximated by the following formula:

$$K_M = 10^{\left(-1.62 + 1.87 \log_{10} \text{Median}\left(\frac{L^{0.44}}{W^{1.45}}\right)\right)} \quad (7.23)$$

where L (µm) is the length and W (µm) is the width of individual fibers, and Median is the Median of the values for a certain set of fibers.

Equation (7.23) fits the K_M values calculated for crocidolite, amosite, anthophyllite, fluoro-edenite, Libby amphiboles, and balangeroite with $R = 0.90$, $R^2 = 0.81$, $p < 0.05$. The potency factors from Berman and Crump (2008a) were used for the Eq. (7.23) fitting, except for fluoro-edenite for which the K_M potency factor was used as in Wylie and Korchevskiy (2023) (by extrapolation from the Hodgson and Darnton R_M value).

Table 7.5 demonstrates the comparison between estimations of potency factors for various size categories of fibers from Berman and Crump, and using the model from Eq. (7.23). It should be noted that Berman and Crump used the PCME

Table 7.5 Potency factors for nonserpentine minerals.

Size category	Mesothelioma potency K_M by Berman and Crump (average, 95% confidence interval) ($\times 10^8$)	Modeled potency K_M ($\times 10^8$) (Eq. 7.23, Wylie and Korchevskiy 2023)
The "thin" fibers: Width <0.4 μm, length >10 μm	30.8 (16.5, 61.5)	24.20
The "thinnest" fibers: Width <0.2 μm, length >10 μm	32.0 (0, 89.9)	75.56
The "all width": Width <3 μm, length >10 μm	13.8 (3.5, 26.3)	7.82
The PCME: Width >0.2 μm, length >5 μm	8.5 (3.5, 19)	0.45
The updated PCME: Width >0.06 μm, Length >5 μm)	—	7.55

category assuming that no fibers thinner than 0.2 μm would be visible on optical microscopy. However, recently, it was suggested that non-serpentine fibers would be visible at the PCM down to the width of 0.06 μm (Lenters et al. 2011). It can be seen from Table 7.5 that for the conditional "PCME" category, the difference in the Berman and Crump estimation of mesothelioma potency and our model can be at least partly explained by the resolved "visibility" issue.

The plot of estimations of total cancer risk for various levels of amosite and chrysotile cumulative exposure, with various combinations of duration and intensity, based on the Berman and Crump coefficients for the Peto model (as in Table 7.3, for "PCME" category) is demonstrated in Figure 7.6 (onset age 30 years). Based on our understanding, the "PCME" potency factors used by Berman and Crump should be applied to actual amphibole and chrysotile exposure category as supposed to be seen by PCM microscopy (fibers longer than 5 μm and wider than 0.06 μm for amphiboles, and fibers longer than 5 μm and wider than 0.15 μm for chrysotile).

In 2011, Berman revisited the differences between the estimations of inhalation unit risk (IUR) for asbestos fibers based on the US EPA IRIS (PCME) metric and Berman and Crump size-specific "protocols" (Berman 2011). Berman updated the IUR values assuming the most up-to-date US mortality statistics for 2000, smoking mortality statistics, and smoking rate estimates. In particular, for "long, thin" fibers, the IUR of 0.05 for chrysotile asbestos and 5.82 for amphibole asbestos were calculated. For the "long, all width" metric, the IUR for chrysotile was 0.03 and for amphiboles 2.75.

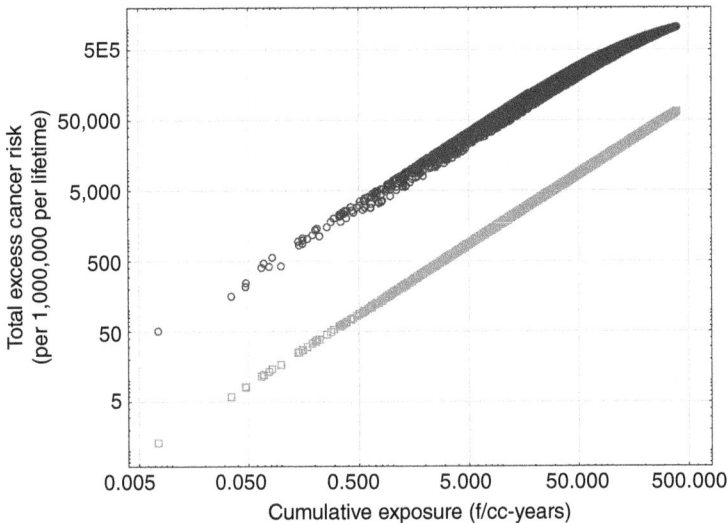

Figure 7.6 Estimations of total cancer risk for various levels of amosite and chrysotile exposure based on the Peto model with Berman and Crump coefficients. The circles – amosite, the squares – chrysotile.

These values can be compared to other published IURs. Korchevskiy et al. (2020) estimated the IUR for chrysotile at the level of 0.025 vs. 1.7 for crocidolite (average between the Berman and Crump, and the Hodgson and Darnton approaches). The US EPA used an IUR value of 0.23 as an average for all types of asbestos fibers (data from Nicholson 1986). The recent risk evaluation for chrysotile determined the upper bound potency for chrysotile at the level of 0.16 (US EPA 2020). However, the US EPA (2014) estimated the IUR for Libby amphiboles at the level of 0.17, virtually identical to their estimation for chrysotile, which apparently contradicts the scientific data (in particular, the Berman and Crump ratio between amphibole and chrysotile asbestos potency).

In essence, Berman and Crump significantly advanced the science of asbestos risk assessment, by updating the parameters of the Peto model and involving additional variables (like fiber length and width) to explain the variability of potency factors. Berman and Crump also tested the Peto model in a detailed epidemiological basis for various cohorts, confirming the model in general, and exploring several additional parameters that can modify the original model (in particular, the parameter α for lung cancer and parameters K and Z for mesothelioma). The data on fiber dimensionality that became available after the Berman and Crump publications allow for potential further improvements in the estimations of the potency factors fully incorporating the mineral type and the length and width of fibers into the general dose–response equation suggested by Peto.

References

Acheson, E. and Gardner, M. (1979). The ill effects of asbestos on health. In: *Asbestos* (ed. W. Simpson). Vol. 2, Final Report of the Advisory Committee on Asbestos. London: HMSO.

Armitage, P. and Doll, R. (1954). The age distribution of cancer and a multi-stage theory of carcinogenesis. *British Journal of Cancer* 8 (1): 1–12. https://doi.org/10.1038/bjc.1954.1.

Berman, D.W. (2011). Apples to apples: the origin and magnitude of differences in asbestos cancer risk estimates derived using varying protocols. *Risk Analysis* 31 (8): 1308–1326. https://doi.org/10.1111/j.1539-6924.2010.01581.x.

Berman, D.W. and Case, B.W. (2012). Over reliance on a single study: There is no real evidence that applying quality criteria to exposure in asbestos epidemiology affects the estimated risk. *Annals of Occupational Hygiene* 56 (8): 869–878. https://doi.org/10.1093/annhyg/mes027.

Berman, D.W. and Crump, K. (2001). Technical support document for a protocol to assess asbestos-related risk. Final Draft, Prepared for US EPA, # DTRS57-01-C-10044, September 4, 2001. Document Display | NEPIS | US EPA.

Berman, D. and Crump, K. (2003). Technical support document for a protocol to assess asbestos-related risk. Final Draft, Prepared for US EPA # 9345.4-06, October, 2003. Document Display | NEPIS | US EPA.

Berman D.W. and Crump, K.S. (2008a). Update of potency factors for asbestos-related lung cancer and mesothelioma. *Critical Reviews in Toxicology* 38 (Suppl 1): 1–47. https://doi.org/10.1080/10408440802276167.

Berman D.W. and Crump, K.S. (2008b). A meta-analysis of asbestos-related cancer risk that addresses fiber size and mineral type. *Critical Reviews in Toxicology* 38 (Suppl 1): 49–73. https://doi.org/10.1080/10408440802273156.

Berman, D.W., Crump, K.S., Chatfield, E.J., et al. (1995). The sizes, shapes, and mineralogy of asbestos structures that induce lung tumors or mesothelioma in AF/HAN rats following inhalation. *Risk Analysis* 15 (2): 181–195. https://doi.org/10.1111/j.1539-6924.1995.tb00312.x.

Billings, W.Z., Clifton, J., Hiller, J., et al. (2022). An axiomatic and contextual review of the Armitage and Doll model of carcinogenesis. *Spora: A Journal of Biomathematics* 8: 7–15. https://doi.org/10.30707/SPORA8.1.1647886301.817127.

Cook, P.J., Doll, R., and Fellingham, S.A. (1969). Mathematical model for the age distribution of cancer in man. *International Journal of Cancer* 4 (1): 93–112. https://doi.org/10.1002/ijc.2910040113.

Crump, K.S. (2005). The effect of random error in exposure measurement upon the shape of exposure response. *Dose Response* 3 (4): 456–464. https://doi.org/10.2203%2Fdose-response.003.04.002.

Dement, J.M. and Harris, R.L. (1979). Estimates of pulmonary and gastrointestinal deposition for occupational fiber exposures. US DHEW, NIOSH Technical Report 79–135, NTIS PB80-149644. US HEW Contract 78-2438. ntrl.ntis.gov/NTRL/dashboard/searchResults.xhtml;jsessionid=598763895e943d8bd787d32d0a70?searchQuery=PB80149644.

Dement, J.M. and Loomis, D. (2023). Manufactured doubt and the EPA 2020 chrysotile asbestos risk assessment. *American Journal of Industrial Medicine*, 66 (7): 543–553. https://doi.org/10.1002/ajim.23476.

Dement, J.M., Kuempel, E.D., Zumwalde, R.D., et al. (2007). Development of a fiber size-specific job-exposure matrix for airborne asbestos fibers. *Occupational and Environmental Medicine* 65 (9): 605–612. https://doi.org/10.1136/oem.2007.033712.

Fisher, J.C. and Hollomon, J.H. (1951). A hypothesis for the origin of cancer foci. *Cancer* 4 (5) 916–918. https://doi.org/10.1002/1097-0142(195109)4:5%3C916::AID-CNCR2820040504%3E3.0.CO;2-7.

Freedman, D.A. and Navidi, W.C. (1989). Multistage models of carcinogenesis. *Environmental Health Perspectives* 81: 169–188. https://doi.org/10.1289%2Fehp.8981169.

Gibbs, G.W. and Hwang, C.Y. (1980). Dimensions of airborne asbestos fibres. In: *Biological Effects of Mineral Fibres* (ed. Wagner, J.C.), 69–78, Lynn: IARC Scientific Publications No. 30. PMID: 7239672.

Gilham, C., Rake, C., Burdett, G., et al. (2016). Pleural mesothelioma and lung cancer risks in relation to occupational history and asbestos lung burden. *Occupational Environmental Medicine* 73 (5): 290–299. https://doi.org/10.1136/oemed-2015-103074.

Green, F.H., Harley, R., Vallyathan, V., et al. (1997). Exposure and mineralogical correlates of pulmonary fibrosis in chrysotile asbestos workers. *Occupational and Environmental Medicine* 54 (8): 549–559. https://doi.org/10.1136/oem.54.8.549.

Harding, C., Pompei, F., and Wilson, R. (2012). Peak and decline in cancer incidence, mortality, and prevalence at old ages. *Cancer* 118 (5): 1371–1386. https://doi.org/10.1002/cncr.26376.

Hobbs, M.S., Woodward, S.D., Murphy, B., et al. (1980). The incidence of pneumoconiosis, mesothelioma and other respiratory cancer in men engaged in mining and milling crocidolite in Western Australia. *IARC Scientific Publications* 30: 615–625. PMID: 7228317.

Hwang, C.Y. and Gibbs, G.W. (1981). The dimensions of airborne asbestos fibres – I. Crocidolite from Kuruman area, Cape Province, South Africa. *Annals of Occupational Hygiene.* 24 (1): 23–41. https://doi.org/10.1093/annhyg/24.1.23.

Korchevskiy, A.A. and Wylie, A.G. (2022). Dimensional characteristics of the major types of amphibole mineral particles and the implications for carcinogenic risk assessment. *Inhalation Toxicology* 34 (1–2): 24–38. https://doi.org/10.1080/08958378.2021.2024304.

Korchevskiy, A., Rasmuson, J.O., Rasmuson, E.J., et al. (2020). Inhalation unit risk (IUR) of asbestos based on available science. *Inhalation Toxicology* 32 (9–10): 372–374. https://doi.org/10.1080/08958378.2020.1829210.

Lash, T.L., Crouch, E.A., and Green, L.C. (1997). A meta-analysis of the relation between cumulative exposure to asbestos and relative risk of lung cancer. *Occupational and Environmental Medicine* 54 (4): 254–263. https://doi.org/10.1136/oem.54.4.254.

Lenters, V., Vermeulen, R., Dogger, S. et al. (2011). A meta-analysis of asbestos and lung cancer: is better quality exposure assessment associated with steeper slopes of the exposure-response relationships? *Environmental Health Perspectives* 119 (11): 1547–1555. https://doi.org/10.1289/ehp.1002879.

Loomis, D, Richardson, D.B., and Elliott, L. (2019). Quantitative relationships of exposure to chrysotile asbestos and mesothelioma mortality. *American Journal of Industrial Medicine* 62 (6): 471–477. https://doi.org/10.1002%2Fajim.22985.

McDonald, J.C. (1998), Unfinished business: The asbestos textile mystery. *Annals of Occupational Hygiene* 42 (1): 3–5. https://doi.org/10.1016/s0003-4878(98)00007-6.

Moolgavkar, S.H., Meza, R., and Turim, J. (2009). Pleural and peritoneal mesotheliomas in SEER: age effects and temporal trends, 1973–2005. *Cancer Causes Control* 20 (6): 935–944. https://doi.org/10.1007/s10552-009-9328-9.

Newhouse, M.L. and Berry, G. (1976). Prediction of mortality from mesothelial tumours in asbestos factory workers. *British Journal of Industrial Medicine* 33 (3): 147–151. https://doi.org/10.1136%2Foem.33.3.147.

NIOSH. (2019). Asbestos and other fibers by PCM. 7400, Issue 3. NMAM METHOD 7400 (cdc.gov).

Nordling, C.O. (1953). A new theory of the cancer-inducing mechanism. *British Journal of Cancer* 7 (1): 68–72. https://doi.org/10.1038/bjc.1953.8.

Peto, J. (1978). The hygiene standard for chrysotile asbestos. *Lancet* 1 (8062): 484–489. https://doi.org/10.1016/s0140-6736(78)90145-9.

Peto, J. (1980). The incidence of pleural mesothelioma in chrysotile asbestos textile workers. *IARC Scientific Publication* 30: 703–711. PMID: 7228327.

Peto, J., Seidman, H., and Selikoff, I.J. (1982). Mesothelioma mortality in asbestos workers: implications for models of carcinogenesis and risk assessment. *British Journal of Cancer* 45 (1): 124–135. https://doi.org/10.1038%2Fbjc.1982.15.

Pira, E., Pelucchi, C., Piolatto, P.G., et al. (2007). First and subsequent asbestos exposure in relation to mesothelioma and lung cancer mortality. *British Journal of Cancer* 97 (9): 1300–1304. https://doi.org/10.1038%2Fsj.bjc.6603998.

Reid, A., de Klerk, N.H., Magnani, C., et al. (2014). Mesothelioma risk after 40 years since first exposure to asbestos: a pooled analysis. *Thorax* 69 (9): 843–850. https://doi.org/10.1136/thoraxjnl-2013-204161.

Rödelsperger, K., Jöckel, K.H., Pohlabeln, H., et al. (2001). Asbestos and man-made vitreous fibers as risk factors for diffuse malignant mesothelioma: results from a German hospital-based case-control study. *American Journal of Industrial Medicine* 39 (3): 262–275. https://doi.org/10.1002/1097-0274(200103)39:3%3C262::aid-ajim1014%3E3.0.co;2-r.

Roggli, V.L. (2013). Fiber analysis vignettes: electron microscopy to the rescue. Roundtable 208, American Industrial Hygiene Conference and Exhibition, Montreal, Canada, May, 2013.

Seidman, H., Selikoff, I.J., and Hammond, E.C. (1979). Short-term asbestos work exposure and long-term observations. *Annals of the New York Academy of Sciences* 330: 61–89. https://doi.org/10.1111/j.1749-6632.1979.tb18710.x.

Selikoff, I.J., Hammond, E.C., and Seidman, H. (1979). Mortality experience of insulation workers in the United States and Canada, 1943–1976. *Annals of the New York Academy of Sciences* 330: 91–116. https://doi.org/10.1111/j.1749-6632.1979.tb18711.x.

US EPA. (1976). Asbestos fibers in natural runoff and discharges from sources manufacturing asbestos products. Final Report. Part II. EPA 560/6-76-020. October, 1976. Document Display | NEPIS | US EPA.

US EPA. (1988). Integrated Risk Assessment System (IRIS). Asbestos (CASRN 1332-21-4). https://iris.epa.gov/static/pdfs/0371_summary.pdf. Accessed April 4th, 2013.

US EPA. (2008). Framework for investigating asbestos-contaminated superfund sites. OSWER Directive #9200.0-68, September, 2008. Framework For Investigating Asbestos-Contaminated Superfund Sites - OSWER 9200.0-68 (FRAMEWORK FOR INVESTIGATING ASBESTOS-CONTAMINATED SUPERFUND SITES - OSWER 9200.0-68 (epa.gov)). Accessed January 24, 2024.

US EPA. (2014). Toxicological review of Libby amphibole asbestos. In Support of Summary Information on the Integrated Risk Information System (IRIS). EPA/635/R-11/002F, December, 2014. https://cfpub.epa.gov/ncea/iris/iris_documents/documents/toxreviews/1026tr.pdf. Accessed January 24th, 2024

US EPA. (2020). Risk evaluation for asbestos. Part 1. Chrysotile asbestos. https://www.epa.gov/assessing-and-managing-chemicals-under-tsca/final-risk-evaluation-asbestos-part-1-chrysotile. Accessed January 24th, 2024

US EPA. Nicholson W. (1986). Airborne asbestos health assessment update. United States Environmental Protection Agency Report 600/884003F. Washington, DC, June, 1986. Document Display | NEPIS | US EPA.

White, M.C., Holman, D.M., Boehm, J.E., et al. (2014). Age and cancer risk: a potentially modifiable relationship. *American Journal of Preventative Medicine.* 46 (3 Suppl 1): S7–15. https://doi.org/10.1016/j.amepre.2013.10.029.

Wittenmore, A. ed. (1977). Environmental health quantitative methods. *Proceedings of a Conference on Environmental Health*, Alta Utah (5–9 July 1976). Philadelphia, Pennsylvania, US: SIAM.

Wylie, A.G. and Korchevskiy, A.A. (2023). Dimensions of elongate mineral particles and cancer: A review. *Environmental Research* 230: 114688. https://doi.org/10.1016/j.envres.2022.114688.

Wylie, A.G., Korchevskiy, A.A., Van Orden, D.R., et al. (2022). Discriminant analysis of asbestiform and non-asbestiform amphibole particles and its implications for toxicological studies. *Computational Toxicology* 23: 100233. https://doi.org/10.1016/j.comtox.2022.100233.

8

The Hodgson and Darnton Approach to Quantifying the Risks of Mesothelioma and Lung Cancer in Relation to Asbestos Exposure

Lucy Darnton

Science Division, Epidemiology and Predictive Modelling, Bootle, Merseyside, UK

Introduction

As early as the 1930s, there had been suspicions about asbestos as a cause of the cancer mesothelioma (McDonald and McDonald 1996). Observations by Wagner published in 1960 of those working or living close to the South African crocidolite mines confirmed the link (Wagner et al. 1960), and just prior to this, it had already been shown without doubt that asbestos could cause lung cancer (Doll 1955). Other early epidemiological studies of asbestos workers in the United States and Britain published in 1964 showed positive findings and led to many further case–control and cohort studies in European and North American workers over the next decade (McDonald and McDonald 1996).

By 1970, it was clear from the emerging epidemiological evidence that crocidolite was particularly hazardous, but it wasn't until the mid-1980s that amosite as well as crocidolite was recognized as a considerably more potent cause of mesothelioma than chrysotile (Acheson et al. 1984; Doll and Peto 1985). This was further highlighted by studies of miners in South Africa and Australia in the late 1980s and early 1990s (Armstrong et al. 1988; Sluis-Cremer et al. 1992). By the mid-1990s, a very large number of epidemiological studies of asbestos-exposed worker cohorts had been published, though only a subset included quantitative estimates of cumulative asbestos exposure.

Aiming to further inform an already long-running debate about the relative potency of the three main commercial asbestos types in relation to the two principal asbestos-related malignancies – mesothelioma and lung cancer – Hodgson and

Health Risk Assessment for Asbestos and Other Fibrous Minerals, First Edition.
Edited by Andrey Korchevskiy, James Rasmuson, and Eric Rasmuson.
© 2024 John Wiley & Sons, Inc. Published 2024 by John Wiley & Sons, Inc.

Darnton embarked on a meta-analysis seeking to quantify these differences (Hodgson and Darnton 2000). This task was considered particularly apposite given clear indication from the literature of the very considerably higher potency of amphiboles than chrysotile in relation to mesothelioma induction that had not been accounted for in risk assessments available at the time. Obvious inconsistencies in the evidence for whether the mineral type of fibers also affected lung cancer risk potencies also called for this to be quantitatively assessed. Around 20 years since the original meta-analysis, the available evidence has been reviewed and an updated assessment was published in 2023 (Darnton 2023).

Overview of the Hodgson and Darnton Approach

The aim of the Hodgson and Darnton meta-analysis was to investigate the consistency of evidence about the potency (risk per unit dose) of the three main commercial asbestos types (crocidolite, amosite, and chrysotile) in relation to mesothelioma and lung cancer. Focus was on the extent to which variation in potency between different studies could be explained by the mineral type of asbestos fibers alone, or whether other factors, such as the industrial process, may also be relevant.

In view of the potentially large number of epidemiological studies, but wide variation in methods, an initial consideration was to choose a meta-analytical method that allowed as much of the available evidence as possible to be considered whilst minimizing the impact of limitations and uncertainties in that evidence.

Internal exposure–response assessments for lung cancer are a common feature of epidemiological studies of asbestos-exposed cohorts where sufficient data on exposure are available. A natural approach is therefore to summarize the available regression slopes in relation to explanatory variables such as the mineral type of fibers and industrial process (Berman and Crump 2008). These kinds of analyses are generally more problematic for mesothelioma given that it tends to be a rare malignancy and numbers of cases have often been relatively low among even among study populations with high asbestos exposures. The number of studies available with internal exposure–response regression analyses based on quantitative assessments of cumulative exposure is relatively small. A further statistical consideration relates to the tendency for bias toward the null in such analyses if exposure is misclassified, and there is clear potential for this given the limitations in the way exposures were assessed (as discussed in Chapter 10 later in this volume).

In the light of these various limitations, Hodgson and Darnton derived cohort-level measures of excess cancer mortality and evaluated these levels against the corresponding cohort average cumulative exposures. This approach is arguably less susceptible than the within-cohort exposure–response analyses to downward bias in the estimated potency (risk per unit cumulative exposure) of different asbestos mineral types. It also meant that some studies without internal assessments of mesothelioma and lung cancer exposure–response could be included in the meta-analysis. The method was applied to published studies of exposed cohorts which assessed both mesothelioma and lung cancer mortality, and which reported – or allowed the derivation of – quantitative estimates of cumulative exposure, as well as information about the mineral type of fibers. Studies where the exposure was exclusively to one kind of mineral type of fiber were particularly central to this approach with studies being classified as "pure fiber" (crocidolite, amosite, or chrysotile) or "mixed fiber" exposures.

Metrics and Data Requirements

Lung Cancer

Cohort studies reporting Standardized Mortality Ratios (SMRs) for lung cancer (SMRLC) allow the excess deaths attributed to asbestos exposure to be determined and this can be expressed as a percentage of expected lung cancer deaths (given the mortality rates for the population from which the cohort came). Dividing by the average cumulative asbestos exposure then gives a standardized measure of the asbestos-related lung cancer mortality per unit of cumulative asbestos exposure (R_L) that can be compared across cohorts:

$$SMR_{LC} = 100 \times \frac{O_{LC}}{E_{LC}}$$

$$R_L = 100 \times \frac{O_{LC} - E_{LC}}{E_{LC} \times CE}$$

Where,

O_{LC} = Observed lung cancer deaths;
E_{LC} = Expected lung cancer deaths; and
CE = Mean cumulative asbestos exposure (f/cc-yrs).

The effect of asbestos exposure is likely to be such that it approximately multiplies the underlying (background) lung cancer risk (produced mainly by cigarette smoking) and this means the relative risk of lung cancer (as characterized by the SMR in cohort studies) for the asbestos exposed vs. those not exposed would be the same in smokers and nonsmokers. Therefore, SMRs from different cohort studies with different smoking prevalences should reflect only the effect of asbestos providing the SMRs themselves are calculated using reference rates from an appropriate population with the same smoking prevalence as the worker cohort.

Mesothelioma

Mesothelioma is relatively rare in general populations even in countries that used asbestos extensively. Though background rates of mesothelioma have been estimated for nonexposed populations, the expected number of mesothelioma deaths in relatively small worker cohorts is typically low and can be difficult to calculate accurately due to a lack of reliable reference mortality rates. Most studies therefore simply reported the number of observed mesothelioma deaths, usually assessed on the best evidence available about the cause of death rather than relying only on death certificate information.

Peto originally described the relationship in which mesothelioma incidence increases in proportion to a power of time (years) since the start of asbestos exposure (Peto et al. 1982). This implies that in the long-term, mesothelioma rates increase rapidly and in proportion to deaths from all causes (see Chapter 7). After sufficient follow-up (around 20 years), the proportion of total expected deaths in a cohort due to mesothelioma will therefore remain broadly constant over time. Assuming workers in different cohorts started their exposure at a similar age (e.g. toward the start of their working life), mesothelioma as a proportion of expected all-cause mortality will constitute a standardized measure of mesothelioma mortality across cohorts. In fact, the average age at the start of exposure was not invariant between cohorts and so an adjustment to standardize the expected all-cause mortality for follow-up starting at age 30 was derived for each using lifetable methods (Hodgson and Darnton 2000). Again, expressing the proportion of mesotheliomas as a percentage and dividing by the estimated average cohort cumulative exposure give the metric R_M as follows:

$$R_M = 100 \times \frac{O_M}{E_{AC(adj)} \times CE}$$

Where

O_M = Observed mesothelioma deaths;
$E_{AC(adj)}$ = Expected all-cause mortality from age 30; and
CE = Mean cumulative asbestos exposure (f/cc-yrs).

Other Data Issues

In some studies, pleural mesotheliomas had been miscoded as lung cancer and these were excluded where possible from the observed lung cancer count prior to calculating the lung cancer excess. Although this could have led to a slight underestimation of the lung cancer SMR (since the exclusion cannot be applied to the denominator), this was preferable to the potentially large overestimation of the SMR if the exclusion had not been applied.

Variation in the values of the metrics R_M and R_L within fiber-type cohort groups were explored using Poisson regression. Where values for specific fiber-type groups were found to be statistically consistent, summary R_M and R_L values with associated 95% confidence intervals were calculated. These can be interpreted as the fiber-specific potency in relation to mesothelioma and asbestos-related lung cancer induction.

Charts and tables in this chapter follow the original Hodgson and Darnton labeling convention for data points, using "o," "a," and "y" for crocidolite, amosite, and chrysotile respectively, with appropriate combinations of these letters for mixed fiber exposure cohorts.

Although cumulative exposure was a natural choice (as a proxy for dose) to relate to cancer risk, nevertheless, since it is a function of both exposure intensity and duration, the Peto model for mesothelioma incidence over time implies that equivalent cumulative exposures accrued over shorter durations will confer higher mesothelioma risks than those accrued over longer durations. This effect of duration was not taken into account in Hodgson and Darnton 2000 since most of the cohorts considered typically involved similar and relatively short durations of exposure (about 2 years on average). However, this effect can become important in application of the resulting risk estimates to situations where the cumulative exposures were accrued over extended periods (i.e. substantially longer than 5 years, say).

Summary of Cohorts Included in the Original and Updated Meta-Analyses

The original Hodgson and Darnton meta-analysis considered 17 published reports of studies with sufficient data to be included in the meta-analysis. These reports allowed the identification of 21 separate cohorts in total. Three of these cohorts entailed exposure exclusively to crocidolite fiber, two to amosite, and six to chrysotile. The remaining 10 cohorts involved exposures to mixtures of more than one fiber type. The recent updated meta-analysis of the "pure fiber" studies (Darnton 2023) included additional follow-up on the crocidolite miners at Wittenoom, Australia (Musk et al. 2008), and two amosite-exposed cohorts not considered in the original analysis: a cohort of textile workers at Tyler, Texas USA (Levin et al. 2016), and a cohort of miners at Libby, Montana, USA, exposed to so-called "Libby amphiboles," a mixture of mineral fibers, including winchite, richterite, and tremolite (Sullivan 2007). Three additional chrysotile studies were also included, two of which were cohorts from China: miners at

Qinghai (Wang et al. 2013a), and factory workers in Chongqing (Wang et al. 2013b); and finally, a cohort of workers manufacturing textiles in North Carolina, USA (Loomis et al. 2009). Brief descriptions of the "pure fiber" cohorts are given in the following section.

Crocidolite Cohorts

The extensively studied cohort of crocidolite miners at Wittenoom, Western Australia, constitutes a key part of the evidence in relation to the cancer risks posed by this mineral fiber. The fact that this cohort has been regularly reviewed based on additional follow-up means that it now contributes a large majority of the mesothelioma and excess lung cancer deaths arising from crocidolite exposures in the meta-analysis. The other two crocidolite studies are the miners of the Transvaal crocidolite in South Africa (Sluis-Cremer et al. 1992) and a small group of factory-based workers manufacturing cigarette filters in Massachusetts, USA (Talcott et al. 1989).

The Wittenoom crocidolite miners cohort included around 6,500 men who had moved to Wittenoom to work at the mine or the related mill at some time between 1943 and 1966 after which the mine closed. Exposure levels were high, but duration of employment was typically short (median four months), with the overall mean cumulative exposure calculated as 23 f/cc-yrs. The original Hodgson and Darnton meta-analysis included data based on follow-up until the end of 1986 by which time there had been 72 mesothelioma deaths ($R_M = 0.52$) and 87 lung cancers vs. 48.7 expected ($R_L = 3.4$). A more recent study by Musk et al. (2008) extended follow-up to the end of year 2000 and reported sufficient data on lung cancer for inclusion in the updated meta-analysis analysis: the 281 lung cancer deaths vs. 136.7 expected from this analysis gives a slightly higher value of $R_L = 4.6$. Subsequently, Berry et al. (2012) reported mesothelioma mortality based on follow-up to the end of 2008, by which time the number of mesothelioma deaths had increased to 316, giving an updated R_M value of 0.51 (very close to the original value). There were also an additional 13 mesothelioma deaths among 419 women also employed by the company, though they did not work in the mine or mill.

The South African crocidolite miners' cohort included 3,430 white men employed in the Transvaal mines between 1925 and 1980. Unfortunately, black and mixed-race men could not be included due to service record and death certificate data being of insufficient quality. Exposures were reportedly particularly high in mine surface-operations, but employment durations were again typically short (median, approx. 1 year) so that the calculated average cumulative exposure was of a similar order to that of Wittenoom (16 f/cc-yrs). Based on follow-up until 1980, there had been 20 mesothelioma deaths ($R_M = 0.59$) and 19 lung cancer deaths vs. 10.2 expected ($R_L = 5.2$): both R_M and R_L are slightly higher than for Wittenoom.

The Massachusetts cigarette filter manufacturers cohort is a small cohort of 33 men employed at the factory at some time during 1953. The process involved the use of crocidolite fiber as well as cotton and acetate fibers by dry mixing, carding, and depositing on crepe paper. The workers were subject to very high levels of exposure to crocidolite fibers for a relatively short period of employment during the 1950s. Although the median duration of employment was less than 2 years (for the 23 workers with information about employment duration), the estimated cumulative exposure was 120 f/cc-yrs, much higher than exposures of the Wittenoom and South African miners. Very high rates of disease due to mesothelioma and lung cancer were seen: by the end of 1988, 28 of the 33 workers had died compared with 8.3 expected from all causes of death; 5 of the deaths were due to mesothelioma ($R_M = 0.68$) and 8 from lung cancer ($R_L = 10$), again higher than for Wittenoom, though these values are clearly associated with very substantial uncertainty.

Meta-analysis using the Hodgson and Darnton approach suggests that for mesothelioma, the R_M values for these three cohorts are statistically homogeneous with a combined summary value of $R_M = 0.52$ (95% CI: 0.47–0.58). This is relatively close to the study-specific value for the Wittenoom cohort, due to the weight this has statistically. Similarly, for lung cancer, the combined overall summary value of $R_L = 4.8$ (95% CI: 3.8–5.9) is statistically homogeneous and relatively close to the Wittenoom value. As will be discussed further (Chapter 10), cumulative exposure estimates for all three of these cohorts have significant uncertainties, which cannot be fully taken into account in the combined summary values. The study-specific R_M and R_L values for Wittenoom are already the lowest of those for the three cohorts, and if the overall cumulative exposure is an underestimate – as suggested by Burdett – these risk estimates are reduced further, which in turn, reduces the overall summary estimates and their statistical consistency. As it will be seen from Chapter 10, the relevance of the risk estimate from the Massachusetts cohort can also be discussed given that the exposure was to Bolivian crocidolite. This is likely to have a different distribution of fiber sizes to other commercial crocidolite (Shedd 1985), as well as different chemical composition (Korchevskiy et al. 2019) that could affect the estimated potency levels. However, given the small size of this study, its data point carries much less weight statistically in the analysis than that for Wittenoom.

Amosite Cohorts

All amosite fiber originated from the mines in the northwest Cape region of South Africa. The original Hodgson and Darnton analysis included the cohort of these miners (Sluis-Cremer et al. 1992) and a cohort of workers from a factory in Paterson, New Jersey, that manufactured insulation products for pipes, boilers, and turbines in US Navy ships (Seidman et al. 1986). In 2023, Darnton included an additional study of a factory in Tyler, Texas USA, which took over production of asbestos insulation products after the Paterson factory closed (Levin et al. 2016; Darnton 2023). A further factory manufacturing amosite containing products in Uxbridge, UK, could not be included in the analysis due to insufficient quantitative exposure data (Acheson et al. 1984).

The South African amosite miners' cohort included 3,212 white men employed in the northwest Cape mines between 1925 and 1980. The pattern of exposure was similar to that at the South African crocidolite mines, and again tended to be intense – particularly in surface operations – but relatively brief; the overall mean cumulative exposure was calculated as 24 f/cc-yrs. By the end of 1980, there had been 4 deaths from mesothelioma ($R_M = 0.06$) and 21 from lung cancer vs. 14.5 expected ($R_L = 1.9$).

The Paterson amosite insulation products factory operated between 1941 and 1954. The cohort included 820 workers who were still alive 5 years after hire. Workers were subject to very intense exposure levels, but the duration of employment was again typically relatively short (625 worked for less than 2 years), with the overall cumulative exposure calculated as 65 f/cc-yrs, substantially higher than the South African mines from which the fiber originated. Follow-up to the end of 1982 revealed 17 mesothelioma deaths ($R_M = 0.12$), a high proportion of which (9/17) were peritoneal cases, and a large excess of lung cancer mortality: 98 observed deaths vs. 20.5 expected ($R_L = 5.8$).

Levin and co-workers reported mortality among 753 white males who had worked at the Tyler plant between 1954 and 1972. Based on follow-up to the end of 1993, there were 222 deaths from all causes vs. 133.6 expected. No overall average cumulative exposure estimate was given, but average fiber concentrations of between 15.9 and 91.4 f/cc were obtained from three surveys carried out between 1967 and 1971 (Ribak and Ribak 2008), as discussed later by Burdett. The average length of employment was 12.7 months. Taken together these data suggest an overall average cumulative exposure of around 50 f/cc-yrs was likely. There were 6 reported deaths from mesothelioma ($R_M = 0.12$), 2 of which were peritoneal cases, and 35 from lung cancer vs. 12.6 expected ($R_L = 2.9$).

Meta-analysis of the above data shows that for mesothelioma, the R_M values for these three amosite cohorts are statistically homogeneous with a combined summary value of $R_M = 0.11$ (95% CI: 0.070–0.15), around one-fifth of the summary value of 0.52 for the three crocidolite cohorts. For lung cancer, the combined data are again statistically homogeneous (summary value of $R_L = 4.0$, 95% CI: 3.3–5.0), but in contrast to mesothelioma, this is much closer in relative terms to the value for the crocidolite cohorts ($R_L = 4.8$).

Other Amphiboles: Vermiculite Miners and Associated Workers, Libby, Montana, USA

McDonald and colleagues reported mortality among a cohort of 406 mine workers exposed to fibrous amphibole material as a contaminant of the mined vermiculite (McDonald et al. 2004). The cohort included men who were hired before 1963 and had worked at the mine for at least one year. Based on follow-up to 1999, there were 285 deaths from all causes vs. 224.4 expected, and 44 lung cancers vs. 18.3 expected. There was some uncertainty in the number of mesotheliomas due to the way the deaths were coded; however, there may have been 12 mesotheliomas in total, 2 of which were peritoneal. The reported average duration of employment (9 years) and fiber concentration (18 f/cc) implies an average cumulative exposure of about 160 f/cc-yrs leading to values of R_M and R_L of 0.041 and 0.93, respectively.

Sullivan reported mortality among a larger cohort of 1,672 vermiculite miners, millers, and processors at Libby (Sullivan 2007). This was an expanded version of a cohort originally studied by NIOSH which itself overlapped with the McDonald study and included all workers employed before 1970 who worked for at least one year at the mine. Sullivan expanded this cohort to include all workers regardless of employment duration (808 workers had worked for less than 1 year). Based on follow-up to 2001, there were 711 deaths from all causes vs. 574 expected, 15 mesotheliomas (14 pleural; 1 peritoneal), and 89 lung cancer deaths against 52.5 expected. An average of exposure category mid-points weighted by the expected lung cancer deaths and expected nonmalignant respiratory disease deaths implies an average cumulative exposure of about 85 f/cc-yrs giving values of R_M and R_L of 0.030 and 0.82, respectively – somewhat lower than those derived from McDonald's results.

In 2010, Larson et al. published a further study that included a slightly larger number of workers ($n = 1,862$) than the Sullivan study – including a small number of female workers (Larson et al. 2010). Based on follow-up to 2006, there were 952 deaths from all causes vs. 732 expected, 19 mesotheliomas, and 104 lung cancer deaths vs. 64.6 expected. The reported median cumulative exposure for the whole cohort was substantially lower than reported by Sullivan (median 4.3 vs. 8.7 f/cc-yrs). The mean cumulative exposure derived from the internal lung-cancer exposure–response analysis (weighting mid-points of the cumulative exposure categories by the expected lung cancer mortality) was approximately 20 f/cc-yrs compared with 85 f/cc-yrs using the same approach in the analyses by Sullivan. This naturally leads to much higher estimates of mesothelioma and lung cancer risk per unit of cumulative exposure ($R_M = 0.13$ and $R_L = 3.0$). However, internal cox-regression analyses show a much shallower exposure–response for lung cancer of 0.1% per f/cc-yr.

Chrysotile Cohorts

Six studies of worker cohorts exposed exclusively to pure chrysotile fiber were included in the original meta-analysis, two of which comprised the male and female workers (considered separately) at the South Carolina textile factory (Dement et al. 1994), and a further cohort which was a subgroup of a New Orleans cement manufacturing plant which also employed other groups of workers exposed to amphibole fiber (Hughes et al. 1987). By far, the largest study of chrysotile-exposed workers was that of around 11,000 miners and millers in Quebec (Liddell et al. 1997). A cohort of chrysotile miners in the Italian Balangero mine (Piolatto et al. 1990) and a cohort of workers at a Connecticut friction products factory (McDonald et al. 1984) were also included as pure chrysotile studies. The updated meta-analysis (Darnton 2023) also considered a further three cohorts: a cohort of chrysotile miners from the Qinghai mine, China (Wang et al. 2013a); a cohort of workers from four textile manufacturing plants in North Carolina, USA (Loomis et al. 2009); and a cohort of factory workers using chrysotile at Chongqing, China (Wang et al. 2013b). Updated data based on additional follow-up on the Balangero miners (Pira et al. 2017), South Carolina textile workers (Hein et al. 2007), and Connecticut friction products manufacturers (Finkelstein and Meisenkothen 2010) were also incorporated.

Results of mortality among the Quebec chrysotile miners cohort were last reported in 1997 (Liddell et al. 1997) based on follow-up to 1992 of this cohort of approximately 10,000 men born between 1891 and 1920. Exposures are likely to have very high and, in contrast to many of the other cohorts considered, exposure durations tended to be longer (typically around 10 years). An overall average cumulative exposure of 200 million particles per cubic foot years (mppfc-yrs) can be derived from the exposure–response analysis presented in Table 8 of Liddell et al. 1997. In Hodgson and Darnton 2000, this was assumed equivalent to a value of 600 f/cc-yrs using a conversion factor of 1 mppcf = 3 f/cc-yrs on the basis of comparisons of later PCM and particle counts (Gibbs and LaChance 1974), but correlations were poor and later reviews have questioned the validity of these global conversions (Gibbs 1994). By 1992, there had been 33 mesothelioma deaths (excluding those from an asbestos factory where amphibole was also used) giving an R_M value of 0.0009 and 587 lung cancers vs. 431.6 expected ($R_L = 0.060$). These are among the lowest values of any of the cohorts considered in this review and if taken these at face value carry considerable weight statistically given the size of the cohort. This is arguably the case even if the overall cumulative exposure amounts to a substantial overestimate, as discussed later.

The Balangero chrysotile miners cohort comprises 1,056 men employed between 1930 and 1990 followed up from 1946. Exposures were high prior to 1968 with an overall average cumulative exposure of 300 f/cc-yrs estimated in the original Hodgson and Darnton meta-analysis. Indeed, Korchevskiy and Wiley recently re-estimated the central tendency exposure level at a considerable higher value of 490 f/cc-yrs (Korchevskiy and Wylie 2023). A recent update of mortality based on follow-up to and including 2014 (Pira et al. 2017) identified 7 mesothelioma deaths ($R_M = 0.0042$, cumulative exposure 300 f/cc-yrs), 2 of which were peritoneal cases, with 53 observed lung cancer deaths vs. 45.5 expected ($R_L = 0.053$, cumulative exposure 300 f/cc-yrs).

Wang and co-workers reported a mortality analysis of a cohort of 1,539 male workers from China's largest chrysotile mine in Qinghai province, employed for at least 1 year in 1981, followed up to the end of 2006 (Wang et al. 2013a). The mine and mill were reported to be very dusty environments, and the workers were generally employed for a number of decades. Based on periodic personal dust samples (with conversion to fiber estimates based on a limited number of parallel samples in 1991), the estimated overall mean cumulative exposure was over 100 f/cc-yrs. No mesotheliomas were observed, but there was a large excess of lung cancer and clear exposure–response relationship with cumulative exposure. Based on calculations applying a 10-year lag, there were 56 lung cancer deaths vs. 11.4 expected giving a value of R_L of 3.3, around 50-fold higher than the R_L values for the Quebec and Balangero mines.

Hughes and co-workers reported the mortality experience of 6,931 employees at two asbestos cement products manufacturing plants mainly using chrysotile fiber (Hughes et al. 1987). However, amosite was also used in plant 1 and crocidolite was reportedly used in the separate pipe production part of the plant 2 site. In the Hodgson and Darnton 2000 meta-analysis, these workers were considered as separate cohorts: mixed exposure to chrysotile and amosite (plant 1), mixed exposure to chrysotile and crocidolite (pipe production at plant 2), and chrysotile only (plant 2 workers who had never worked in pipe production). In the chrysotile-only subgroup, no mesothelioma deaths were observed, but there were 42 lung cancer deaths vs. 32.4 expected giving an R_L value of 1.3 based on an estimated cumulative exposure of 22 f/cc-yrs.

Asbestos textiles were first produced at the South Carolina chrysotile factory in 1909 using chrysotile mainly from Quebec. Though small quantities of crocidolite yarn were reported to be used from 1950 to 1975, chrysotile was the only fiber actually processed at the site. The cohort originally studied by Dement and co-workers included only white male workers employed for at least one month between 1940 and 1965 (Dement et al. 1994), but subsequent updated analyses included black males and white females. The original meta-analysis included 1,241 white males and 1,036 white females as two separate cohorts; black males were not included due to concerns about the consistency of their results, which may have reflected different smoking prevalences and asbestos exposures among white and black males. Asbestos exposures were reported to have been relatively stable from

1940 due to the implementation of control measures by that date, with an overall average cumulative exposure of 28 f/cc-yrs among white males and 26 f/cc-yrs among white females. An update of the South Carolina studies by Hein et al. (2007) slightly expanded the number of subjects in the white male and white female cohorts and extended follow-up to the end of 2001. Three mesotheliomas were identified in the male cohort, one of which was a peritoneal case ($R_M = 0.016$). The large lung cancer excesses identified in earlier results persisted with additional follow-up (males: $R_L = 4.8$; females: $R_L = 4.7$).

The cohort study of workers at four textile plants in North Carolina consists of 3,975 men and 1,795 women employed between 1950 and 1973 and followed up until the end of 2003 (Loomis et al. 2009). Three of the four plants converted raw asbestos – mostly from Quebec – and cotton fibers into yarn and woven materials. The fourth plant manufactured friction products. Studies of available exposure measurements discussed by Burdett later in this volume suggest that exposures were higher at the North Carolina plants than at South Carolina. A reported R_L value of 1.67 (Loomis et al. 2009) implies an overall average cumulative exposure of 68 f/cc-yrs (based on 249 lung cancer deaths vs. 116.4 expected). The reported R_M estimate was 0.010 based on eight deaths from pleural mesothelioma. A further feature of the North Carolina cohort is the large difference between the reported R_L estimate of 1.67 and the implied value from the internal lung cancer regression analysis. For latter, a relative risk of 1.102 per 100 f/cc-yrs was reported, which gives an alternative R_L value of 0.1 (much closer to the R_L values for the Quebec and Balangero mines than to the higher values for the South Carolina cohorts).

The Chongqing asbestos factory has been in operation since 1939 and since the late 1950s had used chrysotile asbestos from mines in the Sichuan region of China in the production of textiles, cement, rubber, and friction products. In 2001, Yano et al. analyzed the long-term mortality of 515 of the male factory workers based on deaths occurring up until the end of 1996 (Yano et al. 2001), and more recently, this study was expanded to additional workers followed-up to the end of 2008 (Wang et al. 2013b). Most workers experienced long-term exposures (at least 15 years) and the overall average cumulative exposure was estimated to be 105 f/cc-yrs. Two mesothelioma deaths (one of which was a peritoneal case) ($R_M = 0.022$) with 53 lung cancer deaths vs. 40 expected ($R_L = 2.9$) among the 586 male workers in the expanded cohort. See Chapter 12 for further discussion on mesothelioma rates for asbestos textile workers in China.

The data used by Hodgson and Darnton 2000 for this cohort of workers from a Connecticut factory that manufactured friction products using chrysotile were extracted from the most recent analysis available at the time in which there were no cases of mesothelioma (McDonald et al. 1984). Finkelstein and Meisenkothen re-examined the incidence of mesothelioma among these workers and discussed information about six individuals with mesothelioma subsequently extracted from litigation case files (Finkelstein and Meisenkothen 2010), and two additional cases identified in a separate study (Teta et al. 1983). They concluded that two of these eight cases met the criteria for inclusion in the original cohort mortality study, but the mesotheliomas occurred after the cut-off date for follow-up. Based on these two cases and an estimate of the expected all-cause mortality to include further follow-up, they then derived an estimate of $R_M = 0.002$. Unfortunately, because the information about the two cases that apparently meet the criteria for inclusion in the cohort study was taken from a separate source, it is not clear whether the individuals were actually among the original cohort members. There is thus at least some potential for the risk estimate to be biased in the upward direction (since it is more likely that any eligible individuals that weren't included in the original study would be subsequently identified if they were cases than if they were noncases).

In the original Hodgson and Darnton 2000 meta-analysis, despite the wide range of study-specific R_M values (ranging from 0.0009 for Quebec to 0.016 for the South Carolina males), a summary value for mesothelioma of $R_M = 0.001$ was statistically homogeneous, but this was largely because of the zero or small numbers of observed mesotheliomas in five of the six chrysotile-exposed cohorts, with 33 or the 37 cases contributed from the Quebec study. In 2023, Darnton reported that a summary value based on the nine cohorts now available ($R_M = 0.0014$) is no longer statistically homogeneous, with a substantially reduced proportion (37/55) of the total cases now contributed by the Quebec study. The four chrysotile textile cohorts all have relatively high study-specific R_M values, and as a subset have a summary value of $R_M = 0.01$ (95% CI: 0.0057–0.017) without significant heterogeneity. This is an order of magnitude higher than a combined summary value for the remaining five cohorts of $R_M = 0.0011$ (95% CI: 0.00079–0.0014). It should be noted that statistically homogeneous estimations of R_M can be obtained if textile cohorts (including Chongqing factory) are considered as a separate category vs. nontextile cohorts.

As in the original meta-analysis for lung cancer, very considerable heterogeneity is present across the chrysotile cohorts with study-specific R_L values ranging from 0.06 for the Quebec study to values some 30–80 times higher for the four textiles cohorts. Whereas in the original meta-analysis, the available mining cohorts showed similar R_L values, this is no longer the case with the inclusion of the Qinghai mine, which has an R_L value of 3.3 and is more in-line with the textiles cohorts.

Summary of Original and Updated Meta-Analyses

Mesothelioma

Figure 8.1 shows the study-specific values of R_M from the updated meta-analysis and summary R_M values for cohorts grouped by fiber type. Mixed exposure cohorts (i.e. which entailed exposure to more than one type of fiber) that were included in

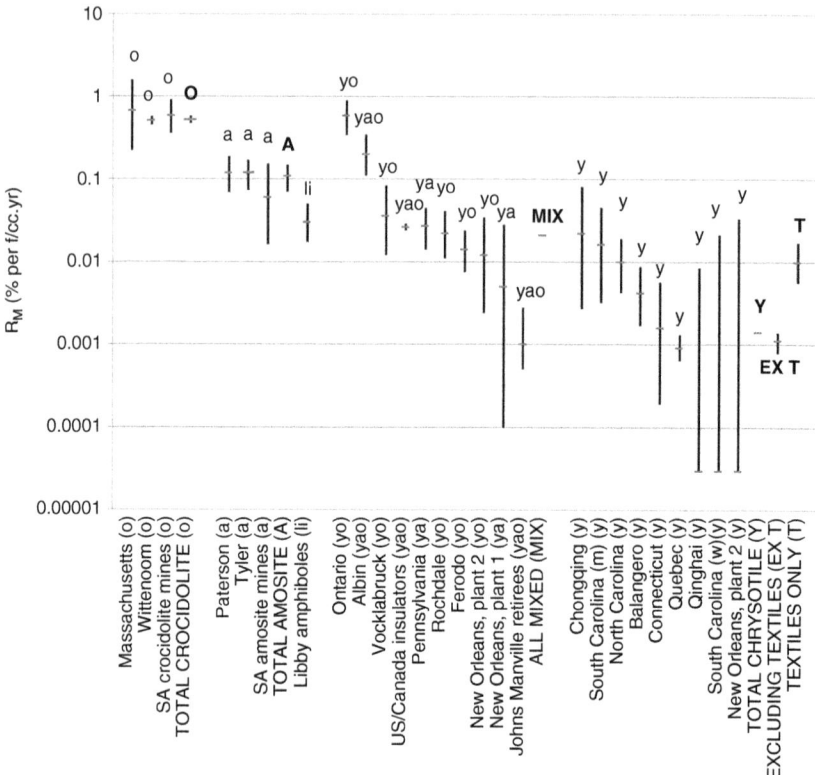

Figure 8.1 Exposure-specific mesothelioma mortality (R_M) by cohort and fiber type groupings, showing 95% confidence intervals. Group means labeled in capitals. Confidence intervals not shown for groups with very significant heterogeneity.

the original 2000 meta-analysis are also reproduced, and the underlying data are also shown in Table 8.1. This shows the wide range of R_M values across cohorts with nonzero counts of mesothelioma deaths, ranging from 0.0009 per f/cc-yrs for the Quebec miners, to value around 750 times higher (0.68) for the Massachusetts factory workers. The results are broadly consistent with the original analysis, with some subtle differences. The clear pattern of risk per unit exposure (i.e. R_M) by fiber type is evident, with the ordering of R_M values from high to low (left to right on the chart) showing a clear tendency for the amphibole cohorts (labeled "o" for crocidolite and "a" for amosite) to appear on the left, chrysotile cohorts (labeled "y") on the right, with mixed fiber cohorts (labeled with combinations of "yao") in-between.

Individual study R_M values remain statistically consistent for the group of crocidolite cohorts ($R_M = 0.52$), and for the group of amosite cohorts ($R_M = 0.11$),

Table 8.1 Summary of mesothelioma mortality data and exposure-specific risk estimates (R_M).

Fiber type	Cohort[a]		Fiber label	Process[b]	Average Cum. Exposure (f/cc-yrs)	Total mesothelioma deaths	(Number Peritoneal)	Expected all-cause mortality	R_M	(95% CI[c])	[p for heterogeneity[c]]
Crocidolite	1	Wittenoom	o	M	23	316	(48)	2617.32	0.51	(0.45, 0.57)	
	13o	SA crocidolite mines	o	M	16.4	20	(2)	223.2	0.59	(0.36, 0.91)	
	14	Massachusetts	o	O	120	5	(3)	8.3	0.68	(0.22, 1.6)	
Total crocidolite									0.52	(0.47, 0.58)	
Amosite	13a	SA amosite mines	a	M	23.6	4	(1)	305.7	0.06	(0.016, 0.15)	
	12	Paterson	a	I	65	17	(9)	355.9	0.12	(0.068, 0.19)	
	—	Tyler	a	I	50	23	(7)	397.4	0.12	(0.073, 0.17)	
Total amosite									0.11	(0.070, 0.15)	
Libby amphibole	3	Libby	li	M	85	15	(1)	574	0.030	(0.017, 0.050)	
Mixed	3	Johns Manville retirees	yao	I	750	8	(2)	762.5	0.001	(0.0006, 0.0028)	
	4	Ontario	yo	C	60	17	(8)	62.2	0.59	(0.34, 0.95)	
	5a	New Orleans (plant 1)	ya	C	79	1	(0)	294.5	0.005	(0.0001, 0.028)	
	5o	New Orleans (plant 2, yo)	yo	C	93	3	(0)	217	0.012	(0.0024, 0.034)	
	7	Vocklabruck	yo	C	25	5	(1)	530.2	0.036	(0.012, 0.084)	
	15	Albin	yao	C	13	13	(0)	493.3	0.20	(0.11, 0.35)	
	8	US/Canada insulators	yao	L	500	453	(282)	3170.6	0.026	(0.024, 0.029)	
	9	Rochdale	yo	T	74	10	(0)	602.5	0.022	(0.011, 0.041)	
	11	Pennsylvania	ya	TF	60	14	(4)	821.1	0.027	(0.014, 0.044)	
	17	Ferodo	yo	F	35	13	(0)	2646.3	0.014	(0.0075, 0.024)	

(*Continued*)

Table 8.1 (Continued)

Fiber type	Cohort[a]		Fiber label	Process[b]	Average Cum. Exposure (f/cc-yrs)	Total mesothelioma deaths	(Number Peritoneal)	Expected all-cause mortality	R_M	(95% CI[c])	[p for heterogeneity[c]]
All mixed									0.021		[p<0.001]
Chrysotile	6	Quebec	y	M	600	33	(0)	5912.7	0.0009	(0.0006,	0.0013)
	10	Balangero	y	M	300	7	(0)	533.4	0.0042	(0.0017,	0.0088)
	—	Qinghai mine	y	M	120	0	(−)	174.2	0	(0,	0.0084)
	5y	New Orleans (plant 2, y)	y	C	22	0	(−)	397.1	0	(0,	0.034)
	2f	South Carolina (women)	y	T	26	0	(−)	549.6	0	(0,	0.021)
	2m	South Carolina (men)	y	T	28	3	(1)	571.1	0.016	(0.0032,	0.045)
	—	North Carolina	y	T	68.3	8	(0)	1275.3	0.010	(0.0042,	0.019)
	—	Chongqing factory	y	T	105.2	2	(1)	197.3	0.022	(0.0027,	0.081)
	16	Connecticut	y	F	46	2	(0)	2800	0.0016	(0.00019,	0.0057)
Total chrysotile									0.0014		[P<0.001]
Excluding textiles									0.0011	(0.00079,	0.0014)
Textiles only									0.010	(0.0056,	0.017)

[a] Cohort name with original numbering from Hodgson and Darnton 2000 where applicable.
[b] M = mining, I = insulation products, C = cement, T = textiles, F = friction products, L = lagging or work with insulation, O = other.
[c] 95% confidence internals are given summary R_L values where groupings do not show significant heterogeneity; otherwise the P-values for heterogeneity are shown.

with values very similar to the original analysis. However, values for the chrysotile cohorts are no longer statistically consistent due to the increased influence of the four chrysotile textile cohorts, which do form a statistically consistent group ($R_M = 0.010$), an order of magnitude higher than the five remaining chrysotile cohorts (which are also statistically consistent with an $R_M = 0.0011$). Nevertheless, the higher of these two chrysotile groups is still an order of magnitude below the amosite summary value. Considered as a whole, the seven amphibole cohorts (including the Libby amphibole cohort) are not statistically consistent, and the lower value of R_M for Libby ($R_M = 0.030$) is not statistically consistent with the amosite cohorts; the latter is around three times that for the chrysotile textiles cohorts. This observation is in-line with the differences in mineral composition and size of fibers for various mineral types of amphiboles (see Chapter 1).

Figure 8.1 also shows the substantial heterogeneity in the R_M values for mixed fiber cohorts with values falling within the range 0.001–0.6 per f/cc-yrs. Whereas the features of the summary data continue to suggest that much of the variation in the potency of asbestos in relation to mesothelioma can be explained by mineral type alone, the updated data for chrysotile cohorts emphasize that other factors associated with the industrial context are also likely to be important, particularly for chrysotile. In this context, it should be noted that statistically consistent estimations of R_M can be obtained if textile cohorts (including Chongqing factory) are considered as a separate category vs. nontextile cohorts. Furthermore, whether the apparent differences in risk per fiber based on PCM for the different amphiboles truly reflect mineral type alone is not entirely clear given the role that fiber size and chemical composition have been shown to have. For example, crocidolite exposures tend to entail a relatively higher proportion of very thin fibers relevant to mesothelioma risk than amosite exposures. If a higher proportion of the thinnest fibers are missed by the PCM metric for crocidolite vs. amosite, then the true difference in potency for the two amphiboles would tend to be smaller than estimated here.

Given the very substantially higher mesothelioma risk per unit of cumulative exposure for amphibole (particularly amosite and crocidolite) vs. chrysotile, even small amounts of amphibole present in the exposures for cohorts categorized as chrysotile in this analysis could bias the risk attributed to chrysotile upward, potentially by a considerable amount. Unfortunately, small amounts of amphiboles may have been present alongside chrysotile either due to their natural occurrence in the ore or because they were intentionally used in certain industrial circumstances. The extent to which this was the case may not be documented or well understood. An example of a cohort where the exposure was predominantly to chrysotile but where varying small amounts of amphibole may have been present is the North Carolina textiles cohort. Some have referred to evidence pointing to the likelihood of amphibole exposure at two of the four plants that contributed most to the available person-time for this study (Garabrant 2020), although the counterargument that most of the available evidence suggests that neither amosite nor crocidolite was likely to have been processed has also been made strongly (Dement and Loomis 2023). Reclassifying this particular cohort as a mixed mineral fiber type study and removing it from the meta-analysis has only a marginal effect (the chrysotile summary R_M estimate based on the remaining eight cohorts is reduced from 0.0014 to 0.0012 and is still associated with substantial heterogeneity); nevertheless, the potential for varying degrees of concomitant amphibole exposure being at least part of the explanation for the heterogeneity among the chrysotile cohorts remains. A further example is the presence of a naturally occurring amphibole-like fiber – balangeroite – potentially influencing the risk attributed to chrysotile among the Balangero miners (Korchevskiy and Wylie 2023), though again the risk levels from balangeroite and potentially tremolite contamination of Balangero chrysotile remains uncertain.

Lung Cancer

Figure 8.2 shows the study-specific values of R_L from the updated Hodgson and Darnton meta-analysis and summary R_L values for cohorts grouped by fiber type. As for mesothelioma, the mixed exposure cohorts (i.e. which entailed exposure to more than one type of fiber) included in the original 2000 meta-analysis are also reproduced, and the underlying data are also shown in Table 8.2.

As for mesothelioma, a very wide range of R_L values is evident among those studies ranging (for those studies with a positive lung cancer excess) from 0.053 for the Balangero miners to a value of 10 – nearly 200-fold higher – for the Massachusetts factory workers. However, there is clearly more heterogeneity present in the lung cancer than the mesothelioma data, particularly for the chrysotile cohorts. Figure 8.2 broadly preserves the left-to-right ordering used in Figure 8.1 for mesothelioma. While the amphibole cohorts (shown on the left) have a clear tendency to have higher R_L values, the chrysotile cohorts have R_L values spanning nearly the full range: for example, 0.06 for the Quebec mining cohort to 4.8 for the South Carolina male textiles factory workers who processed the same fiber mined by the Quebec workers. Mixed fiber cohorts again generally fall between the two extremes.

Ostensibly, the data still suggest that the mineral type of fibers is an important determinant of lung cancer risk, but that other factors not summarized within Figure 8.2 – such as the industrial process – may potentially play as large a role. As in the original 2000 meta-analysis – and in contrast to the results for mesothelioma – the crocidolite and amosite cohorts are statistically consistent when considered as a single group with a summary estimate of $R_L = 4.3$ (95% CI: 3.7–5.0). This is slightly lower than that for the five amphibole cohorts included

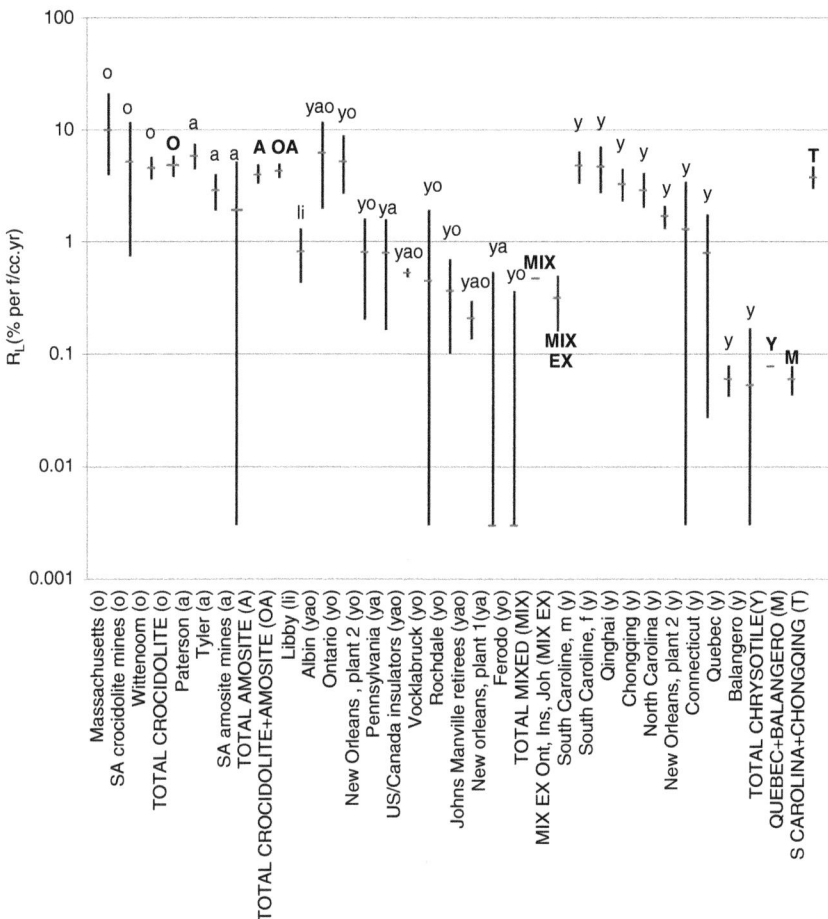

Figure 8.2 Exposure-specific excess lung cancer mortality (R_L) by cohort and fiber type groupings, showing 95% confidence intervals. Group means labeled in capitals. Confidence intervals not shown for groups with very significant heterogeneity.

originally ($R_L = 4.8$, 95% CI: 3.9–5.8). However, the lower value for the Libby amphiboles cohort ($R_L = 0.82$) is not consistent statistically with the other amphibole data (once again, demonstrating that amphibole category is not homogenous for cancer potency).

For mesothelioma, even the highest study-specific R_M values among the chrysotile cohorts are substantially lower than the amphibole values (and most of the values for mixed cohorts), but this is not the case for lung cancer. Here, study-specific R_L values span the full range of results (with the exception of Massachusetts),

Table 8.2 Summary of lung cancer mortality data and exposure-specific risk estimates (R_L).

Fibre type	Cohort[a]		Fibre label	Process[b]	Average Cum. exposure (f/cc-yr)	Observed lung cancer	Expected lung cancer	Excess	R_L	(95% CI[c])		[p for heterogeneity[c]]
Crocidolite	1	Wittenoom	o	M	23	281	136.7	144.3	4.6	(3.6,	5.7)	
	13o	SA crocidolite mines	o	M	16.4	19	10.2	8.8	5.2	(0.74,	12)	
	14	Massachusetts	o	O	120	8	0.6	7.4	10	(4.0,	21)	
Total crocidolite									4.8	(3.8,	5.9)	
Amosite	13a	SA amosite mines	a	M	23.6	21	14.5	6.5	1.9	−(0.44,	5.1)	
	12	Paterson	a	I	65	98	20.5	77.5	5.8	(4.4,	7.4)	
	—	Tyler	a	I	50	89	36.5	52.5	2.9	(1.9,	4.0)	
Total amosite									4.0	(3.3,	4.9)	
Total crocidolite and amosite									4.3	(3.7,	5.0)	
Libby amphibole	—	Libby	la	M	85	89	52.5	36.5	0.82	(0.43,	1.3)	
Total amphibole (including Libby)									2.7			[p<0.001]
Mixed	3	Johns Manville retirees	yao	I	750	73	28.4	44.6	0.21	(0.14,	0.30)	
	4	Ontario	yo	C	60	22	5.3	16.7	5.2	(2.7,	8.8)	
	5a	New Orleans (plant 1)	ya	C	79	21	22.5	−1.5	0	−(0.53,	0.54)	
	5o	New Orleans (plant 2, yo)	yo	C	93	31	17.7	13.3	0.81	(0.20,	1.6)	
	7	Vocklabruck	yo	C	25	47	42.2	4.8	0.45	−(0.73,	1.9)	
	15	Albin	yao	C	13	35	19.4	15.6	6.2	(2.0,	12)	
	8	US/Canada insulators	yao	L	500	934	256.8	677	0.53	(0.48,	0.57)	

#	Cohort	age	Type	col1	col2	col3	col4	col5	CI
9	Rochdale	yo	T	138	56	37.1	18.9	0.37	(0.10, 0.70)
11	Pennsylvania	ya	TF	60	50	33.8	16.2	0.80	(0.16, 1.6)
17	Ferodo	yo	F	35	241	242.5	−1.5	0	−(0.36, 0.36)
	All mixed							0.47	[p < 0.001]
	Mixed excl. Ontario, Insulators, and JM (lung cancer)							0.33	0.53
6	Quebec	y	M	600	587	431.6	155	0.060	(0.042, 0.079)
10	Balangero	y	M	300	53	45.7	7.3	0.053	−(0.044, 0.17)
—	Qinghai mine	y	M	120	56	11.4	44.6	3.3	(2.3, 4.5)
5y	New Orleans (plant 2, y)	y	C	22	42	32.4	9.6	1.3	−(0.30, 3.4)
2f	South Carolina (women)	y	T	26	61	27.5	33.5	4.7	(2.7, 7.1)
2m	South Carolina (men)	y	T	28	116	49.6	66.4	4.8	(3.3, 6.4)
—	North Carolina	y	T	68.3	249	116.4	132.6	1.7	(1.3, 2.1)
—	Chongqing factory	y	T	105.2	53	13	40	2.9	(2.0, 4.1)
16	Connecticut	y	F	46	49	35.8	13.2	0.80	(0.027, 1.8)
	Total chrysotile							0.078	[p < 0.001]
	Quebec and Balangero only		M					0.060	(0.043, 0.078)
	Mining cohorts		M					0.064	[p < 0.001]
	Textile cohorts		T					2.2	[p < 0.001]
	South Carolina and Chongqing only		T					3.8	(3.0, 4.7)

[a] Cohort name with original numbering from Hodgson and Darnton 2000 where applicable.
[b] M = mining, I = insulation products, C = cement, T = textiles, F = friction products, L = lagging or work with insulation, O = other.
[c] 95% confidence internals are given summary R_L values where groupings do not show significant heterogeneity; otherwise the P-values for heterogeneity are shown.

and those for the South Carolina cohorts are similar to many of the amphibole values. These high values of R_L are around 80-fold higher than that for Quebec where the raw fiber that was processed in South Carolina originated, an observation that has been the subject of extensive debate since over several decades.

The original Hodgson and Darnton 2000 meta-analysis largely set aside the South Carolina lung cancer observations on the basis that they appear to be untypical of most situations where chrysotile is the predominant exposure. This argument appealed not only to the evidence from other chrysotile cohorts, but also to that from the mixed exposure cohorts shown in Figure 8.2. The weight of evidence from these cohorts is that R_L values are usually substantially lower than those of South Carolina even in the context of textile production, such as at Rochdale, UK, which had an R_L value of 0.37, over an order of magnitude lower than South Carolina. Taken together with the consistently high R_L values for the crocidolite and amosite cohorts, this suggests that amphibole is playing a relatively more important role in lung cancer risk than chrysotile. If true, and given an overall summary R_L estimate for mixed exposure settings of 0.32, this suggests that a value of R_L for chrysotile for general risk estimation purposes would be substantially lower still – perhaps closer to that of the Quebec study ($R_L = 0.06$), and implying that the South Carolina values are high for reasons other than chrysotile exposure per se.

However, the inclusion of additional chrysotile studies with high values of R_L – such as the Qinghai miners and the Chongqing factory workers – in the updated meta-analysis suggests that this argument is now less easy to sustain and tends to weaken the case for a summary R_L value that is close to that of the Quebec study. On the other hand, updated results from the Balangero miners cohort do tend to corroborate the Quebec R_L value and suggest that chrysotile exposure does constitute a much lower lung cancer risk per fiber than amphibole at least in certain contexts. If the higher estimate of cumulative exposure at Balangero estimated by Korchevskiy and Wiley is adopted (Korchevskiy and Wylie 2023), the lung cancer risk estimate is reduced by a factor 1.6 returning the updated R_L value close to its original value at about half that of Quebec. The additional data therefore emphasize the need to determine those factors in addition to the mineral type of fibers – such as fiber size – that substantially modify the risk in certain settings.

The semiquantitative argument described above from the original Hodgson and Darnton 2000 meta-analysis led to the Quebec study playing a central role in determining the value of the R_L appropriate for general risk estimation for chrysotile and to downplaying the relevance of the South Carolina results for such purposes. However, another lung cancer meta-analysis of largely the same set of cohorts used methods that resulted in the opposite conclusion, leading some to a view that the evidence does not support there being a difference in potency between amphibole and chrysotile in the causation of lung cancer (Lenters et al. 2012).

Lenters et al. found that the summary lung cancer risk was higher for the subgroups of studies satisfying each of five separate criteria related to study quality, and that this was independent of fiber type. Series of summary risk values determined by excluding studies after the application of the five criteria in turn (and in different orders) showed monotonic increases. This led to the conclusion that study quality itself is an important determinant of lung cancer risk and that the risk is best characterized on the basis of a much-restricted pool of studies that meet all or most of these criteria (the two studies meeting all five criteria being the South Carolina and the Libby amphibole cohorts).

Hodgson examined these arguments in detail and noted that while the association between risk and quality is clearly evident, the effect is not necessarily as strong as presented and that fiber type remains a strong determinant of risk alongside study quality. Excluding studies that do not meet certain quality criteria therefore amounts to an arbitrary prioritizing of this effect over the effect of fiber type. The key issue is not that the association between quality and risk exists, but how to take it into account in summarizing the weight of evidence (Hodgson 2013). Completely excluding some – or indeed most – studies is one approach, but this makes the strong claim that the effect of exposure on risk can only be quantified based on a certain subset of studies. How far down the path of excluding studies is sufficient to ensure that this signal is correctly characterized remains a matter of subjective judgment, and the approach does not address the potential for substantial bias to be introduced by ignoring any genuine signal between exposure and effect from lower quality studies, which could still be strong.

This can be illustrated by considering the Quebec study, where the assumed cumulative exposure value of 600 f/cc-yrs in the meta-analysis can reasonably be debated and sometimes is suggested to be potentially overestimated (as discussed later in Chapter 10). If so, the R_L value would in turn have been underestimated. Taking an extreme case that the cumulative exposure was overestimated by as much as a factor of 5 (bringing the estimate into line with that of the Qinghai mine), the resulting fivefold higher R_L value would still be an order of magnitude below the R_L value for the Qinghai mine. Given its weight in statistical terms, and despite its deficiencies and resulting uncertainties, the Quebec study does therefore provide a strong signal about the lung cancer risk per fiber and whether this should simply be dismissed entirely is certainly open to debate. These arguments are particularly important in the context of the particular set of cohorts in this meta-analysis given that the studies are different important ways other than quality, including fiber type and industrial context. Ultimately, the quality criteria approach does not answer the question of whether the South Carolina study has a much higher R_L value than Quebec because of it is higher quality, or that it simply provides a good estimate of the risk in the very particular exposure setting studied, which may or may not be appropriate for risk estimation in other exposure settings. Also, it should be noted the potential overestimation of the cumulative exposure in Quebec workers remains a judgment. As discussed in Chapter 10, the exposure intensity in Quebec miners was of the order of about 100 f/cc at least at the influential period of the mine's history, and many of the workers with asbestos-associated disease could be exposed to even higher levels. In general, our analysis with the inclusion of the Quebec study seems to be reasonable and valid.

Nonlinear Exposure–Response Relationship

The observation from the cohort studies summarized in the above meta-analyses reflect the effect of high cumulative asbestos exposures. The average cumulative exposures of the included studies ranged from tens of f/cc-yrs to many hundreds of f/cc-yrs. These exposures were generally accrued over a number of years of working in conditions where fiber levels were orders of magnitude higher than required by standards of control typically expected today. Nowadays, risk estimation may therefore often be focused on much lower cumulative exposures and require a very substantial extrapolation below the range of data for which empirical observations exist. Exposures of interest might, for example, include workers subject to high fiber levels but for a very limited time-period as a result of poorly controlled asbestos abatement work, or very long-term exposures at very low fiber levels that might be found in certain environments such as buildings containing asbestos materials in poor condition.

The simplest approach to estimating the risk for low-level exposures is to apply the summary values of R_M and R_L within a relevant population to derive lifetime risks, which in effect assumes a simple linear extrapolation to low cumulative exposures (see the final section of this chapter). In 2000, Hodgson and Darnton discussed certain features of the cohort study data that might draw this assumption into question.

The main motivation for exploring the possibility of a nonlinear exposure–response was that, although the total mesothelioma risk looks to increase in a broadly linear fashion with cumulative exposure within the range of the available data, the relationship looks different for pleural and peritoneal mesothelioma when considered separately. In the original meta-analysis, peritoneal mesotheliomas were not seen among the chrysotile cohorts (with the exception of a possible single case in the South Carolina textiles cohort). Among the cohorts with amphibole exposure, the proportion of cases that were peritoneal tends to increase as the average cumulative exposure increases. For example, in the Wittenoom study (cumulative exposure: 23 f/cc-yrs) peritoneal cases accounted for 15% of mesotheliomas, whereas in the Massachusetts cohort (cumulative exposure: 120 f/cc-yrs), three of the five cases were peritoneal, and in the Johns Manville cohort (cumulative exposure: 750 f/cc-yrs) 62% were peritoneal.

Examination of pleural and peritoneal mesothelioma as a proportion of expected all-cause mortality in cohorts with pure crocidolite or pure amosite exposure (regardless of whether quantitative exposure estimates were available) suggested a nonlinear relationship in which the peritoneal mesothelioma rate was proportional to at least the square (perhaps as much as the cube) of the pleural rate. The implication of this is that the apparent linear relationship between total mesothelioma and cumulative exposure in the observed data masks two different exposure–response relationships for pleural and peritoneal mesothelioma. Despite the fact that both pleural and peritoneal mesothelioma reflect the effect of the same carcinogen on the same kind of tissue, a different dose-response might be expected depending on how asbestos fibers are distributed around the body following inhalation.

The updated meta-analysis (Darnton 2023) revisited the nonlinear models developed originally, and considered the impact of incorporating the new data. The general form of these models is such that the percentage excess risk of each outcome (i.e. pleural mesothelioma, peritoneal mesothelioma, or lung cancer) is proportional to cumulative exposure to a specific type of asbestos raised to some power. The power parameter – which we call the "slope" – introduces the nonlinearity to the exposure–response. After exploring a range of these kinds of models, Hodgson and Darnton 2000, adopted a form with a common slope for all fiber types (but different proportionality coefficients), which can be written mathematically as follows:

For pleural mesothelioma: $P_r = A_{(fib)} \cdot X^r$

For peritoneal mesothelioma: $P_t = B_{(fib)} \cdot X^t$

For lung cancer: $P_l = C_{(fib)} \cdot X^l$

where,

P_r = percent excess pleural mesothelioma mortality;
P_t = percent excess peritoneal mesothelioma mortality;
P_l = percent excess lung cancer mortality;
$A_{(fib)}$, $B_{(fib)}$, and $C_{(fib)}$ are regression coefficients for fiber-type factors;
X = cumulative asbestos exposure (f/cc-yrs); and
r, t, and l are the fitted slope parameters that describe a sublinear exposure–response if less than 1 and a supra-linear relationship if greater than 1.

Pleural Mesothelioma

The best fitting model for pleural mesothelioma from the original 2000 analysis – which was sublinear in form with slope parameter $r = 0.75$ (and fiber coefficients

0.93, 0.13, and 0.0047 for crocidolite, amosite, and chrysotile, respectively) no longer provides and adequate fit to the updated data (excluding the Libby amphibole cohort data point). Reoptimizing the parameters to the new data set (and allowing a separate factor for Libby) produces a model with more extreme sublinearity ($r = 0.48$), but this does not constitute an adequate fit to the data. This lower slope value is driven by the chrysotile data, and in particular, the presence of both the North Carolina and Quebec data points, which are influential statistically. If the chrysotile data is excluded, the best-fitting slope returns to a value close to the original ($r = 0.77$), but a model with a linear slope ($r = 1$) does not amount to a significant worsening of the fit statistically (Model deviance $D = 2.14$ on 3 degrees of freedom for $r = 0.77$ vs. $D = 2.66$ for $r = 1$). The position of the data points shown in Figure 8.3 is suggestive of a model with two separate chrysotile coefficients. This model – indicated by the lines in Figure 8.3 (the grey lines determined by the "high" and "low" chrysotile coefficients) – does provide an adequate fit with a slope of $r = 0.82$. Overall, the pleural mesothelioma data are still consistent

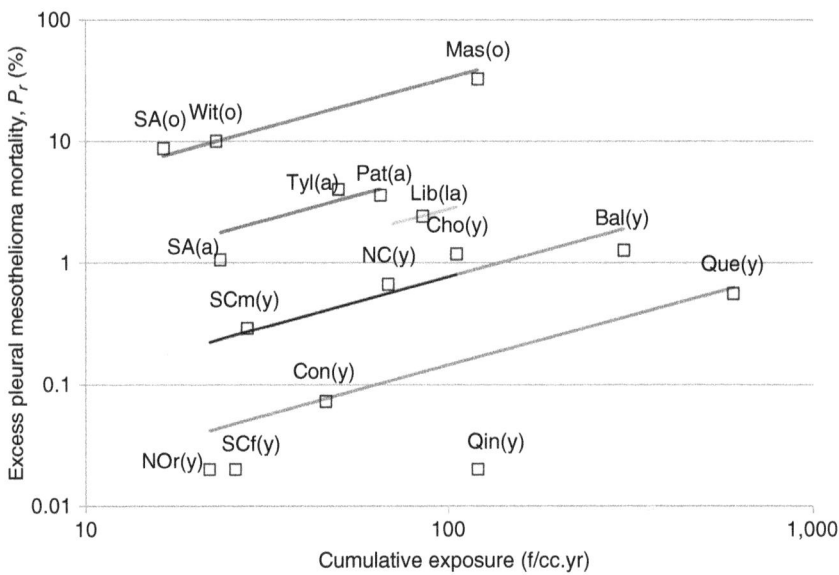

Figure 8.3 Excess pleural mesothelioma mortality by cumulative asbestos exposure. Model coefficients are: $r = 0.82$, $A_{(o)} = 0.77$ (top line), $A_{(a)} = 0.13$ (second line), $A_{(la)} = 0.064$ (third line), $A_{(y,high)} = 0.018$ (second from bottom line), and $A_{(y,low)} = 0.0034$ (bottom line). Key to data points: Wit(o) Wittenoom; SA(o) SA crocidolite mines; Mas(o) Massachusetts; SA(a) SA amosite mines; Pat(a) Paterson; Tyl(a) Tyler; Lib(la) Libby amphiboles; Que(y) Quebec; Bal(y) Balangero; Qin(y) Qinghai mine; Nor(y) New Orleans, plant 2; SCf(y) South Carolina (women); SCm(y) South Carolina (men); NC(y) North Carolina; Cho(y) Chongqing factory; Con(y) Connecticut. Zero values are plotted at 0.02 on the vertical axis.

with a sublinear exposure response, although the extent of this nonlinearity is dependent on the treatment of the chrysotile data.

The sublinear form of these models is unusual. Progressing down the cumulative exposure scale, the predicted risk from these models departs increasingly (in an upward direction) from a simple linear extrapolation to the origin. For example, for amosite exposure at 1 f/cc-yrs, the nonlinear model shown in Figure 8.3 (with $r = 0.82$) gives a predicted risk estimate of 0.13% vs. an estimate of 0.11% for a linear extrapolation (i.e. $r = 1$). At 0.1 f/cc-yrs – that is a factor of 10 lower – the predicted risk for the nonlinear model is reduced by a factor of 6.5 to a value of 0.02%, whereas the linear extrapolation reduces by a factor 10 to 0.011%. If true, this nonlinearity has important implications for low-dose extrapolations, which are often the focus of risk estimation in contemporary exposure settings.

Given uncertainties in the underlying data, and other available evidence about mesothelioma exposure–response, the veracity of sublinear (at higher doses) relationship seen here can be questioned. Korchevskiy and Korchevskiy (2022) demonstrated that the Peto model of age-dependent mesothelioma incidence (Chapter 7) could lead to an apparently sublinear (at higher doses) relationship in the presence of unaccounted for variation in exposure duration between studies. At the same time, a recent British study of mesothelioma lifetime risk in relation to amphibole asbestos lung burden showed evidence of a linear dose-response (Gilham et al. 2016) (though the linearity of mesothelioma risk by lung burden in this study is also associated with considerable uncertainty).

Peritoneal Mesothelioma

A nonlinear exposure–response for peritoneal mesothelioma is less easy to dismiss based on the available data – and particularly given the observation of a generally increasing proportion of peritoneal cases in studies with higher exposures (of a given fiber type) – see Table 8.1. If the pleural mesothelioma exposure–response is linear or mildly sublinear in form, this implies that the peritoneal mesothelioma exposure–response must be supra-linear.

In both the original and updated meta-analyses modeling was restricted to the amphibole cohorts given the small number of peritoneal cases in chrysotile exposure settings. Figure 8.4 shows the model for excess peritoneal mesothelioma mortality. A good fit to the data is achieved with a slope parameter of $t = 2.1$ – the same as in the original analysis – and with similar fiber-type coefficients (a third coefficient has been introduced in the update to incorporate the Libby amphibole data).

These models imply that the peritoneal mesothelioma risk increases with more than the square of cumulative exposure. However, the fiber-specific regression coefficients have small values compared to those of the model for pleural

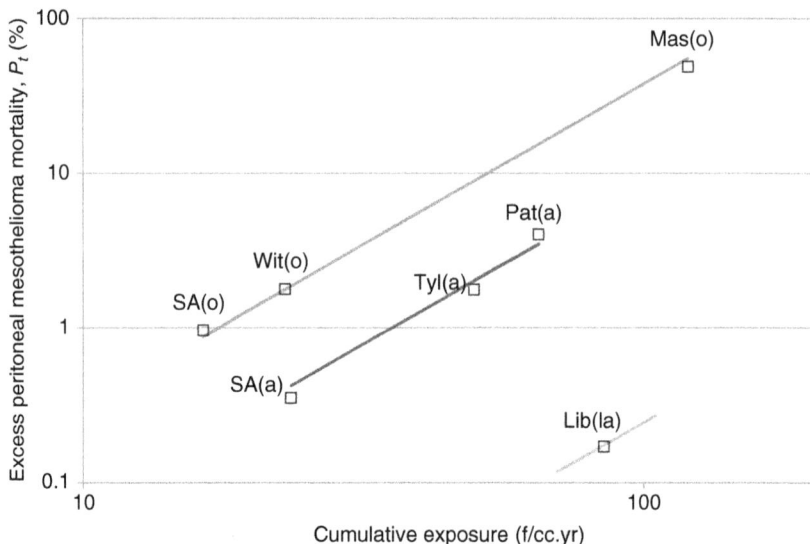

Figure 8.4 Excess peritoneal mesothelioma mortality by cumulative asbestos exposure. Model coefficients are: $t = 2.1$, $B_{(o)} = 0.0026$ (top line), $B_{(a)} = 0.00059$ (second line), $B_{(la)} = 0.000017$ (bottom line). Key to data points: Wit(o) Wittenoom; SA(o) SA crocidolite mines; Mas(o) Massachusetts; SA(a) SA amosite mines; Pat(a) Paterson; Tyl(a) Tyler; Lib(la) Libby amphiboles.

mesothelioma. This explains why when considering mesothelioma in total (pleural plus peritoneal cases), peritoneal cases only become the dominant component at very high cumulative exposures. The implication for low-dose extrapolations is that below about 1 f/cc-yrs, the total mesothelioma risk is effectively entirely determined by pleural cases.

Lung Cancer

The presence of a nonlinear exposure–response for mesothelioma – at least for peritoneal mesothelioma, if not both the pleural and peritoneal components – raises the question of whether such a relationship also exists for lung cancer.

The amphibole data are shown in Figure 8.5 and the chrysotile data in Figure 8.6. The latter are treated separately in the modeling due to the very considerable uncertainties in the data points.

In the original meta-analysis, models with a wide range of slopes with $l > 1$ and a single coefficient for both crocidolite and amosite combined were consistent with the five available amphibole data points. It was not possible to decide between these to determine a best-fitting slope on statistical grounds alone. A preferred slope of $l = 1.3$, which is toward the lower end of the full range of possible slopes,

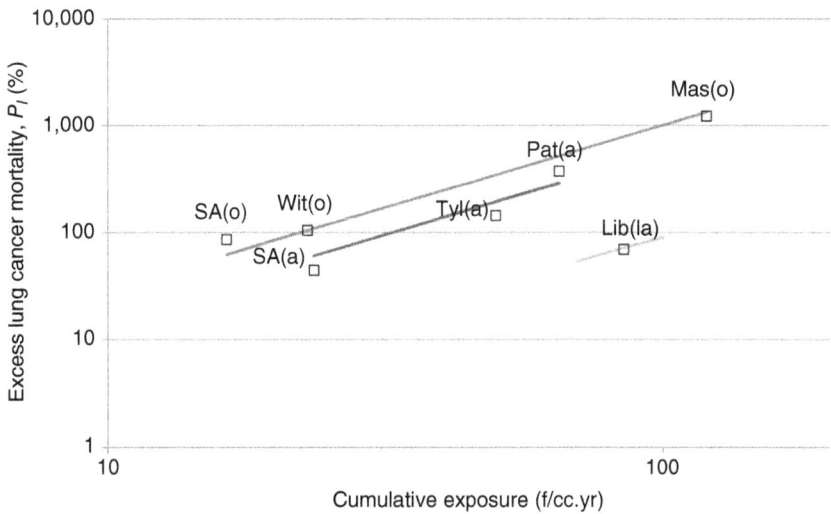

Figure 8.5 Excess lung cancer mortality by cumulative asbestos exposure (amphibole cohorts). Model coefficients are: $l = 1.54$, $C_{(o)} = 0.84$ (top line), $C_{(a)} = 0.47$ (second line), and $C_{(la)} = 0.07$ (bottom line). Key to data points: Wit(o) Wittenoom; SA(o) SA crocidolite mines; Mas(o) Massachusetts; SA(a) SA amosite mines; Pat(a) Paterson; Tyl(a) Tyler; Lib(la) Libby amphiboles. Zero values are plotted at 0.02 on the vertical axis.

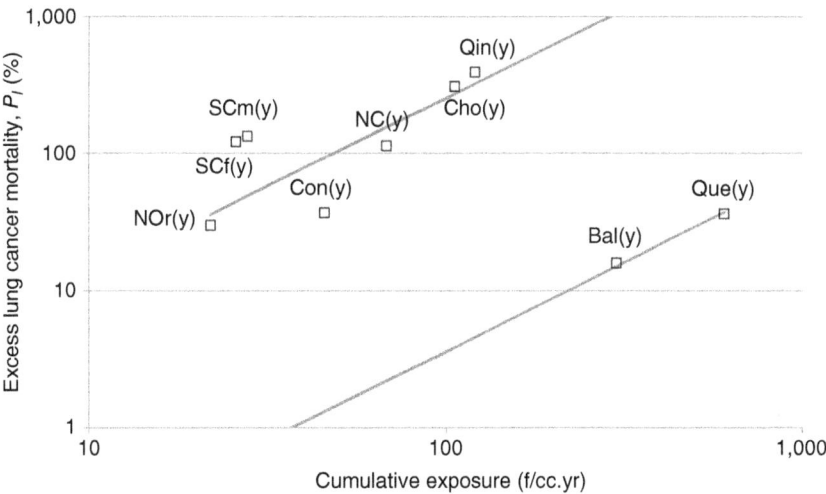

Figure 8.6 Excess lung cancer mortality by cumulative asbestos exposure (chrysotile cohorts). Model coefficients are: $l = 1.3$, $C_{(y\ tex)} = 0.64$ (top left line), $C_{(y\ nontex)} = 0.009$ (bottom right line). Key to data points: Que(y) Quebec; Bal(y) Balangero; Qin(y) Qinghai mine; Nor(y) New Orleans, plant 2; SCf(y) South Carolina (women); SCm(y) South Carolina (men); NC(y) North Carolina; Cho(y) Chongqing factory; Con(y) Connecticut.

was justified by arguing that while the data presented in this way are suggestive of a supra-linear relationship, an upward bias could have been introduced (particularly by the Massachusetts cohort), and noting that within study exposure–response relationships of this form have rarely been considered in other research. Treating all the amphibole data as a whole (refitting with a single fiber coefficient as originally) does not result in an adequately fitting model with the updated data. Instead, a model with separate coefficients for crocidolite, amosite, and Libby amphiboles is suggested by the data in Figure 8.5, but this has stronger supra-linearity ($l = 1.54$) and constraining the slope to lower values such as $l = 1.3$ (the original best slope) or $l = 1$ (linear) degrades the fit statistically.

Treating the chrysotile data shown in Figure 8.6 as a single group (with a single fiber coefficient) is not possible as this produces a negative slope. Two separate lines are suggested by the data: a "high chrysotile" line through most of the data points including the textiles cohorts, and a "low chrysotile" line determined by the Quebec and Balangero cohorts. If this form is adopted, modeling is then suggestive of lower slope values than for the amphibole data with a linear model within the range of possibilities – although the fit of all these models is poor statistically. Figure 8.6 shows the version in which the slope is constrained to a value or $l = 1.3$ (the preferred model from the original analysis).

Summary

Clearly, there are considerable uncertainties associated with the evidence presented here. We can be most confident about the conclusions for peritoneal mesothelioma in relation to amphibole exposure for which the data strongly suggest a supra-linear exposure response. A supra-linear relationship (though less extreme) is also suggested for lung cancer in relation to amphibole exposure, but whether some lesser degree of nonlinearity is present in the exposure–response for lung cancer in relation to chrysotile is not clear from the data. The pleural mesothelioma exposure–response is perhaps the most challenging to decide upon since this has important implications for low-dose extrapolations used in risk assessment. Taken at face value, a model with a moderately sublinear form is supported by the data and would lead to conservative risk estimates when applied at low exposures.

Application of Hodgson and Darnton for Risk Assessment

The fiber-specific potency values for mesothelioma (R_M) and excess lung cancer due to asbestos (R_L), and their equivalent forms (P_r, P_t, and P_l) based on the non-linear models developed by Hodgson and Darnton, can be used to estimate the lifetime risk of each outcome for a given level of cumulative exposure (in f/cc-yrs).

Since it expresses the proportion of total mortality due to mesothelioma (as a percentage), the metric R_M approximates to the lifetime risk under the assumption that the long-term pattern of increase in incidence over time is the same as for total mortality. However, for very long follow-up at old ages, the overall mortality rate eventually becomes very high, and it is not clear whether mesothelioma incidence keeps track with this pattern. Indeed, some studies have suggested mesothelioma incidence eventually reduces at very long follow-up times, although this too is uncertain (McDonald et al. 2006; Harding et al. 2009). In light of this, Hodgson and Darnton in 2000 suggested truncating the risk estimation at age 80 (i.e. 50 years after exposure starting at age 30). The average male lifetable used to adjust the expected all-cause mortality in each cohort in the meta-analysis to age 30 predicts that 70% of all-cause deaths occur by age 80. Thus, lifetime risk estimates based on a truncated view of the long-term mesothelioma risk were multiplied by 0.7. This factor applies the R_M metric as derived in its historical context of the average overall mortality of the populations from which the various studies were part. Different adjustments may be appropriate for the application of the risk models to more recent populations with different expected long-term total morality patterns (see Chapter 12).

The predicted number of mesothelioma cases for cumulative exposure starting at age a (M_a) can be summarized as:

$$M_a = R_M \times \text{CE} \times N \times 0.7 \times F_a / 100$$

where

R_M is the fiber-type specific potency factor (for the linear model);
CE is the average cumulative exposure in f/cc-yrs (assuming exposure during working hours);
N is the size of the group for which the lifetime risk is being estimated; and
F_a – age-specific adjustment factor derived using the long-term pattern of mesothelioma incidence from different starting ages using the Peto model.

These risk estimates will be valid for cumulative exposures accrued over relatively short periods (up to about 5 years). Further adjustment is required for cumulative exposures accrued over extended exposure periods given that the Peto model implies that a given cumulative exposure accrued over a long period will confer a lower risk that the same cumulative exposure accrued over a short period.

Similarly, the projected number of lifetime excess lung cancer cases (L) in a cohort of N workers, with average cumulative exposure to asbestos CE (assuming exposure during regular occupational working year) (f/cc-years) can be estimated as:

$$L = \left(R_L \times \text{CE} + 1\right) \times E_L \times N / 100$$

where

R_L is the fiber-type specific potency factor (for the linear method) and
E_L is the total expected lung cancer mortality in corresponding nonasbestos-exposed reference population (deaths per person).

In spite of the significant limitations, this approach can serve as a basis for risk prediction for various exposure scenarios. In particular, the linear approximation of risks at low doses fits the overall methodological vision of different regulatory agencies, including the US EPA (US EPA 2005). Application of the nonlinear model for mesothelioma to lower dose scenarios will tend to lead to conservative risk estimates.

Conclusions

Mortality studies of asbestos-exposed worker cohorts show that asbestos mineral fiber type is a major determinant of potency, particularly in relation to mesothelioma induction.

Quantitative differences in mesothelioma potency can be inferred from mortality and exposure data for cohorts categorized as mainly involving exposure to particular asbestos mineral fiber types. Average lifetime mesothelioma risks per unit of cumulative exposure are highest for cohorts categorized as crocidolite only: about 4 times higher than amosite, 17 times higher than the Libby mixed amphiboles, and up to 500 times higher than cohorts categorized as chrysotile.

Three features of the data in this analysis continue to support the conclusion that amosite and crocidolite asbestos usually have a higher potency than chrysotile in relation to lung cancer: (1) amosite- and crocidolite-exposed cohorts have consistent risks per unit exposure, which are at the upper end of values across all the cohorts considered, (2) while some of the chrysotile cohorts (typically, for asbestos textile workers) have risks per unit cumulative exposure as high as amphiboles, very substantial heterogeneity is present across the group of chrysotile cohorts, with some having much lower values than crocidolite and amosite, and, (3) cohorts with mixed mineral fiber type exposures also show very substantial heterogeneity with risks spanning the same range as the chrysotile cohorts, and again many of these are much lower than for crocidolite or amosite only. Nevertheless, the additional evidence available since the original Hodgson and Darnton meta-analysis tends to strengthen the argument that chrysotile exposure in textile manufacturing may pose a lung cancer risk per cumulative exposure of a similar order to amphiboles. Other subtypes of chrysotile asbestos tend to have lower lung cancer potency factors (about an order of magnitude less than asbestos textile cohorts). This emphasizes the need to determine those factors in addition to the mineral type of fibers – such as the distribution of fiber size, habit of the mineral, and potentially other factors – that may substantially modify the risk in certain settings.

Inter-study Poisson regression analyses, which allow the exposure–response to be nonlinear, show the peritoneal mesothelioma risk is related to about the square of cumulative exposure. The exposure–response for pleural mesothelioma and lung cancer may also be nonlinear by cumulative exposure, though the evidence for these outcomes is less clear-cut and any nonlinearity is likely to be less extreme in extent than for peritoneal mesothelioma, and also can be driven by other factors (like exposure duration).

There are additional arguments supporting the conclusions of our meta-analysis. The proportional difference between mesothelioma potency for various mineral types of asbestos can be efficiently reconstructed from data on the chemical composition and dimensionality of fibers. Remarkably, even assuming that exposure intensity measurements are very uncertain in the mining cohorts, the potency factors per year of exposure duration for the Wittenoom crocidolite cohort, South Africa amosite cohort, and Quebec mining cohort are estimated at 100 : 10 : 1 for lung cancer and 1,000 : 10 : 1 for mesothelioma, again showing significant difference by mineral type (A. Korchevskiy, personal communication).

When this book was being prepared for publication, a new paper was published by Schüz et al. regarding a cohort study of workers from the Uralasbest chrysotile asbestos mine in Russia (Schüz et al. 2024). In the cohort of 30,445 workers, 13 mesothelioma cases along with around 45 excess lung cancer cases were found. The authors estimated the overall average cumulative exposure of workers as 33.6 f/cc-yrs, an order of magnitude lower than that estimated for the Quebec and Balangero chrysotile miners' cohorts. Further analysis is needed to incorporate these results into the meta-analysis, but preliminary calculations suggest an R_M value of a similar order to that seen for the Balangero miners and an R_L value around an order of magnitude higher (which is still substantially lower than values typical of the chrysotile textiles cohorts). This suggests that inclusion of these new data points would not significantly impact on the meta-analytical values proposed in this chapter.

The results of the meta-analysis based on the Hodgson and Darnton method create a basis for practical risk assessment, especially when stochastic uncertainties in the data are taken into account. Further studies would be needed to better understand the role of additional characteristics that influence the difference in potency for various cohorts and types of cancer.

The views expressed in this chapter are those of the author and not necessarily the ones of Health and Safety Executive (HSE).

References

Acheson, E.D., Gardner, M.J., Winter, P.D., and Bennett, C. (1984). Cancer in a factory using amosite asbestos. *International Journal of Epidemiology* 13(1): 3–10. https://doi.org/10.1093/ije/13.1.3.

Armstrong, B.K., de Klerk, N.H., Musk, A.M., and Hobbs, M.S. (1988). Mortality in miners and millers of crocidolite in Western Australia. *British Journal of Industrial Medicine* 45(1): 5–13. https://doi.org/10.1136%2Foem.45.1.5.

Berman, D.W. and Crump, K.S. (2008). A meta-analysis of asbestos-related cancer risk that addresses fiber size and mineral type. *Critical Reviews in Toxicology* 38(S1): 49–73. https://doi.org/10.1080/10408440802273156.

Berry, G., Reid, A., Aboagye-Sarfo, P., et al. (2012). Malignant mesotheliomas in former miners and millers of crocidolite at Wittenoom (Western Australia) after more than 50 years follow-up. *British Journal of Cancer* 106(5): 1016–1020. https://doi.org/10.1038/bjc.2012.23.

Darnton, L. (2023). Quantitative assessment of mesothelioma and lung cancer risk based on Phase Contrast Microscopy (PCM) estimates of fibre exposure: an update of 2000 asbestos cohort data. *Environmental Research* 230: 114753. https://doi.org/10.1016/j.envres.2022.114753.

Dement, J.M., Brown, D.P., and Okun, A. (1994). Follow-up study of chrysotile asbestos textile workers: cohort mortality and case-control analyses. *American Journal of Industrial Medicine* 26(4): 431–447. https://doi.org/10.1002/ajim.4700260402.

Dement, J.M. and Loomis, D. (2023). Manufactured doubt and the EPA 2020 chrysotile asbestos risk assessment. *American Journal of Industrial Medicine* 66(7): 543–553. https://doi.org/10.1002/ajim.23476.

Doll, R. (1955). Mortality from lung cancer in asbestos workers. *British Journal of Industrial Medicine* 12(2): 81–86. https://doi.org/10.1136%2Foem.12.2.81.

Doll, R. and Peto, J. (1985). *Asbestos: Effects in Health of Exposure to Asbestos: Health and Safety Commission*. London: Her Majesty's Stationary Office.

Finkelstein, M.M. and Meisenkothen, C. (2010). Malignant mesothelioma among employees of a Connecticut factory that manufactured friction materials using chrysotile asbestos. *The Annals of Occupational Hygiene* 54(6): 692–696. https://doi.org/10.1093/annhyg/meq046.

Garabrant, D. (2020). Garabrant comments on US EPA Draft Risk Evaluation for Asbestos, dated March 2020 (EPA Document # EPA-740-R1-8012). May 2020. Docket identification (ID) number EPA-HQ-OPPT-2019-0501. https://www.regulations.gov/document/EPA-HQ-OPPT-2019-0501-0034. Accessed September 9th, 2022.

Gibbs, G.W. and LaChance, M. (1974). Dust-fiber relationships in the Quebec chrysotile industry. *Archives of Environmental Health* 28(2): 69–71. https://doi.org/10.1080/00039896.1974.10666439.

Gibbs, G.W. (1994). The assessment of exposure in terms of fibres. *The Annals of Occupational Hygiene* 38(4): 477–487. https://doi.org/10.1093/annhyg/38.4.477.

Gilham, C., Rake, C., Burdett, G., et al. (2016). Pleural mesothelioma and lung cancer risks in relation to occupational history and asbestos lung burden. *Occupational and Environmental Medicine* 73(5): 290–299. https://doi.org/10.1136/oemed-2015-103074.

Harding, A.-H., Darnton, A., Wegerdt, J., and McElvenny, D. (2009). Mortality among British asbestos workers undergoing regular medical examinations (1971–2005). *Occupational and Environmental Medicine* 66(7): 487–495. https://doi.org/10.1136/oem.2008.043414.

Hein, M.J., Stayner, L.T., Lehman, E., et al. (2007). Follow-up study of chrysotile textile workers: cohort mortality and exposure-response. *Occupational and Environmental Medicine* 64(9): 616–625. https://doi.org/10.1136%2Foem.2006.031005.

Hodgson, J.T. and Darnton, A. (2000). The quantitative risks of mesothelioma and lung cancer in relation to asbestos exposure. *The Annals of Occupational Hygiene* 44(8): 565–601.

Hodgson, J.T. (2013). Quality of evidence must guide risk assessment of asbestos, by Lenters, V; Burdorf, A; Vermeulen, R; Stayner, L; Heederik, D. *Annals of Occupational Hygiene* 57(5): 670–674. https://doi.org/10.1093/annhyg/met016.

Hughes, J.M., Weill, H., and Hammad, Y.Y. (1987). Mortality of workers employed in two asbestos cement manufacturing plants. *British Journal of Industrial Medicine* 44(3): 161–174. https://doi.org/10.1136%2Foem.44.3.161.

Korchevskiy, A., Rasmuson, J.O., and Rasmuson, E.J. (2019). Empirical model of mesothelioma potency factors for different mineral fibers based on their chemical composition and dimensionality. *Inhalation Toxicology* 31(5): 180–191. https://doi.org/10.1080/08958378.2019.1640320.

Korchevskiy, A.A. and Korchevskiy, A. (2022). Non-linearity in cancer dose-response: the role of exposure duration. *Computational Toxicology* 22: 100217. https://doi.org/10.1016/j.comtox.2022.100217.

Korchevskiy, A.A. and Wylie, A.G. (2023). Toxicological and epidemiological approaches to carcinogenic potency modeling for mixed mineral fiber exposure: the case of fibrous balangeroite and chrysotile. *Inhalation Toxicology* 35(7–8): 185–200. https://doi.org/10.1080/08958378.2023.2213720.

Larson, T.C., Antao, V.C., and Bove, F.J. (2010). Vermiculite worker mortality: estimated effects of occupational exposure to Libby amphibole. *Journal of Occupational and Environmental Medicine* 52(5): 555–560. https://doi.org/10.1097/jom.0b013e3181dc6d45.

Lenters, V., Burdorf, A., Vermeulen, R., et al. (2012). Quality of evidence must guide risk assessment of asbestos. *Annals of Occupational Hygiene* 56(8): 879–887. https://doi.org/10.1093/annhyg/mes065.

Levin, J.L., Rouk, A., Shepherd, S., et al. (2016). Tyler asbestos workers: a mortality update in a cohort exposed to amosite. *Journal of Toxicology and Environmental Health, Part B* 19(5–6): 190–200. https://doi.org/10.1080/10937404.2016.1195319.

Liddell, F.D., McDonald, A.D., and McDonald, J.C. (1997). The 1891–1920 cohort of Quebec chrysotile miners and millers: development from 1904 and mortality to 1992. *Annals of Occupational Hygiene.* 41(1): 13–36. https://doi.org/10.1016/s0003-4878(96)00044-0.

Loomis, D., Dement, J.M., Wolf, S.H., et al. (2009). Lung cancer mortality and fibre exposures among North Carolina asbestos textile workers. *Occupational and Environmental Medicine* 66(8): 535–542. https://doi.org/10.1136/oem.2008.044362.

McDonald, A.D., Fry, J.S., Woolley, A.J., et al. (1984). Dust exposure and mortality in an American chrysotile asbestos friction products plant. *British Journal of Industrial Medicine* 46: 151–157. https://doi.org/10.1136%2Foem.41.2.151.

McDonald, J.C. and McDonald, A.D. (1996). The epidemiology of mesothelioma in historical context. *European Respiratory Journal* 9(9): 1932–1942. https://doi.org/10.1183/09031936.96.09091932.

McDonald, J.C., Harris, J., and Armstrong, B. (2004). Mortality in a cohort of vermiculite miners exposed to fibrous amphibole in Libby, Montana. *Occupational and Environmental Medicine* 61: 363–366. https://doi.org/10.1136%2Foem.2003.008649.

McDonald, J.C., Harris, J.M., and Berry, G. (2006). Sixty years on: the price of assembling military gas masks in 1940. *Occupational and Environmental Medicine* 63(12): 852–855. https://doi.org/10.1136%2Foem.2006.028258.

Musk, A.W., de Klerk, N.H., Reid, A., et al. (2008). Mortality of former crocidolite (blue asbestos) miners and millers at Wittenoom. *Occupational and Environmental Medicine* 65(8): 541–543. https://doi.org/10.1136/oem.2007.034280.

Peto, J., Seidman, H., and Selikoff, I.J. (1982). Mesothelioma mortality in asbestos workers: implications for models of carcinogenesis and risk assessment. *British Journal of Cancer* 45(1): 124–135. https://doi.org/10.1038%2Fbjc.1982.15.

Piolatto, G., Negri, E., La Vecchia, C., et al. (1990). An update of cancer mortality among chrysotile asbestos miners in Balangero, northern Italy. *British Journal of Industrial Medicine* 47(12): 810–814. https://doi.org/10.1136%2Foem.47.12.810.

Pira, E., Romano, C., Donato, F., et al. (2017). Mortality from cancer and other causes among Italian chrysotile asbestos miners. *Occupational and Environmental Medicine* 74(8): 558–563. https://doi.org/10.1136/oemed-2016-103673.

Ribak, J. and Ribak, G. (2008). Human health effects associated with the commercial use of grunerite asbestos (amosite): Paterson, NJ; Tyler, TX; Uxbridge, UK. *Regulatory Toxicology and Pharmacology* 52(1, Suppl): S82–S90. https://doi.org/10.1016/j.yrtph.2007.10.002.

Schüz, J., Kovalevskiy, E., Olsson, A., et al. (2024). Cancer mortality in chrysotile miners and millers, Russian Federation: main results (Asbest Chrysotile

Cohort-Study). *JNCI: Journal of the National Cancer Institute*, djad262. https://doi.org/10.1093/jnci/djad262.

Seidman, H, Selikoff, I.J., and Gelb, S.K. (1986). Mortality experience of amosite asbestos factory workers: dose–response relationships 5 to 40 years after onset of short-term work exposure. *American Journal of Industrial Medicine* 10(5–6): 479–514. https://doi.org/10.1002/ajim.4700100506.

Shedd, K.B. (1985). Fiber dimensions of crocidolites from Western Australia, Bolivia, and the Cape and Transvaal Provinces of South Africa. United States Department of the Interior. Bureau of Mines Report of Investigations No. 8998.

Sluis-Cremer, G.K, Liddell, F.D., Logan, W.P. et al. (1992). The mortality of amphibole miners in South Africa, 1946–1980. *British Journal of Industrial Medicine* 49(8): 566–575. https://doi.org/10.1136/oem.49.8.566.

Sullivan, P.A. (2007). Vermiculite, respiratory disease, and asbestos exposure in Libby, Montana: update of a cohort mortality study. *Environmental Health Perspectives* 115(4): 579–585. https://doi.org/10.1289%2Fehp.9481.

Talcott, J.A, Thurber, W.A, Kantor, A.F., et al. (1989). Asbestos associated diseases in a cohort of cigarette-filter workers. *New England Journal of Medicine* 321(18): 1220–1223. https://doi.org/10.1056/nejm198911023211803.

Teta, M.J., Lewinsohn, H.C., Meigs, J.W., et al. (1983). Mesothelioma in Connecticut, 1955–1977. Occupational and geographic associations. *Journal of Occupational Medicine* 25(10): 749–756.

US EPA. (2005). Guidelines for Carcinogen Risk Assessment, EPA/630/P-03/001F.

Wagner, J.C., Sleggs, C.A., and Marchand, P. (1960). Diffuse pleural mesothelioma and asbestos exposure in the North Western Cape Province. *British Journal of Industrial Medicine* 17(4): 260–271. https://doi.org/10.1136/oem.17.4.260.

Wang, X., Lin, S., Yu, I., et al. (2013a). Cause-specific mortality in a Chinese chrysotile textile worker cohort. *Cancer Science* 104(2): 245–249. https://doi.org/10.1111%2Fcas.12060.

Wang, X., Yano, E., Lin, S., et al. (2013b). Cancer mortality in Chinese chrysotile asbestos miners: exposure-response relationships. *PLoS One* 8(8): e71899. https://doi.org/10.1371/journal.pone.0071899.

Yano, E., Wang, Z.M., Wang, X.R., et al. (2001). Cancer mortality among workers exposed to amphibole-free chrysotile asbestos. *American Journal of Epidemiology* 154(6): 538–543. https://doi.org/10.1093/aje/154.6.538.

9

Prediction of Mesothelioma Mortality in the Context of Country-wide Risk Evaluation

Lucy Darnton

Science Division, Epidemiology and Predictive Modelling, Bootle, Merseyside, UK

Mesothelioma mortality projection is an important exercise that contributes to our understanding of the overall health impact of historical patterns of asbestos usage and subsequent strategies to control exposures – such as bans and strict measures for asbestos abatement and management – at the national level. The purpose of this chapter is to describe the methodological approach underpinning the latest projections of national mesothelioma mortality in Great Britain, published as National Statistics by the Health and Safety Executive (2020), and to demonstrate how this approach can be applied to assess the extent to which different population asbestos exposure scenarios lead to different health impacts.

Predictions of mesothelioma mortality have been made in various countries, some of which are described below. In Britain, the rapid increases in annual mesothelioma mortality occurring during the 1980s and 1990s led to early modeling and projections of future mortality based on a simple age-birth cohort methodology (Peto et al. 1995). These results highlighted for the first time the likely full impact of widespread past asbestos usage in Britain. This drew attention to the need to carry out focused research to fully understand the past sources of mesothelioma risk in order to inform the development of the most effective regulation and control approaches.

Examples of mesothelioma projection modeling in countries other than Great Britain include work by Segura, Burdorf, and Looman who predicted the expected number of mesothelioma deaths in the Netherlands from 2000 to 2028 (Segura et al. 2003). Through an age-period-cohort modeling technique, age-specific mortality rates and cohort-relative risks by year of birth were calculated from the

Health Risk Assessment for Asbestos and Other Fibrous Minerals, First Edition.
Edited by Andrey Korchevskiy, James Rasmuson, and Eric Rasmuson.
© 2024 John Wiley & Sons, Inc. Published 2024 by John Wiley & Sons, Inc.

mortality of pleural mesothelioma in 1969–1998. Numbers of deaths for both sexes were predicted for 2000–2028, taking into account the most likely demographic development. The modeling demonstrated that men in the oldest age group had the highest age-specific death rates (79 per 100,000 person-years in the age group 80–84 years) and the highest relative risks were estimated for the birth cohorts of 1938–1942 and 1943–1947. The most plausible scenario predicts an increase in pleural mesothelioma mortality up to 490 cases per year in men, with a total death toll close to 12,400 cases during 2000–2028. However, using different assumptions, this death toll could rise to nearly 15,000 in men (20% increase). Mortality among women remained low, with a total death toll of about 800 cases. For 2018, four prediction models tested by the authors demonstrated mortality in men from 450 to 900 cases. The National Cancer Register of the Netherlands, at the same time, reported 445 mesothelioma death cases among males in 2018, generally confirming the less conservative trend of mortality (Integraal Kankercentrum Nederland 2020).

Oddone et al. (2020) discussed the prediction of malignant pleural mesothelioma (MPM) mortality in Italy after the ban on asbestos use. The authors described the trends of MPM deaths in Italy (1970–2014) and predicted the future number of cases in both sexes (2015–2039), with consideration of the national asbestos ban that was issued in 1992. For each 5-year period from 1970 to 2014, mortality rates were calculated and age-period-cohort Poisson models were used to predict future burden of MPM cases until 2039. During the period of 1970–2014, a total number of 28,907 MPM deaths were observed. MPM deaths increased constantly over the study period, ranging from 1,356 cases in 1970–1974 to 5,844 cases in 2010–2014. The peak of MPM cases is expected to be reached in the period of 2020–2024 (about 7,000 cases). Based on the developed model, the decrease will be slow: about 26,000 MPM cases are expected to occur in Italy during the next 20 years (2020–2039). The authors emphasized that the MPM epidemic in Italy is far from being concluded despite the national ban implemented in 1992, and the peak is expected in 2020–2024, in both sexes.

Lin et al. (2019) attempted to predict the MPM incidence in the next 30 years for Taiwan based on historical asbestos consumption. Annual data on local asbestos consumption during 1939–2015 was collected and sex-specific incidence of pleural cancer as a proxy for MPM during 1979–2013 was used. The authors applied Poisson log-linear models to predict future MPM numbers under the assumption that latency periods between asbestos exposure and MPM incidence were between 25 and 45 years. Asbestos consumption in Taiwan reached a peak in the 1980s, with a total of 668 thousand metric tons during 1939–2015. The observed number of MPM cases increased by nine-fold and six-fold in males and females respectively during 1979–2013, with a cumulative total of 907. Assuming a latency period of 31 years, MPM incidence was expected to peak at around 2012–2016 for males and 2016–2020 for females. In 2017–2046, the predicted total number of new MPM was 659 cases (95% confidence interval = 579–749); the male-to-female ratio was predicted to range from 1.8 to 2.8.

In the United States, Price and Ware (2009) examined trends in mesothelioma incidence based on data from the National Cancer Institute (NCI) Surveillance, Epidemiology, and End Results (SEER) database and made projections of future cases separately for men and women. Price recently updated these analyses based on more recent SEER data which includes 28% coverage of the US population, although this may disproportionately include populations with a higher likelihood of historic asbestos exposure in the United States. An age-birth cohort model was again applied to describe the observed data and provide a basis for future projections (Price 2022). The updated analyses confirmed previous conclusions that mesothelioma incidence attributed to past asbestos exposure is expected to decline during the 2020s and 2030s in both men and women, and of the cases occurring after 2040, the vast majority will be background cases.

The first projections of annual mesothelioma mortality in Great Britain were based on an age-birth cohort model fitted to male mortality data for 1968–1991 (Peto et al. 1995). These results suggested that annual male mesothelioma deaths would peak at between 2,700 and 3,300 deaths around the year 2020. In the underlying model, the pattern of increase in mesothelioma incidence with age is assumed to be the same in each period of birth. Observations of male mortality after 2000 suggested this may not be the case, with more recent birth cohorts tending to show a somewhat less steep rate of increase in incidence with age. The simple age-birth cohort approach could therefore lead to an overprediction of the risk in more recent birth cohorts and thus future mortality overall. These concerns led to the development of an alternative model in which annual mesothelioma mortality is modeled in terms of the past pattern of population exposure by calendar year, the propensity for exposure at different ages, and applying the relationship between time since exposure and incidence described by Peto (Peto et al. 1982; Hodgson et al. 2005; Tan et al. 2010). The original motivation was to assume that annual asbestos consumption (as measured by annual import tonnages) would be a proxy for annual population asbestos exposure. However, the available data on total annual imports (all asbestos mineral types combined) did not provide a sufficiently well-fitting model for predicting observed annual deaths within this framework. Instead, a model based on constructed annual exposure index was used. This index was found to correlate only with the amphibole (crocidolite and amosite) data (Hodgson et al. 2005). Other evidence also suggests that past amphibole exposures are the main past source of mesothelioma risk in Britain (Gilham et al. 2018 and that the reason why Britain has among the highest mesothelioma rates worldwide is due to particularly extensive use of amphibole asbestos in construction materials (Rake et al. 2009).

The model developed by Hodgson et al. and further refined by Tan and Warren remains the basis for the British mesothelioma projections, with the most recent results based on observations of mortality up to and including year 2017 (Health and Safety Executive 2023). The model expresses national mesothelioma incidence in terms of the sum of the risks produced by average population asbestos exposure in each past calendar year (with each risk assumed to increase with a power of the number of years since each took place). Population asbestos exposure is modeled as a single value for each calendar year – the "exposure index" – which represents the time-specific exposure averaged over the whole population. Varying propensities to be subject to this year-specific exposure are allowed for different population age categories.

The mesothelioma risk for an individual of a particular age (A) in a particular year (T) is given by the sum of the risk components due to exposure in all previous years of their lifetime (excluding the most recent 10 years), with each component for a given year of their lifetime – for example the year in which their age was $A - t$ years – calculated as the product of the exposure propensity appropriate to their age in that year, the exposure index value for that year, and the number of years t since that earlier year (i.e. the time since exposure lagged 10 years) raised to the power k.

For example, the component of the risk for someone aged 40 in year 1980 due to exposure in the year in which they were aged 20 (i.e. 1960) will be:

$$W_{20} D_{1960} \{20+1-10\}^k \tag{9.1}$$

where W_{20} is the relative propensity for exposure at age 20, D_{1960} is the exposure index for value for 1960, and k is the power of the time (20 years) since the exposure lagged by 10 years.

Such an individual will in fact have risk components from each year between 1940 and through to 1970:

$$W_0 D_{1940} \{40+1-10\}^k + W_1 D_{1941} \{39+1-10\}^k + \ldots + W_{30} D_{1970} \{10+1-10\}^k \tag{9.2}$$

In general terms, we write the risk component at age A and year T due to exposure in the year t years earlier as:

$$W_{A-t} D_{T-t} \{t+1-10\}^k \left(\text{for } t >= 10; \{\ \} \text{ set to zero otherwise}\right) \tag{9.3}$$

The predicted number of mesothelioma deaths at age A in year T is then proportional to the sum of these risk contributions multiplied by the total population aged A in year T (i.e. the person-years for age A and year T). The relationship is one of proportionality rather than equality since the annual exposure index describes the shape of the past population exposure in relative terms over time rather than, say, the actual airborne exposures in each past year. Thus, the predicted annual deaths from the model are rescaled so that the total fitted number of mesothelioma deaths over the period for which observed deaths are available is equal to the total observed number.

The model also incorporates parameters to account for the potential clearance of asbestos fibers from the lung and a "diagnostic trend" toward more complete recording of deaths over time.

The complete model can be represented mathematically as follows:

$$F_{A,T} = \frac{\left[\sum_{l=0}^{A+1} W_{A-l} D_{T-l} \{l+1-10\}^{k} \times c \right] \times d_T \times P_{A,T}}{\sum_{A,T} \left\{ \left[\sum_{l=0}^{A+1} W_{A-l} D_{T-l} \{l+1-10\}^{k} \times c \right] \times d_T \times P_{A,T} \right\}} \times (M-B) + B_{A,T}$$

(9.4)

where

$F_{A,T}$ = number of deaths at age A in year T;
W_A = age-specific exposure propensity at age A;
D_T = overall population exposure index for year T;
k = exponent of time representing the increase of risk with increase of time since exposure;
$P_{A,T}$ = person-years at risk for age A in year T;
M = total observed mesothelioma deaths in the overall observation period;
c = clearance of asbestos fibers from the lungs, with $c = (1/2)^{l/H}$, so that H is the clearance half-life in years. Clearance is indexed by the number of years from exposure, l, and the parameter H is a fitted parameter within the model such that large H leads to less clearance.
d_T = term to estimate a linear trend in diagnosis, that is the proportion of mesothelioma deaths in year T that are recorded; and
$\{\}$ = zero when negative.

The summations indexed by l represent the cumulative effect at age A of the exposures at earlier ages; l indexes years lagged from the risk year.

B = total "background" mesothelioma deaths in the observation period, equal to the sum of $B_{A,T}$, the background cases at age A and year T. This background describes the incidence of cases that would have occurred anyway in the British population in the absence of any commercial asbestos importation and usage. The background rate is a fitted parameter in the model and assumed to increase rapidly with age: it is proportional to $(A-10)^k$, with k taking the same value as the exponent of time since asbestos exposure.

The age-specific exposure propensity, W_A, is defined by assigning nine parameters to the age groups 0–4, 5–15, 16–19, 20–29, 30–39, 40–49, 50–59, 60–64, and 65+ years. The overall population exposure distribution, D_T, is parameterized by defining growth and decline rates for years in multiples of 10 before and after the maximum exposure year (in which exposure growth/decline is zero), with growth rates for years intermediate between the 10-yearly values were determined by linear interpolation.

In reality, the average population exposure arises from a complicated distribution of exposures accrued by workers, and others, across a wide range of settings. In the past this will have included traditional exposed industrial settings such as shipbuilding, asbestos product manufacturing, and asbestos lagging and construction activities. Following the cessation of new use of asbestos, this will have shifted to comprise asbestos removal workers and building maintenance workers. Within these groups, there will have been considerable variation in the number of workers exposed and the extent of their exposure on any given day. The model does not tell us anything about these complicated underlying distributions. Rather, the exposure index, D_T, just expresses the mean effective carcinogenic dose of asbestos to the population delivered in a given year. Summarizing the exposure in this way is justifiable given that the dose-response relationship is likely to be linear, or approximately so, so that the mean dose of a group (or population as a whole) will reliably predict the mean risk among that group (or population).

The results of the model-fitted data for males are shown in Figure 9.1a and for females in Figure 9.1b below. Models were fitted separately to the national mesothelioma data for males and females given the differing past patterns of population asbestos exposure in each case.

The profile of annual population exposure, D, implied by the model in each case is shown by the black dotted line in Figure 9.1a and b, with the actual annual deaths to 2017 shown by the black squares, the predicted annual mortality shown by the solid black lines, and the uncertainty interval implied by the particular statistical model used in each case shown by dashed black lines. For both males and females, the implied profile of annual population exposure, D, rises to a peak during the mid-1960s before reducing rapidly after this and during the 1970s. In broad terms, the consequent pattern of annual mesothelioma mortality lags this exposure profile by around 50 years with predicted peaks in annual mortality occurring in the latter part of the 2010s at around 2,100 deaths per year in males and 450 per year in females. The scaling of exposure profile is such than in females the peak is one-fifth of the male value and the shape of the subsequent decline in exposure results in the predicted level of annual mortality in the long-term (i.e. by 2050) being similar in both males and females (around 200–300 deaths per year in each case). The fitted background rate is similar in both males (1.25 per million per year) and females (1.30 per million per year), which is broadly equivalent to about 30 deaths per year in each sex in 2020.

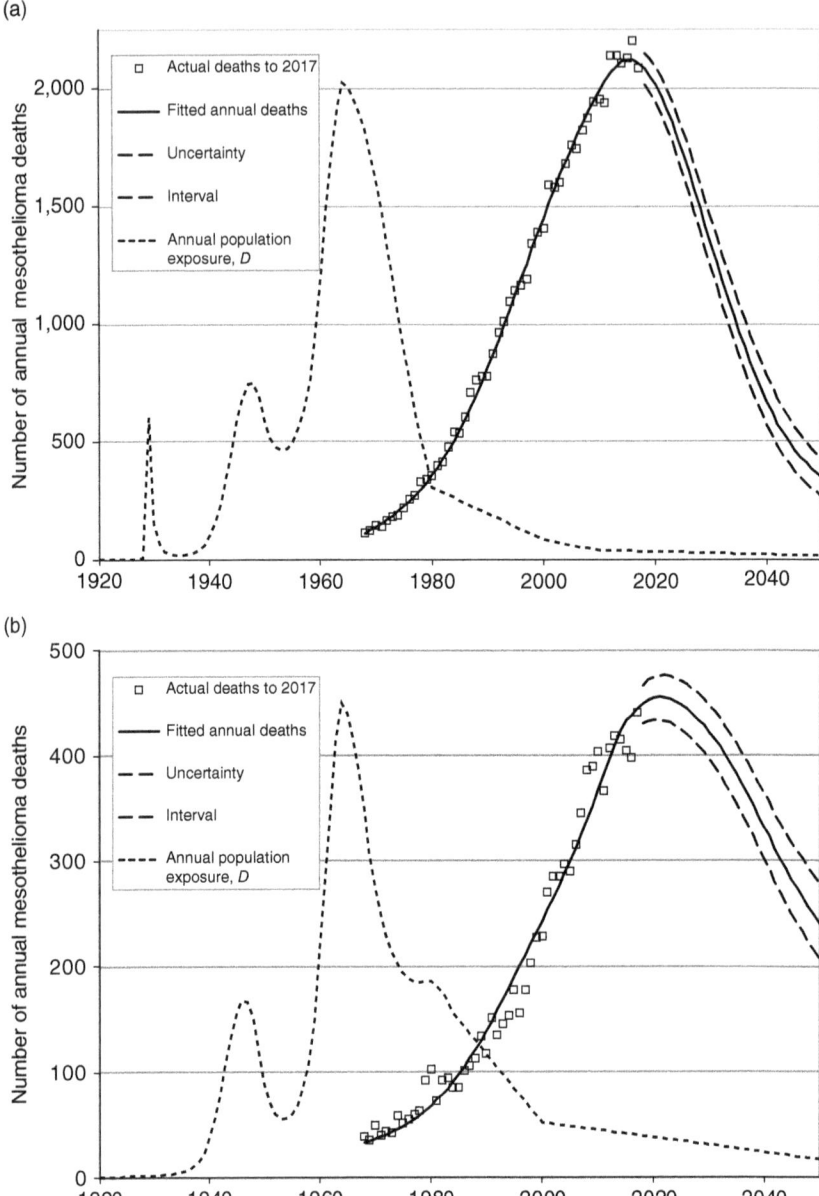

Figure 9.1 (a) Observed and predicted annual mesothelioma mortality, with annual population exposure, males aged 20–94. Fitted model parameters: $k = 2.547$; peak exposure year = 1964; no lung clearance; fitted background rate = 1.25 per million per year; no diagnostic trend; age-specific exposure propensity, W_A = 0.19, 1, 1.65, 1.35, for ages 16–19, 20–29, 30–39, 40–49, respectively, zero otherwise. (b) Observed and predicted annual mesothelioma mortality, with annual population exposure, females aged 20–94. Fitted model parameters: $k = 3.0$; peak exposure year = 1964; no lung clearance; fitted background rate = 1.30 per million per year; no diagnostic trend; age-specific exposure propensity, W_A = 0.01, 0.15, 0.95, 1.0, 0.40, for ages 5–15, 16–19, 20–29, 30–39, 40–49, respectively, zero otherwise.

The models for both males and females provide strong evidence that annual exposure peaked during the 1960s and then reduced rapidly during the 1970s. However, a consequence of the latency between exposure and mortality is that the strength of inferences that can be drawn about D falls rapidly beyond 1980 and by the late 1980s, the value of D is essentially undetermined. We will refer to this part of the exposure as the profile "tail." A wide range of scenarios for the profile tail is possible without adversely affecting the fit of the model. The application of different profile tails provides a means of comparing the predicted long-term mesothelioma health impact of counterfactual scenarios in reference to default assumption that there was a continued reduction in exposure during the 1980s and 1990s followed by a more gradual reduction beyond 2000. This approach has a number of potential applications, such as assessing the historic value of regulation to reduce worker and wider population exposure to asbestos, and assessing the potential impact of future initiatives to modify exposure control requirements.

The exposure profile tails presented in Figure 9.1a and b – showing a linear reduction in D between 1980 and 2000, followed by a more gradual decline thereafter – were developed based on assumptions and semi-quantitative arguments drawing on a range of evidence sources about the most likely situation in Britain between 1980 and the present time, and what might reasonably be expected to occur to year 2050 (Hodgson et al. 2005). These profiles represent default scenarios for making projections beyond the peak mortality in around year 2020.

Figure 9.2 replicates the male projections based on the default exposure profile tail shown in Figure 9.1a but with three additional possible tails also shown, two of which are broadly consistent with the observations of mesothelioma mortality to 2017. Grey shaded lines (dashed, solid and dotted/dashed lines) represent the population exposure D with the alternative tails, and lines with the same shading to the right of these show the consequences of those exposure profile tails in terms of predicted annual mesothelioma deaths to 2050 based on the model.

Two scenarios illustrating a more rapid decline in population exposure during the 1980s than the default scenario are shown. In the "rapid decline" scenario (light grey dashed line), exposure continues to reduce at the rate of decline inferred for the 1970s so that exposure is zero by 1983. However, this is in fact inconsistent statistically with the observed data: the fit of the model is significantly worse than for the default scenario and annual deaths in peak years are substantially underestimated. This suggests that asbestos exposure must have continued well into the 1980s. The "fastest arguable decline" scenario (solid dark grey line) represents the steepest linear decline in exposure that can be assumed from 1980 without a statistically significant worsening of model fit. The "no decline" scenario represents a leveling off of the exposure, D, at the highest level possible for the earliest year after 1980 (namely, 1986) without a statistically

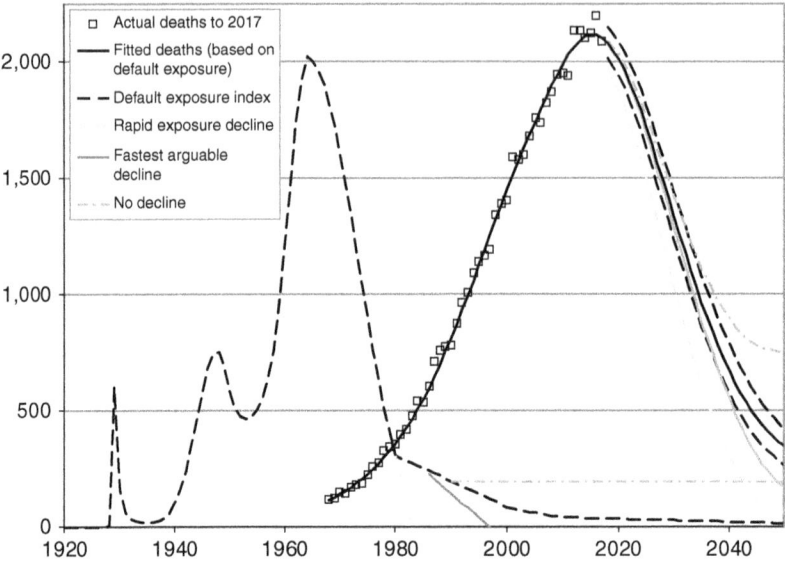

Figure 9.2 Observed and predicted annual mesothelioma mortality for various possible annual population exposure profiles from 1980, males aged 20–94.

significant worsening of model fit (grey dotted/dashed line). This model clearly predicts much higher numbers of mesothelioma deaths in the long-term which, following an initial decline after year 2020, eventually starts to increase again during the 2040s due to gradual predicted changes in population demographics.

The default scenario was determined by assuming a linear decline in the value of D between 1980 and 2000, with the value for year 2000 estimated using other data sources, and further assuming that exposure beyond 2000 would decline in proportion to size of the remaining stock of asbestos-containing buildings during the period from 2000 to 2050. The index of population exposure used in the model is scaled arbitrarily: it is the *shape* of the exposure profile over time that is used to predict subsequent patterns in mesothelioma mortality rather than the absolute value of D in any given year. Estimating the absolute scale of D from other data sources therefore requires calibration of the predictions of the mesothelioma model against predictions based on these other sources.

The particular approach adopted to estimate the value of D in 2000 was first to consider evidence about the distribution of exposure levels for key subgroups of the working and wider population of Britain on any given day in order to then estimate the overall average concentration of the population at that given point in time (i.e. in year 2000), using the standard metric for airborne asbestos concentrations (f/ml). Next, published information about the exposure-response model for mesothelioma (expressed in terms of cumulative exposure using the same metric) was used to estimate the long-term mesothelioma mortality that would be expected for continued exposure at this average level. Then, the value of the exposure index D in the mesothelioma projections model was set to predict this same level of future deaths. This process led to an estimated value of D in year 2000 of 4.2% of the peak value in 1964.

In the absence of any explicit regulatory requirements to actively manage asbestos remaining in British buildings, values of D beyond year 2000 were assumed to reduce from the level in 2000 in proportion to the expected rate of demolition of existing buildings with a high probability of containing asbestos materials. This was derived by applying a demolition rate of 1% of current building stock per year in 2000, rising gradually to reach 2% after 25 years, and then accelerating to reach 4% by year 2050. These assumptions led to an overall average demolition rate of just over 2% per year, and imply numbers of annual demolition jobs that were broadly consistent with annual asbestos notifications for licensed work in Britain.

The reduction in exposure during the period from 1980 to 2000 – driven by the estimate of D for 2000 of 4.2% – is a key feature of the default scenario, particularly in the context of the "no decline" scenario, which is statistically consistent with observations of mesothelioma mortality to 2017 but which predicts much higher levels of future mortality. However, the calculations to estimate the value of D in year 2000 encompass uncertainties that are difficult to quantify but are likely to be considerable. The plausibility of the default and other scenarios can nevertheless be considered informally with reference to an alternative source of evidence about population asbestos exposure, which has become available since the default scenario was developed, namely estimates of the asbestos lung content of the general population from a recent British study (Gilham et al. 2018).

Gilham et al. demonstrated a strong relationship between mesothelioma risk and amphibole asbestos lung burden based on both case–control analyses and correlations between overall average population lung burdens and national mesothelioma rates in successive birth cohorts (Gilham et al. 2016, 2018). Amphibole asbestos lung burden is reflection of cumulative exposure to asbestos – that is it reflects mainly the duration and the average intensity (i.e. the airborne concentration) of exposure, assuming that there will be minimal clearance of bio-persistent amphibole fibers from the lung over time. However, since lung tissue samples in the lung burden study were obtained during a fairly short time window (of a few years) but reflect a much wider range of ages, the period of birth comparisons presented by Gilham et al. are influenced by the amount of time available for those born in successive birth periods (i.e. of different ages) to be exposed, as well as changes in the average exposure intensity over time. For example, those born in the 1950s will, on average have started work by the early 1970s and had the potential for around 40 years of exposure by the time the lung samples were collected, whereas those born in the 1970s will have had the potential for only around 20 years exposure on average at working ages. The exposure metric in the mesothelioma projections model, on the other hand, reflects the average annual intensity of population exposure. However, the equivalent pattern of cumulative exposure implied by the population exposure profile, D, and age-specific exposure propensity factors from projections model can be calculated, based on exposures accrued up to the point when the lung burden samples were collected. Comparisons between the implied birth-cohort-specific cumulative exposures for various exposure profile tails can then be made with the results from the lung burden data to informally assess the plausibility of each.

Figure 9.3a reproduces the overall average male lung burden by 5-year birth cohort from Table 1 of Gilham (dashed black line) and also shows the implied average cumulative exposures for three of the exposure profile tails shown in Figure 9.2 (solid grey, black, and dotted/dashed lines). Given the scaling of the implied cumulative exposure calculated from the projections model is arbitrary, the chart presents the changes over time relative to the 1940–1944 birth cohort. The time periods shown in brackets on the horizontal scale of Figure 9.3a give an indication of when men born in each birth cohort would have started work and thus have the highest potential for exposure to asbestos.

Figure 9.3a suggests that the Default scenario from the projections model matches the lung burden results more closely than the "no decline" scenario (which implies higher cumulative exposures) and the "Fastest arguable decline" scenario (which implies lower cumulative exposures).

The implied cumulative exposures calculated from the projections model presented in Figure 9.3a, assume that the age-specific exposure potential (factor W_A in the model) remains fixed for all time with relative values of 0.19, 1.0, 1.65, and 1.35 at ages 16–19, 20–29, 30–39, and 40–49 years respectively. Estimation of these parameters is influenced mainly by exposures prior to 1980 when "consumption" rather than releases from the stock of existing asbestos materials in buildings was the dominant exposure source. However, it might be argued that those below working age have a greater potential for exposures from the stock of existing materials than implied by these values, and if so this would tend to decrease the rate of decline of the cumulative exposure in successive birth cohorts. To illustrate this effect, the calculations underpinning Figure 9.3a can be re-run after arbitrarily assuming higher alternative relative exposure propensity values of 0.25 and 0.5 at ages 5–15 and 16–19. The results of this exercise are shown in Figure 9.3b. Here the scenario of a faster decline in exposure post 1980 produces cumulative exposures that most closely match the lung burden values.

Although not definitive, the results presented in Figure 9.3a and b add weight to the evidence that population exposures continued to reduce during the 1980s

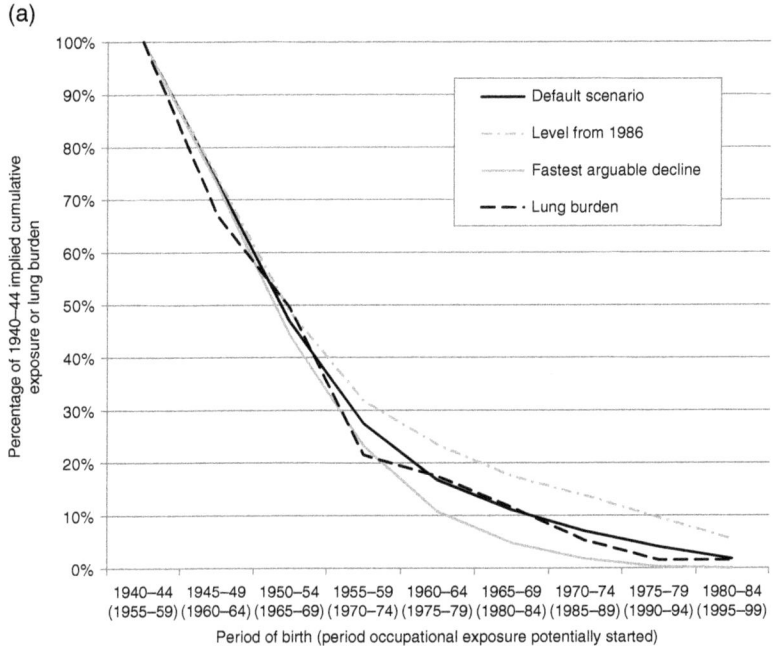

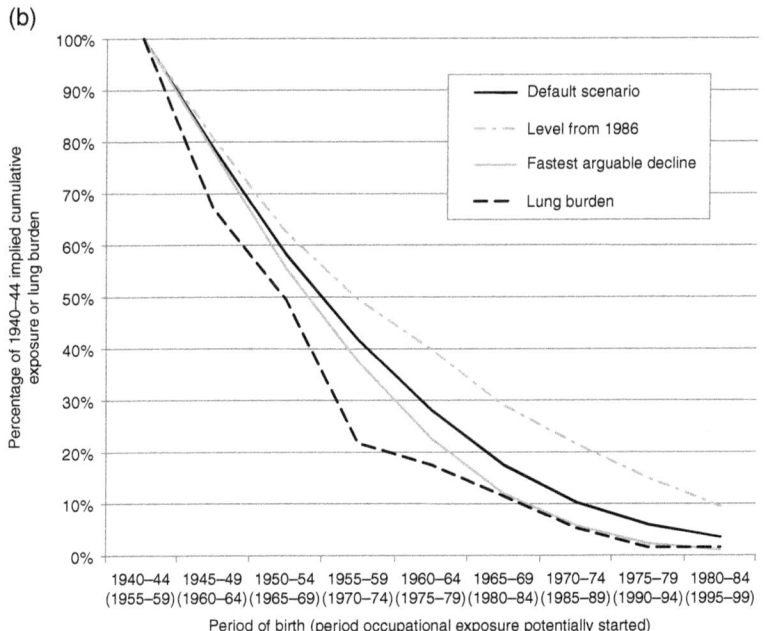

Figure 9.3 (a) Average population lung burdens of amphibole asbestos in males by period of birth and implied cumulative exposures under three exposure tail scenarios. (b) Average population lung burdens of amphibole asbestos in males by period of birth and implied cumulative exposures with higher exposure potential below working age. (c) Average population lung burdens of amphibole asbestos in females in Great Britain by period of birth and implied cumulative exposures.

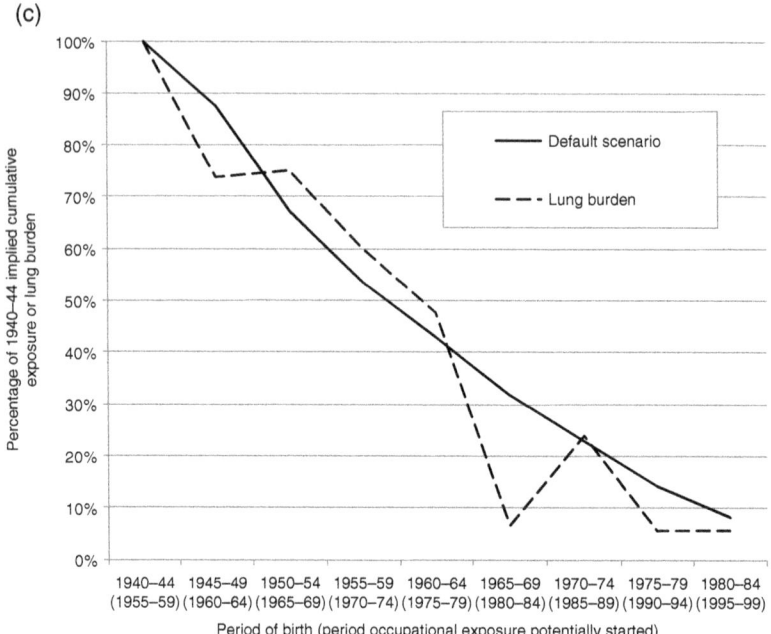

Figure 9.3 (Continued)

and 1990s in Britain, at least to some extent and that the original default profile tail remains a reasonable estimate of the likely pattern in this period.

Alternative exposure profile tails have not been explored to the same extent in females. Nevertheless, as described earlier, the default tail for females results in similar long-term predictions of annual mortality in females as in males. This is consistent in general terms with the lung burden research results which imply that as the effect of heavy past occupational exposures wanes – which particularly affected males – numbers of future cases in both men and women will tend to converge. A similar analysis of the implied pattern of cumulative exposures based on just the default exposure tail from the female projections also shows good agreement with the female lung burden results (see Figure 9.3c).

More recent observations of annual mesothelioma mortality in Great Britain beyond 2017 generally confirm the validity of models presented here. Figure 9.4a and b show an additional four data points for observed male and female mesothelioma deaths at ages 20–94 in Great Britain (solid black squares). These data points are broadly consistent with the predictions shown based on observed data to 2017 (solid black line).

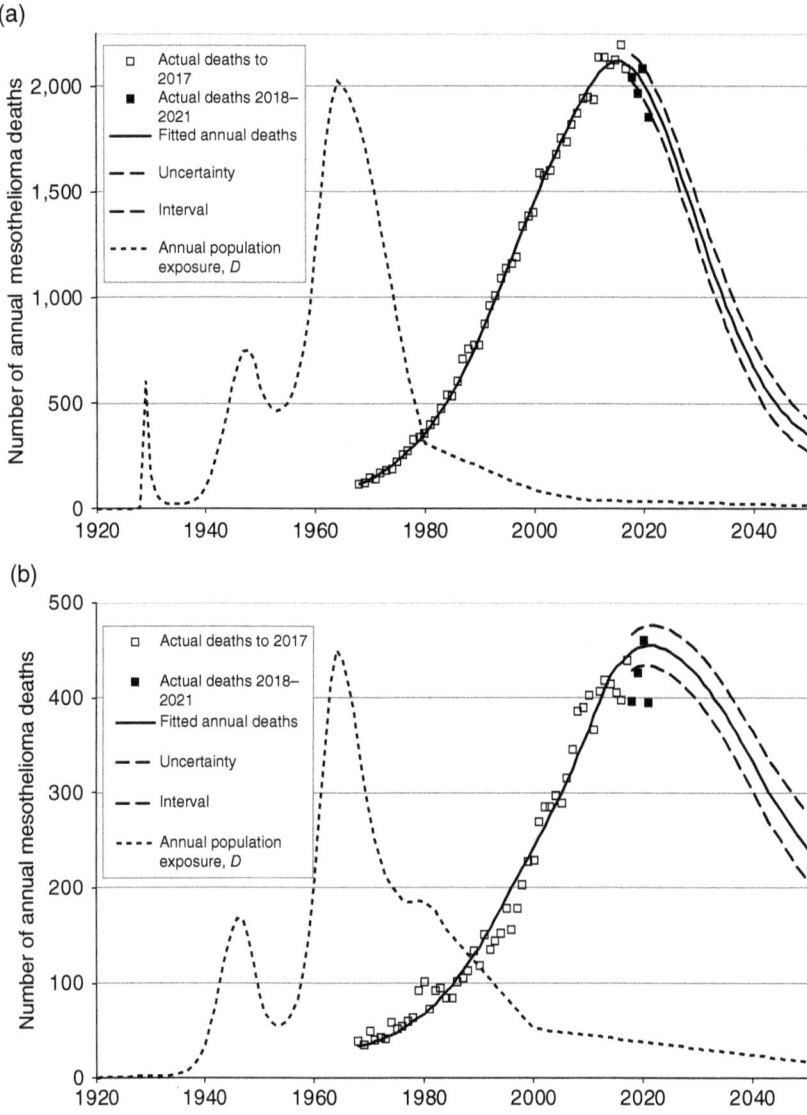

Figure 9.4 (a) Observed and predicted annual mesothelioma mortality in males, with additional annual mortality data from 2018 to 2021. (b) Observed and predicted annual mesothelioma mortality in females, with additional annual mortality data from 2018 to 2021.

Conclusions

The results presented here demonstrate the feasibility of producing projections of mesothelioma mortality in terms of the past pattern of asbestos exposure at a national level. Whilst providing confidence about the likely future pattern of the overall burden of mesothelioma, the utility of this particular approach of inferring the past temporary pattern of exposure is also demonstrated in terms of then allowing health impact assessment based on different assumed recent and future exposure scenarios. This will be most informative where the assumed exposure index can be calibrated against other sources of data on population asbestos exposure. This was possible in the British context with reference to information on population lung burden by birth cohort.

Additional monitoring of trends in national mesothelioma mortality will provide a basis for further assessment and potential refinement of the developed model, which could also be utilization in various countries. However, the poor correlation between the inferred population exposure index and total Britain imports of all asbestos mineral types combined emphasizes the need for caution when making international comparisons of country-level data on asbestos consumption and mesothelioma outcomes. While the temporal pattern of asbestos usage within a country may allow good predictions of mesothelioma mortality, international comparisons are confounded by development status and asbestos mineral type. In the British context, whereas chrysotile imports account for the vast majority of historical import tonnages, the relatively heavy past usage of amosite asbestos compared with other countries with similar development status was a key factor in producing the higher rates of mesothelioma observed to date (Darnton et al. 2014).

The views expressed in this chapter are those of the author and not necessarily the Health and Safety Executive.

References

Darnton, A., Gilham, C., and Peto, J. (2014). *Epidemiology of Malignant Pleural Mesothelioma in Europe*. Bentham Science, 33–52, (20). https://doi.org/10.2174/9781681081946116010008; https://benthamscience.com/public/chapter/8397.

Gilham, C., Rake, C., Burdett, G., et al. (2016). Pleural mesothelioma and lung cancer risks in relation to occupational history and asbestos lung burden. *Occupational and Environmental Medicine* 73(5): 290–299. https://doi.org/10.1136/oemed-2015-103074.

Gilham, C., Rake, C., Hodgson, J., et al. (2018). Past and current asbestos exposure and future mesothelioma risks in Britain: The Inhaled Particles Study (TIPS). *International Journal of Epidemiology* 47(6): 1745–1756. https://doi.org/10.1093/ije/dyx276.

Health and Safety Executive. (2020). Table MESO06: Projections of annual male and female mesothelioma deaths in Great Britain. https://www.hse.gov.uk/statistics/tables/meso06.xlsx. Accessed November 6th, 2023.

Health and Safety Executive. (2023). Mesothelioma statistics for Great Britain, 2023. November 2023. https://www.hse.gov.uk/statistics/assets/docs/mesothelioma.pdf

Hodgson, J., McElvenny, D., Darnton, A., et al. (2005). The expected burden of mesothelioma mortality in Great Britain from 2002 to 2050. *British Journal of Cancer* 92(3): 587–593. https://doi.org/10.1038/sj.bjc.6602307.

Integraal Kankercentrum Nederland. (2020). "Cancer statistics," NKR Cijfers (https://iknl.nl/en/ncr/ncr-data-figures). Accessed September 21st, 2020.

Lin, R.-T., Chang, Y.-Y., Wang, J.-D., and Lee, L. J.-H. (2019). Upcoming epidemic of asbestos-related malignant pleural mesothelioma in Taiwan: a prediction of incidence in the next 30 years. *Journal of the Formosan Medical Association* 118(1), Part 3: 463–470. https://doi.org/10.1016/j.jfma.2018.07.013.

Oddone, E., Bollon, J., Nava, C.R., et al. (2020). Predictions of mortality from pleural mesothelioma in Italy after the ban of asbestos use. *International Journal of Environmental Research and Public Health* 17(2): 607. https://doi.org/10.3390/ijerph17020607.

Peto, J., Seidman, H., and Selikoff, I. (1982). Mesothelioma mortality in asbestos workers: implications for models of carcinogenesis and risk assessment. *British Journal of Cancer* 45(1): 124–135. https://doi.org/10.1038/bjc.1982.15.

Peto, J., Hodgson, J.T., Matthews, F.E., and Jones, J.R. (1995). Continuing increase in mesothelioma mortality in Britain. *Lancet* 345(8949): 535–539. https://doi.org/10.1016/s0140-6736(95)90462-x.

Price, B. (2022). Projection of future numbers of mesothelioma cases in the US and the increasing prevalence of background cases: an update based on SEER data for 1975 through 2018. *Critical Reviews in Toxicology* 52(4): 317–324. https://doi.org/10.1080/10408444.2022.2082919.

Price, B. and Ware, A. (2009). Time trend of mesothelioma incidence in the United States and projection of future cases: an update based on SEER data for 1973 through 2005. *Critical Reviews in Toxicology* 39(7): 576–588. https://doi.org/10.1080/10408440903044928.

Rake, C., Gilham, C., Hatch, J., et al. (2009). Occupational, domestic and environmental mesothelioma risks in the British population: a case-control study. *British Journal of Cancer* 100:1175–1183.

Segura, O., Burdorf, A., and Looman, C. (2003). Update of predictions of mortality from pleural mesothelioma in the Netherlands. *Occupational and Environmental Medicine* 60:50–55. http://doi.org/10.1136/oem.60.1.50.

Tan, E., Warren, N., Darnton, A.J., and Hodgson, J.T. (2010). Projection of mesothelioma mortality in Britain using Bayesian methods. *British Journal of Cancer* 103(3): 430–436. http://doi.org/10.1038/sj.bjc.6605781.

10

Implications of Exposure Measurement Methodologies for Dose–Response Assessment in Asbestos Worker Cohorts

Garry Burdett

Research and Measurement Scientist (Retired), UK Health and Safety Executive

This chapter focuses on the air measurement data in historical asbestos worker cohorts, which forms an essential part of the epidemiological assessment of cancer risks in those settings. Most studies rely on a combination of air measurements, local employer knowledge, and records to estimate the cumulative exposure for each worker. This is usually in the form of a matrix of the estimated average fiber concentration in fibers/milliliter (f/ml) of air in a work area or job category multiplied by the number of years worked at a particular time. If several different jobs are held during employment, these will be added together to give the cumulative exposure in terms of (f/ml-years). The number, type, and relevance of the air concentration measurements available for each asbestos cohort varies widely and presents a major challenge when comparing cohorts. For instance, for the Wittenoom occupational cohort, exposures were calculated from 87 job categories graded from 0 to 10 for the relative dust concentrations based on the knowledge of an ex-supervisor. However, only one set of fiber concentration measurements is available. These were carried out using a long-running thermal precipitator (TP) just before the mine closed and these only survived as a summary table of the average concentration in six positions.

In meta-analyses of such epidemiological studies, formal synthesis of the data relating to cancer risk per unit exposure typically focuses on uncertainties associated with the health outcomes, whereas the potential impact of the substantial uncertainties in the exposure data may either be ignored or investigated only in qualitative or semiquantitative terms, leaving the interpretation of such analyses open to debate. For many cohorts, the relevant exposures occurred many decades ago and so measurement approaches and metrics are in terms of historic methods such as particle count (see chapter 5). The conversion between measurement

Health Risk Assessment for Asbestos and Other Fibrous Minerals, First Edition.
Edited by Andrey Korchevskiy, James Rasmuson, and Eric Rasmuson.
© 2024 John Wiley & Sons, Inc. Published 2024 by John Wiley & Sons, Inc.

indices to the current index of exposure and when this was carried out is thus a key source of uncertainty in the estimated cumulative exposure.

As cancer risks have typically been characterized in terms of exposures using the metric of Phase Contrast Microscopy (PCM) – that is counts of fibers >5 µm long as the "index of exposure" – this chapter first summarizes the information available from cohorts for the:

- airborne fiber size distributions from analytical electron microscopy (EM) measurements.
- analysis of the types of fiber present.
- PCM conversion factors used from earlier sampling and measurement methods.

The air concentration measurements that are available for the main crocidolite-only, amosite-only, and chrysotile-only asbestos cohorts used in various meta-analysis are reviewed, for the:

- Extent of data available,
- Representativeness of data to the historical conditions, and
- Reliability of the conversion of data to PCM fiber counts.

The amosite, crocidolite, and chrysotile-only cohorts in the Hodgson and Darnton meta-analyses are the starting point for this review (Hodgson and Darnton 2000; Darnton 2023) due to the relationship of lung disease to mineral fiber type and fiber dimensions and our current knowledge of the biological mechanisms that give rise to the risk (Gualtieri 2023; Wylie and Korchevskiy 2023).

Electron Microscopy Fiber Size Distribution for Different Cohorts and Their Relationship to PCM Fiber Counts

Until recently, only the transmission electron microscope (TEM) was capable of giving an accurate fiber size distribution and count of all the asbestos fibers present on a membrane filter (MF) sample and scanning electron microscopy (SEM) was used to assess the same population of >5 µm long fibers that were likely to be counted by phase contrast microscopy (PCM). The MF-PCM index of exposure usually represents <10% of the total numbers of fibers (i.e. particles with aspect ratios >3 : 1) present in chrysotile and crocidolite. Both the fiber width/diameter and the fiber refractive index difference from the mount will determine whether the fibers are visible and included in the PCM count (Kenny et al. 1987; Rooker et al. 1982). The differences in the early EM fiber size distributions – on the samples for which MF-PCM conversions were made – are examined in relation to their effect on the PCM counts and the effect on the counts of longer fibers.

TEM Fiber-Size-Distribution in Cohorts from Mines and Mills

TEM, SEM, and PCM measurements of the airborne fiber size distributions at the South African and Canadian mines were reported in the early 1980s (Gibbs and Hwang 1980; Hwang and Gibbs 1981; Walton 1982; Hwang 1983). They used TEM micrographs of ×15,000 to size the fibers, but due to the small size of the TEM grid openings (45 μm) and the micrograph magnification, the maximum length of fiber that could be measured was truncated. Also, as all fiber sizes were measured, the numbers of longer fibers counted were limited so do not have good precision. However, using their published data for >5 μm and >10 μm long fibers, it can be seen (Table 10.1) that there are substantial differences between fiber types in what will be counted by PCM. Some 21% of all amosite fibers were countable by PCM compared with around 2% of the crocidolite fibers.

The PCM limit of visibility ~0.2 μm width, has been applied to the mining data for three main commercial asbestos types, for example it was reported as 0.21 μm for crocidolite (Hwang and Gibbs 1981). Table 10.1 shows that due to the finer fibers present, only about a quarter of the >5-μm-long crocidolite fibers and about half of the chrysotile fibers will be visible, compared with the amosite fibers (as indicated by the ratios in the bottom row of the table). In a direct comparison of different areas of the same filter by PCM vs. TEM, it was estimated that just over one-third of the TEM >5-μm-long crocidolite fibers were counted by the MF-PCM method.

While longer fibers (e.g. >10 or >40 μm long) are considered likely to be more carcinogenic, the percentage of >10-μm-long fibers in mine bagging samples, was found to represent ~1% of the total numbers of airborne fibers, except for amosite (see Table 10.1). This reduction in the ratio of fibers >5 μm long: fibers >10 μm long is a factor of 8, 3.3, and 4 for crocidolite, amosite, and chrysotile, respectively. For >40-μm-long fibers, the TEM data are too sparse to give reliable comparisons, but the light microscope data showed that 2% of the >5-μm-long crocidolite fibers were >40 μm long. The low numbers of long fibers have important implications

Table 10.1 Summary of fiber size distribution data from asbestos mines and mills for >5 and >10-μm-long fibers of different widths (Walton 1982; Hwang 1983).

Length (μm)	Crocidolite		Amosite		Chrysotile	
	% >5	% >10	% >5	% >10	% >5	% >10
All widths	7.1	0.89	24.5	7.5	4.12	1.04
Widths >0.2 (μm)	1.85	0.22	20.89	6.8	2.01	0.55
Width ratio all / >0.2	3.84	4.05	1.17	1.10	2.05	1.89

for their statistical precision and stratified counting is needed to additionally count a larger number of fields of view (i.e. an increased filter area) for longer fibers only.

Substantial differences between the mining areas and the bagging areas of the milled asbestos ore were found in terms of the >5 μm long fibers as a percentage of the TEM fiber size distribution (see Table 10.2). The milling operations separate the fibers from the host rock to produce fibers of different length grades (e.g. the long fiber Quebec chrysotile grade 3 was used for textiles manufacture).

TEM Size Distributions from Manufacturing Cohorts

Due to the voluntary and formal prohibition on the use of amphiboles in the manufacturing industry during the 1980s, it is difficult to find historic TEM size distributions of amosite and crocidolite in manufacturing settings and most of the data in the literature relates to chrysotile. Table 10.3 summarizes the size distribution for fibers >0.5 μm long from a chrysotile textiles factory in Rochdale, UK, which used textile-grade chrysotile from the Canadian mines (Rood and Streeter 1984).

Table 10.2 Percentage of >5-μm-long fibers in the TEM size distributions at mines and mills.

Asbestos type	Crocidolite		Amosite		Chrysotile	
Site	Mining	Bagging	Mining	Bagging	Mining	Bagging
% of >5-μm-long fibers	4.1	7.1	12.4	24.5	1.2	4.1

Source: Reproduced with permission from Gibbs and Hwang (1980) / Europe PMC

Table 10.3 Examples of the effect of fiber visibility on counts of >5-μm-long fibers from monitoring different processes at a chrysotile textile factory in Rochdale, UK).

	Percentage of >5-μm-long fibers with diameters > (μm)			
Process/diameter	0.1	0.2	0.3	0.4
Carding	77	49	26	20
Spinning	69	31	19	14
Weaving	78	38	26	24

Source: Adapted from Rood and Streeter (1984)

Only a few percent of the fibers sized were >5 μm long and the size distributions from the carding, spinning, and weaving areas for the factory were similar. The TEM size distribution showed that the number of >5-μm-long fibers with >0.1 μm diameter, was some ×3 greater than for >0.3 μm diameter, so the microscope set-up and visibility of fibers will have an important effect on the PCM fiber count. A similar chrysotile size distribution was found at a UK chrysotile friction products plant from the preform ring and cure pressing areas, but the yarn dressing area size distribution showed fibers were significantly shorter and thinner (Rood and Scott 1989). However, as relatively few >5-μm-long fibers were present in the size distributions in these papers, it is difficult to draw robust conclusions and the use of a stratified count to include greater number of longer fibers is required. Both of these studies were at factories that used textile-grade fibers (e.g. grade 3 long fiber compared to the shorter fibers used for asbestos cement and papers grades 4 and 5).

Particular attention has been given to fiber size distributions for two US chrysotile textile manufacturing cohorts, to determine if the differences in the lung cancer rates can be reconciled by the differences in fiber size. One of the most studied cohorts is the South Carolina textile workers, who used the raw fiber supplied from the Quebec asbestos mines. As the slope of the lung cancer exposure–response for this cohort is a factor of 80 higher than for Quebec miners (Lash et al. 1997), it was argued that the longer grades of fibers required for textile manufacture are potentially one of the explanations for the different disease rates. Differences in the fiber size distribution have long been suspected as an important determinant of the different disease rates (i.e. potency) observed for amphibole fibers (e.g. Berman and Crump 2008a, 2008b; Lippmann 1990; Loomis et al. 2012).

TEM analysis of 84 archived MFs collected between 1964 and 1968 from chrysotile textile manufacturing in South Carolina has been reported (Dement et al. 2008). Based on a stratified count of 18,824 fibers and fiber bundles, some 16.5% of the fibers were >5 μm long, a much greater percentage than found at the UK factories and Canadian mines. Some 6.6% of all fibers had dimensions equivalent to fibers that would be expected to be counted by PCM (i.e. a particle >5 μm length and >0.25 μm width with an aspect ratio >3 : 1), but some 9.2% of the fiber population were <0.25 μm diameter giving an overall adjustment factor of 2.4 for all >5-μm-long respirable fibers.

Fiber counts were available for 10 production process categories (see Table 10.4). Between 4.5% and 11.4% of the fibers in different zones were >5 μm long but too thin to be visible by PCM. The main purpose of the analysis was to use the TEM size distribution to apply "adjustment factors" to the PCM counts. However, as the TEM analysis sized a much larger number of fibers, it also allowed other size ranges to be assessed, which have been suggested to be more biologically active, for example "Stanton" fibers (>8 μm long and >0.25 μm wide) and >40-μm-long

Table 10.4 Results from TEM analysis of 84 filters collected at a South Carolina chrysotile textile factory for general area personnel in 1965 and calculated adjustment factors to the PCM count for fibers >5 μm long and >0.25 μm width.

Exposure zone	Associated process/ department	Mean PCM exposure (f/ml)	PCM adjustment factor	Mean TEM exposure of PCME fibers (f/ml)
1	Preparation	5.8	1.02	5.92
2	Carding	2.4	1.15	2.76
3	Ring spinning	8.2	1.31	10.74
4	Mule spinning	6.3	1.33	8.38
5	Foster winding	4.2	1.32	5.54
6	Twisting	5.4	2.17	11.72
7	Universal winding	4.1	1.13	4.63
8	Heavy weaving	2.6	1.34	3.48
9	Light weaving	2.7	1.69	4.56
10	Finishing	0.2	2.15	0.43

PCME = PCM equivalent fibers (>0.25 μm diameter and >5 μm long with >3 : 1 aspect ratio).
Source: Dement et al. 2008

Table 10.5 Summary of percentage of various categories of biologically active fiber sizes in the TEM analysis of 84 filters collected at a South Carolina chrysotile textile factory for general area personnel in 1965.

Fiber width (μm)	Fiber length (μm)		
	>5	>15	>40
<0.25	9.2	2.5	0.4
PCM equivalent (PCME) >0.25–<3	6.6	2.0	0.5
>3	0.7	0.4	0.1
Total	16.5	4.9	1

Source: Dement et al. 2008

fibers (Stanton et al. 1981; Berman et al. 1995). The results for these have been extracted from the data and are summarized in Table 10.5. It can be seen that >40 μm long fibers represented 1% of TEM size distribution, but still only half of these long fibers were likely to be counted by PCM.

A similar exercise was carried out for North Carolina chrysotile textile plants using a random stratified sample of 77 MFs from 333 stored filters collected by the

US Public Health service between 1964 and 1971 (Loomis et al. 2010). The TEM size data and exposure from 22,766 fibers were used to estimate the fiber concentration from different bivariate (length and width) ranges and then used to assess the dose–response relationships for lung cancer based on 181 deaths (mesothelioma deaths were too few to use). The dose–response model based on the bivariate distribution, which most closely represents the PCM estimate of exposure provided a better fit than models for distributions associated with TEM exposure indicators. This finding suggests that even if most of the risk resides in the longer fibers, the >5-µm-long PCM index does not exclude them unless they are too thin to be visible. Moreover, the statistical analysis found that, "the best fit model for lung cancers were associated with cumulative exposures to fibers 20–40 µm in length" but other size distributions (fiber lengths >10 µm, 5–40 µm, and >40 µm for various width categories) also correlated well in the statistical analysis.

Further analyses have been carried out using combined data for employees at the North and South Carolina asbestos textile mills using TEM size data (Loomis et al. 2012). They found that lung cancer exposure–response models based on fibers >5 µm long and <0.25 µm in diameter provided the best fit to the data, while models based on fibers 5–10 µm long and <0.25 µm in diameter showed the strongest associations with lung cancer.

SEM Size Distributions from Manufacturing Cohorts

SEM size analyses from an operational chrysotile manufacturing plant in Chongqing, China have been reported (Courtice et al. 2016). The plant had a range of asbestos production including textile and asbestos-reinforced rubber products. The chrysotile size distribution based on 8,846 primary chrysotile fibers and 4,584 chrysotile fiber bundles and matrices (component fibers) found that overall, 11% were >0.25 µm and longer than 5 µm (PCM countable fibers); 3% were longer than 10 µm; and 0.5% were longer than 20 µm. The fibers were also observed to become wider as they became longer. The SEM analysis suggests that the percentage of PCM countable fibers compared with all asbestos fibers counted was greater at Chongqing than in other textile factories (e.g. 5.0–6.4% for North Carolina and 4.5–11.4% for South Carolina), but this is more likely to be due to the difference between the visibility of short and thin chrysotile fibers in the SEM compared with the TEM analysis used for other chrysotile factories. Also, a relatively large pore size 0.8 µm polycarbonate filter was used for sampling the Chongqing samples, which may not have collected the shorter fibers efficiently.

The premise that airborne chrysotile fiber dimensions in textiles change with the stages of processing (e.g. the raw fiber is carded, spun, doubled before being woven into textiles) was not statistically supported by the Chongqing SEM fiber size measurements. Surprisingly, the rubber workshop had larger proportions of

20–40 µm long fibers and >40 µm long fibers, than the weaving workshop. Using the same nested statistical analysis methods as used for the South Carolina data, no overall significant differences in either the length or fiber size distributions were found in the various workshops sampled.

SEM and PCM size distributions in two UK factories that manufactured amosite-containing insulating board (Cape Uxbridge and Cape Glasgow) were reported (Beckett and Jarvis 1979). No significant differences in the length distributions were found for the PCM samples at 8 and 13 locations at Uxbridge and Glasgow, respectively; two positions at each site with high airflows did give differences, but these were due to the large differences in the airflow and sampling velocities (nonisokinetic sampling). Fibers sampled at side-by-side locations onto 0.8 µm pore size polycarbonate filters were measured on the SEM screen at ×10,000 magnification and only fibers >0.2 µm long with >3 : 1 aspect ratio (i.e. >0.6 µm long) were evaluated. While the sanding operation gave more short fibers, their >5 µm long fiber distributions were similar to samples taken elsewhere in both of the factories. It was concluded for amosite that the PCM counts were "not preferentially missing long but very fine fibers."

Overall, the early comparisons between the MF-PCM and by both TEM and SEM showed that the PCM was reasonably able to monitor the >5-µm-long fiber concentrations in air but with some underestimation of chrysotile (e.g. a factor of 1–2.2 : 1) between work areas. While crocidolite was not directly compared, its size distribution shows that it is likely to have a similar or higher ratio than chrysotile, while the broader width distribution of amosite gives close to a 1 : 1 ratio.

Also, as shown by the TEM size data in Tables 10.1 and 10.2, the proportion of visible PCM countable fibers >5 µm long is ×3.85 higher for amosite than crocidolite. This size difference and the greater visibility of amosite fibers could account for some of the apparent potency difference between amosite and crocidolite in relation to mesothelioma (Hodgson and Darnton 2000; Berman and Crump 2008a).

EM Determinations of Asbestos Fiber Types in Asbestos Industry Cohorts

The MF-PCM method only counts fibers and does not give information on the type of asbestos present, or whether the fiber is an asbestos fiber. Whilst this produces a conservative estimate of exposure (i.e. all fibers counted are assumed to be the asbestos type in use), this is however limited if a mixed-exposure to chrysotile and amphiboles occurs, where there is considered to be a large difference in potencies between fiber types. The most recent meta-analyses of mining and manufacturing industry cohorts (Garabrant and Pastula 2018; ECHA 2021; Darnton 2023), which include the three main industrially used asbestos types, showed that an approximately 500-fold difference in potency exists between chrysotile and crocidolite asbestos for mesothelioma. This implies that a presence of just 0.2% of these >5 µm long fibers in air would have the same risk for mesothelioma as 99.8% of the >5 µm long chrysotile fibers. Therefore, the presence of amphibole asbestos fibers in chrysotile-exposed cohorts is of particular relevance. There are three situations where the analysis of fiber type is important: the natural occurrence of amphiboles in the chrysotile deposit, the use of more than one asbestos type in the production process (i.e. mixed exposure), and situations where the cohort may be unduly exposed to external sources of amphibole fibers.

Natural Occurrence

The presence of tremolite (a regulated amphibole asbestos) in chrysotile mines has long been suggested as a possible cause of the different disease rates within chrysotile-only exposed cohorts (Rowlands et al. 1982). The greater bio-durability of tremolite fibers in the lung, compared with chrysotile was first observed in lung sample analysis. Often tremolite was found to be a major component of the lung burden remaining on autopsy, although >99% of the fiber extracted was chrysotile (Sebastien et al. 1989). The presence of tremolite in the bulk chrysotile is often only a fraction of a percent and below the conventional bulk analysis methods such as X-ray diffraction or Infra-red spectroscopy. A chrysotile digestion procedure was used to concentrate the tremolite before analysis (Addison and Davies 1990); of the 81 bulk chrysotile samples analyzed, 28 had detectable tremolite, the highest being 0.6%. Only one of the eight Quebec samples had detectable tremolite (0.24%). As tremolite in the Quebec chrysotile mining areas is usually present as seams of a slightly different color, the results depend on the area being mined, whether the tremolite seams were avoided, as well as the representativeness of the subsample analyzed.

The largest chrysotile mine in China, located in Qinghai Province, was reported (Wang et al. 2013b) to have tremolite below the limit of detection by XRD (<0.1%). Two other chrysotile mines in Sichuan Province, which supplied the Chongqing factory, also had amphibole contents of below the limit of detection (<0.001%) by TEM analysis (Yano et al. 2001; Courtice et al. 2016); however a further TEM assessment reported tremolite in bulk samples taken from six Sichuan mines 0.002–0.312% (Tossavainen et al. 2000). The largest chrysotile mine in Russia is also reported to have a low tremolite content (Tossavainen et al. 2000; Kashansky et al. 2001; Kovalevskiy et al. 2021).

Although some attempts have been made to assess airborne chrysotile for tremolite by analytical electron microscopy, the tremolite fibers are rarely present at concentrations that are detectable in the microscopic analysis of a single sample. For example, of the 22,776 fibers and fiber bundles analyzed by analytical TEM for the South Carolina cohort, only 16 had a tremolite or actinolite composition (0.07%) (Dement et al. 2011). The SEM analysis of the Chongqing chrysotile factory air samples (Courtice et al. 2016) found 13,435 chrysotile fibers and bundles and 1,075 tremolite fibers (7.5%). In order to obtain greater numbers of counts for the much rarer longer fibers, a stratified analysis was used to scan the sample three times at ×10,000 magnification for three different size categories, so some care with the interpretation is needed. A comparison of fiber length and width frequencies by fiber type showed tremolite is generally shorter and thicker than chrysotile in the air samples analyzed and no tremolite fibers >30 µm long were found. The adjusted tremolite percentage was 1.5% of all airborne asbestos fibers analyzed and for fibers longer than 5, 10, and 20 µm, comprised some 2.3%, 1.6%, and 0.5% of total fibers, respectively. These analyses suggest that the tremolite component in the airborne fibers was around 100-fold higher in the Chongqing cohort than the South Carolina cohort, as a similar stratified analysis was used, as reported by Courtice et al. (2016). This may account for the twofold greater rate of lung cancer and the relative higher rate of mesothelioma reported in this relatively small prospective cohort of 577 workers compared with the larger ($n = 3,072$) well-established historic South Carolina textile cohort.

Tremolite – actinolite asbestos is not the only fibrous amphibole known to occur with chrysotile; the Italian chrysotile mine at Balangero has been found to contain a new end member of the amphibole Calcic series. It was reported that fibrous balangeroite accounts for 0.2–0.5% of the total mass of the chrysotile produced and has been associated with some excess risk of laryngeal cancer and pleural mesothelioma (Piolatto et al. 1990; Pira et al. 2009).

Mixed-Use

The use of chrysotile with lesser amounts of amosite and/or crocidolite in manufacturing cohorts was widespread in textile, cement, and board manufacturing. While historic records of feedstock are the usual way of establishing the types of fibers used and their relative quantities by year, these may not be used equally across the worker cohort and different areas/processes may need to be evaluated. This relies on suitable historic samples being available for analysis. This is particularly important for cases of mesothelioma in chrysotile cohorts.

A good example of this is the North Carolina cohort, which in the recent US EPA chrysotile risk assessment, was considered as only one of the two available chrysotile-only cohorts with good data quality (US EPA 2020). However, approximately 96% of the person-time accrued for this cohort was from workers at plants three and four (Paustenbach et al. 2021), which had documented use of amosite and crocidolite asbestos. These two plants accounted for the four pleural and four mesotheliomas recorded in the follow-up of this cohort (Loomis et al. 2019).

External Sources

Chrysotile-only cohorts are also potentially confounded by exposure to amphibole asbestos outside of the study workplace. The most frequent cause of these exposures is previous or later employment where exposure to amphibole asbestos takes place. This can only be considered if the analysis and follow-up of the cohorts are sufficiently detailed. Again, using the North Carolina cohort as an example, it has been argued that as short-term workers (i.e. those with <30 days employment in the North Carolina cohort) account for 42% of the total person years of observation, any asbestos exposure from the rest (i.e. >99.8%) of their employment history is unaccounted for. Also, the employment periods and possibility of other sources of exposures for the four pleural and four mesothelioma cases are not discussed (Paustenbach et al. 2021).

Other potential sources of nonoccupational exposure can also occur from living in areas where there is naturally occurring amphibole asbestos or by living close to industries that have used large quantities of amphibole asbestos (e.g. Naval Shipyards). Para-occupational exposures can also occur if living in a household where one member works with amphibole asbestos.

Lung Burden Analysis

The retained dose after death in a worker's lung can provide useful evidence as to the types of asbestos to which they have been exposed. The retention of fibers in the lung will change with time since the last exposure as the bio-persistence of fibers will vary by both fiber type and size. Other factors, such as whether fibrosis is present, also needs to be considered. This evidence is particularly important for chrysotile cohorts to help assess the amount of amphibole asbestos present but can also be used to help supplement the established degree of exposure when air sampling information is limited or lacking.

Conversions of Historic Cohort Measurement Indices to MF-PCM Fiber Counts

The introduction of MF-PCM method for exposure assessment coincided with the establishment of quantitative epidemiological studies for many of the historic mining and manufacturing cohorts available for use by meta-analyses. As the past dust emissions in these industries were often less controlled, this was also when often much higher airborne exposures were commonplace. These early exposures therefore have a large influence on the incidence of asbestos-related cancers in these retrospective cohorts. Therefore, the conversion of early measurements to estimates to MF-PCM airborne fiber concentrations is both important but an additional source of uncertainty.

The relationship between two different measurement indices was usually established by paired (i.e. side-by-side sampling) of static samplers, if it was carried out at all. The individual paired values often differ substantially and had poor correlation. This meant that conversions were usually calculated using logarithmically transformed data to give a geometric mean for the whole cohort, or at best several individual values for the main processes/workshop areas. The numbers of paired samples, when and where they were taken and how the conversions were made, will have a large influence on the form and gradient of the dose–response curve(s). These limitations in converting the data to the current index of exposure were recognized by those carrying out the data conversions, but when it comes to performing meta-analyses for risk assessment, the underlying bias and lack of precision are usually represented as part of the overall uncertainty interval around the central estimate of the risk.

The conversion factors used in the Health Council of the Netherlands meta-analysis (HCN 2010) are given in Table 10.6. As over half of the cohort studies in this meta-analysis were from the United States and Canada, the conversion from impinger particle counts to MF-PCM fiber counts was a key influence on estimating past exposure to chrysotile. The conversions used related back to the assessment derived for the US National Research Council report (NRC 1984).

Conversion from Impinger Counts to MF-PCM

The US public health service (Lynch et al. 1970) carried out a detailed study to convert impinger measurements. Comparisons of MF-PCM counts of >5-μm-long fibers on the same sample at ×450 and ×970, found a factor of 2 increase over a

Table 10.6 Factors for the conversion of atmospheric fiber concentrations obtained by means of the various measurement methods used to measure workplace asbestos concentrations (NRC 1984; HCN 2010).

	Equivalent value for alternative method			
Original measurement method	Impinger (Million particles per cubic foot)	MF-PCM (>5-μm-long fibers/ml)	TEM (s/ml)	Gravimetric (mg/m^3)
Impinger (mppcf)	1	6	360	0.2
PCM (>5 f/ml)	0.17	1	60	0.03
TEM (s/ml)	0.0028	0.017	1	0.0005
Gravimetric (mg/m^3)	5	30	2,000	1

wide range of chrysotile factory samples, but the counting at ×5,000 in the TEM did not give further increase. Owing to the high number of shorter fibers (<1 μm long), it was concluded that the MF-PCM count at ×450 for >5-μm-long fibers would be used in the United States.

A comparison between 800 pairs of impinger: MF-PCM samples taken side-by-side in US textile, friction products, and asbestos cement pipe manufacturing factories were compared and gave impinger: MF-PCM conversion ratios for 1 million particles per cubic foot (mppcf) as equivalent to 5.9, 2.2, and 1.9 f/ml, respectively. The authors noted that the correlation coefficients were always <0.6, but the cement pipe plants were not significantly different from zero (i.e. no linear correlation), and stated that "...the hypothesis that the measurements of exposure are unrelated cannot be rejected." In general, <20% of the impinger particles counted were fibers. It was concluded that workshop/area-specific ratios should be derived for conversions to the MF-PCM method.

Other conclusions from the analysis of nearly 10,000 air samples were also made. The ratio between MF-PCM counts of >5 and >10-μm-long fibers was close to two for most of the processes monitored in the textile, friction products, pipe, and insulation plants. The authors concluded that the different grades of long and short chrysotile fibers used for the different manufacturing processes had no apparent effect on the length distribution of the fibers present in the air samples. Also, the different processes for secondary dispersion of particles (release of dust and fibers shaken loose from the raw asbestos) were found to be not substantially different from primary dispersion processes (where the dust is being created by processes such as cutting and finishing). The MF-PCM results suggested that the impinger-based US Threshold Limit Value (TLV) of 5 mppcf was equivalent to 30 f/ml in the chrysotile textile plants.

The midget impinger and MF-PCM conversions were investigated (Gibbs and LaChance 1974) in surveys in August 1971 and 1972 at up to nine selected sites in each of the five mines and mills of the Quebec chrysotile mining and milling industry. A total of 87 pairs of static MF samples and midget impinger samples were collected side-by-side. The linear correlation coefficient was poor (0.32 for 56 samples with counts of >20 fibers and 0.03 for 31 samples with counts <20 fibers) and it was concluded that the comparison did not give a reliable single conversion of the dust-to-fiber concentration and even doubted whether the impinger particle number data could be converted to an MF-PCM count unless workshop specific conversion factors were used. However, the poor correlation coefficients observed were highly influenced by the low fiber counts from the MF-PCM method used (e.g. sampling times of 5–30 minutes and 20 graticule areas counted on a 37 mm diameter filter) and the underlying statistical (Poisson) distribution (e.g. the 95% confidence interval for a count of 11 fibers has a range of 5.5–19.7). A later review (Gibbs 1994) concluded that no overall single factor could be derived for the

Quebec mining and milling industry. However, it has been possible to derive conversion factors for concentrations at the individual mill and work areas based on a ratio of 3.6 f/ml to 1 mppcf instead of the 6 : 1 ratio used previously (NRC 1984) and in the Health Council of the Netherlands meta-analysis (HCN 2010).

For the South Carolina textile plant (Dement et al. 2008) workshop specific linear correlations were conducted. A conversion factor of 7.8 f/ml to 1 mppcf was calculated for jobs involving the highest asbestos exposure (preparation). For the rest of the jobs, a mean conversion factor of 2.9 f/ml to 1 mppcf was used. More detailed assessment for the conversion of impinger measurements using additional information provided by the TEM analysis of stored MF samples have also been calculated for the North and South Carolina textile cohorts (Dement et al. 2009, 2011). For asbestos cement manufacture (Lash et al. 1997), a conversion factor was reported of 0.7 f/ml to 1 mppcf, approximately an order of magnitude lower than value often used for textile manufacture (see Table 10.6).

As discussed in the previous chapter, there were at least three types of impingers, and all had very poor collection and counting efficiency for <1 µm particles. The index of measurement was also nonspecific to fibers (all visible particles) and unstable. As particles were collected into a liquid, chrysotile would start to divide with the individual fibers, bundles, and clusters breaking down into many more fibers. If left for any significant time before transferring and counting them at the bottom surface of a liquid sedimentation cell, they would also start to reagglomerate.

Conversions from Other Particle Counting Methods

Whilst impinger measurements were largely confined to North America, TPs and konimeter sampling were much more widespread geographically. However, each of these instruments also had at least three or more versions, often with the different versions favored in one or more countries. Both instruments had one advantage over impingers in that it was a direct method of sampling, so the sample collected would not change appreciably with time after sampling.

Good evidence (see previous chapter) suggests that the conversion factor for the various TP counts of >5-µm-long fibers to MF-PCM fiber counts were likely to be reasonably close to ~1 : 1. However, users of the original standard thermal precipitator (STP) commented that counts of >10-µm-long fibers were curtailed (Holmes 1965). Conversion from konimeter particle counts is more problematic and relies on specific side-by-side measurements in the workshop/area being sampled.

Conversions from Gravimetric Measurement

Only a few cohorts have exposure measurements in terms of mass concentration of airborne dust. In the United States, when the conversion of the current impinger exposure limit of 2 mppcf (i.e. the tolerance level value [TLV]) to

MF-PCM fiber count was being considered, it also included a comparison with the total and respirable dust concentrations based on the magnesium content (Lynch et al. 1970). Despite the magnesium content not discriminating between chrysotile fibers and the nonfibrous antigorite particles in the chrysotile-bearing serpentine ore, the gross weight ratios for textiles, friction products, and pipes were found to be approximately 2, 1 and 0.5 mg/m^3, respectively. This ratio range for total dust conversion (×4) was only slightly greater than the ratio (×3) for the MF-PCM conversion factors. The percentage magnesium contents of the respirable fraction of the total dust were determined as 31, 29, 44, and 22 for textile, friction products, pipe, and insulation, respectively. Taking into consideration that the magnesium content in the total dust decreased from 68% to 4% and for respirable dust from 64% to 6% for the same product categories, the gravimetric magnesium content did not appear to be too far adrift of the MF-PCM method, for use as an index of exposure.

Another attempt to measure mass of chrysotile in a sample by XRD and relate it to PCM fiber concentrations was used to monitor the filter outlets from three Italian chrysotile plants (Puledda and Marconi 1991). The pooled analysis divided samples into four categories with three ranges of conversion factors per process: splitting (0.4–0.7), oven drying (1.8–3.4), silo sorting (4.0–4.9), and packing (1.7–2.1). The limitations of XRD to discriminate between the nonfibrous antigorite host rock and chrysotile fibers probably accounted for part of the order of magnitude range in the ratios.

The rapid decline in asbestos mining and manufacture in the West, means that Russia and China are now by far the biggest producers and manufacturers of asbestos. This means any new cohorts and knowledge will come from these countries, which use static gravimetric air measurements to assess controls. All gravimetric conversions are dependent on their being a relationship between the mass of the nonfibrous and fibrous particles and the number PCM fibers. This requires that the particle size of the airborne dust and the percentage of asbestos it contains is relatively constant between similar processes over time. However, the percentage of asbestos fibers present in the material will change as the ore is refined during the milling, processing, and downstream manufacturing. The gravimetric measurements will also be directly influenced by what other materials are added during the manufacturing process (e.g. asbestos cement, asbestos board, and insulation manufacturing).

Particle size and the equivalent aerodynamic diameter are particularly important for weight conversions. For example, if typical dimensions of a countable chrysotile fiber were ~0.2×10 μm, this has a volume of 0.3146 μm^3, while a 4 μm spherical particle (i.e. a particle with a 50% probability of penetrating into the alveolar region) has a volume of 25.2 μm^3 and, if of equal density, is equivalent to 80 such fibers. In total dust samples, much larger particles will deposit on the filter and a single large nonrespirable particle can represent many thousands of such fibers. Therefore, the ratio of the particulate mass: MF-PCM fiber count can vary substantially with respirable mass and much more so for total mass measurements.

For the Qinghai cohort, the conversion to PCM fiber counts was based on 35 paired samples taken in 1991 in the mill and the mine. Based on linear regression of the logarithmic data, the following relationship was determined: mass concentration (mg/m^3) = 3.29356 × fiber concentration (f/ml)-1.0945, with a correlation coefficient of 0.88. However, this applied only to the conversions from 1984 to 2006. The geometric mean airborne concentrations were estimated at ~240 f/ml in the mill and 40 f/ml in the mines. This conversion was supported by the high number of asbestosis cases ($n = 87$) in the mill workers.

The same chrysotile fiber was used by the Chongqing manufacturing cohort which was divided into seven workshop areas. The average dust concentrations in the main workshops reached 146.2 mg/m^3 prior to the 1960s, but despite improvements, have continued to far exceed the Chinese national standard (Deng et al. 2012). From 1999, air samples of both dust and fibers were taken in each workshop. In total, 223 MF-PCM fiber counts were available, but only 90 samples were selected for the years 1999, 2002, and 2006 and paired with the static mass concentrations to calculate the conversion factors. Multiple linear regression analysis of the logarithmic data showed that fiber concentrations were significantly correlated ($p < 0.001$) with the dust concentration in the individual workshops. The estimated geometric mean fiber concentration was 13.8 f/ml (SD 17.3) during the period from 1970 to 1990.

A new retrospective study established for workers in a large Russian chrysotile-producing mine operated by Uralasbest has over 100,000 measurements of the gravimetric (total) dust concentrations at static sampling locations from 1951 onward (Schonfeld et al. 2017; Schuz et al. 2024). Attempts to correlate with MF-PCM fiber counts have been described (Feletto et al. 2017). Three different sampling campaigns were used to compare side-by-side static measurements (1995, 2007, and 2013/2014) in the mine and two mills (enrichment plants) but, owing to the high dust concentrations prevalent, very short and often different durations of sampling periods and sampling rates were used (e.g. 10–107 minutes for the dust sampling and 1–69 minutes for MF-PCM). A total of 620 daily median fibers-to-dust ratios were derived and grouped together by production unit. Using the log-transformed fiber-to-dust ratios, linear mixed models were built to estimate the log-transformed fiber-to-dust ratio separately for the mine and factories as a function of year of sampling, season, and unit. Several fixed and random variables were used in the modeling and the data were censored to exclude outliers (e.g. outside a total dust range of 0.1–15 mg/m^3). All measurements were undertaken by the central laboratory for the Uralasbest mining company.

In the two mills, modeled ratios varied by unit, generally increasing along the stages of asbestos enrichment (i.e. as the nonfibrous host rock is increasingly separated from the ~2.3% chrysotile content of the ore). In the large open-cast mine, ratios were higher in winter compared with summer. Overall, an approximate 1 : 1 ratio between the medians of the two measurement indices was obtained (i.e. 1 f/ml = 1 mg/m^3 of total dust), but the ratios showed a strong negative dependency with increasing dust concentration. The authors considered that "Extrapolating to earlier years should be done with caution, especially considering that process automation and the introduction of more modern technologies, changed mining and asbestos processing practices and allowed for enrichment of ore with lower chrysotile content. We have no ability to account for these changes in our model." The authors felt that the higher ratio of fiber:mass conversions obtained in the Chinese and South Carolina studies were not comparable due to the chrysotile being more refined for asbestos textile manufacture and/or the effects from undercounting of fibers, as many of the MF-PCM sample filters were overloaded.

Examples of gravimetric to PCM conversions used by various sources is collated in Table 10.7 and showed that ratios of between 0.56 and 50 have been used for total dust measurements and 0.5–4.7 for filtered stack emissions and respirable dust.

Crocidolite Cohort Exposures

Wittenoom Occupational

Crocidolite asbestos was mined and milled at Wittenoom in Western Australia between 1943 and 1966 and male workers have high rates of mesothelioma (>5%). Exposure estimates are complicated as several mines and mills are involved. Low-level "pick and shovel" mining activity started in 1936 and the Wittenoom Gorge mine and mill operated from 1943 to 1958 but also hired previously exposed workers from the Yampire Gorge crocidolite mine and mill, which operated from 1937 to 1946. A new mill was constructed in 1949 and various modifications and attempts at dust control were made until its closure in 1958, but there was little change in dust concentrations in this period (Armstrong et al. 1988). The Colonial Gorge mine commenced operation in 1953 and was closed in 1966 for economic reasons. A new mill was commissioned in September 1957 close to the new mine.

The only available airborne fiber concentration data (Major 1968; Rogers and Major 2002) was collected in 1966 at the Colonial Gorge mine. Thirty-eight long-running thermal precipitator (LRTP) static samples were collected about 2 months before it closed. The results were reported only in very general terms based on

Table 10.7 Examples of gravimetric to MF-PCM fiber conversions for asbestos mining and manufacturing.

Source	Type of plant	Type of activity	Sample number	MF-PCM (f/ml)/ gravimetric (mg/m^3)
EEC directive (1987)	All	Environmental discharges	Not given	20
Bauer et al. (1997)	All	All	Not given	50
HCN (2010)	All	All	Not given	30
Institute of Occup. Safety, Maribor, SI (1986)	Asbestos cement	All	Not given	10
	Asbestos-only	All	Not given	50
Puledda and Marconi (1991)	Asbestos cement	Environmental stack discharges Sheet production	11	0.7–4 (Filtered air)
		Environmental stack discharges pipe production	6	0.5–3.5 (Filtered air)
Dodič-Fikfak (2007)	Asbestos cement	All activities	67	0.7 (respirable) Range 0.04–25
		Pipe all activities	7	4.7 (respirable)
		Sheet all activities	4	1.6 (respirable)
Feletto et al. (2017)	Mining chrysotile	Mining	60	1.09 (0.15–5.3)
	Milling chrysotile factories 4 and 6	All units	381	0.56–1.8
		Packaging only	12	1.09 (0.21–5.18)
Deng et al. (2012)	Manufacturing textiles, asbestos cement, and rubber	Raw material	28	4.78
		Carding and spinning	37	6.65
		Weaving	16	6.65
		Maintenance, cement, and rubber	9	3.92

Gravimetric refers to total dust sampled by the instrument unless stated otherwise.

some measurements over 12 shifts with the number of fibers expressed as "about 10%" for the mine samples. Although no details of the analysis are available, it is likely that the ARC method (Holmes 1965) was used based on ×500 bright field microscopy. Results have been summarized in Table 10.8.

The use of static sampling has to be taken into account and the way the ore was mined suggests that the personal exposure would have been considerably higher. It is unusual that the control platform should give the highest fiber counts and suggests oversampling may have occurred causing undercounting in other samples. This is also supported by the sampling times being reported as usually 4–5 hours (Armstrong et al. 1988). The relationship between LRTP fiber counts and the early PCM method was taken as 1 : 1 (Walton 1982). As the LRTP samples were taken at the very end of production, these measurements are likely to underestimate the airborne fiber concentrations during previous years. However, it was stated that the facility was in full production at the time of sampling. The underestimation of the fiber concentrations and the use of a median time-weighted average over 23 years for broad categories of exposure such as 50 f/ml for "ever mill workers," 12 f/ml for "ever mine workers but with no mill exposure," and 5 f/ml for "other workers" was judged to be "somewhat low" by the occupational hygienist that collected the samples, who also considered they should be regarded as "guestimates" (Rogers and Major 2002). Unfortunately, no opinion was expressed as to what they might be, but even a simple average of the mill results in 1966, suggests that 150 f/ml could be argued.

Table 10.8 Summary of the only existing data on the fiber concentration at Wittenoom mines and mills (Colonial Gorge) taken in 1966 just before the mine closed.

Process	Particles >0.5 μm wide (p/cc)	All fibers lengths (s/cc)	>5-μm-long fibers (f/ml)
Colonial mine			
Miner and scraper operator	1,500	120	20
Ross ore feeder	100	10	~1–3
Picking belt and drier area	200	20	~1
Colonial mill			
Plant operator control platform	3,000	670	270
Hand bagging	3,000	600	100
Mechanical bag press operator	2,000	330	80

Sampling carried out by the Mines Department of Western Australia prior to 1966, used a konimeter to give particle counts. Between 1948 and 1958, periodic measurements were made. As many of the results recorded were >1,000 p/cc, representing overloaded samples, results that were of limited value other than showing that airborne concentrations were consistently high (Rogers and Major 2002). It has been reported (Reid et al. 2008) that between 1948 and 1951, the results were some six to eight times above "safe levels" but also mentioned was anecdotal evidence suggesting that operations were shut down before inspections commenced (Armstrong et al. 1988). While other konimeter measurements were taken for in-house mine operation and ventilation purposes, these were no longer available.

To verify the LRTP concentrations, the lung burden of fibers from deceased workers (de Klerk et al. 1996) was measured. A log v log plot of some 90 mesothelioma cases showed an overall trend between estimated cumulative exposure and lung burden. However, as individual points of cumulative fiber exposure were often spread over some 2–4 orders of magnitude, this did not provide good evidence that the lung burden data supported previous estimates of the airborne exposure. Based on various assumptions, the residence time in the lung was estimated to have a half-life of 92 months. This constant removal of fibers was suggested as the reason for a reduction in risk from lung cancers, long after exposure had ceased. The intensity of exposure was expressed in terms of a geometric mean of 20 f/ml.

Due to the absence of PCM data from Wittenoom, it is of interest to compare the PCM and TEM data from an active crocidolite mine in Cochucumba, Bolivia (Ilgren et al. 2015). The chemistry of the Bolivian crocidolite deposit differs markedly from that found in Wittenoom and the South Africa crocidolite mines (Shedd 1985). This has resulted in a much coarser airborne fiber width distribution compared with other crocidolite deposits (e.g. 0.30–0.48 µm compared with 0.06–0.17 µm for Cape crocidolite (Ilgren et al. 2015). However, in the fiberizing section of the mill, individual personal exposures of nearly 400 f/ml have been recorded with an average of 269.5 f/ml for shoveling in a semienclosed work area with natural ventilation. It should be noted that exposures from the MF-PCM count are likely to be significantly underestimated at such high concentrations and a TEM analysis showed the PCM-sized fiber concentrations were substantially higher (e.g. up to 826 f/ml (Ilgren et al. 2015)). Again, this suggests that airborne concentrations in the Wittenoom mill with much finer crocidolite fibers were likely to be in the range of several hundreds of f/ml.

Wittenoom Environmental

The original town of Wittenoom was established 1 km from the Wittenoom Gorge mine and mill but was moved 12 km away in 1957, so it was closer to the site of the Colonial Gorge mine and mill. Mine tailings containing crocidolite were widely used in its infrastructure (e.g. roads, playgrounds, backyards, school running track, etc.) giving rise to widespread environmental exposure. The environmental conditions when the mine was operating, due to traffic movements, transport of tailings and fugitive emissions, were likely to be very different from the situation after the mine closure in December 1966. Most cases of mesothelioma amongst nonemployees have been associated with para-occupational exposures from employed family members. In a study of 2,608 females (Reid et al. 2008) covering the period from 1960 to 2005, about 2.5% of the workers compared to 1% of the residents had mesothelioma (total 47) based on a median duration of exposure of 1.3 years. These mesothelioma rates were ~77 times higher than the background rate for Western Australia.

Environmental measurement data were collected on several occasions, the first being collected by LRTP in 1966 (Rogers and Major 2002), which found concentrations of 2 and 0.5 f/ml some one hundred meters outside the mill. Subsequent environmental data was obtained from MF sampling and PCM analysis of personal and static samples, in 1973, 1977, 1978, 1980, 1984, 1986, and 1992. Based on these measurements in the epidemiological study, people not working directly with asbestos were assigned an intensity of exposure of 1.0 PCM (f/ml) of air from 1943 to 1957 (when the new mill was commissioned), and then 0.5 f/ml between 1958 and 1966, when the mining operations ceased. Interpolation between surveys using personal monitors assigned exposures from 0.5 f/ml in 1966 to 0.01 f/ml in 1992.

As personal PCM measurements were only taken in 1977, 1978, and 1980, and a high percentage of these were overloaded, it is likely that the mean is an underestimate. The static samples gave much lower results than the personal samples with the highest personal sample being 0.88 f/ml. As MF-PCM results were only taken over 10 years after mining ceased and the population was much reduced, it is unknown what airborne concentrations the population was exposed to during the time the mine and mill were operational. Some 80% of the environmental exposed cohort started their exposure when the mine was operating. Anecdotal information suggests that there was likely to be a contribution from the mine and mill to the environmental and para-occupational exposures when it was operational. Pilots flying into Wittenoom were guided by the blue haze visible from a long distance away and clerical staff in the company office, well away from the production areas, were able to write in the fibrous dust that settled on their desks (Reid et al. 2008). There are no measurements of para-occupational exposures from Wittenoom.

South African Mines and Mills

Commercial amphibole asbestos production in South Africa started in the 1890s. The Cape crocidolite and the Transvaal crocidolite and amosite mines and surface processing mills were air sampled infrequently from about 1940. Most of the employees were nonwhite and they were mainly itinerant workers who worked for 1 year or less and no epidemiological or medical follow-up is available. Therefore, the only epidemiology relates to a minority of white workers who often had the less dusty jobs and only <1% of the cohort reported had exposures before 1940, and 6.4% were exposed before 1950 when the airborne concentrations were much higher (Sluis-Cremer et al. 1992). Over the next three decades, about one-third each of the remaining cohort was first exposed in the 1950s, 1960s, and 1970s, when air measurements were consistently lower (see Figure 10.1).

The occupational measurements reported have inherent problems as they were often averaged for the mine and the mill, and then for groups of mines, or over time. At varying times measurements were made using two types of konimeters and two types of TPs. Furthermore, it was not always clear what was being measured, viz. particles and/or fibers. The definition of a fiber is determined by the ratio of its length to diameter and the measurement of the diameter itself. This definition was not constant; the fiber length: diameter ratio was changed from two to three in 1965, and the diameter from 5 to 3 µm in 1970 (du Toit 1989). As different optics were used for the microscopical evaluation (e.g. ×150 dark field and ×1,000 bright field) and various sample treatments (e.g. acid washing and incineration) were used at different times, it is difficult to establish a single reliable index of exposure. An early PCM method (ARC 1971) was introduced for health surveillance in 1975, although reference is made to a large PCM survey in 1970. A summary of the early measurements is given in Figure 10.1.

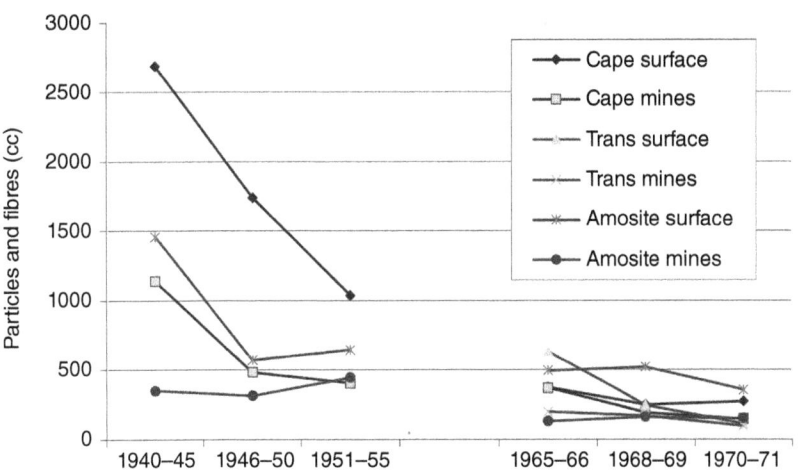

Figure 10.1 Summary of the historical airborne particle concentrations made in the main South African Amphibole mining areas (Cape and Transvaal) using the konimeter (1940–1955) and thermal precipitator methods (1965–1971).

It was reported that, in general, the limit of visibility for particles in a ×150 dark field count with a konimeter could be equated to a TP count of >0.5 μm using ×1,000 bright field microscopy (du Toit and Gilfillan 1977).

Given that about 94% of the cohort was exposed in the 1950–1970s, it should be possible to give a reasonable estimate of their average of >5-μm-long PCM fiber exposure from side-by-side comparisons of the monitoring techniques. Also, some occasional fiber counting took place from the 1950s. Although comparing particle counts to fiber counts by separating out the fibers should be relatively straightforward process, the attempts to do this for the MF-PCM method in the South African mines are difficult to reconcile. It was also reported that the air samples were originally collected on 8 μm pore size filters instead of the recommended 0.8 μm pore size (du Toit and Gilfillan 1977; du Toit 1989). The conversion factors are summarized in Table 10.9, along with a later set of values produced in cooperation with the Asbestos International Association (AIA) (du Toit et al. 1983), which after detailed statistical analysis were found to be around five times lower. However, this last set of values was extremely dependent on one high data point and the slope could change from 1.09 to 2.68 without this point (Burdett 1998).

The estimates of the actual fiber concentrations used for the epidemiology are given in Table 10.10. However, given the uncertainties between conversion between indices and methods, it is perhaps only possible to conclude that recorded concentrations of airborne dust in the Cape crocidolite mines and mills and the Transvaal amosite mills fell by about a factor of 3 between 1940 and 1945 to 1950 and 1955, but over the same period, those in the amosite mines increased by about a half. A similar trend was seen from 1965 to 1971 when airborne dust concentrations in both the Cape and Transvaal crocidolite mines and mills fell, usually by around one-half. This reduction coincided with an increase in production and a much greater number of people being exposed. The crocidolite cohort of 3,430 men

Table 10.9 Summary of South African mine conversion factors in terms of the expected membrane filter fiber count from a concentration of 1 f/ml measured by the konimeter and the (standard thermal precipitator).

Filter pore size used (μm) (publication)	Cape crocidolite	Transvaal crocidolite	Chrysotile	Amosite
8[a]	0.75 (0.9)	2.2 (1.9)	4.6 (7.3)	4.4 (5.7)
0.8[b]	4.3 (4.5)	6.8 (5.7)	8.1 (9.8)	7.9 (8.8)
0.8[c]	1 (0.53)	1 (0.53)		1 (0.53)

[a] du Toit and Gilfillan (1977).
[b] du Toit and Gilfillan (1979).
[c] du Toit et al. (1983).

Table 10.10 Overall estimated average in South African mines (underground) and mills (surface) in terms of PCM f/ml (Sluis-Cremer et al. 1992).

Year	1945	1960	1970
Amosite surface	150	56	40
Amosite underground	14	6	4
Crocidolite surface	30–160	10–50	8–30
Crocidolite underground	2–60	2–6	2–6

had an estimated average exposure of 9.6 f/ml, which was multiplied by the number of years employed to give the cumulative exposure.

By the early 1980s, when MF PCMs were being collected, the results from two laboratories for 10 crocidolite samples from the mills ranged from 0.5 to 57.2 f/ml and three amosite samples were between 20.4 and 41.2 f/ml. Only one mine/underground sample was available from each site with the highest laboratory count being 0.75 and 0.3 f/ml, respectively (du Toit et al. 1983). This suggests air concentrations in the mills remained much higher than the mine. Although several other mines were present in this region, no PCM data is available for comparison.

Compared with the Wittenoom crocidolite mines and mills, the airborne concentrations at the South African crocidolite mines are substantially lower, even though, from the fiber width distributions, Cape blue is nearly as fine.

Massachusetts Cigarette Filter Manufacturing

There was one set of impinger measurements from this small facility that operated from 1952 to 1957 (Talcott et al. 1989). It produced filters for Kent cigarettes using Bolivian crocidolite; the MF-PCM concentrations were estimated in an appendix in the Hodgson and Darnton meta-analysis (Hodgson and Darnton 2000). This involved the conversion of impinger measurements of ~2.3 mppcf. Further evidence was obtained from the reported lung burdens from two mesothelioma cases from a plant in Kentucky (Dodson et al. 2003), which processed the crocidolite-containing filters made at the Massachusetts facility. One case employed for <2 years had cut the lengths of filter plugs received to the correct size. The work area was described as very dusty and the lung burden of 77 million fibers/g of dried lung (note: 1 million fibers/g of dried lung is usually taken as evidence of an occupational exposure) showed that this must have been the case. Nearly all fibers found were high magnesium crocidolite typical of the Bolivian crocidolite also used specifically by the Massachusetts cigarette filter manufacturer. The original conversion (Hodgson and Darnton 2000) was based on the crocidolite in use being similar to South African Cape crocidolite. It was not until much later (Ilgren et al. 2015) that the larger width distribution was reported. This suggests the original estimate was likely to be too high (e.g. factor of 2–3).

UK Gas Mask Workers

No air concentration measurements are available from the UK gas mask assembly facilities (Birmingham, Blackburn, Leyland, and Nottingham), which were operating before and during the Second World War. Acheson et al. (1982) reported on the civilian gas masks workers at Blackburn, and the military gas workers at Leyland. Military gas masks used pads manufactured elsewhere with a mixture of ~20% crocidolite (mainly South African) and either chrysotile or merino wool. Much higher incidences of mesothelioma were found in this workforce than found in civilian gas mask production, which only used chrysotile asbestos. The order of exposure estimate made in the Hodgson and Darnton analysis was based solely on lung analysis of predominantly mesothelioma victims and therefore represents a selective group of diseased workers.

Morgan and Holmes (1982) also reported lung burdens from selected cases at the Blackburn, Leyland, and Nottingham factories by MF-PCM analysis (fibers >2.5 m long) of digested material. Results ranged from 0 to 0.07 to 216 million fibers/dry gram (mfpdg), but there was no clear correlation between concentration and duration of exposure. Lungs of workers who were not employed on the assembly line also had significant fiber concentrations. The geometric mean fiber burden for the Leyland cases (8.6 mfpdg) was about twice that for the Nottingham cases (4.6 mfpdg) and the Blackburn cases were lower with a range of 0.5–6.5 mfpdg. It was considered that the higher geometric means for Leyland were consistent with the reported lack of control measures.

TEM measurements of the concentrations of various types of asbestos in the lungs of 22 Nottingham gas-mask factory workers (Jones et al. 1980) found that crocidolite fibers predominated with a median of 40 mfpdg, which was the equivalent to around 5 mfpdg of PCM countable fibers. A further study (Berry et al. 2009) analyzed lung tissue samples from 50 deaths from mesothelioma and 20 other causes at the Nottingham factory. They reported TEM analyzed lung burdens of up to 1,949 mfpdg for the Nottingham workers with medians of 67 mfpdg for those working less than 2 years and 93 mfpdg for those working for more than 2 years. There were clearly high airborne concentrations present, but the measured burdens were lower than those of the Wittenoom miners and millers who had shorter exposure periods; some 77% worked <1 year (Armstrong et al. 1988).

Other Cohorts Exposed to Crocidolite

While the gas mask workers using crocidolite were separated by geographical location and the operation took place over a specific time period, the lack of detail of personal exposure history for those exposed to some crocidolite in the textile and asbestos cement manufacturing presented difficulties for estimating the dose–response by fiber type. Crocidolite was used in varying amounts and times in many textile cohorts and the epidemiologists overseeing the cohort may or may not class it as a trivial or nontrivial mixed exposure (e.g. US chrysotile textile cohorts).

Asbestos cement product manufacture was by far the largest user of asbestos. For instance, a large, pooled analysis of Italian asbestos cement cohorts who used mixtures of chrysotile, crocidolite, and amosite (Luberto et al. 2019) with airborne measurement data derived from various sources, reported airborne concentrations were very high before 1980 (0.2–45 f/ml), intermediate between 1980 and 1989 (0.2–11 f/ml) and lower after 1989 (<0.1–0.3 f/ml).

Crocidolite Summary

Table 10.11 summarizes the crocidolite exposure of various cohorts. Overall, the average assessed cumulative exposure at Wittenoom seems low, given the probable airborne concentrations and conditions. However, this reflects the limited time people worked at the site and is not dissimilar from the white South African experience. The scant information from the small Massachusetts cigarette filter cohort does not allow any real understanding of the risk from Bolivian as opposed to other commercial crocidolite. The use of crocidolite in textiles and asbestos cement, produces vastly different health outcomes for mesothelioma compared with that for chrysotile alone, but there are few airborne concentration measurements and even records of when it was used are rare.

Amosite Cohort Exposures

Amosite was mined in one region of South Africa and the raw fiber was used to manufacture insulation products. Three cohort studies have been reported: Patterson, Tyler, and Uxbridge. Uxbridge used a mixture of mainly amosite and some chrysotile. The three manufacturing cohorts have been previously reviewed by Ribak and Ribak (2008).

South African Amosite Mining

South Africa was the sole commercial source of grunerite (amosite) asbestos. The Penge, Weltevrede and Kromellemboog mines in the Transvaal, Limpopo

Table 10.11 Summary of crocidolite cohort exposure and mesothelioma rate.

Cohort dates reference to latest update	Estimated mean airborne concentration in terms of PCM (f/ml) for date/period stated	Calculated average cumulative exposure (f/ml-years)	Size of cohort	Comments on fiber measurement biases
Wittenoom Workers 1943–1966 Musk et al. (2020)	1966: Mills >250 Mines 15–40 No PCMs but 28 LRTPs taken in 1966	23*	6895	LRTP: PCM conversions are usually assumed to be 1 : 1. All samples collected just before mine and mill closed and likely to underestimate the historical fiber levels.
South African mines and mills 1893–1992 Sluis-Cremer et al. (1992)	1945: mills 30–160, mines 2–60 1960: mills 10–50, mines 2–6 1970: mills 8–30, mines 2–6 (PCMs from 1975)	16.4 * 9.6	3430	Conversions from historical measurement methods to MF-PCM are difficult to reconcile (see Table 10.9).
Wittenoom domestic/environmental 1943–1993 Musk et al. (2020)	1943–57: 1.0 1958–66: 0.5 1966–94: 0.1 (PCMs from 1973)	5.4 mean 15.2 median	5042	No measurements relevant to high early exposures when mine was open. The high estimated average cumulative exposure due to more hours of exposure than workers.
Massachusetts 1952–1957 Talcott et al. (1989)	One set of impinger measurements, from which PCM concentration estimated at 60	120*	55	Likely an overestimate of the fiber concentration and cumulative exposure as the original conversion did not account for subsequent size data.
UK gas mask workers 1939–1945 Berry et al. (2009)	No air measurements	Lung burdens less than Wittenoom mine and mill workers	1600	No air measurements taken, so best estimate based on lung burdens

* Value used by Hodgson and Darnton (2000).

province being the main source of production. The ease of mining (only Penge mine had underground shafts) meant that amosite was the main fiber type produced from South Africa up to 1960. However, epidemiological investigations are hampered by the fact that crocidolite asbestos was also present in this region and at some places occurred alongside or interwoven with the amosite deposits. Many of the workers at the mines were transient, with staff turnover approaching 100% per annum among Penge miners. Owing to this, white miners are the only group studied; yet in 1963, Penge had 360 white and 6,500 black workers (McCulloch 2002).

In addition to the two mills at Penge mine, there were many small mills in the Penge area. They were of basic design and caused extensive environmental pollution in the valleys (McCulloch 2002). The workers themselves were exposed to extremely high fiber concentrations, as sweepers, sorters, and packers. Cobbing, whereby the fiber was loosened from the rock, was done primarily by women and children at Penge and other smaller operations. In 1940, 25% of the Penge workforce were boys younger than 16 (McCulloch 2002). The employment of children younger than 16 years of age was only prohibited in 1973 and hand cobbing was reported to have continued past 2002.

Despite the limited number of exposure measurements, anecdotal evidence and the amount of nonmalignant lung disease clearly indicates that the concentrations of asbestos fibers in and around the Penge mining area were high (Murray and Nelson 2007). A study (Davies et al. 2001) of the prevalence of asbestos disease in 770 women who had worked in asbestos mining (80% were cobbers) in the Pietersburg asbestos field from 1929 to 1980 found that 96% had pleural and/or parenchymal asbestosis. They ascribed the high prevalence due to exposure to high concentrations of amphibole asbestos dust, predominantly amosite. Autopsies of Black Penge miners who died while employed, reported 80% with at least mild asbestosis. Widespread environmental contamination has been described around the mines and downstream. In 1988, some 12 years after the closure Felix (1997) reported environmental concentrations, with children in the area having the highest exposures (mean 0.02 f/ml: range 0.002–0.090 f/ml).

In 1947, workplace static sampler particle measurements in the Penge mills ranged from 162 to 720 p/cc, and underground, from 80 to 228 p/cc: concentrations appeared to increase in the 1950s, probably due to mechanization, when 30% of the counts in the mills were above 780 p/cc (Sluis-Cremer 1965). Using personal samples, Rendall (1971, *Unpublished data*) recorded fiber counts of up to 327 f/ml in the mills and more than 100 f/ml in a range of other surface jobs: underground concentrations were <5 f/ml. In the 1980s, the fiber counts at Penge ranged from 0.1 to 6.5 f/ml in the mills and from 0.2 to 1.0 f/ml underground (Murray and Nelson 2007). Table 10.12 details the available air concentration

Table 10.12 Summary of measured air concentrations for amosite mining at Penge.

			Air concentration	
Source reference	Year	Site	Particles/cc	Fibers/ml
Sluis-Cremer et al. (1992)	1945	Surface		150
		Underground		14
Sluis-Cremer (1965)	1947	Mills	162–720	
		Underground	80–228	
	1951	Mills	30% >780	
Sluis-Cremer et al. (1992)	1960	Surface		56
		Underground		6
Sluis-Cremer et al. (1992)	1970	Surface		40
		Underground		4
Rendall (1971, *Unpublished data*)	1970	Mills		1.4–326.7
		Other surface jobs		0.2–113.4
		Underground		<5
SA Dept. of minerals and energy: personal communication	1986	Mills		1.1–6.5
		Underground		0.2–1.0

Source: From Murray and Nelson 2007 / Elsevier

measurements for amosite mining at Penge. It should be noted that conditions in the amosite mines were particularly hot and humid making MF-PCM sampling problematic.

Patterson, New Jersey

The first study to establish the carcinogenicity of amosite was conducted on a cohort of 820 workers from a plant in Paterson, New Jersey. The plant which operated from 1941 to 1954 has been reported to have utilized almost exclusively amosite asbestos. The exposure period for the cohort was limited to the workers who started work between June 1941 and December 1945. There is no record of any air sampling measurements from Patterson and the airborne fiber concentrations at the factory were assumed to be similar to those monitored at Tyler, some 25 years later. Both facilities were operated by the same company and in some cases, used the very same machinery and processes. Based on Tyler, a range of between 15.9 and 91.4 f/ml was estimated (Levin et al. 1998), a median exposure of 50 f/ml was estimated (Seidman et al. 1986) and 30 f/ml was used as a lower limit for the average exposure in a risk assessment (Nolan et al. 1999).

Tyler, Texas

The Paterson factory closed in 1954 and moved to Tyler, Texas where a cohort of 1,130 workers fabricated amosite containing products until February 1972, when it was closed after failing to control air concentrations to the hygiene limit (Levin et al. 1998). Some 170 personal air samples were collected and analyzed in 1970 and 1971 some 13 years after the plant opened with over 70% of them collected in 1971, shortly before the plant closed. The concentration range was reported to be from 15.9 to 91.4 f/ml (Levin et al. 1998). More detailed exposure data specific to various tasks that the workers performed are available for 1967, 1970, and 1971, and are shown in Table 10.13 (Hurst et al. 1979; Johnson et al. 1982). Since these air samples were taken in the last phase of operation at the plant where it was carrying out modifications to meet the new industrial hygiene limit of 12 f/ml, the values in Table 10.13 are more likely to represent the lowest concentrations achievable rather than the mean airborne concentrations over the previous 13 years.

Although the airborne concentrations at Tyler are high, there were only three asbestosis deaths (representing 1.4% of the total mortality) from the 18 years it operated (Levin et al. 1998). This meant that there was some 3.7-fold less asbestosis mortality than reported for the Patterson cohort (Seidman et al. 1986), which had a maximum exposure duration of 13 years. This suggests the air concentrations were several times higher at Patterson.

Uxbridge

The use of amosite began in 1947 and ended when the factory closed in 1979. Between 1946 and 1973, chrysotile asbestos was also used but generally constituted <3% by weight of the board material being manufactured, compared with

Table 10.13 Results of personal air sampling for five different functions at the amosite factory in Tyler, Texas.

Operation	PCM air concentrations (f/ml)			
	1967	1970	1971	Mean
Milling/fiberizing	163.5	36.2	74.4	91.4
Forming	33.3	25.7	50.6	36.5
Curing	2.5	31	14.4	15.9
Finishing	44.6	34.8	39.5	39.6
Packing	16.7	17.9	22.8	19.1

Source: From Ribak and Ribak (2008) / Elsevier

an average of ~20% amosite. The fiber content was reduced over time for both economic and other reasons before being replaced by a nonasbestos board. Exposures prior to 1964 were thought to be as high as 100 f/ml, particularly on the "beater floor" where the raw asbestos bales were unpacked and opened. This process was removed to a new building and enclosed in 1966, which resulted in significant dust reduction. Four of the five mesothelioma cases were exposed to the dustiest operations such as milling or fiberizing of asbestos before 1960.

PCM air measurements were made from 1969, after the improvements had been made and concentrations of around 30 f/ml were reported (Acheson et al. 1984). Evidence provided to the UK Advisory Committee on Asbestos (Simpson/ACA 1977) showed that for the first half of 1976, the average fiber concentration was 1.26 f/ml with 10.1% between 2 and 4 f/ml and 1.3% above 4 f/ml. There were 11 respirator areas in the 90 stations monitored, showing that meeting the 2 f/ml exposure limit was difficult.

The estimate of the early airborne concentrations, ~100 f/ml for opening and fiberizing (Acheson et al. 1982, 1984), are similar to the average of 91.4 f/ml reported for milling/fiberizing at the Tyler factory (Hurst et al. 1979). One worker with 4 months of exposure had both asbestosis and pleural mesothelioma, suggesting that airborne fiber concentrations were very much higher in some areas. The lung burdens from 43 autopsy cases from Cape Uxbridge (Gibbs et al. 1994) showed that amosite was the predominant fiber found, with a mean of 784.9 million fibers per dry gram (mfpdg) with two subjects having no exposure and the highest being 7852.9 mfpdg. These are very high lung burdens and again suggest very high airborne fiber concentrations existed in the factory until 1964. Crocidolite was the next most common fiber type in the lungs with a mean of 22 mfpdg: the highest burden was 72.7 mfpdg and 19 subjects had no known exposure. The chrysotile mean burden was 12.1 mfpdg, the highest being 24.3 mfpdg and there were 14 subjects with no known exposure. Cases of lung cancer and mesothelioma had by far the highest lung burdens, respectively 1,434 and 1,001 mfpdg and accounting for 14 and 5 of the deaths, respectively. Other cancers have 24 associated deaths, but lung burdens were an order of magnitude lower (297 mfpdg). There were strong correlations between the lung burden and the grade of asbestosis.

Amosite Summary

All three factories are similar in that they produced high airborne fiber exposures during their early use (Table 10.14). The Uxbridge factory reduced airborne fiber concentrations significantly after 1966. Overall, the cancer mortality rates between the three manufacturing cohorts are similar, but the asbestosis rate is significantly different: with the Uxbridge cohort having double the rate and

Table 10.14 Summary of amosite cohort exposures.

Cohort dates reference to the latest update	Estimated mean airborne concentration in terms of PCM (f/ml) for date/period stated	Calculated average cumulative exposure (f/ml-years)	Size of cohort	Comments on fibers measurement biases
South African mines and mills 1914–1992 Murray and Nelson (2007)	1914–1945: mills 150, mines 14 1960: mills 56, mines 6 1970: mills 40, mines 4 1992: mills <6.5, mines <5	23.6* 15.2	3212	Conversions from historical measurement methods to MF-PCM are difficult but exposures higher than for S.A. crocidolite
Patterson, New Jersey 1941–1954 Ribak et al. (1989); Ribak and Ribak (2008)	No air concentration measurements, estimated from Tyler, Texas	65* Median 50	820	Equipment moved to Tyler and air concentrations estimated from Tyler plant
Tyler, Texas 1954–1972 Levin et al. (1998)	1954–1966: 60 1966–1972: 40 (Range 15.9–91.4)	50* 35	1130	Most PCM measurements in last few years of operation with focus on controls.
Uxbridge 1946–1979 Acheson et al. (1981, 1984)	1946–1965: 60–100 1964–1969: 30 1969–1979: 2	50 1	654 early 4166 late	Early exposures due to factory changes are estimates. ~100 PCMs between 1969 and 1976

* Value used by Hodgson and Darnton (2000).

Patterson quadruple the rate associated with the Tyler cohort. If these were indicative of the exposures, this would suggest that the Tyler cohort had the lowest exposures and the Patterson cohort the highest.

Chrysotile Mining and Milling Cohort Exposures

The Quebec, Balangero, and Qinghai Province, China, cohorts are the three main quantitative cohort studies of chrysotile mining and milling. Exposure measurement records exist from 1948 onward for Quebec and from 1969 for Balangero and from 1984 from Qinghai, China, so the historic exposure in these cohorts relies heavily on reconstructed estimates of the earlier exposures. A further Russian cohort is currently under study (Schüz et al. 2013) and exposure data is available, but the mortality rates have only recently been published (Schuz et al. 2024). Several chrysotile cohorts were reviewed in detail by (Bernstein et al. 2013).

Quebec, Canada

The Quebec cohort of ~11,000 men born between 1891 and 1921 was established in 1966 from mainly chrysotile mines and mills at Asbestos and Thetford (Liddell et al. 1997; McDonald et al. 1993). Cumulative exposures were based on time employed (>1 month) between 1904 and 1966 and the impinger measurements collected. Some 4,500 impinger dust measurements were taken across all companies in the cohort during annual surveys between 1948 and 1966. These exposures were analyzed in 10 exposure groups from <3 to >1,000 mppcf-years (i.e. assessed as <11.2–3,600 f/ml-years). The impinger measurements fell from an average at about 75 mppcf in 1948 to about 10 mppcf in 1968 when the PCM conversions were made: 10 mppcf equivalent to 30 f/ml (Gibbs and LaChance 1974). As discussed, the conversion was based on highly scattered data with poor correlation coefficients and the robustness of estimated historic exposure in 1948 of 225 f/ml is debatable. A later review (Gibbs 1994) concluded that no overall single factor could be derived for the Quebec mining and milling industry. However, it was possible to derive conversion factors from the airborne concentrations at the individual mill and work areas based on a ratio of 3.6 f/ml to 1 mfpcf.

After the 1960s conditions improved considerably, and exposures were considered to be relatively inconsequential compared with previous exposures. The annual trends in dust concentration were estimated from 1967 onward for the ~2,500 men in the cohort employed after November 1966 by McDonald et al. (1993).

Balangero, Italy

The Balangero chrysotile mine and mill opened in 1917 and closed in 1990, after operating for 5 years at reduced capacity. The mine had between 200 and 400 employees between 1942 and 1989. No dust controls were in place before 1968 and the operation was very dusty. For example, the crushed ore arriving at the mill was fed into rotary and tower ovens against a counterflow of hot air, but the introduction of dust filters on the air exhausts did not start until 1970. No airborne sampling or measurement before 1967 appears to have taken place. The first airborne sampling of work areas was carried out by industrial hygienists from the Occupational Health Department of the University of Milan who, from 1967 to 1970, conducted four short surveys. The fiber concentrations were estimated by work area from very low-volume, short-term static area samples of 1–2 l of ambient air collected over 10–20 minutes. The first set of samples did not use the standard PCM magnification or aspect ratio for counting the fibers (a magnification of ×250 and an aspect ratio of >8 : 1 was used) but reported air concentrations in the mill of 80.30, 51.00, 50.90, 29.65, and 17.50 f/ml for crushing, drying, beneficiation, mixing and recovery, respectively. Two years later in 1969, the same areas gave results of 11.3, 13.0, 11.2, 10.1, and 28.6 f/ml, but it is unclear whether additional controls were in place to account for this reduction.

The in-house laboratory carried out static area air sampling measurements from 1975 to 1981. Since 1981, personal air samples were also collected. In total, there were 1,099 area and 484 personal measurements. Sampling duration was usually short (10 minutes) for area samples, with two samples taken in each surveyed area per day. Even when sampling duration was longer, it rarely exceeded 30–40 minutes in the 1970s, whilst most samples lasted 60–120 minutes in the 1980s. The in-house measurements showed that the average fiber concentration fell from around 10 f/ml in 1970 to 1 f/ml in 1980 and 0.5 f/ml in the mid-1980s when mining ceased.

Most of the exposure to the cohort of 974 workers was from before any measurements were taken (pre-1967) with some 50% estimated as having cumulative exposures of between 100 and 4,146 f/ml-years. The historical asbestos concentrations were estimated in a study conducted during 1976 and 1977 by the head of the in-house laboratory, simulating working conditions in the absence of local exhausts, and using obsolete machinery and plant. The estimated average air concentrations across the site were thought to have reduced from 72 f/ml in 1930s to 11 f/ml by the end of the 1960s, when the first static air samples were collected and analyzed for fiber concentration. The highest exposures were attributed to the beneficiation area, where the estimated concentrations were some three times higher than the site average until the mid-1950s. The average fiber concentrations for each decade are summarized in Table 10.15.

Qinghai, China

China's largest chrysotile mine is in Qinghai province. The mine and mill opened in 1958 and is still operational. The cohort includes 1,539 workers (of which 400 were office workers) and covers the period from 1981 to 2006. During the follow-up period, over 50% of the workers had retired from the mine. The exposures were based on a measurement taken from 1984 of the total dust in the air using static gravimetric samples with a flow rate of 5 l/min. Samples were taken over four hours and positioned close to the workers' breathing zone. The conversion from mass to fiber concentration was based on 35 side-by-side samples taken in 1991 (Wang et al. 2013b). There was an order of magnitude difference between

Table 10.15 Summary of the average estimated air concentration by decade for the main work areas at the Balangero mine and mill).

	Estimated PCM fibers concentration (f/ml)							
Decade	Mine	Crushing	Drying	Beneficiation	Mixing and packaging	Dumping site	Offices and labs	Average
1920s	16.10	39.58	72.00	250.00	70.30	26.36	21.00	71.90
1930s	16.78	29.79	72.00	175.00	74.20	17.89	16.53	60.60
1940s	22.47	20.77	72.00	175.00	82.87	17.91	16.21	61.89
1950s	10.83	28.41	72.00	136.00	53.67	14.96	13.08	45.35
1960s	3.85	40.26	53.70	40.82	52.59	10.62	9.01	26.89
1970s	1.47	3.24	14.52	5.47	7.24	2.64	1.48	4.53
1980s	0.37	0.38	1.01	1.22	0.95	0.09	0.24	0.77

Source: Adapted from Silvestri et al. (2020)

the mining and milling areas for the gravimetric airborne dust concentrations. It was stated that "average dust concentrations in the mine dropped from 800 mg/m^3 in the 1980s to about 140 mg/m^3 in the 1990s, some 400 to 70 times the previously applied national standard (2 mg/m^3)."

The current conversions based solely on the 35 MF-PCM samples from 1991 suggest that the MF-PCM exposures in the 1980s were around 250 f/ml, not dissimilar to what had been estimated for historic exposure concentrations for Canadian miners. This fell to 44 f/ml in the 1990s. An additional set of gravimetric samples taken in 2006 confirmed the particle concentration for total dust remained high with a calculated geometric mean of 15.9 f/ml (Table 10.16).

The gravimetric measurements showed that the exposures to workers in these areas remain high with the geometric and arithmetic cumulative mean exposures estimated at 57.7 and 212.4 f/ml-years, respectively. These relatively high estimates of airborne concentrations are supported by the high number of asbestosis cases ($n = 87$) in the cohort. However, it has been reported (Yano 2018), that attempts to establish the PCM concentration by internationally recognized methods were thwarted and the actual PCM fiber concentrations may be different. Other environmental/office worker exposures have been measured around the mines but to date, have not been used for the epidemiological assessment. Yano also suggested that the mine conversions may have been overestimated and this accounts for some of the differences in the dose responses between mines and textile workers.

Table 10.16 Airborne exposure concentrations at Qinghai mine and mills in 2006 based on the 1991 side-by-side total mass and MF-PCM measurements (Wang et al. 2013b).

Workshop	Number of samples	Geometric mean (range)	
		Dust (mg/m^3)	PCM fibers (f/ml)
Main workshops	13	47.1	15.9
Mining area	5	5.3 (0.6–17.3)	1.9 (0.5–5.6)
Milling plant 1	4	48.2 (12.5–196.7)	15.2 (4.1–63.8)
Milling plant 2	4	84.3 (31.5–128.0)	26.0 (9.9–47.4)
Other areas	15	3.6	1.6
Milling plant office	3	9.7 (6.3–21.3)	3.7 (2.3–6.8)
Package site	3	8.4 (4.3–13.7)	2.9 (1.6–4.5)
Maintenance area	6	2.4 (0.3–5.8)	1.1 (0.4–2.1)
Transportation site	3	2.4 (0.8–5.1)	1.1 (0.6–1.9)

Uralasbest, Russia

The Bazhenovskoye chrysotile asbestos deposit has been exploited since 1896. Gravimetric air monitoring started in 1950 and, to avoid filter overloading (i.e. >200 mg), several samples were often taken for 15–30 minutes during the day, at each sampling point. A large database ($n = 89{,}200$) of monthly static dust concentrations in the mine and mills has allowed a detailed analysis of temporal trends in asbestos-containing dust between 1951 and 2001. The gravimetric data are consistent with strong downward trends in the 1950s and 1960s, which then slowed down (and in some areas flattened off) in more recent decades (Schonfeld et al. 2017).

The geometric means of 1,457 monthly averaged concentrations in the mine from 1964 to 2001 showed (Schonfeld et al. 2017) that the annual mean dust concentration was between 2.7 and 7.8 mg/m^3 in the 1960s and 2.5 and 4.7 mg/m^3 in the 1970s and early 1980s. There were overall decreases from the mid-1980s to the end of the 1990s and early 2000s with annual mean dust concentrations in these later years ranging from 1.8 to 2.2 mg/m^3. Based on modeling described for conversion of the Russian gravimetric data into MF-PCM f/ml in the section above, the overall daily median fiber concentration was estimated at 0.80 f/ml for 2007 and 0.63 f/ml in 2013/2014 (Feletto et al. 2017). However, dry periods can result in fiber concentrations >2.7 f/ml. Overall, it appears that the estimated geometric mean fiber concentration in the mines was much lower (a few f/ml) than other mines.

The calculated overall daily median fiber concentration for the two mills that were operating was 0.92 f/ml in 1995, 2.50 f/ml in 2007, and 2.22 f/ml in 2013/2014. Based on the gravimetric sampling record, it was thought that the geometric mean concentration in the mills never exceeded 5 f/ml from the mid-1950s and was between 2 and 4 f/ml from the mid-1970s onward (Feletto et al. 2017).

As the arithmetic mean concentration is used for cohort exposures, the geometric mean and median values need further interpretation. At present, no arithmetic mean values are available. An approximate estimate based on what was found at the Qinghai mines suggests the arithmetic means could be some four times higher. Therefore, the Russian measurement data equates to arithmetic mean airborne concentrations in the mill as <20 f/ml in the 1950s and between 8 and 16 f/ml from the mid-1970s, while the mine remains relatively unchanged since the 1950s at around 8 f/ml. The 2 : 1 ratio between the air concentrations in the mills and mines appears to be significantly lower than reported for the other chrysotile producers, as do the estimates of the airborne fiber concentrations.

Chrysotile Mining Summary

The chrysotile disease rates are much lower than those for the crocidolite and amosite mines and mills (Table 10.17) although facilities at both Quebec and Balangero have similar average airborne concentrations. The Quebec cohort has an overall SMR = 1.07 with asbestosis rates of 0.9%. The Balangero cohort had an SMR = 1.28 with asbestosis rate of 3.7% (Ferrante et al. 2020). The Chinese Qinghai cohort has an overall SMR = 2.46 with higher rates of lung cancers (SMR = 4.69) and asbestosis rates (5.7%) despite the reported average cumulative exposures being ~×5 and ×10 lower than Balangero and Quebec (Wang et al. 2013a, 2013b). The Uralasbest mine in Russia has even lower calculated fiber concentrations than those at Qinghai, the latest information shows that there were 13 mesotheliomas and increases in lung cancers among men with increasing cumulative exposure (Schuz et al. 2024).

Chrysotile Textiles

In general, the chrysotile textile cohorts have the best historical measurement data and are some of the most studied. Much of the interest arises from the higher risk estimates for lung cancers compared with the Canadian mines and mills that supplied most of the feedstock. The importance of the difference was amplified by the recent decision by the US EPA to determine chrysotile risk for all industrial

Table 10.17 Summary of chrysotile mining and milling cohorts.

Cohort dates reference to the latest update	Estimated mean airborne concentration in terms of PCM (f/ml) for date/period stated	Calculated average cumulative exposure (f/ml-years)	Size of cohort	Comments on fiber measurement biases
Quebec 1930–1992 Liddell et al. (1997)	1948: 270 1968: 36	600[a]	10981	Very poor correlations with PCM for early impinger data.
Balangero 1930–1990 Ferrante et al. (2020)	No data before 1967. 1967: mill 17.5–80.3 1969: mill 10.1–28.6	300[a] 261	974	Heavily dependent on reconstruction estimates using old machinery etc. See Table 10.15.
Qinghai, China 1981–2006 Wang et al. (2011, 2013b)	No data before 1984. 1980s: 250 1990s: 44 2006: 58 (15.7 GM)	120[a] 212.4 (57.7 GM)	1539	Measurements made in terms of mass concentration, conversion to PCM fibers based on 35 side-by-side PCM and mass static samples in 2006.
Uralasbest, Russia Feletto et al. (2017) Schuz et al. (2024)	Mills: 1955–1975 <5[b] 1975–2001 2–4[b] Mine: 1955–2001 2[b]	Not available	30,445	Results in terms of Median and GMs make it difficult to convert to the arithmetic mean values used for epidemiology. Air concentrations seem very low compared to other sites.

[a] From Hodgson and Darnton (2000); and Darnton (2023).
[b] Geometric mean (GM) values only-suggested arithmetic means based on Qinghai results would be ×3.8.

sectors based on linear modeling of the two US textile cohorts in South and North Carolina (US EPA 2020). The two cohorts differ markedly when the central integrated unit risk (IUR) for lung cancer and mesothelioma was determined from linear models, 0.201 and 0.065 (×3.1 difference).

South Carolina Textile Workers

A cohort of 3,072 workers exposed to chrysotile in a South Carolina asbestos textile plant (1916–1977) was followed up for mortality (Hein et al. 2007). A detailed study of plant processes and dust control methods over the period from 1930 to 1975 was also conducted. Linear statistical models for reconstructing historic dust exposure concentrations were developed, taking the following into account: textile processes, dust control measures, and job assignments (Dement et al. 1983). Parameters of these statistical models were estimated using 5,952 industrial hygiene sampling measurements covering the period from 1930 to 1975. For most textile operations, exposure concentrations were significantly reduced by about 1940, when most engineering dust control measures were in place. Results of the exposure estimates indicated "precontrol" exposure concentrations that ranged from 3 to 78 f/ml with typical concentrations well above 10 f/ml. After textile operations were provided with dust control measures, estimated exposure concentrations ranged from 3 to 17 f/ml and were usually in the range of 5–10 f/ml. The Job exposure matrix was broken down into 10 areas (see Table 10.4). Clearly, most of the cohort's exposure was before the MF-PCM method was available, so the conversion of the historic impinger measurements is key to the exposure–response estimates.

Air concentrations were monitored from 1930 to 1939 on five occasions by the Metropolitan Life insurance company using an impinger, with analysis by counting all particles visible at ×100 magnification at the bottom of a sedimentation cell and results expressed as millions of particles per cubic foot (mppcf). The plant also used impinger monitored on an infrequent basis from the 1930s and routinely from 1956. The US public health service also monitored in 1968 and 1971 using both PCM and impingers. The conversion from impinger measurements to MF-PCM fiber counts is crucial and was based on the PCM method described by Edwards and Lynch (1968).

McDonald et al. (1983) investigated a parallel cohort of 2,543 workers at the same factory employed for 1 month or more between 1938 and 1958 to investigate a 50-fold increase in risk compared with the Canadian miners. They considered various factors for the impinger to PCM conversions and reported ratios of between 1.3 and 10 f/ml = 1 mppcf with an average of about 6 f/ml for 1 mppcf. This is close to the overall ratio (5.9) reported by the US Public Health studies (Ayer et al. 1965) and twice the ratio used by Dement (Dement et al. 1983) except for the zone 1, where 7.9 f/ml was used. Although McDonald felt that the impinger results were extraordinarily low compared with those obtained at the Canadian mines and mills, they found the concentrations were comparable or lower than those of other textile plants in the United States (e.g. Connecticut).

Another area where some of the differences may arise was the exclusion of several very dusty practices which took place as "overtime" activities, which were not included in the air monitoring in South Carolina. While this meant that some people had extremely high exposures that went unrecorded and have not been included in the analysis, these were thought unlikely to exceed a twofold underestimate of the cumulative exposure estimates. So overall, a factor of 10 difference would still exist in the dose–response relationship for lung cancer between the South Carolina textile manufacturing and the mining and milling of the material supplied from Quebec (McDonald et al. 1983).

When MF-PCM analysis was introduced, portions of the sampled filters were stored and were able to be analyzed by TEM at a later date. This analysis allowed conversion factors to be applied to the original PCM counts and to further analyze the size distribution of the airborne fibers in the factory to determine whether the longer textile-grade fibers were an additional factor. The mean cumulative exposure for the cohort was 28.2 (range <1–700) f/ml-years (Elliott et al. 2012).

North Carolina Textile Workers

A cohort of 5,397 chrysotile asbestos textile workers employed between 1950 and 1973 in any of four manufacturing plants in North Carolina, USA that produced asbestos textile products were followed up to 2003 for lung cancer (Loomis et al. 2009). Historical exposures to asbestos fibers were estimated from work histories and the 3,578 industrial hygiene measurements taken between 1935 and 1986. The three larger factories have also been followed up for mortality from mesothelioma and pleural cancer (Loomis et al. 2019).

Air monitoring started in 1935 and used midget impinger samples to measure particle concentrations. Measurement of fiber concentrations by MF sampling and phase-contrast microscopy (MF-PCM) was introduced in 1964 and both methods were used until 1971, after which time MF-PCM was used exclusively. Data from approximately 1,000 paired and concurrent samples obtained by both methods were used to estimate plant and period-specific factors to convert measurements of dust concentration to estimated MF-PCM-equivalent fiber concentrations. A total of 3,420 air samples were analyzed using multivariable mixed models to estimate average fiber concentrations by plant, department, job, and period (Table 10.18). These data were linked to individual work-history records to estimate the cumulative exposure to asbestos for each worker. Further details of the exposure assessment methods and results have been discussed earlier in this chapter and are similar to that described for the South Carolina cohort. .

Studies for a pooled analysis (Dement et al. 2011) found that exposure concentrations in the North Carolina plants were in excess of 50 f/cc for many operations through to about 1955 owing to a lack of dust control measures in early years, whereas concentrations in the South Carolina plant were generally less than 10 f/cc by about 1950. The high concentrations of dust in North Carolina meant that

Table 10.18 Arithmetic mean fiber PCM fiber concentrations (AM) by plant, exposure zone, and calendar period for North Carolina asbestos textile plants.

Period	1935–1946		1947–1970		1971–1986	
Plant 1	Number	AM	Number	AM	Number	AM
All combined	129	65.3	419	9.5	Closed 1970	
Preparation	25	139.2	43	15.8		
Carding	35	110.4	76	13.4		
Spinning	17	18.9	106	9.7		
Twisting	18	21.6	66	9.8		
Weaving	17	10.4	75	3.4		
Winding	10	14.7	41	7.0		
Finishing	2	5.2	2	1.7		
All others	5	6.5	10	7.2		
Plant 3	Number	AM	Number	AM	Number	AM
All combined	279	100.5	787	10.1	831	5.1
Preparation	59	182.3	124	11.1	116	7.0
Carding	60	128.9	158	14.7	187	7.7
Spinning	35	16.0	133	4.0	99	4.2
Twisting	21	22.4	118	6.5	93	4.7
Weaving	65	104.5	153	14.4	170	3.6
Winding	18	54.2	30	7.5	56	4.0
Finishing	5	31.0	16	2.1	44	1.0
All others						
Plant 4	Number	AM	Number	AM	Number	AM
All combined	16	46	514	21.2	456	3.5
Preparation	—	—	52	25.9	52	3.1
Carding	—	—	94	21.4	116	6.1
Spinning	—	—	83	19.5	27	1.1
Twisting	—	—	55	29.5	25	6.5
Weaving	14	47.7	147	21.3	80	2.5
Winding	—	—	39	16.5	51	1.7
Finishing	2	33.9	19	9.2	15	3.6
All others	—	—	25	12.7	83	2.1

Source: Dement et al. 2009

workers were withdrawn from the workforce based on chest X-ray and the cumulative exposures (f/ml-years) were estimated at 80.4 for all plants with 174.5 (<1–1,297), 62.4 (<1–2,944), and 147.5 (<1–1,271) at plants 1, 3, and 4, respectively (Elliott et al. 2012).

Chongqing Chrysotile Cohort

The factory at Chongqing was established in 1939 and only ever used chrysotile from two mines in Sichuan Province. Chrysotile samples analyzed by TEM were below the limit of detection (<0.001%) for amphiboles (Courtice et al. 2016; Yano et al. 2001), but a further assessment found small amounts of tremolite in six Sichuan mines of 0.002–0.312% (Tossavainen et al. 2000). From 1958, the main products were textiles, rubber products, and asbestos cement. Cement and textile production was greatly reduced in the 1990s and had ceased in 2002, but in the 1980s, major innovations to improve the workplace environment had been made (Yano et al. 2001; Yano 2018). This included the installation of local ventilation systems throughout the factory, although from the beginning of the 1990s, the systems were no longer in use to cut costs. Additionally, a small hut with a scrubber for the final treatment of exhaust air had been turned into a storage area for dumping sacks. As most staff lived on or near the premises, there was considerable environmental exposure to the 1,200 employees and their families.

A cohort of 586 males and 279 females (females employed since 1970) were followed prospectively for 37 years from the start of 1972 until the end of 2008. The measured median cumulative fiber exposure for the cohort was 132.6 f/ml-years (IQR 89.3–548.4). Exposure–response relationships demonstrated that at the highest exposure concentration, there was a near sixfold increase in the risk for lung cancer and a threefold increase for asbestosis.

Exposure assessment was based on gravimetric sampling of total dust from 556 area measurements taken every 4–5 years from 1970 to 2006 (Deng et al. 2012). From 1999, paired dust and fiber concentration samples were taken in collaboration with Japanese health scientists. A total of 223 measurements of fiber concentration by MF-PCM were available. The paired dust and fiber samples from 1999 to 2006 were used to estimate the dust to MF-PCM fiber-equivalent concentrations for the 1970–1994 exposures, using a similar approach to that used for South Carolina (Dement et al. 2009).

The concentrations of dust measured at different workshops periodically were generally far higher than the Chinese national standards (Wang et al. 2013a). Yano et al. (2001) reported fiber concentrations in the raw material section and the textile section of the plant were 7.6 and 4.5 f/ml, respectively. Measurements conducted in 2002 indicated that the asbestos fiber concentrations in air samples were 18 f/ml in the raw material section and 6 f/ml in the textile section; the fiber

Table 10.19 Summary of the concentration of fibers and dusts for workers in major sections of the Chongqing asbestos plant by job category 1999).

Job category	PCM fiber concentration (f/ml)	Dust concentration (mg/m^3)
Raw material opening	6.5 (5.8–7.5)	8.8 (6.1–12.3)
Raw material bagging	12.6 (5.2–58.4)	18.2 (14.5–22.4)
Rubber plate	2.8 (2.6–3.1)	237.5 (176.0–320.5)
Textile	4.5 (0.7–17.1)	22.4 (15.8–35.5)
Asbestos cement	0.1	22.3

Source: Adapted from Yano (2018)

concentrations for personal samples were 6 and 8 f/ml in the two sections, respectively (Wang et al. 2012). A comparison of the side-by-side dust concentration vs. fiber concentration (Yano 2018) showed wide variations (up to an order of magnitude) by workshop. A summary of air concentrations is given (Table 10.19).

Chrysotile Textiles Summary

The chrysotile textile air concentrations are summarized in Table 10.20. As explained in the introduction, the wide variation between the Quebec mines/mills and the Carolina textile factories represents the two extremes in the lung cancer dose–response relationships (about a factor of 80) for chrysotile. Small numbers of mesotheliomas have been reported in both chrysotile mining and textile manufacture, but the influence of the small amphibole components cannot be ruled out entirely. There is some evidence that the Quebec cohort conversions from impinger measurements to fiber concentration may be too high, while that for the Carolina's is too low. This observation would bring the dose–response relationships closer, but an approximate order of magnitude difference in lung cancer is likely to remain. Variations in the percentage of long and thin fibers may be another factor to explain the differences.

Other Chrysotile Cohorts

Asbestos cement cohorts generally involve exposure to mixed asbestos types. Two New Orleans plants that opened in the 1920s were investigated in part to compare chrysotile and mixed exposures. Plant 2 consisted of four separate production buildings; the one involved in pipe manufacture from 1946 used both crocidolite and chrysotile. The three other buildings were described as chrysotile-only (Hughes et al. 1987). Plant 2 had some 300 workers in the 1940s and over 900 in

Table 10.20 Summary of chrysotile textile cohorts.

Cohort dates reference to the latest update	Estimated mean airborne concentration in terms of PCM (f/ml) for date/ period stated	Calculated average cumulative exposure (f/ml-years)	Size of cohort	Comments on fiber measurement biases
South Carolina 1916–1977 Elliott et al. (2012)	1930–1940: >10 1940–1965: 5–10 1965–1971: 0.2–8.4 by work area (Table 10.4)	28 males[a] 26 females[a] 28.2	3072	Early data based on impinger particle counts. PCM-TEM conversion based on >0.25 µm fiber width, which may underestimate the PCM count.
North Carolina 1935–1986 Elliott et al. (2012) Loomis et al. (2009, 2019)	Plants(P): 1/3 and 4 1935–1946 65.3/100.5/>46 1947–1970 9.5/10.1/21.2 1971–1986 Closed/5.1/3.5	68.3[b] 80.4 $P_1 = 174.5$ $P_3 = 62.4$ $P_4 = 147.5$	5397	Early data based on impinger particle counts. PCM-TEM conversion based on >0.25 µm fiber width, which may underestimate the PCM count.
Chongqing 1972–2008 Wang et al. (2012)	1999: 4.5–12.6 By job category (Table 10.19)	105.2[b] 132.6 (median)	586 males 279 females	Some issues remain with the mass to PCM conversions of the 1999 data.

[a] Hodgson and Darnton (2000).
[b] Darnton (2023).

1950s and the cohort exposure was between 1937 and 1969, so measurements were largely based on impinger data. A detailed review of available measurements was carried out: some 100 in total for Plant 1 and 1664 for Plant 2. The converted estimated average fiber concentration was 10.4 f/ml for Plant 2, but this was thought to be an underestimate, especially during the 1940s and 1950s when most of the cohort were employed, as anecdotal evidence suggests that the average concentrations may be a factor of 2 higher. The average cumulative exposure was estimated to be 22 f/ml-year (Hodgson and Darnton 2000).

A friction products factory that opened in Connecticut in 1913 used chrysotile only until 1937. A cohort of 3,641 men employed for >1 month between 1938 and 1958 were followed. From the late 1930s, dry molded back brake linings and die-cast molded clutch linings were manufactured. During the 1940s automatic transmission discs and bands were also produced. Prior to 1970, air monitoring was conducted by the Metropolitan Life Insurance Company on only four occasions between 1930 and 1939. An impinger was used to measure particle number concentrations of between 1 and 5 mppcf (McDonald et al. 1984). No conversions to f/ml were attempted. It was found that the average dust concentration of 1.84 mppcf was similar to that found at the South Carolina textile plant. About 400 lbs of crocidolite was used experimentally in the laboratory between 1942 and 1972 and from 1957, some anthophyllite was also used in the production. No cases of asbestosis were mentioned on the death certificates, and 9 of the 12 cases of pneumoconiosis had exposures of <10 mpcf-years, with half of the cases working for <1 year. The average cumulative exposure was estimated to be 58 f/ml per year (Hodgson and Darnton 2000).

Discussion and Outlook

Since the original Hodgson and Darnton meta-analysis (2000), there have been several further meta-analyses that have also found differences between the three main asbestos types (chrysotile, crocidolite, and amosite) (e.g. Berman and Crump 2003; Berman and Crump 2008b; HCN 2010; Lenters et al. 2011; Garabrant and Pastula 2018; ECHA 2021). The latest update of the Hodgson and Darnton analysis (Darnton 2023) found similar differences in the risk potency factors by asbestos type as the original study and again highlighted the difference between chrysotile textiles and chrysotile mining and milling and the much larger volume of nontextile manufacturing. While the updates have sought to improve the precision of the exposure–response estimates and further investigate the role of mineral fiber type and size in cancer risk, there have been only a small number of additional asbestos worker cohort data sets to consider, including the Qinghai chrysotile miners and millers and the Chongqing factory workers.

Bans on the processing and use of asbestos across an increasing number of countries have resulted in most of the remaining asbestos mining and manufacture of asbestos goods taking place in Russia and China, where static gravimetric sampling of total dust remains the air monitoring method of choice. This suggests that any additional knowledge of cohort exposures may also have limitations unless extensive collaboration and the use of side-by-side and personal sampling assessment have been implemented. Assessing the consistency of cancer risks per unit exposure from any new studies from these countries with the evidence from existing meta-analyses may need to rely on semiquantitative judgments. The recent epidemiological results from the Uralasbest mining and mill cohort appear to be particularly challenging not only due to the type and conversion of the exposure data but also due to the relatively short life expectancy of the cohort and the high incidence of alcohol related deaths.

Early air measurement data reviewed in this chapter were often collected before the PCM-MF fiber counting became established as the index of exposure for asbestos epidemiology in the late 1960s. Assessments of early exposures therefore rely on conversions from different measurement methods and indices. Due to the long latency of asbestos-related cancers, epidemiological assessments of the exposure–response from cohort studies are typically heavily influenced by the earlier part of study exposure periods when PCM measurements were unavailable and when air measurements were typically highest due to poorer standards of control. The data available suggest that air measurements for mining and milling in the early (less well-controlled) period of production were likely to be in excess of 100 f/ml for milling and somewhat lower, that is several tens of f/ml in the mines. The exception to this seems to be the relatively new measurement data from the Uralasbest mine. Many asbestos manufacturing plants were subject to increasing controls from the 1930s but were still of the order of several tens of f/ml and either decreased over time in order to meet the airborne exposure limits or closed. Specific areas, such as bag opening and fiberizing of the raw fiber were particularly dusty and if not well controlled would have produced airborne levels not dissimilar to the mills, which produced and bagged the fibers.

It is clear from the summary tables in the chapter that the amount of measurement information for amosite- and crocidolite-exposed cohorts is much more limited than for chrysotile-only cohorts. This is partly a reflection that more than 95% of commercial asbestos production was chrysotile. Epidemiological comparisons of cancer risk in relation to the different types of asbestos therefore necessarily rely on a more uncertain body of evidence about exposures in relation to amphiboles than chrysotile. Epidemiological analyses often do not take into account the uncertainty in the underlying exposure measurements (although potential biases in analysis due to exposure misclassification are well known). In this context, it is perhaps striking that assessments of both mesothelioma and lung cancer risk per unit of cumulative exposure are reasonably consistent across the amphibole-exposed cohorts, whereas consistency is generally much poorer within the chrysotile-exposed cohorts, as described by Hodgson and Darnton (2000) and Darnton (2023). While there are different levels of uncertainty between and within individual cohorts, a consistency across the air measurement by main activity at different sites gives some assurance that the dose–response relationships determined by major meta-analyses (Hodgson and Darnton 2000; Berman and Crump 2008a, 2008b; Darnton 2023) are reasonably robust.

The difference in the potency of asbestos types in relation to mesothelioma and lung cancer induction has important ramifications in a number of areas, including setting occupational exposure limits, cost-benefit analyses, and policy development for the management of *in-situ* asbestos in buildings, its removal and disposal, as well as the development and remediation of brownfield land. As the current and future management of asbestos materials remaining in many settings will account for billions of dollars of expenditure worldwide, adopting an informed science-led, risk-based policy approach is the best way to meet societal need. However, a "science-led" approach still has challenges and limitations, particularly in relation to assessing the implications of uncertainty in the evidence base. The conversion of early exposure measurements to the current MF-PCM "index of exposure" remains a key source of uncertainty in that evidence that is likely to affect some studies more than others. However, any approach to assessing the validity of the quantitative evidence about historic exposures reviewed here will necessarily involve some reliance on qualitative arguments about the value of parts of the evidence that are more or less limited in nature.

There is no straightforward solution to dealing with this issue for any epidemiological assessment of the risk, and different approaches could lead to conclusions that are biased in different directions. One such approach is to exclude large parts of the evidence deemed to be more uncertain and therefore of less value but given the very wide variation in risk estimates from different studies (particularly for lung cancer in relation to chrysotile), this approach seems particularly susceptible to bias. The exclusion of the mining and milling cohorts, which as a group are relatively consistent in their estimated airborne concentrations, seems unwise, particularly as it was these cohorts that produced the fiber that many of the manufacturing plants were using. For example, exclusion of the Quebec cohort does not seem reasonable given the plausible extent to which this data point could be errant (many thousands of air measurements were made), and doing so would likely bias the chrysotile risk estimation upward considerably. Similarly, it seems unreasonable to rely only on a very restricted number of studies, such as those relating to asbestos textiles, where there are questions about the generalizability of the exposure scenarios and where differences in fiber sizes could affect the potency factors significantly (Berman and Crump 2008b; Wylie and Korchevskiy 2023). It must also be considered that the Canadian chrysotile mines were a major source for many of the asbestos manufacturing cohorts and that chrysotile textiles represent only a very small amount of the legacy asbestos in buildings, which is now often the main focus of asbestos policy and concern. It is the asbestos cement, asbestos insulating boards, asbestos insulation, and asbestos floor tile that building occupants, maintenance workers, and asbestos removal operatives are mainly interfacing and/or disturbing, and it is important that risk estimation should be relevant to these materials and the type of asbestos they contain.

Questions remain as to how well the PCM fiber count represents the risk, particularly if thinner fibers are implicated with lung diseases. This has led the EU to recently update the asbestos workers protection directive (EU 2023). This requires that EU member states should reduce the current PCM occupational exposure limit (OEL) of 0.1 f/ml by an order of magnitude by 21st December 2025 and by 21st December 2029 implement an overall x50 reduction to 0.002 f/ml. A six-year transition period has also been given for transitioning to a new OEL where "fibres with a breadth of less than 0,2 micrometres shall also be taken into consideration". This will require the use of high magnification TEM or field emission gun (FEG) SEM microscopy to count the fibres but the OEL for the new index of exposure will be set at 0.01 f/ml. This effectively assumes that five times more fibres will be visible using EM methods but this is clearly unfounded given the measurements made nearly half a century ago in Table 10.1. Even textile grade chrysotile manufacture supports only a factor of 2-3 increase in fibre numbers (Table 10.3) and seems to undermine the basis of the risk assessment behind the OEL

This change to an EM index of exposure will create a further epidemiological measurement and conversion challenge, as each asbestos mineral type, its source, and its subsequent processing will have a significant influence on the proportion of finer fibers present and what proportion would be visible by PCM analysis. The assumption that a PCM can only render fibres above 0.2 µm width visible has also been shown to be questionable.

The reduction in the OEL based on PCM fiber counts (values have fallen from 12 f/ml in the 1960s to an EU proposed value of 0.002 f/ml by 2030; a ×6,000 reduction), means that some difficult sampling, analysis and risk-assessment challenges will need to be resolved, along with extensive advances in reducing fibre release during asbestos maintenance and removal operations. Without such advances much greater use of air supplied respirators, or similar, seem the most likely response the new EU directive.

Acknowledgement

Jean Prentice for comments.

References

Acheson, E.D., Bennett, C., Gardner, M.J., et al. (1981). Mesothelioma in a factory using amosite and chrysotile asbestos. *The Lancet* 318(8260–8261): 1403–1406. 556.

Acheson, E.D., Gardner, M.J., Pippard, E.C., et al. (1982). Mortality of two groups of women who manufactured gas masks from chrysotile and crocidolite asbestos: a

40-year follow up. *British Journal of Industrial Medicine* 39(4): 344–348. https://doi.org/10.1136/oem.39.4.344.

Acheson, E.D., Gardner, M.J., Winter, P.D., et al. (1984). Cancer in a factory using amosite asbestos. *International Journal of Epidemiology* 13(1): 3–10. https://doi.org/10.1093/IJE/13.1.3.

Addison, J. and Davies, L.S.T. (1990). Analysis of amphibole asbestos in chrysotile and other materials. *The Annals of Occupational Hygiene* 34(2): 159–175. https://doi.org/10.1093/annhyg/34.2.159.

Armstrong, B.K., de Klerk, N.H., Musk, A.W., et al. (1988). Mortality in miners and millers of crocidolite in Western Australia. *British Journal of Industrial Medicine* 45(1): 5–13. https://doi.org/10.1136/OEM.45.1.5.

Asbestosis Research Council (ARC). (1971). Technical note 1. The measurement of airborne asbestos dust by the membrane filter method' and provisional notes for guidance on various aspects of asbestos safety. Rochdale, England.

Ayer, H.E., Lynch, J.R., and Fanney, J.H. (1965). A comparison of Impinger and membrane filter techniques for evaluating air samples in asbestos plants. *Annals of the New York Academy of Sciences* 132(1): 274–287. https://doi.org/10.1111/J.1749-6632.1965.TB41108.X.

Bauer, H.D., Blome, H., and Gelsdorf, H. (1997). Umrechnungsfaktoren der Meßverfahren (Conversion factors of measuring methods). *BK Report Faserjahre* 1: 64–65.

Beckett, S.T. and Jarvis, J. L. (1979). A study of the size distribution of airborne amosite fibres in the manufacture of asbestos insulation boards. *The Annals of Occupational Hygiene* 22(3): 273–284. https://doi.org/10.1093/annhyg/22.3.273.

Berman, D.W. and Crump, K.S. (2003). Technical support document for a protocol to assess asbestos-related risk. Washington, DC, 20460.

Berman, D.W. and Crump, K.S. (2008a). A meta-analysis of asbestos-related cancer risk that addresses fiber size and mineral type. *Critical Reviews in Toxicology* 38(suppl 1): 49–73. https://doi.org/10.1080/10408440802273156.

Berman, D.W. and Crump, K.S. (2008b). Update of potency factors for asbestos-related lung cancer and mesothelioma. *Critical Reviews in Toxicology* 38(suppl 1): 1–47. https://doi.org/10.1080/10408440802276167.

Berman, D.W., Crump, K.S., Chatfield, E.J., et al. (1995). The sizes, shapes, and mineralogy of asbestos structures that induce lung tumors or mesothelioma in AF/HAN rats following inhalation. *Risk Analysis* 15(2): 181–195. https://doi.org/10.1111/J.1539-6924.1995.TB00312.X.

Bernstein, D., Dunnigan, J., Hesterberg, T., et al. (2013). Health risk of chrysotile revisited. *Critical Reviews in Toxicology* 43(2): 154–183. https://doi.org/10.3109/10408444.2012.756454.

Berry, G., Pooley, F., Gibbs, A., et al. (2009). Lung fiber burden in the Nottingham gas mask cohort. *Inhalation Toxicology* 21(2): 168–172. https://doi.org/10.1080/08958370802291304.

Burdett, G. (1998). A comparison of historic asbestos measurements using a thermal precipitator with the membrane filter-phase contrast microscopy method. *The Annals of Occupational Hygiene* 42(1): 21–31. https://doi.org/10.1016/s0003-4878(97)00048-3.

Courtice, M.N., Berman, D.W., Yano, E., et al. (2016). Size- and type-specific exposure assessment of an asbestos products factory in China. *Journal of Exposure Science and Environmental Epidemiology* 26(1): 63–69. https://doi.org/10.1038/jes.2015.46.

Darnton, L. (2023). Quantitative assessment of mesothelioma and lung cancer risk based on phase contrast microscopy (PCM) estimates of fibre exposure: an update of 2000 asbestos cohort data. *Environmental Research* 230: 114753. https://doi.org/10.1016/J.ENVRES.2022.114753.

Davies, J.C.A., Williams, B.G., Debeila, M.A., et al. (2001). Asbestos-related lung disease among women in the Northern Province of South Africa. *South African Journal of Science* 97: 87–93. https://journals.co.za/doi/pdf/10.10520/EJC97294.

Dement, J.M., Harris, R.L., Symons, M.J., et al. (1983). Exposures and mortality among chrysotile asbestos workers. Part I: exposure estimates. *American Journal of Industrial Medicine* 4(3): 399–419. https://doi.org/10.1002/AJIM.4700040303.

Dement, J.M., Kuempel, E.D., Zumwalde, R.D., et al. (2008). Development of a fibre size-specific job-exposure matrix for airborne asbestos fibres. *Occupational & Environmental Medicine* 65(9): 605–612. https://doi.org/10.1136/oem.2007.033712.

Dement, J.M., Myers, D., Loomis, D., et al. (2009). Estimates of historical exposures by phase contrast and transmission electron microscopy in North Carolina USA asbestos textile plants. *Occupational and Environmental Medicine* 66(9): 574–583. https://doi.org/10.1136/oem.2008.040410.

Dement, J.M., Loomis, D., Richardson, D., et al. (2011). Estimates of historical exposures by phase contrast and transmission electron microscopy for pooled exposure-response analyses of North Carolina and South Carolina, USA asbestos textile cohorts. *Occupational and Environmental Medicine* 68(8): 593–598. https://doi.org/10.1136/oem.2010.059972.

Deng, Q., Wang, X., Wang, M., et al. (2012). Exposure-response relationship between chrysotile exposure and mortality from lung cancer and asbestosis. *Occupational and Environmental Medicine* 69(2): 81–86. https://doi.org/10.1136/OEM.2011.064899.

Dodič-Fikfak, M. (2007). An experiment to develop conversion factors to standardise measurements of airborne asbestos. *Arhiv Za Higijenu Rada i Toksikologiju* 58(2): 179–185. https://doi.org/10.2478/v10004-007-0003-9.

Dodson, R.F., O'Sullivan, M., Brooks, D.R., et al. (2003). Quantitative analysis of asbestos burden in women with mesothelioma. *American Journal of Industrial Medicine* 43(2): 188–195. https://doi.org/10.1002/ajim.10164.

Edwards, G.H. and Lynch, J.R. (1968). The method used by the US public health service for enumeration of asbestos dust on membrane filters. *The Annals of Occupational Hygiene* 11(1): 1–6. https://doi.org/10.1093/annhyg/11.1.1.

EEC. (1987). Council of the European Communities. Council Directive of 19 March 1987 on the prevention and reduction of environmental pollution by asbestos. Official Journal of the European Community (87/217/EEC) (OJ L 85, 28.3.1987, p. 40). https://eur-lex.europa.eu/legal-content/EN/TXT/PDF/?uri=CELEX:01987L0217-20180704.

Elliott, L., Loomis, D., Dement, J., et al. (2012). Lung cancer mortality in North Carolina and South Carolina chrysotile asbestos textile workers. *Occupational and Environmental Medicine* 69(6): 385–390. https://doi.org/10.1136/OEMED-2011-100229.

European Chemicals Agency (ECHA). (2021). ECHA Scientific report for evaluation of limit values for asbestos at the workplace. European Chemicals Agency, Helsinki, Finland. https://echa.europa.eu/documents/10162/d5f8d584-5e7d-bc97-3a98-4e9a39715f41.

European Commission (EU). (2023). Directive 2009/148/EC on the protection of workers from the risks related to exposure to asbestos at work. Official Journal of the European Union OJ L, 30 November 2023. http://data.europa.eu/eli/dir/2023/2668/oj.

Feletto, E., Schonfeld, S.J., Kovalevskiy, E.V., et al. (2017). A comparison of parallel dust and fibre measurements of airborne chrysotile asbestos in a large mine and processing factories in the Russian Federation. *International Journal of Hygiene and Environmental Health* 220(5): 857–868. https://doi.org/10.1016/j.ijheh.2017.04.001.

Felix, M.A. (1997). Environmental asbestos and respiratory disease in South Africa. PhD Thesis. Faculty of Medicine. University of the Witwatersrand, Johannesburg. p. 281.

Ferrante, D., Mirabelli, D., Silvestri, S., Azzolina, D., Giovannini, A., Tribaudino, P., & Magnani, C. (2020). Mortality and mesothelioma incidence among chrysotile asbestos miners in Balangero, Italy: a cohort study. *American Journal of Industrial Medicine*, 63(2), 135–145.

Garabrant, D.H. and Pastula, S.T. (2018). A comparison of asbestos fiber potency and elongate mineral particle (EMP) potency for mesothelioma in humans. *Toxicology and Applied Pharmacology* 361: 127–136. https://doi.org/10.1016/j.taap.2018.07.003.

Gibbs, G.W. (1994). The assessment of exposure in terms of fibres. *The Annals of Occupational Hygiene* 38(4): 447–487. https://europepmc.org/article/med/7978969.

Gibbs, G.W. and Hwang, C.Y. (1980). Dimensions of airborne asbestos fibres. *IARC Scientific Publications* 30: 69–78. http://europepmc.org/abstract/MED/7239672.

Gibbs, G.W. and LaChance, M. (1974). Dust-fiber relationships in the Quebec chrysotile industry. *Archives of Environmental Health: An International Journal* 28(2): 69–71. https://doi.org/10.1080/00039896.1974.10666439.

Gibbs, A.R., Gardner, M.J., Pooley, F.D., et al. (1994). Fiber levels and disease in workers from a factory predominantly using amosite. *Environmental Health Perspectives* 102(Suppl. 5): 261–263. https://doi.org/10.1289/EHP.94102S5261.

Gualtieri, A.F. (2023). Journey to the centre of the lung. The perspective of a mineralogist on the carcinogenic effects of mineral fibres in the lungs. *Journal of Hazardous Materials* 442: 130077. https://doi.org/10.1016/j.jhazmat.2022.130077.

HCN. (2010). Asbestos: risks of environmental and occupational exposure. Health Council of the Netherlands. Advisory Report 03-06-2010. https://www.healthcouncil.nl/documents/advisory-reports/2010/06/03/asbestos-risks-of-environmental-and-occupational-exposure (accessed 08 September 2023).

Hein, M.J., Stayner, L.T., Lehman, E., et al. (2007). Follow-up study of chrysotile textile workers: cohort mortality and exposure-response. *Occupational and Environmental Medicine* 64(9): 616–625. https://doi.org/10.1136/OEM.2006.031005.

Hodgson, J. and Darnton, A. (2000). The quantitative risks of mesothelioma and lung cancer in relation to asbestos exposure. *The Annals of Occupational Hygiene* 44(8): 565–601. https://doi.org/10.1093/annhyg/44.8.565.

Holmes, S. (1965). Developments in dust sampling and counting techniques in the asbestos industry. *Annals of the New York Academy of Sciences* 132(1): 288–297. https://doi.org/10.1111/j.1749-6632.1965.tb41109.x.

Hughes, J.M., Weill, H., and Hammad, Y.Y. (1987). Mortality of workers employed in two asbestos cement manufacturing plants. *Occupational and Environmental Medicine* 44(3): 161–174. https://doi.org/10.1136/OEM.44.3.161.

Hurst, G.A., Spivey, C.G., Matlage, W.T., et al. (1979). The Tyler asbestos workers program. I. A medical surveillance model and method. *Archives of Environmental Health: An International Journal* 34(6): 432–439. https://doi.org/10.1080/00039896.1979.12088652.

Hwang, C.Y. (1983). Size and shape of airborne asbestos fibres in mines and mills. *British Journal of Industrial Medicine* 40(3): 273–279. https://doi.org/10.1136%2Foem.40.3.273.

Hwang, C.Y. and Gibbs, G.W. (1981). The dimensions of airborne asbestos fibres – I. Crocidolite from the Kuruman area, Cape Province, South Africa. *The Annals of Occupational Hygiene* 24(1): 23–41. https://doi.org/10.1093/annhyg/24.1.23.

Ilgren, E.B., van Orden, D.R., Lee, R.J., et al. (2015). Further studies of Bolivian crocidolite–Part IV: Fibre width, fibre drift and their relation to mesothelioma induction: preliminary findings. *Epidemiology Biostatistics and Public Health* 12(2): 1–11. https://doi.org/10.2427/11167.

Institute of Occupational Safety. (1986). *Maribor Centre for the Protection of Environment. The Measurements of Dust Concentration and the Concentration of Asbestos Fibres in Salonit Anhovo, Sheets Subsidiary*. Maribor: Institute of Occupational Safety.

Johnson, W.M., Lemen, R.A., Hurst, G.A., et al. (1982). Respiratory morbidity among workers in an amosite asbestos insulation plant. *Journal of Occupational Medicine* 24(12): 994–999. https://journals.lww.com/joem/Abstract/1982/12000/Respiratory_Morbidity_Among_Workers_in_an_Amosite.13.aspx.

Jones J.S., Smith P.G., Pooley F.D., et al. (1980). The consequences of exposure to asbestos dust in a wartime gas–mask factory. *IARC Scientific Publications* 30: 637–653. https://pubmed.ncbi.nlm.nih.gov/7228319/.

Kashansky, S.V., Domnin, S.G., Kochelayev, V.A., et al. (2001). Retrospective view of airborne dust levels in workplace of a chrysotile mine in Ural, Russia. *Industrial Health* 39(2): 51–56. https://doi.org/10.2486/INDHEALTH.39.51.

Kenny, L.C., Rood, A.P., and Blight, B.J.N. (1987). A direct measure of the visibility of amosite asbestos fibres by phase contrast optical microscopy. *The Annals of Occupational Hygiene* 31(2): 261–264. https://doi.org/10.1093/annhyg/31.2.261.

de Klerk, N.H., Musk, A.W., Williams, V., et al. (1996). Comparison of measures of exposure to asbestos in former crocidolite workers from Wittenoom Gorge, W. Australia. *American Journal of Industrial Medicine* 30(5): 579–587. https://doi.org/10.1002/(sici)1097-0274(199611)30:5%3C579::aid-ajim5%3E3.0.co;2-o.

Kovalevskiy, E.V., Schüz, J., Bukhtiyarov, I.V., et al. (2021). Experience of cohort formation and data collection in a retrospective cohort epidemiological study. *Russian Journal of Occupational Health and Industrial Ecology* 61(4): 253–266. https://doi.org/10.31089/1026-9428-2021-61-4-253-266.

Lash, T.L., Crouch, E.A., and Green, L.C. (1997). A meta-analysis of the relation between cumulative exposure to asbestos and relative risk of lung cancer. *Occupational and Environmental Medicine* 54(4): 254–263. https://doi.org/10.1136/oem.54.4.254.

Lenters, V., Vermeulen, R., Dogger, S., et al. (2011). A meta-analysis of asbestos and lung cancer: is better quality exposure assessment associated with steeper slopes of the exposure-response relationships? *Environmental Health Perspectives* 119(11): 1547–1555. https://doi.org/10.1289/ehp.1002879.

Levin, J.L., McLarty, J.W., Hurst, G.A., et al. (1998). Tyler asbestos workers: mortality experience in a cohort exposed to amosite. *Occupational and Environmental Medicine* 55(3): 155–160. https://doi.org/10.1136/OEM.55.3.155.

Liddell, F.D., McDonald, A.D., and McDonald, J.C. (1997). The 1891–1920 birth cohort of Quebec chrysotile miners and millers: development from 1904 and mortality to 1992. *The Annals of Occupational Hygiene* 41(1): 13–36. https://doi.org/10.1016/s0003-4878(96)00044-0.

Lippmann, M. (1990). Effects of fiber characteristics on lung deposition, retention, and disease. *Environmental Health Perspectives* 88: 311–317. https://doi.org/10.1289/ehp.9088311.

Loomis, D., Dement, J.M., Wolf, S.H., et al. (2009). Lung cancer mortality and fibre exposures among North Carolina asbestos textile workers. *Occupational and Environmental Medicine* 66(8): 535–542. https://doi.org/10.1136/oem.2008.044362.

Loomis, D., Dement, J., Richardson, D., et al. (2010). Asbestos fibre dimensions and lung cancer mortality among workers exposed to chrysotile. *Occupational and Environmental Medicine* 67(9): 580–584. https://doi.org/10.1136/OEM.2009.050120.

Loomis, D., Dement, J.M., Elliott, L., et al. (2012). Increased lung cancer mortality among chrysotile asbestos textile workers is more strongly associated with exposure to long thin fibres. *Occupational and Environmental Medicine* 69(8): 564–568. https://doi.org/10.1136/OEMED-2012-100676.

Loomis, D., Richardson, D.B., and Elliott, L. (2019). Quantitative relationships of exposure to chrysotile asbestos and mesothelioma mortality. *American Journal of Industrial Medicine* 62(6): 471–477. https://doi.org/10.1002/ajim.22985.

Luberto, F., Ferrante, D., Silvestri, S., et al. (2019). Cumulative asbestos exposure and mortality from asbestos related diseases in a pooled analysis of 21 asbestos cement cohorts in Italy. *Environmental Health* 18: 71. https://doi.org/10.1186/s12940-019-0510-6.

Lynch, J.R., Ayer, H.E., and Johnson, D.J. (1970). The interrelationships of selected asbestos exposure indices. *American Industrial Hygiene Association Journal* 31(5): 598–604. https://doi.org/10.1080/0002889708506298.

Major G. (1968). Asbestos dust exposure. In: *The First Australian Pneumoconiosis Conference* (ed. G. Major), 467–474. Joint Coal Board Sydney.

McCulloch, J. (2002). Asbestos blues: labour, capital, physicians, and the state in South Africa (African issues). James Currey, January 1, 2002.

McDonald, A.D., Fry, J.S., Woolley, A.J., et al. (1983). Dust exposure and mortality in an American chrysotile textile plant. *Occupational and Environmental Medicine* 40(4): 361–367. https://doi.org/10.1136/oem.40.4.361.

McDonald, A.D., Fry, J.S., Woolley, A.J., et al. (1984). Dust exposure and mortality in an American chrysotile asbestos friction products plant. *Occupational and Environmental Medicine* 41(2): 151–157. https://doi.org/10.1136/OEM.41.2.151.

McDonald, J.C., Liddell, F. D., Dufresne, A., et al. (1993). The 1891–1920 birth cohort of Quebec chrysotile miners and millers: mortality 1976–88. *In British Journal of Industrial Medicine* 50(12): 1073–1081. https://doi.org/10.1136/oem.50.12.1073.

Morgan, A. and Holmes, A. (1982). Concentrations and characteristics of amphibole fibres in the lungs of workers exposed to crocidolite in the British gas-mask factories, and elsewhere, during the second world war. *Occupational and Environmental Medicine* 39(1): 62–69. https://doi.org/10.1136/oem.39.1.62.

Murray, J. and Nelson, G. (2007). Health effects of amosite mining and milling in South Africa. *Regulatory Toxicologu and Pharmacology* 52(1, Supplement): S75–S81 https://doi.org/10.1016/j.yrtph.2007.09.011.

Musk, A.W., Reid, A., Olsen, N., et al. (2020). The Wittenoom legacy. *International Journal of Epidemiology* 49(2): 467–476. https://doi.org/10.1093/IJE/DYZ204.

National Research Council (NRC). (1984). *Asbestiform Fibers: Nonoccupational Health Risks*. Washington, DC: The National Academies Press. https://doi.org/10.17226/509.

Nolan, R.P., Langer, A.M., and Wilson, R. (1999). A risk assessment for exposure to grunerite asbestos (amosite) in an iron ore mine. *Proceedings of the National Academy of Sciences of the United States of America* 96(7): 3412–3419. https://doi.org/10.1073/PNAS.96.7.3412.

Paustenbach, D., Brew, D., Ligas, S., et al. (2021). A critical review of the 2020 EPA risk assessment for chrysotile and its many shortcomings. *Critical Reviews in Toxicology* 51(6): 509–539. https://doi.org/10.1080/10408444.2021.1968337.

Piolatto, G., Negri, E., La Vecchia, C., et al. (1990). An update of cancer mortality among chrysotile asbestos miners in Balangero, northern Italy. *Occupational and Environmental Medicine* 47(12): 810–814. https://doi.org/10.1136/oem.47.12.810.

Pira, E., Pelucchi, C., Piolatto, P.G., et al. (2009). Mortality from cancer and other causes in the Balangero cohort of chrysotile asbestos miners. *Occupational and Environmental Medicine* 66(12): 805–809. https://doi.org/10.1136/oem.2008.044693.

Puledda, S. and Marconi, A. (1991). Study of the count-to-mass conversion factor for asbestos fibres in samples collected at the emissions of three industrial plants. *The Annals of Occupational Hygiene* 35(5): 517–524. https://doi.org/10.1093/annhyg/35.5.517.

Reid, A., Heyworth, J., de Klerk, N.H., et al. (2008). Cancer incidence among women and girls environmentally and occupationally exposed to blue asbestos at Wittenoom, Western Australia. *International Journal of Cancer* 122(10): 2337–2344. https://doi.org/10.1002/IJC.23331.

Ribak, J. and Ribak, G. (2008). Human health effects associated with the commercial use of grunerite asbestos (amosite): Paterson, NJ; Tyler, TX; Uxbridge, UK. *Regulatory Toxicology and Pharmacology* 52(1 Suppl): S82–S90. https://doi.org/10.1016/j.yrtph.2007.10.002.

Ribak, J., Seidman, H., and Selikoff, I.J. (1989). Amosite mesothelioma in a cohort of asbestos workers. *Scandinavian Journal of Work, Environment and Health* 15(2): 106–110. https://doi.org/10.5271/sjweh.1877.

Rogers, A. and Major, G. (2002). The quantitative risks of mesothelioma and lung cancer in relation to asbestos exposure: the Wittenoom data. *Annals of Occupational Hygiene* 46(1): 127–128. https://doi.org/10.1093/annhyg/mef002.

Rood, A.P. and Scott, R.M. (1989). Size distributions of chrysotile asbestos in a friction products factory as determined by transmission electron microscopy. *The Annals of Occupational Hygiene* 33(4): 583–590. https://doi.org/10.1093/annhyg/33.4.583.

Rood, A.P. and Streeter, R.R. (1984). Size distribution of occupational airborne asbestos textile fibres as determined by transmission electron microscopy. *The Annals of Occupational Hygiene* 28(3): 333–339. https://doi.org/10.1093/annhyg/28.3.333.

Rooker, S.J., Vaughan, N.P., and Le Guen, J.M. (1982). On the visibility of fibers by phase contrast microscopy. *American Industrial Hygiene Association Journal* 43(7): 505–515. https://doi.org/10.1080/15298668291410125.

Rowlands, N., Gibbs, G.W., and McDonald, A.D. (1982). Asbestos fibres in the lungs of chrysotile miners and millers – a preliminary report. *Inhaled Particles V*: 411–415. https://doi.org/10.1016/B978-0-08-026838-5.50032-5.

Schonfeld, S.J., Kovalevskiy, E.V., Feletto, E., et al. (2017). Temporal trends in airborne dust concentrations at a large chrysotile mine and its asbestos-enrichment factories in the Russian Federation during 1951–2001. *Annals of Work Exposures and Health* 61(7): 797–808. https://doi.org/10.1093/annweh/wxx051.

Schuz JNCI (2024). https://doi.org/10.1093/jnci/djad262.

Schüz, J., Schonfeld, S.J., Kromhout, H., et al. (2013). A retrospective cohort study of cancer mortality in employees of a Russian chrysotile asbestos mine and mills: study rationale and key features. *Cancer Epidemiology* 37(4): 440–445. https://doi.org/10.1016/j.canep.2013.03.001.

Sebastien, P., McDonald, J.C., McDonald, A.D., et al. (1989). Respiratory cancer in chrysotile textile and mining industries: exposure inferences from lung analysis. *Occupational and Environmental Medicine* 46(3), 180–187. https://doi.org/10.1136/oem.46.3.180.

Seidman, H., Selikoff, I.J., and Gelb, S.K. (1986). Mortality experience of amosite asbestos factory workers: dose-response relationships 5 to 40 years after onset of short-term work exposure. *American Journal of Industrial Medicine* 10(5–6): 479–514. https://doi.org/10.1002/AJIM.4700100506.

Shedd, K.B. (1985). Fiber dimensions of crocidolites from Western Australia, Bolivia and the Cape and Transvaal provinces of South Africa, US Bureau of Mines Investigation 8998. US Department of the Interior. Fiber Dimensions of Crocidolites From Western Australia, Bolivia, and the Cape and Transvaal Provinces of South Africa (umd.edu).

Silvestri, S., Ferrante, D., Giovannini, A., et al. (2020). Asbestos exposure of chrysotile miners and millers in Balangero, Italy. *Annals of Work Exposures and Health* 64(6): 636–644. https://doi.org/10.1093/ANNWEH/WXAA045.

Simpson, W.J. (1977). Selected written evidence provided to the advisory committee on asbestos, 1976–1977. Advisory Committee on Asbestos (ACA), Health and Safety Executive HMSO. London. 011883004X, 9780118830041.

Sluis-Cremer, G.K. (1965). Asbestosis in South Africa – certain geographical and environmental considerations. *Annals of the New York Academy of Sciences* 132(1): 215–234. https://doi.org/10.1111/j.1749-6632.1965.tb41103.x.

Sluis-Cremer, G. K., Liddell, F.D. Logan, W.P., et al. (1992). The mortality of amphibole miners in South Africa, 1946–80. *Occupational and Environmental Medicine* 49(8): 566–575. https://doi.org/10.1136/OEM.49.8.566.

Stanton, M.F., Layard, M., Tegeris, A., et al. (1981). Relation of particle dimension to carcinogenicity in amphibole asbestoses and other fibrous minerals. *Journal of the National Cancer Institute* 67(5): 965–975. https://doi.org/10.1093/jnci/67.5.965.

Talcott, J.A., Thurber, W.A., Kantor, A.F., et al. (1989). Asbestos-associated diseases in a cohort of cigarette-filter workers. *New England Journal of Medicine* 321(18): 1220–1223. https://doi.org/10.1056/NEJM198911023211803.

du Toit, R.S.J. (1989). The estimation of the cumulative fibre exposure of persons employed on South African asbestos mines. National Institute of Occupational Health. NCOH Report No. 4, Johannesburg.

du Toit, R.S.J. and Gilfillan, T.C. (1977). Simultaneous airborne dust samples with konimeter, thermal precipitator and dosimeter in asbestos mines. *The Annals of Occupational Hygiene* 20(4): 333–344. https://doi.org/10.1093/ANNHYG/20.4.333.

du Toit, R.S.J. and Gilfillan, T.C. (1979). Conversion of the asbestos fibre concentration recorded by means of the konimeter and the thermal precipitator to that expected by means of the membrane filter method. *The Annals of Occupational Hygiene* 22(1): 67–83. https://doi.org/10.1093/annhyg/22.1.67.

du Toit, R.S.J., Isserow, L.W., Gilfillan, T.C., et al. (1983). Relationships between simultaneous airborne dust samples taken with five types of instruments at South African asbestos mines and mills. *The Annals of Occupational Hygiene* 27(4): 373–387. https://doi.org/10.1093/annhyg/27.4.373.

Tossavainen, A., Kovalevskiy, E., Vanhala, E., et al. (2000). Pulmonary mineral fibers after occupational and environmental exposure to asbestos in the Russian chrysotile industry. *American Journal of Industrial Medicine* 37(4): 327–333. https://doi.org/10.1002/(sici)1097-0274(200004)37:4%3C327::aid-ajim1%3E3.0.co;2-1.

US EPA. (2020). Final risk evaluation for asbestos, Part 1: chrysotile asbestos. https://www.epa.gov/sites/default/files/2020-12/documents/1_risk_evaluation_for_asbestos_part_1_chrysotile_asbestos.pdf.

Walton, W.H. (1982). The nature, hazards and assessment of occupational exposure to airborne asbestos dust: a review. *The Annals of Occupational Hygiene* 25(2): 117–119. https://doi.org/10.1093/annhyg/25.2.117.

Wang, X., Lin, S., Yano, E., et al. (2011). Mortality in a Chinese chrysotile miner cohort. *International Archives of Occupational and Environmental Health* 85(4): 405–412. https://doi.org/10.1007/S00420-011-0685-9.

Wang, X., Yano, E., Qiu, H., et al. (2012). A 37-year observation of mortality in Chinese chrysotile asbestos workers. *Thorax* 67(2): 106–110. https://doi.org/10.1136/thoraxjnl-2011-200169.

Wang, X., Lin, S., Yu, I., et al. (2013a). Cause-specific mortality in a Chinese chrysotile textile worker cohort. *Cancer Science* 104(2): 245–249. https://doi.org/10.1111/cas.12060.

Wang, X., Yano, E., Lin, S., et al. (2013b). Cancer mortality in Chinese chrysotile asbestos miners: exposure-response relationships. *PLoS One* 8(8): e71899. https://doi.org/10.1371/journal.pone.0071899.

Wylie, A.G. and Korchevskiy, A.A. (2023). Dimensions of elongate mineral particles and cancer: a review. *Environmental Research* 230: 114688. https://doi.org/10.1016/J.ENVRES.2022.114688.

Yano, E. (2018). Adverse health effects of asbestos: solving mysteries regarding asbestos carcinogenicity based on follow-up survey of a Chinese factory. *Environmental Health and Preventive Medicine* 23: 35. https://doi.org/10.1186/s12199-018-0726-z.

Yano, E., Wang, Z.-M., Wang, X.-R., et al. (2001). Cancer mortality among workers exposed to amphibole-free chrysotile asbestos. *American Journal of Epidemiology* 154(6): 538–543. https://doi.org/10.1093/AJE/154.6.538.

11

Mathematical Modeling of Cancer Potency for Various Fibrous Minerals

Andrey Korchevskiy, James Rasmuson, and Eric Rasmuson

Chemistry & Industrial Hygiene, Inc., Lakewood, CO, USA

Quantitative structure–activity relationship (QSAR) modeling gradually becomes one of the most important and promising directions of the toxicological and risk assessment sciences (Cherkasov et al. 2014). For example, Barratt (2000) emphasized that "the prediction of toxicity from chemical structure can make a valuable contribution to the reduction of animal usage in the screening out of potential toxic chemicals at the early stage and in providing data for making positive classifications of toxicity."

The Organization for Economic Co-operation and Development (OECD 2007) developed validation principles for QSAR modeling applied to toxicological assessment and regulations, including

1) Defined endpoint.
2) Unambiguous algorithm.
3) Defined domain of applicability.
4) Appropriate measures of goodness-of-fit, robustness, and predictability.
5) Mechanistic interpretation.

However, while numerous models were developed to approximate various toxic characteristics of chemical agents based on their chemical structure, much less is known about the approaches for predicting cancer potency of particles, in general, and asbestiform and nonasbestiform fibers, in particular.

Some advances were made in this direction. For example, Puzyn et al. (2011) developed a "nano-QSAR" model to predict the cytotoxicity of metal-oxide nanoparticles. Based on the toxicity data and structural descriptors, a "nano-QSAR" equation was proposed to predict the cytotoxicity (denoted EC50 – the effective

Health Risk Assessment for Asbestos and Other Fibrous Minerals, First Edition.
Edited by Andrey Korchevskiy, James Rasmuson, and Eric Rasmuson.
© 2024 John Wiley & Sons, Inc. Published 2024 by John Wiley & Sons, Inc.

concentration of a compound that brings about a 50% reduction in bacteria viability) of the metal-oxide nanoparticles:

$$\log\left(\frac{1}{EC50}\right) = 2.59 - 0.50\ \Delta H_{Me+} \quad (11.1)$$

where the descriptor ΔH_{Me+} represented the enthalpy of formation of a gaseous cation having the same oxidation state as that in the metal oxide structure.

It is well known that various asbestos minerals demonstrate different propensities to cause both lung cancer and mesothelioma. For example, mesothelioma mortality in human cohorts exposed to chrysotile asbestos is significantly lower than in cohorts exposed to amphibole asbestos – crocidolite, amosite, tremolite, and others. Moreover, it is clear now that asbestos toxicity should not be considered in isolation, but in the wider context of the carcinogenic potential of elongate mineral particles expressing asbestiform habits, not necessarily corresponding to the commercial or regulatory definition of asbestos.

The difference in carcinogenic potency of mineral fibers was demonstrated in toxicological studies. Wagner et al. (1985) emphasized the enhanced potential for mesothelioma induction of erionite compared with crocidolite. Hodgson and Darnton (2000) demonstrated the statistical differences in mesothelioma potencies not only among chrysotile and amphibole minerals in general but also between various amphiboles, with the following order of potencies: crocidolite>amosite>chrysotile (see Chapter 8). Berman and Crump (2008a) showed a similar, but even more striking proportion between the potencies of the various asbestos mineral classes (amphiboles vs. chrysotile) (see Chapter 7). Also, there is an inherited difference in potency of asbestiform and nonasbestiform elongate mineral particles, as was demonstrated in many toxicological studies (Mossman 2008; Wylie and Korchevskiy 2023).

Different parameters and characteristics can explain the observed variation in asbestos potency demonstrated in the epidemiological studies. For example, the Agency for Toxic Substances and Disease Registry (ATSDR 2010) mentioned that the three main determinants of asbestos toxicity are fiber size, durability, and iron content.

Korchevskiy et al. (2019) attempted to develop QSAR models that would quantitatively characterize the mesothelioma potency of fibers based on their chemical composition and dimensions. Only the cohorts where workers or populations were exposed directly to naturally occurring fibrous minerals, and not to manufactured, fiber-containing products, were included in the modeling process. The published mesothelioma potency factors were utilized for crocidolite cohorts from Australia and South Africa, an amosite cohort from South Africa, the Libby amphibole asbestos cohort from the United States, and the Canadian chrysotile cohort. In addition, the estimated mesothelioma potency factors for Russian anthophyllite and Turkish erionite were included.

We updated the results from Korchevskiy et al. (2019) based on the up-to-date epidemiological data as reported in Chapter 8. We also included lung cancer potencies in our modeling process.

Values of R_M and R_L that were utilized for modeling are summarized in Tables 11.1 and 11.2.

Table 11.1 Published mesothelioma (R_M) potency factors.

Asbestos mineral type (location)	R_M, average (%) (95% CI)	Reference
Chrysotile, Quebec	0.0009 (0.0006, 0.0013)	Darnton (2023)
Amosite, South Africa	0.06 (0.016, 0.15)	Darnton (2023)
Crocidolite, South Africa	0.59 (0.36, 0.91)	Darnton (2023)
Crocidolite, Australia	0.51 (0.45, 0.57)	Darnton (2023)
Libby amphiboles, Libby, MT	0.030 (0.017, 0.05)	Darnton (2023)
Russian anthophyllite, Ekaterinburg	0.056 (0.003, 0.14)	Calculated based on Korchevskiy et al. (2013) in Korchevskiy et al. (2019)
Turkish erionite, Karain	4.67 (1.84, 7.51)	Calculated based on Simonato et al. (1989) in Korchevskiy et al. (2019)

Table 11.2 Published lung cancer (R_L) potency factors.

Asbestos mineral type (location)	R_L, average (%) (95% CI)	Reference
Chrysotile, Quebec	0.060 (0.042, 0.079)	Darnton (2023)
Amosite, South Africa	1.9 (−0.44, 5.1)	Darnton (2023)
Crocidolite, South Africa	5.2 (0.74, 12)	Darnton (2023)
Crocidolite, Australia	4.6 (3.6, 5.7)	Darnton (2023)
Libby amphiboles, Libby, MT	0.82 (0.43, 1.3)	Darnton (2023)
Russian anthophyllite, Ekaterinburg	1.8 (N/A)	Wylie and Korchevskiy (2023)
Balangeroite, Italy	1.28 (95% CI 0.75, 1.77)	Korchevskiy and Wylie (2023)
Fluoro-edenite	2.34 (N/A)	Wylie and Korchevskiy (2023)

The potency factors used in this study were derived from various sources.

As shown in Table 11.1, in addition to the potencies published by Darnton (2023), the mesothelioma potency for anthophyllite ($R_M = 0.056\%$, 95% CI 0.003, 0.14) was estimated based on a study of mesothelioma among the population around a Russian anthophyllite mine (Korchevskiy et al. 2013). The lung cancer potency for Russian anthophyllite was assumed to be 1.8%, with an unknown confidence interval, as in Wylie and Korchevskiy (2023).

Despite the significant interest in Turkish erionite as apparently one of the most toxic minerals on earth, there is scarce quantitative information about its mesothelioma potency in humans. An IARC publication from 1989 concluded that a cumulative dose of 1 f/cc-year of erionite induced a pleural mesothelioma rate of 996 per 100,000 person-years in exposed populations of Karain and Sarihidir villages in Cappadocia, Turkey (Simonato et al. 1989; Dogan 2005). Based on the IARC data, the average level of R_M for erionite can be estimated as 4.67% (95% CI 1.84, 7.51).

Fluoro-edenite was used for validation purposes in mesothelioma modeling and as a data point in lung cancer modeling. Balangeroite was also used for lung cancer modeling and for validation in mesothelioma.

The chemical composition of various minerals and aspect ratios (ARs) of the fibers used in the modeling process were as in Korchevskiy et al. (2019). In addition, chemical composition for balangeroite was used as in Wylie and Korchevskiy (2023). The median AR of fibers determined as fiber length divided by fiber width was used as a quantitative characteristic of their dimensionality. The median ARs were selected rather than mean values because AR, in general, appeared to be more log-normally distributed than normal.

As in 2019, we assumed that the model should have a log–log regression shape. The choice of a "log–log" shape for the regression model was determined by statistically testing the distribution of all data points for each of the variables. For example, a set of R_M parameters, included in the modeling process for the seven cohorts (Table 11.1), is distributed log–normally (p-value for Shapiro–Wilk test 0.699), but significantly deviates from normality ($p = 0.001$). Similarly, the AR of fibers for the seven cohorts is log-normally distributed ($p = 0.604$), as well as specific surface area ($p = 0.877$), MgO content ($p = 0.79$), Fe_2O_3 content ($p = 0.76$), and other components of the chemical composition of the fibers. The use of log-normality test in this case can be considered empirical because the origin of individual data points can be a source of heterogeneity. However, we used it as in Korchevskiy et al. (2019) to demonstrate the origin of our log–log regression approach.

The methodology and principles of the QSAR modeling were described in Korchevskiy et al. (2019). For the update, we utilized the "best subset" procedure for the regression analysis instead of the "forward stepwise" method, assuming that it would avoid the loss of some preferable combinations. Both approaches ("best subset" and "forward [or backward] stepwise") are widely used in the regression analysis literature. The purpose of both approaches is a search for the best set of independent variables predicting the specific outcome of the dependent variable. In our case, we searched for a combination of chemical and dimensional predictors, to reconstruct mesothelioma and lung cancer potency factors. When the "forward stepwise" procedure is used, the first independent variable is determined by correlation analysis (by selecting the best single predicting variable). Then, the next variable is selected, to maximize the correlation again. The process is stopped when adding another variable would not improve the correlation. The "best subset" procedure, on the contrary, analyses all possible combinations of the variables. Then, the model with the best predictive power is selected. The "best subset" is more complex computationally, but it avoids the possibilities of the "best" model to be missed.

The models derived for mesothelioma as a result of the "best subset" selection are demonstrated in Table 11.3. In the table, the partial correlation relates to a correlation of the dependent variable with each of the independent variables, with the assumption that other variables are set constant.

In both models, SiO_2, AR, and iron (ferrous or ferric) content of the fibers have positive coefficients (positively correlate with mesothelioma potency). However, MgO content is negatively correlated with potency. The toxicological meaning of this finding will be discussed further.

The regressions between the predicted and published levels of average potency factors are illustrated in Figures 11.1a and b. Both models have high correlation coefficients although the model with Fe_2O_3 seems to be in slightly better agreement with the data, and accordingly, it was selected by the "best subset" modeling procedure. Below, we use the two models interchangeably, but specify the model under discussion.

Sensitivity analysis demonstrated that the proposed model with Fe_2O_3 is prevailingly affected by MgO (72% of the variance) and AR (16% of the variance), and to a lesser extent SiO_2 (6% of the variance) and Fe_2O_3 (5% of the variance). At the same time, for example, for the Fe_2O_3 model, doubling SiO_2 increases R_M by approximately a factor of 4, doubling AR increases R_M by a factor of 2.5, doubling Fe increases R_M by a factor of 1.2, and doubling MgO decreases the mesothelioma potency by a factor of 2.6. For the model with total Fe, the variance of the model is affected by MgO (75%), AR (18%), SiO_2 (5.5%), and Fe (1.5%).

The results of both models for various fiber types are listed in Table 11.4. For some of the fiber types, the table shows the comparison between modeled values for R_M with previously published levels. For other fiber types, however, the potency has not been published, and the model provides an estimate of the unknown R_M values.

Table 11.3 The derived models for mesothelioma potency and their characteristics.

Model	Statistical characteristics
$\log_{10} R_M = -4.76 + 2.12\log_{10}\text{SiO}_2 + 1.31\log_{10}(AR - 3) + 0.36\log_{10}\text{Fe}_2\text{O}_3 - 1.51\log_{10}\text{MgO}$, (11.2) where R_M – mesothelioma potency (% per f/cc-years), SiO_2 – silicon dioxide content in the chemical composition of the fibers (%), Fe_2O_3 – ferric oxide content in the chemical composition of the fibers (%), MgO – magnesium oxide content in the chemical composition of the fibers, AR – medial aspect ratio of the fibers.	$R = 0.998$, $R^2 = 0.995$, adjusted $R^2 = 0.986$, $p = 0.0093$, goodness-of-fit criteria $F(4,2) = 106.45$, number of observations 7, residuals normally distributed with mean 0, and standard deviation 0.08, no outliers. Partial correlations for the model were found as 0.59 for SiO_2, 0.90 for Fe_2O_3, −0.989 for MgO, and 0.88 for aspect ratio.
$\text{Log}_{10} R_M = -4.75 + 2.07\log_{10}\text{SiO}_2 + 1.44\log_{10}(AR - 3) + 0.24\log_{10}\text{Fe} - 1.60\log_{10}\text{MgO}$, (11.3) where R_M – mesothelioma potency (% per f/cc-years), SiO_2 – silicon oxide content in the chemical composition of the fibers (%), Fe – total iron oxide content in the chemical composition of the fibers (%), MgO – magnesium oxide content in the chemical composition of the fibers, AR – medial aspect ratio of the fibers.	$R = 0.993$, $R^2 = 0.987$, adjusted $R^2 = 0.961$, $p = 0.026$, goodness-of-fit criteria $F(4,2) = 38.00$, number of observations 7, residuals normally distributed with mean 0, and standard deviation 0.13, no outliers. Partial correlations for the model were found as 0.39 for SiO_2, 0.70 for Fe, −0.96 for MgO, and 0.14 for aspect ratio.

For the additional validation of the models, we utilized the "leave-one-out" method (Cheng et al. 2017). The "leave-one-out" method confirms that the model is not fully dependent on single values used for its development, but represents some "natural" objective laws, captured in various values of the variables. The approach allows reconstruction of the potency factors "blindly," excluding them from the initial development of the model and its coefficients. To demonstrate this method, we eliminated the R_M value, systematically and separately, one at a time, for each of the mineral types shown in Table 11.1. Then, with the seven separate regression equations, we solved for the R_M that was eliminated in each instance. The results of this "blinded" modeling are shown in Table 11.5 (for simplicity, only the Fe_2O_3- model is used). The blinded model results are generally within less than a factor of about two to three from the published values.

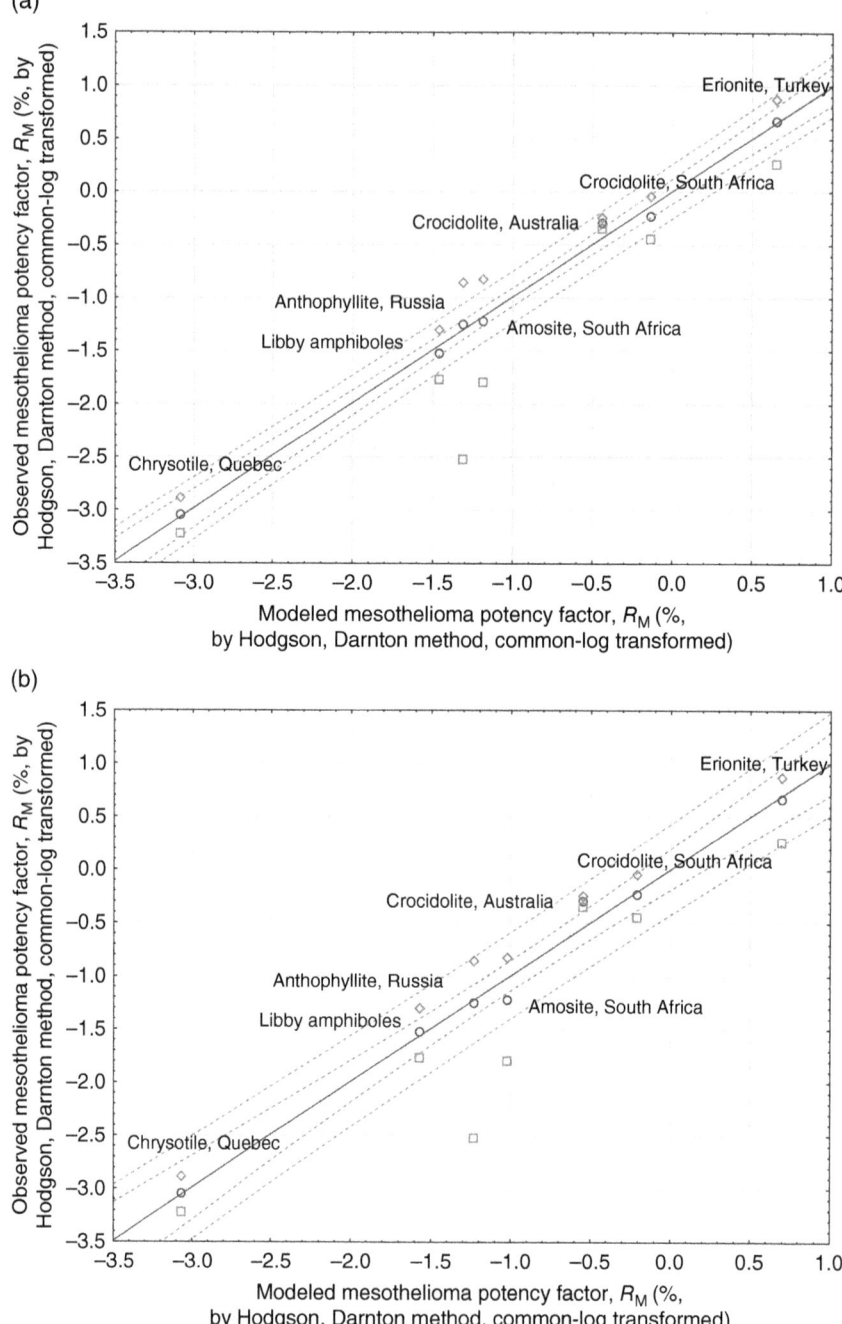

Figure 11.1 (a) Relationship between modeled and observed potency factors. Solid line – linear regression, dotted grey line – confidence interval bands, dotted black lines – prediction interval bands (model with Fe^{3+}). Dots – average published values, squares – lower confidence interval, rhombuses – upper confidence interval for published values. (b) Relationship between modeled and observed potency factors. Solid line – linear regression, dotted black lines – prediction interval bands (model with total iron). Dots – average published values, squares – lower confidence interval, rhombuses – upper confidence interval for published values.

Table 11.4 Modeled and published mesothelioma potency factor (R_M) for selected minerals.

Mineral type	Location	Average R_M, %		
		Published	Modeled, Fe_2O_3	Modeled, total Fe
Chrysotile	Quebec	0.0009^a	0.0008	0.0009
	Zimbabwe	—	0.0010	0.0012
	Russia	—	0.0018	0.0015
Amosite	South Africa	0.06^a	0.065	0.095
Tremolite	Pakistan	—	0.014	0.023
Crocidolite	Cape Province	0.59^a	0.73	0.62
	Transvaal	—	0.49	0.42
	Bolivia	—	0.09	0.06
	Australia	0.51^a	0.36	0.28
Libby amphiboles	Libby, Montana	0.03^a	0.035	0.027
Erionite	Turkey	4.67^b	4.49	4.98
Balangeroite	Italy	0.04^c	0.01	0.01
Anthophyllite	Russia	0.056^b	0.05	0.06
	Finland	—	0.04	0.03
Actinolite	Susa Valley	—	0.024	0.024
Fluoro-edenite	Biancavilla, Italy	0.12^b	0.11	0.14

[a] Darnton (2023). Specific cohort values of potency factors are used (and not meta-analytical averages for the entire mineral type).
[b] Korchevskiy et al. (2019).
[c] Korchevskiy and Wylie (2023).

Table 11.5 The reconstruction of mesothelioma potency factors based on a consequent exclusion of items from the original data set.

Fiber type	Published R_M, %	R_M predicted with a "blinded" model, %
Quebec chrysotile	0.0009	0.0006
Amosite	0.06	0.07
Crocidolite, Cape Province	0.59	1.2
Crocidolite, Australia	0.49	0.26
Libby amphibole	0.03	0.07
Russian anthophyllite	0.056	0.017
Turkish erionite	4.67	2.58

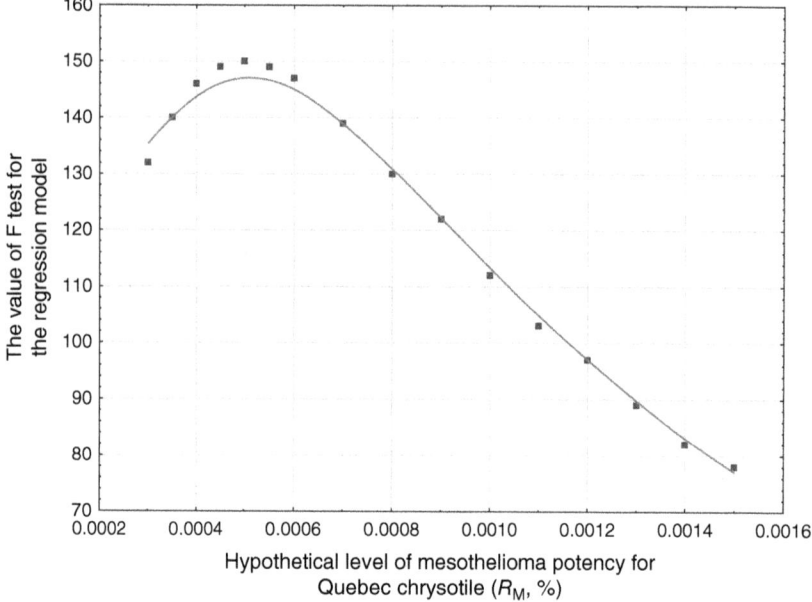

Figure 11.2 The F test for the regression model assuming various levels of Quebec chrysotile (grey line – polynomial approximation).

To understand better the application of the "blind" approach in Table 11.5, the readers should interpret each of the values in the third column as a result of the modeling (prediction) for the model developed just with six and not seven data points (for example, the prediction 0.0006% for chrysotile is made by the model developed with exclusion of the chrysotile data point, etc.).

Also, when R_M for Turkish erionite is varied systematically in small increments from the IARC-derived value of 4.67%, and F (goodness of fit criteria) is maximized in the Fe_2O_3 version of the model, a value of 3.99% is obtained. In a similar fashion, Figure 11.2 illustrates how the potency level of $R_M = 0.0005\%$ for Quebec chrysotile maximizes F criteria in the Fe^{3+} version, compared with the published R_M of 0.0009%. In the total iron version (which apparently is less robust), F increases with chrysotile potency approaching 0 and, separately, with erionite potency unlimitedly increasing.

The same parameters as in the models (Table 11.3) can be used to develop a predictive model for lung cancer potency of mineral fibers. In particular, we used SiO_2, AR, Fe, and MgO as independent variables.

The following model was constructed:

$$\log_{10} R_L = -2.06 + 0.74 \log_{10} SiO_2 + 1.12 \log_{10}(AR - 3) \\ + 0.53 \log_{10} Fe - 0.70 \log_{10} MgO, \tag{11.4}$$

where

R_L – mesothelioma potency (%),
SiO_2 – silicon oxide content in the chemical composition of the fibers (%),
Fe – total iron oxide content in the chemical composition of the fibers (%),
MgO – magnesium oxide content in the chemical composition of the fibers,
AR – medial aspect ratio of the fibers.
$R = 0.997$, $R^2 = 0.993$, adjusted $R^2 = 0.984$, $p = 0.0013$, goodness-of-fit criteria $F(4,3) = 111.31$, number of observations 8, residuals normally distributed with mean 0, and standard deviation 0.12, no outliers.

Partial correlations for the model were found as 0.75 for SiO_2, 0.96 for Fe, −0.95 for MgO, and 0.99 for aspect ratio.

The results of the model for various mineral types are listed in Table 11.6.

Table 11.6 Modeled and published lung cancer potency factor (R_L) for selected minerals.

Mineral type	Location	Average R_L (%)	
		Published	Modeled
Chrysotile	Quebec	0.06[a]	0.06
	Zimbabwe	—	0.07
	Russia	—	0.12
Amosite	South Africa	1.9[a]	2.11
Tremolite	Pakistan	—	0.55
Crocidolite	Cape Province	5.2[a]	5.87
	Transvaal	—	3.84
	Bolivia	—	1.76
	Australia	4.6[a]	3.61
Libby amphiboles	Libby, Montana	0.82[a]	0.76
Erionite	Turkey	—	3.30
Balangeroite	Italy	1.28[b]	1.23
Anthophyllite	Russia	1.8[c]	1.88
	Finland	—	1.37
Actinolite	Susa Valley	—	0.68
Fluoro-edenite	Biancavilla, Italy	2.34[c]	2.29

[a] Darnton (2023). Specific cohort values of potency factors are used (and not meta-analytical averages for the entire mineral type).
[b] Korchevskiy and Wylie (2023).
[c] Wylie and Korchevskiy (2023).

For the lung cancer model, the sensitivity analysis demonstrates that the variance is mostly affected by MgO (76%), AR (19%), Fe (3%), and SiO_2 (2%).

The proposed models have three main points of justification: (1) they are based on a meaningful choice of variables from proposed mechanisms of action in the literature and a general mathematical form, (2) they fit published mesothelioma potency factors well, and (3) they yield reasonable estimations of potency factors for fibers with absent epidemiological information, as demonstrated below.

For example, the ratio between the modeled mesothelioma potency of crocidolite (average between South African and Australian), South African amosite, and Quebec chrysotile, is 500 : 105 : 1 for the Fe model and 681 : 81 : 1 for the Fe_2O_3 model. The corresponding ratio between the mesothelioma potencies for the same cohorts in Darnton (2023) is 611 : 67 : 1. The modeled ratio is also of the same order as the estimation proposed by Hodgson and Darnton (2000) of 500 : 100 : 1 for meta-analytical potency factors of crocidolite, amosite, and chrysotile. The meta-analysis performed by Darnton in 2003 yields the ratio of 372 : 122 : 1. While the model shows reasonable estimations for the data points used to fit its parameters (as for crocidolite, amosite, Libby amphiboles, chrysotile, and erionite), it is interesting to observe what estimations that model would produce for the mineral types not included in the initial modeling process.

For example, we did not use any estimations of lung cancer erionite potency for the initial building of the model. However, our modeling shows erionite lung cancer potency to be lower than for crocidolite, $R_L = 3.31\%$ for erionite vs. about 4.8% for crocidolite, as in Darnton, 2023. Our modeling, in addition, demonstrates that one erionite-related excess lung cancer case is expected to correspond to about 20 cases of malignant mesothelioma (depending on the background risk of lung cancer). For comparison, the epidemiological statistics on lung cancer incidence in the populations exposed to fibrous erionite is quite limited, in particular, because of the unknown cigarette smoking-related lung cancer mortality in erionite-exposed populations. For example, Baris et al. (1996) reported 157 cases of malignant mesothelioma deaths (both pleural and peritoneal) in the village of Karain, Turkey, but only four lung cancers (from a total of 305 deaths). In the village of Tuzkoy, 184 cases of malignant mesothelioma were reported, vs. 14 lung cancer cases (from the total of 519 deaths). The expected number of lung cancer cases in this population is unknown and probably would be affected by an unknown level of smoking-related mortality. Also, the quality of lung cancer records is unknown, and the effect of competing causes of death, e.g. mesothelioma, is likely significant. The low number of observed lung cancer cases compared with the very high number of mesothelioma cases generally confirms our modeled result conclusion that the ratio of lung cancer incidence compared with mesothelioma incidence should be low, but quantitative confirmation is not directly possible.

In addition, Wagner et al. (1985) found a substantial number of mesothelioma cases, but no lung cancer cases in rats exposed to erionite by inhalation. This result is generally consistent with our low ratio between predicted lung cancer cases to mesothelioma cases in humans, because of the limited number of test animals in such studies. Similarly, for crocidolite, they found only one case. Further studies on erionite potency for lung cancer would be beneficial.

In another example, based on our models, R_M of Bolivian crocidolite ranged from 0.06% to 0.09%. That is approximately six times lower than for averaged Australian and South African varieties. This is generally in line with the observations of Ilgren et al. (2012) that magnesium-rich Bolivian crocidolite (magnesioreibeckite) apparently did not cause significant elevation of mesothelioma mortality in workers and the population. However, it should be noted that Ilgren's assessment might not be comprehensive, and mesothelioma could likely still occur, though at a lower level than for the Australian and South African varieties.

The modeled mesothelioma and lung cancer potencies of Canadian chrysotile are remarkably close to the published result. The modeled Russian chrysotile R_M appears to be about twice that of the Quebec variety (because of the apparent differences in chemical composition or, potentially, the dimensional characteristics).

Our models predict a rather low level of mesothelioma potency for Italian balangeroite: 0.01%. For lung cancer, the projected potency is 1.23%. Korchevskiy and Wylie (2023) used more specific dimensionality-based models to demonstrate for balangeroite, the mesothelioma potency factor $R_M = 0.04\%$ (95% CI 0.0058, 0.16) and lung cancer potency factor $R_L = 1.28\%$ (95% CI 0.75, 1.77). However, the models in this chapter (Table 11.3) predict a mesothelioma potency of balangeroite at a level within the confidence interval of the dimensionality-based potency. If additional models are considered (and averaged), as was proposed in Korchevskiy and Wylie (2023), the mesothelioma potency of balangeroite would be slightly lower than initially predicted in our 2019 publication: $R_M = 0.034\%$, but still fully in line with the conclusions of Korchevskiy and Wylie (2023).

Fibrous minnesotaite and other phyllosilicates were suspected to increase the risk of mesothelioma in taconite workers of the Western Mesabi Range in Minnesota (NRRI 2019). Minnesotaite contains around 2% of Fe_2O_3, 6% of magnesium, and 51.3% of SiO (Gruner 1944). Although it appears that generally minnesotaite does not contain long and thin asbestiform fibers (NRC 1984), hypothetically, if some portion of airborne minnesotaite is of a fibrous nature with a median AR of at least 5 : 1, its R_M based on our models could be as high as 0.010–0.015%, lower than that observed for Libby amphiboles, but still significant from risk assessment viewpoint.

It is important to assess for toxicological plausibility of the developed models. In Chapter 2, Mossman discusses the role of iron, particularly in amphibole asbestos, in producing free radicals. In particular, iron "overload" was suggested to play an important role in asbestos fiber toxicity (Quinlan et al. 1994). Iron is a stoichiometric component (about 27%, w/w) in amphiboles such as crocidolite and amosite, while in chrysotile it is a contaminant (about 1–6%, w/w) replacing isomorphous magnesium and silicon (Schwarz and Winer 1971; De Waele et al. 1984). Retained in the lungs, asbestos fibers adsorb iron from the cellular environment creating an iron-containing coating around the fibers. Some view the iron coating as a step in the mechanism of toxicity while others believe the coating is protective (Gilmour et al. 1997; Jaurand et al. 1997; Xu et al. 1999). Fe^{3+} was demonstrated to have a more pronounced effect on reduction of human enzyme activity than Fe^{2+} (Karami et al. 2010). Fisher et al. (1998) emphasized the role of Fe^{3+} in fiber toxicity. Ghio et al. (1992) determined that generation of hydroxyl radicals by crocidolite asbestos is proportional to surface Fe^{3+}. Toyokuni (1996) found that different compounds of Fe^{3+} are carcinogenic in animals. For example, intraperitoneal injections of ferric sacharate followed by administration of nitrilotriacetate (NTA) caused a high incidence of diffuse mesotheliomas. Intraperitoneal injections of Fe-NTA can induce peritoneal mesotheliomas in rats as well (Okada et al. 1989; Toyokuni 1996). Ghio et al. (2008) suggested that differences between serpentine and amphibole fibers in oxidative stress and biological effects do not reflect the inclusion of iron in the crystal lattice, but rather larger numbers of surface functional groups (first, silanol) in amphiboles which can mobilize iron from the host. It is possible either or both iron in fibers and extracellular iron may play a mechanistic role in the carcinogenicity process.

Different authors have also emphasized the possible role of silicon oxide content in carcinogenicity mechanisms (Fubini et al. 1990; The National Academy of Science 2006; Aust et al. 2011).

It should be noted that Ghio et al. (2023) proposed a theory of iron deficiency in cellular structures caused by asbestos fiber intervention. According to the authors, the role of iron in the asbestos lattice is minimal for the toxic effects. On the contrary, "the higher percentage SiO_2 in the lattice of amphibole asbestos can result in higher surface silanol numbers" that Ghio et al. see as a predictor of greater biological effects for amphiboles as compared to chrysotile. However, the QSAR modeling that we performed indicates that iron content in the lattice of fibers can be one of the quantitative predictors of potency. In reality, the iron content of the fibers might be an indirect parameter, reflecting some of the unknown mechanisms of carcinogenicity that should not necessarily be directly related to one or both of the theories: "iron in the lattice causes cancer" vs. "iron deficiency in the cells causes cancer."

In our estimation of the chemical composition of various fibers, we indicated a very low, but not zero level of iron oxide content in erionite fibers. For example, USGS (Lowers et al. 2010) indicated that there is a low iron content in Turkey erionite vs. other deposits where iron was not found. Gualtieri et al. (2016) demonstrated that for Nevada erionite, iron is apparently concentrated on the surface, and not in the mineral lattice, of the fibers. Further studies are needed to better understand the role of iron in asbestos toxicity, and the reasons why it plays a role in the QSAR models we developed.

Also, the size and shape of asbestos fibers are known predictors of their carcinogenicity (Berman et al. 1995; Berman and Crump 2008b; Lippmann 2009; Loomis et al. 2010; Barlow et al. 2017; Wylie and Korchevskiy 2023). AR can be viewed as one of the possible characteristics, useful, even if not universal, to generalize the dimensional characteristics of fibers (Boulanger et al. 2014). It is obvious that using the AR of fibers (and not separate length and width, along with other parameters) as a surrogate dimensionality characteristic has its limitations. However, it was suggested recently that AR can be an important toxicological characteristic not only for asbestos fibers but also for nanoparticles (Tran et al. 2008). Korchevskiy and Wylie (2021) demonstrated that AR should be "corrected" to correlate with mesothelioma potency (length and width in the ratio to be taken in different powers). However, AR in our model seems to fit well the purpose of supporting the fiber potency predictions.

The magnesium content in the chemical composition of fibers seems to emerge as one of the strongest predictors of the potency difference between fibers. Similarly, erionite seems to be producing the highest risk of mesothelioma, and chrysotile the lowest. It is possible that MgO in fibers is a predictor of the biopersistence of fibers (the higher the MgO content, the higher the dissolution rate, and the lower the biopersistence). This correlates with low MgO content in erionite and high MgO content in chrysotile.

It seems significant that the ratio between the half-life of crocidolite and amosite fibers with lengths of 5–20 µm in animals is generally in agreement with human data. For example, de Klerk estimated the average half-life of crocidolite clearance in human lungs as 92 months (de Klerk et al. 1996). Du Toit estimated the lung clearance half-life of Cape Province crocidolite in human lungs as six years (Du Toit 1991). At the same time, Churg and Vedal (1994) evaluated the clearance half-life of amosite fibers as 20 years. Churg (1994) stated:

> The available data suggest that chrysotile is deposited in the parenchyma but is cleared extremely rapidly, with the vast bulk of fibres removed from human lungs within weeks to months after inhalation; by comparison, amphibole clearance half-lives are of the order of years to decades.

Korchevskiy and Wylie (2022) attempted to model the elimination rate of various mineral types of fibers demonstrated based on the proposed model that the average elimination coefficient was estimated for crocidolite as 0.099 vs. the average published value of 0.092, for amosite as 0.169 vs. 0.19, and for chrysotile as 6.45 vs. the average published value of 6.36 (years^{-1}).

The elimination coefficient λ can be found from the first-order kinetic elimination formula:

$$S = S_0 e^{-\lambda t} \tag{11.5}$$

where S_0 – initial number of fibers per weight of lung tissue,
S – number of fibers per weight of lung tissue at the moment t (years).

At the same time, Bernstein et al. (2005) suggested that long tremolite fibers are expected to have an "infinite" half-life in animal lungs.

Korchevskiy and Wylie (2022), however, demonstrated that the elimination rate of tremolite from human lungs can potentially be 0.14 (years^{-1}). This estimation is more in line with the majority of tremolite asbestos subcategories that are not the most carcinogenic among amphibole asbestos, with mesothelioma potency about 1/10th of the crocidolite variety (see Chapter 8). It is also remarkable that the US EPA estimated the best-fitting half-life of Libby amphiboles from Libby, MT as being in the range from 5 to 10 years (corresponding to the elimination rate from 0.13 to 0.07 [years^{-1}]).

General estimations of fiber biopersistence (or half-life in human lungs), as derived from Wylie and Korchevskiy (2022), are demonstrated in Table 11.7.

It can be demonstrated that log-transformed biopersistence of fibers positively and statistically significantly correlates with the log-transformed average mesothelioma potency factors as modeled above ($R = 0.93$ for the Fe^{3+} model and 0.92 for the total iron model, $p < 0.05$).

The following model can be derived:

$$\log_{10} R_M = -3.2 + 1.36 \log_{10}\left(\text{Fiber Half-Life in Lungs}\right) \tag{11.6}$$

where Fiber Half-Life in Lungs is an assumed half-life of various mineral types of fibers in lungs, measured in months.

$\left(R = 0.92, R^2 = 0.84, \text{adjusted } R^2 = 0.79, F(1, 3) = 16.3, p < 0.027\right).$

Table 11.7 Asbestos fiber elimination rate and half-life in human lungs.

Asbestos mineral type	Elimination rate (years^{-1})	Half-life (months)
Chrysotile, Quebec	6.45	1.3
Amosite	0.169	49
Crocidolite	0.099	84
Libby amphiboles	0.14	59

This model allows to reconstruct the half-life for fibers in lungs if their mesothelioma potency is known. For example, based on the regression equation, the half-life of Turkish erionite in human lungs can be estimated as about 57 years (virtually, "infinite" biopersistence). This estimation is in line with the assessment of Gualtieri (2018) who found fibrous erionite to have acellular dissolution about 600 times slower than for chrysotile that should correspond to 65 years of half-life in the human lungs.

The log-transformed level of biopersistence also correlates with variables such as SiO_2 ($R = 0.92$, $p < 0.05$) and Fe_2O_3 ($R = 0.90$, $p < 0.05$), showing that a fraction of iron and silicon can be predictive of the fiber burden and longevity in lungs.

It was also interesting to juxtapose the results of the potency modeling with information on *in vitro* behavior of asbestos fibers. In Table 11.8, determined human membranolytic activities of fibers (HC_{50}) are listed for various fiber types according to Nolan et al. (1991). The HC_{50} is defined as the concentration of particles (given in mg/ml) required to lyse 50% of the erythrocytes in a suspension containing 1.8×10^8 cells/ml. It is important to remember that high HC_{50} values indicate lower toxicity (membranolytic activity).

We determined that our estimations of mesothelioma potencies (total iron model) for different fiber types are highly correlated with HC_{50} (both log-transformed) and that the relationship is linear:

$$\log_{10} R_M^{modeled} = -1.70 + 1.15 \log_{10}\left(HC_{50}\right) \tag{11.7}$$

$\left(\text{log–log slope} = 1.15 \pm \text{standard error of } 0.27\right).$

An intriguing aspect of the results is that fibers with the highest hemolytic activity (and lower HC_{50}) appear to have the lowest cancer potency for mesothelioma. Recently, Nagai et al. (2011) stated that "silica exhibited the most potent

Table 11.8 Criteria of membranolytic (hemolytic) activity of fibers.

Fiber type and location	HC_{50} (mg/ml)
Na–Ca amphiboles, Montana	2.07 ± 1.72
Tremolite, India[a]	2.81 ± 0.33
Crocidolite, Cape Province	4.05 ± 1.22
Amosite, Transvaal	1.76 ± 0.28
Anthophyllite, Finland	3.76 ± 1.14
Chrysotile, Canada	0.08 ± 0.02
Chrysotile, Zimbabwe	0.06 ± 0.00

[a] Used in our comparison for tremolite, Pakistan.

hemolytic activity, followed by chrysotile. Crocidolite and amosite were also hemolytic but with a much lower activity (approximately 200-fold less)."

This order, again, is very noteworthy because crystalline silica, for example, is obviously not a mesothelial carcinogen. It is also known that hemolytic potential is not a determinant of fibrogenicity of mineral dust (Hemenway et al. 1993). Also, Blum et al. (2015) and Henzi et al. (2009) promoted the idea that factors protecting mesothelial cells from asbestos cytotoxicity (like overexpression of calretinin) actually may contribute to mesothelioma pathogenesis. However, this apparent "competition" of potential cell toxicity vs. carcinogenicity should be explored further for a better understanding of potential asbestos toxicological mechanisms of action.

It should be added that during the last year the QSAR methodology for asbestos fibers was explored from the position of the dimensionality parameters as a predictor of carcinogenic potency. In particular, Wylie et al. (2020) demonstrated that for amphiboles, mesothelioma potency factors can be modeled as a function of the fraction of very thin fibers in the exposure. However, chemical composition parameters are needed to predict potency of wider classes of fibers. In Wylie and Korchevskiy (2023), it is demonstrated that the dimensional-specific models for chrysotile can be similar to amphibole models, if only an additional correction coefficient would be introduced (this coefficient can be interpreted as a "chemical composition" part of the model in this chapter). Apparently, similar approaches can be proposed for interpretation of other types of minerals (like, for erionite). Further studies are needed to develop additional models that would fully consider not only mineral type, chemical composition, and dimensions of elongate mineral particles but also their habit (asbestiform vs. non-asbestiform variety of amphiboles, as in Wylie et al. 2022).

References

[ATSDR] Agency for Toxic Substances and Disease Registry. (2010). *Case studies in environmental medicine (CSEM): Asbestos toxicity*. Course WB/CB1093. https://www.atsdr.cdc.gov/hec/csem/asbestos/docs/asbestos.pdf. Accessed December 20, 2023.

Aust, A.E., Cook, P.M., and Dodson, R.F. (2011). Morphological and chemical mechanisms of elongated mineral particle toxicities. *Journal of Toxicology and Environmental Health; Part B, Critical Reviews* 14(1–4): 40–75. https://doi.org/10.1080/10937404.2011.556046.

Bariş, B., Demir, A.U., Shehu, V., et al. (1996). Environmental fibrous zeolite (erionite) exposure and malignant tumors other than mesothelioma. *Journal of Environmental Pathology Toxicology and Oncology* 15(2–4): 183–189. MID: 9216804.

Barlow, C.A., Grespin, M., and Best, E.A. (2017). Asbestos fiber length and its relation to disease risk. *Inhalation Toxicology* 29(12–14): 541–554. https://doi.org/10.1080/08958378.2018.1435756.

Barratt, M.D. (2000). Prediction of toxicity from chemical structure. *Cell Biology and Toxicology* 16(1): 1–13. https://doi.org/10.1023/a:1007676602908.

Berman, D.W. and Crump, K.S. (2008a). Update of potency factors for asbestos-related lung cancer and mesothelioma. *Critical Reviews in Toxicology* 38(Sup 1): 1–47. https://doi.org/10.1080/10408440802276167.

Berman, D.W. and Crump, K.S. (2008b). A meta-analysis of asbestos-related cancer risk that addresses fiber size and mineral type. *Critical Reviews in Toxicology* 38(Suppl 1): 49–73. https://doi.org/10.1080/10408440802273156.

Berman, D.W., Crump, K.S., Chatfield, E., et al. (1995). The sizes, shapes, and mineralogy of asbestos structures that induce lung tumors or mesothelioma in AF-HAN rats following inhalation. *Risk Analysis* 15(2): 181–195. https://doi.org/10.1111/j.1539-6924.1995.tb00312.x.

Bernstein, D.M., Chevalier, J., and Smith, P. (2005). Comparison of Calidria chrysotile asbestos to pure tremolite: final results of the inhalation biopersistence and histopathology examination following short-term exposure. *Inhalation Toxicology* 17(9): 427–449. https://doi.org/10.1080/08958370591002012.

Blum, W., Pecze, L., Felley-Bosco, E., et al. (2015). Overexpression or absence of calretinin in mouse primary mesothelial cells inversely affects proliferation and cell migration. *Respiratory Research* 16: 153–172. https://doi.org/10.1186/s12931-015-0311-6.

Boulanger, G., Andujar, P., Pairon, J.-C., et al. (2014). Quantification of short and long asbestos fibers to assess asbestos exposure: a review of fiber size toxicity. *Environmental Health* 13: 59. https://doi.org/10.1186/1476-069X-13-59

Cheng, H., Garrick, D.J., and Fernando, R.L. (2017). Efficient strategies for leave-one-out cross validation for genomic best linear unbiased prediction. *Journal of Animal Science and Biotechnology* 8(38). https://doi.org/10.1186/s40104-017-0164-6.

Cherkasov, A., Muratov, E.N., Fourches, D., et al. (2014). QSAR Modeling: Where have you been? Where are you going to? *Journal of Medicinal Chemistry* 57(12), 4977–5010. https://doi.org/10.1021/jm4004285.

Churg, A. (1994). Deposition and clearance of chrysotile asbestos. *Annals of Occupational Hygiene* 38(4): 625–633. https://doi.org/10.1093/annhyg/38.4.625.

Churg, A. and Vedal, S. (1994). Fiber burden and patterns of asbestos-related disease in workers with heavy mixed amosite and chrysotile exposure. *American Journal of Respiratory Critical Care Medicine* 150(3): 663–669. https://doi.org/10.1164/ajrccm.150.3.8087335.

Darnton, L. (2023). Quantitative assessment of mesothelioma and lung cancer risk based on Phase Contrast Microscopy (PCM) estimates of fibre exposure: an update of 2000 asbestos cohort data. *Environmental Research* 230: 114753. https://doi.org/10.1016/J.ENVRES.2022.114753.

De Klerk, N.H., Musk, A.W., Williams, V., et al. (1996). Comparison of measures of exposure to asbestos in former crocidolite workers from Wittenoom Gorge, W. Australia. *American Journal of Industrial Medicine* 30(5): 579–587. https://doi.org/10.1002/(sici)1097-0274(199611)30:5%3C579::aid-ajim5%3E3.0.co;2-o.

De Waele, J., Luys, M., Vansant, E., et al. (1984). Analysis of chrysotile asbestos by LAMMA and Mossbauer spectroscopy: a study of the distribution of iron. *Journal of Trace Microprobe Techniques* 2(2): 87–102.

Dogan, A. (2005). Malignant mesothelioma and erionite. In: *Malignant Mesothelioma: Advances in Pathogenesis, Diagnosis, and Translational Therapies* (eds. H.I. Pass, N.J. Vogelzang, and M. Carbone), 242–258. USA: Springer New York, NY. https://link.springer.com/book/10.1007/0-387-28274-2#:~:text=https%3A//doi.org/10.1007/0%2D387%2D28274%2D2

Du Toit, R.S. (1991). An estimate of the rate at which crocidolite asbestos fibres are cleared from the lungs. *The Annals of Occupational Hygiene* 35(4): 433–438. https://doi.org/10.1093/annhyg/35.4.433.

Fisher, C., Brown, D., Shaw, J., et al. (1998). Respirable fibres: surfactant coated fibres release more Fe^{3+} than native fibres at both pH 4.5 and 7.2. *The Annals of Occupational Hygiene* 42(5): 337–345. https://doi.org/10.1016/s0003-4878(98)00022-2.

Fubini, B., Giamello, E., Volante, M., et al. (1990). Chemical functionalities at the silica surface determining its reactivity when inhaled. Formation and reactivity of surface radicals. *Toxicology and Industrial Health* 6(6): 571–598. PMID: 1965871.

Ghio, A.J., Zhang, J., and Piantadosi, C.A. (1992). Generation of hydroxyl radical by crocidolite asbestos is proportional to surface [Fe^{3+}]. *Archives of Biochemistry and Biophysics* 298(2): 646–650. https://doi.org/10.1016/0003-9861(92)90461-5.

Ghio, A.J., Stonehuerner, J., Richards, J., et al. (2008). Iron homeostasis in the lung following asbestos exposure. *Antioxidants & Redox Signaling* 10(2): 371–377. https://doi.org/10.1089/ars.2007.1909.

Ghio, A.J., Stewart, M, Sangani, R.G., et al. (2023). Asbestos and iron. *International Journal of Molecular Sciences* 24(15): 12390. https://doi.org/10.3390/ijms241512390.

Gilmour, P.S., Brown, D.M., Beswick, P.H., et al. (1997). Free radical activity of industrial fibers: role of iron in oxidative stress and activation of transcription factors. *Environmental Health Perspectives* 105 (Suppl 5): 1313–1317. https://doi.org/10.1289/ehp.97105s51313.

Gruner, J.W. (1944). The composition and structure of Minnesotaite, a common iron silicate in iron formations. *American Mineralogist: Journal of Earth and Planetary Materials* 29(9-10): 363–372. http://www.minsocam.org/msa/collectors_corner/amtoc/toc1944.htm.

Gualtieri, A.F., Gandolfi, N.B., Pollastri, S., et al. (2016). Where is iron in erionite? A multidisciplinary study on fibrous erionite-Na from Jersey (Nevada, USA). *Scientific Reports* 6, 37981. https://doi.org/10.1038/srep37981.

Gualtieri, A.F., Pollastri, S., Gandolfi, N.B., Gualtieri, M.L. (2018). *In vitro* acellular dissolution of mineral fibres: a comparative study. *Scientific Reports* 8:7071. https://doi.org/10.1038/s41598-018-25531-4

Hemenway, D.R., Absher, M.P., Fubini, B., and Bolis, V. (1993). What is the relationship between hemolytic potential and fibrogenicity of mineral dusts? *Archives of Environmental Health: An International Journal* 48(5): 343–347. https://doi.org/10.1080/00039896.1993.9936723.

Henzi, T., Blum, W.-V., Pfefferli, M., et al. (2009). SV40-induced expression of calretinin protects mesothelial cells from asbestos cytotoxicity and may be a key factor contributing to mesothelioma pathogenesis. *American Journal of Pathology* 174(6): 2324–2336. https://doi.org/10.2353/ajpath.2009.080352.

Hodgson, J.T. and Darnton, A. (2000). The quantitative risks of mesothelioma and lung cancer in relation to asbestos exposure. *Annals of Occupational Hygiene* 44(8): 565–601. PMID: 11108782.

Ilgren, E., Ramirez, R., Claros, E., et al. (2012). Fiber width as a determinant of mesothelioma induction and threshold – Bolivian Crocidolite: epidemiological evidence from Bolivia – mesothelioma demography and exposure pathways. *Annals of Respiratory Medicine* (online first). https://www.researchgate.net/publication/274697356_Fiber_Width_as_a_Determinant_of_Mesothelioma_Induction_and_ThresholdBolivian_Crocidolite_Epidemiological_Evidence_from_BoliviaMesothelioma_Demography_and_Exposure_Pathways

Jaurand M.C. (1997). Mechanisms of fiber-induced genotoxicity. *Environmental Health Perspectives* 105(Suppl 5): 1073–1084. https://doi.org/10.1289%2Fehp.97105s51073.

Karami, M., Ebrahimzadeh, M.A., Mahdavi, M.R., et al. (2010). Effect of Fe^{2+} and Fe^{3+} ions on human plasma cholinesterase activity. *European Review for Medical and Pharmacological Sciences* 14(10): 897–901. PMID: 21222379.

Korchevskiy, A.A. and Wylie, A.G. (2021). Dimensional determinants for the carcinogenic potency of elongate amphibole particles. *Inhalation Toxicology* 33(6–8), 244–259. https://doi.org/10.1080/08958378.2021.1971340.

Korchevskiy, A.A. and Wylie, A.G. (2022). Asbestos exposure, lung fiber burden, and mesothelioma rates: mechanistic modelling for risk assessment. *Computational Toxicology* 24:100249. https://doi.org/10.1016/j.comtox.2022.100249.

Korchevskiy, A.A. and Wylie, A.G. (2023). Toxicological and epidemiological approaches to carcinogenic potency modeling for mixed mineral fiber exposure: the case of fibrous balangeroite and chrysotile. *Inhalation Toxicology* 35(7–8): 185–200. https://doi.org/10.1080/08958378.2023.2213720.

Korchevskiy, A.A., Rasmuson, E.J., Rasmuson, J.O., and Strode, R.D. (2013). Asbestos mining in Russia: approaches to public health risk assessment. Preprint 13-117 in:

Mining: It's All About People. Pre-prints of the Society for Mining, Metallurgy & Exploration Annual Meeting; February 24-27; Denver, Colorado.

Korchevskiy, A., Rasmuson, J.O., and Rasmuson, E.J. (2019). Empirical model of mesothelioma potency factors for different mineral fibers based on their chemical composition and dimensionality. *Inhalation Toxicology* 31(5), 180–191. https://doi.org/10.1080/08958378.2019.1640320.

Lippmann, M. (2009). *Environmental Toxicants: Human Exposures and Their Health Effects*, 3rd Edition. New Jersey: John Wiley & Sons.

Loomis, D., Dement, J., Richardson, D., et al. (2010). Asbestos fibre dimensions and lung cancer mortality among workers exposed to chrysotile. *Occupational and Environmental Medicine* 67(9): 580–584. https://doi.org/10.1136/OEM.2009.050120.

Lowers, H.A., Adams, D.T., Meeker, G.P., and Nutt, C.J. (2010). Chemical and morphological comparison of erionite from Oregon, North Dakota, and Turkey. Reston, Virginia: USGS. Open-File Report 2010–1286.

Mossman, B.T. (2008). Assessment of the pathogenic potential of asbestiform vs. nonasbestiform particulates (cleavage fragments) in in vitro (cell or organ culture) models and bioassays. *Regulatory Toxicology and Pharmacology* 52(1 Suppl): S200–S203. https://doi.org/10.1016/j.yrtph.2007.10.004.

Nagai, H., Ishihara, T., Lee, W.-H., et al. (2011). Asbestos surface provides a niche for oxidative modification. *Cancer Science* 102(12): 2118–2125. https://doi.org/10.1111/j.1349-7006.2011.02087.x.

[NAS/NRC] National Academy of Sciences, National Research Council. (1984). Asbestiform fibers: nonoccupational health risks. Washington, DC: National Academy Press.

[NAS] National Academy of Sciences. (2006). *Asbestos: Selected Cancers*. Washington, DC: National Academies Press. https://doi.org/10.17226/11665.

[NRRI] Natural Resource Research Institute. (2019). Environmental study of airborne particulates in Mesabi Iron Range communities and Taconite Processing Plants – Mesabi Iron Range Community Particulite Matter Collection and GravimetricAnalysis. Submitted by: Stephen D. Monson Geerts et al., Report Number: NRRI/Ri-2019/30; Project No. 1806-10416-20080. December 2019.

Nolan, R.P., Langer, A.M., Oechsle, G.W., et al. (1991). Association of tremolite habit with biological potential: preliminary report. In: *Mechanisms in Fibre Carcinogenisis* (eds. R.C. Brown, J.A. Hoskins, and N.F. Johnson). NATO ASI Series, Vol. 223. Boston, MA: Springer. https://doi.org/10.1007/978-1-4684-1363-2_21.

Okada, S., Hamazaki, S., Toyokuni, S., et al. (1989). Induction of mesothelioma by intraperitoneal injections of ferric saccharate in male Wistar rats. *British Journal of Cancer* 60(5): 708–711. https://doi.org/10.1038%2Fbjc.1989.344.

[OECD] Organization for Economic Co-operation and Development. (2007). Guidance Document on the Validation of (Quantitative) Structure-Activity

Relationship ((Q)SAR) Models. ENV/JM/MONO(2007)2. 30-Mar-2007. ENVJmMONO_2007_2olis.doc

Puzyn, T., Rasulev, B., Gajewicz, A., et al. (2011). Using nano-QSAR to predict the cytotoxicity of metal oxide nanoparticles. *Nature Nanotechnology* 6(3): 175–178. https://doi.org/10.1038/nnano.2011.10.

Quinlan, T.R., Marsh, J.P., Janssen, Y.M., et al. (1994). Oxygen radicals and asbestos-mediated disease. *Environmental Health Perspectives* 102(Suppl 10): 107–110. https://doi.org/10.1289/ehp.94102s10107.

Schwarz, E.J. and Winer, A. (1971). Magnetic properties of asbestos, with special reference to determination of absolute magnetite contents. *Canadian Institute of Mining and Metallurgy Petroleum, CIM Bulletin* 64: 55–59.

Simonato, L., Baris, R., Saracci, R., et al. (1989). Relation of environmental exposure to erionite fibers to risk of respiratory cancer. *IARC Scientific Publications* (90): 398–405. PMID: 2545613.

Toyokuni, S. (1996). Iron-induced carcinogenesis; the role of redox regulation. *Free Radical Biology & Medicine* 20(4): 553–566. https://doi.org/10.1016/0891-5849(95)02111-6.

Tran, L., Hankin, S. Ross, B., et al. (2008). An outline scoping study to determine whether High Aspect Ratio Nanoparticles (HARN) should raise the same concerns as do asbestos fibers. Report on Project CB0406. https://www.researchgate.net/publication/235223737_An_outline_scoping_study_to_determine_whether_high_aspect_ratio_nanoparticles_HARN_should_raise_the_same_concerns_as_do_asbestos_fibres

Wagner, J.C., Skidmore, J.W., Hill, R.J., and Griffiths, D.M. (1985). Erionite exposure and mesotheliomas in rats. *British Journal of Cancer* 51(5): 727–730. https://doi.org/10.1038/bjc.1985.108.

Wylie, A.G. and Korchevskiy, A.A. (2023). Dimensions of elongate mineral particles and cancer: a review. *Environmental Research* 230: 114688. https://doi.org/10.1016/j.envres.2022.114688.

Wylie, A.G., Korchevskiy, A., Segrave, A.M., and Duane, A. (2020). Modeling mesothelioma risk factors from amphibole fiber dimensionality: mineralogical and epidemiological perspective. *Journal of Applied Toxicology* 40: 515–524. https://doi.org/10.1002/jat.3923.

Wylie, A., Korchevskiy, A., Van Orden, D., and Chatfield, E. (2022). Discriminant analysis of asbestiform and non-asbestiform amphibole particles and its implications for toxicological studies. *Computational Toxicology* 23:100233. https://doi.org/10.1016/j.comtox.2022.100233.

Xu, A., Wu, L., Santella, R.M., and Hei, T.K. (1999). Role of oxyradicals in mutagenicity and DNA damage induced by crocidolite asbestos in mammalian cells. *Cancer Research* 59(23): 5922–5926. PMID: 10606236.

12

Theoretical and Practical Aspects of Asbestos Dose–Response Assessment

Andrey Korchevskiy and James Rasmuson

Chemistry & Industrial Hygiene, Inc., Lakewood, CO, USA

General Considerations and Model of Asbestos Dose–Response Assessment

As a rule, asbestos dose–response assessment requires knowledge of the exposure scenario and characteristics of fibers in the exposure. The exposure scenario is characterized by the level of exposure concentrations, along with duration and exposure onset age. A typical hypothesis is that an exposure concentration is constant for the duration of exposure or follows an assumed distribution. In many cases, average exposure concentration is used for risk assessment purposes instead of the concentration changing with time. In some instances, the cumulative exposure metric is used, with cumulative exposure equal to the product of average exposure concentration times exposure duration either based on a 40-hour week ("occupational cumulative exposure") or a continuous average exposure occurring 24 hours per day for a set time ("environmental cumulative exposure").

Two methods are considered the best developed in asbestos risk assessment. The first can be defined as the Peto method (as described in Chapter 7), and the second is the Hodgson and Darnton method (as described in Chapter 8).

The dose–response relationships for the methods can be summarized as follows:

Method I. The Peto Method

The mesothelioma excess incidence rate t years after the onset of exposure is calculated as

$$I_M(t) = 3 \times K_M \times \int_0^{t-10} E(u)(t-u-10)^2 \, du \tag{12.1}$$

Health Risk Assessment for Asbestos and Other Fibrous Minerals, First Edition.
Edited by Andrey Korchevskiy, James Rasmuson, and Eric Rasmuson.
© 2024 John Wiley & Sons, Inc. Published 2024 by John Wiley & Sons, Inc.

where K_M is the potency factor and $E(u)$ is the exposure level at the time moment u (f/cc).

The lung cancer excess incidence rate t years after the onset of exposure is calculated as

$$I_L(t) = I_E(t) \times \left[1 + K_L \times CE_{10}\right] \tag{12.2}$$

where K_L is the potency factor, CE_{10} is the cumulative exposure with lag 10 years at the moment t, and $I_E(t)$ is the baseline cancer incidence.

The total lifetime excess risk is calculated as

$$\text{Risk} = \sum_{t=10}^{S} \left(I_m(t) + I_L(t)\right) l_{A+t} \tag{12.3}$$

where S is the time between the onset of exposure and the maximum age included in the calculation and l_{A+t} is the number of the population survivors to the age of $A + t$.

Method II. The Hodgson and Darnton Method

Because the Hodgson and Darnton linear method is most precise with the assumption of 30 years onset age and 5 years duration (see Chapter 8), the mesothelioma and lung cancer risk for specific levels of exposure were initially calculated assuming this artificial exposure scenario with conversion from the raw mortality data to the indicated artificial exposure/risk scenario.

Linear Model

The mesothelioma lifetime excess cancer risk for the artificial exposure scenario (30 years onset age, 5 years duration) is calculated as

$$\text{Risk}_M^* = R_M \times CE \times 100 \times N \times 0.7 \tag{12.4}$$

and the lung cancer lifetime excess risk as

$$\text{Risk}_L^* = \left(R_L \times CE \times 100 + 1\right) \times B_L \tag{12.5}$$

where R_M and R_L are potency factors for mesothelioma and lung cancer, respectively, for the artificial exposure scenario in which

B_L is the baseline lung cancer risk,
CE is the cumulative exposure (f/cc-years),
and N is the initial size of the population, for example, 1,000,000, which would result in lifetime cases per million. In the formulas above, the asterisks indicate the artificial exposure scenario. The coefficient 0.7 was used for mesothelioma risk as recommended by Hodgson and Darnton (2000).

Nonlinear Model

The mesothelioma lifetime excess risk for the artificial exposure scenario (30 years onset age, 5 years duration) is calculated as

$$\text{Risk}_M^* = \left(A_{pl} \times \text{CE}^r + A_{per} \times \text{CE}^t\right) \times 0.7 \times 100 \times N \tag{12.6}$$

and the lung cancer lifetime excess risk as

$$\text{Risk}_L^* = \left(A_L \times \text{CE}^s \times 100 + 1\right) \times B_L \tag{12.7}$$

where A_{pl} is the nonlinear potency factor for pleural mesothelioma,
A_{per} is the nonlinear potency factor for peritoneal mesothelioma,
A_L is the nonlinear potency factor for lung cancer,
r is the power value for pleural mesothelioma,
t is the power value for peritoneal mesothelioma,
s is the power value for lung cancer,
B_L is the background lung cancer risk,
CE is the cumulative exposure (f/cc-years), and
N is the initial size of the population.

The values of Risk_M and Risk_L are corrected for the true exposure duration and onset age by coefficients derived from the Peto model, with the use of Berman and Crump coefficients (Berman and Crump 2008).

If $\text{Risk}(P)_M(E, D, A)$ denotes the mesothelioma risk calculated by the Peto method for exposure level of E (f/cc), duration D (years), and onset age A, the Hodgson and Darnton mesothelioma risk Risk_M is calculated as

$$\text{Risk}_M = \text{Risk}_M^* \times \frac{\text{Risk}(P)_M(E, D, A)}{\text{Risk}(P)_M(E, 5, 30)} \tag{12.8}$$

Similarly, if $\text{Risk}(P)_L(E, D, A)$ denotes the lung cancer risk calculated by the Peto method for exposure level of E (f/cc), duration D (years), and onset age A, the Hodgson and Darnton lung cancer risk Risk_L is calculated as

$$\text{Risk}_L = \text{Risk}_L^* \times \frac{\text{Risk}(P)_L(E, D, A)}{\text{Risk}(P)_L(E, 5, 30)} \tag{12.9}$$

The coefficients used for the duration and onset age correction in our methodology are close to the similar coefficients estimated by the United States Environmental Protection Agency (US EPA) as in Chapter 7. We can construct a combined mesothelioma and lung cancer correction coefficient as

$$\left(\text{Risk}_L + \text{Risk}_M\right) / \left(\text{Risk}_L^* + \text{Risk}_M^*\right)$$

and compare it with the US EPA exponential coefficient as quoted in Chapter 7.

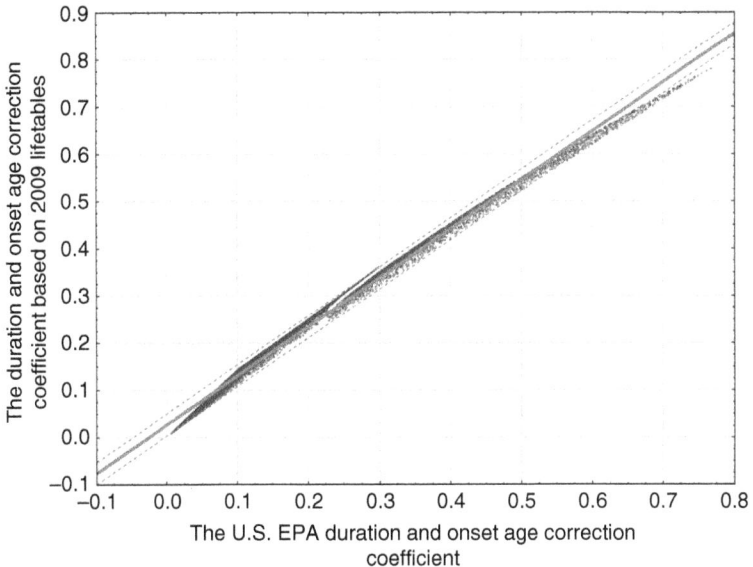

Figure 12.1 The comparison between duration and onset age coefficients. Dotted lines – prediction interval.

To illustrate this fact, Figure 12.1 demonstrates a comparison between the correction coefficients from the US EPA and our study (based on 2009 lifetables) for a range of duration (1–30 years) and onset time values (0–40 years); the average between lung cancer and mesothelioma coefficients was used. The correlation between two estimations of the coefficients is determined at the level of $R = 0.997$, $R^2 = 0.995$, $p < 0.000001$.

It should be noted that we applied the Peto-based correction coefficient as described above to compensate for both linear and nonlinear scenarios in the risk assessment by the Hodgson and Darnton method. It is not fully clear if additional correction would be needed for the nonlinear scenario. However, the applied formulas are fully intuitive, and it is quite reasonable to apply them for practical purposes.

The coefficients R_M, R_L, K_M, and K_L for risk assessment purposes should be used for various mineral types of fibers based on recently published values.

Table 12.1 contains R_M, R_L, K_M, and K_L coefficients for several mineral types of asbestos fibers.

Table 12.2 contains published estimations of A_{pl}, A_{per}, and A_L, as well as r, t, and s for the nonlinear Hodgson and Darnton method, including the data from the original Hodgson and Darnton (2000) paper, and the updates as published by Darnton (2023).

Table 12.1 Published R_M, R_L, K_M, and K_L coefficients for several mineral types of asbestos fibers (for PCM Exposure Equivalent).

Fiber type	Mesothelioma		Lung cancer	
	R_M (%)	K_M (10^8)	R_L (%)	K_L (%)
Crocidolite	0.52^a	8.5^b	4.8^a	1.4^b
Amosite	0.11^a	8.5^b	4.0^a	1.4^b
Libby amphiboles	0.03^a	0.5^c	0.82^a	0.24^c
Chrysotile	0.0014^a	0.009^b	0.078^a	0.2^b

[a] Darnton (2023).
[b] Berman and Crump (2008).
[c] Moolgavkar et al. (2010).

Table 12.2 Published estimations of A_{pl}, A_{per}, and A_L, as well as r, t, and s (mesothelioma and lung cancer power values) for the nonlinear Hodgson, Darnton method.

(a) Original (*Source:* Adapted from Hodgson and Darnton (2000))

Mineral type	A_{pl} (r = 0.75)	A_{per} (t = 2.1)	A_L (s = 1.3)
Crocidolite	0.94	0.0022	1.6
Amosite	0.13	0.0006	1.6
Chrysotile	0.0047	—	0.028

(b) Updated (*Source:* Adapted from Darnton (2023))

Fiber type	A_{pl} (r = 0.48)[a]	A_{pl} (r = 0.78)[a]	A_{per} (t = 2.1)	A_L (s = 1.54 for amphiboles, s = 1.3 for chrysotile)
Crocidolite	2.00	0.89	0.0026	0.84
Amosite	0.45	0.16	0.00059	0.47
Libby amphiboles	0.26	0.079	0.000017	0.07
Chrysotile	0.0028	0.0056 (excluding Quebec)	—	0.64 (textile and Chinese mine)
		0.0044 (excluding North Carolina)		0.009 (asbestos mining and general industry)

[a] Both fits for pleural mesothelioma were estimated by Darnton (2023) as not fully satisfactory. See Darnton (2023) and Chapter 8 for additional explanation.

Please refer to the discussion on the updated coefficients for the nonlinear model in Chapter 8.

Relationship Between Different Estimation of Mesothelioma and Lung Cancer Potency Factors

The estimations of potency for various cohorts and mineral types of fibers based on the Hodgson and Darnton method (R_M and R_L) and based on the Berman and Crump coefficients for the Peto model (K_M and K_L) are related to each other.

In particular, Wylie and Korchevskiy (2023) determined the following relationships:

$$K_L(\%) = 0.241\, R_L(\%) \left(R = 0.986,\, R^2 = 0.943,\, P < 0.000001 \right) \quad (12.10)$$

and

$$K_M(\times 108) = 23.59\, R_M(\%) \left(R = 0.997,\, R^2 = 0.994,\, P < 0.000001 \right) \quad (12.11)$$

Figure 12.2 demonstrates the behavior of dose–response models for mesothelioma and lung cancer, as calculated by the Berman and Crump, and the Hodgson and Darnton methods, for amosite exposure, assuming 10 years of exposure starting at the age of 20. For mesothelioma, the coefficients from Hodgson and Darnton (2000) are used. (For amosite, there is only a 10% increase in the updated values in Darnton 2023.)

It is expected that extremely differing biopersistence as well as differing chemical composition and mineral habit between asbestos mineral types would have a significant effect on the potency factors for chrysotile and amphibole elongate mineral particles (EMPs). Chrysotile, in particular, is known to be quickly dissolved or broken by macrophages, especially when such fibers are shorter in length. Amphiboles, on the contrary, are iron rich, biopersistent, and able to remain in the lungs for a long time (see Section General Considerations and Model of Asbestos Dose-Response Assessment). It is important to note that different estimations of potency factors, published by various research teams, are consistent in the determination of the ratio between amphibole and chrysotile potency factors. This is especially pronounced in mesothelioma risk. This consistency is illustrated in Figure 12.3 for three independent sources of calculated mesothelioma potency. We compared the mesothelioma potency factor ratios between chrysotile and amphiboles for updated values published by Darnton (2023), Berman and Crump (2008), and Garabrant and Pastula (2018).

Figure 12.3 compares the ratios between PCMe mesothelioma potency factors only and should not be confused with the actual potency factors. In particular,

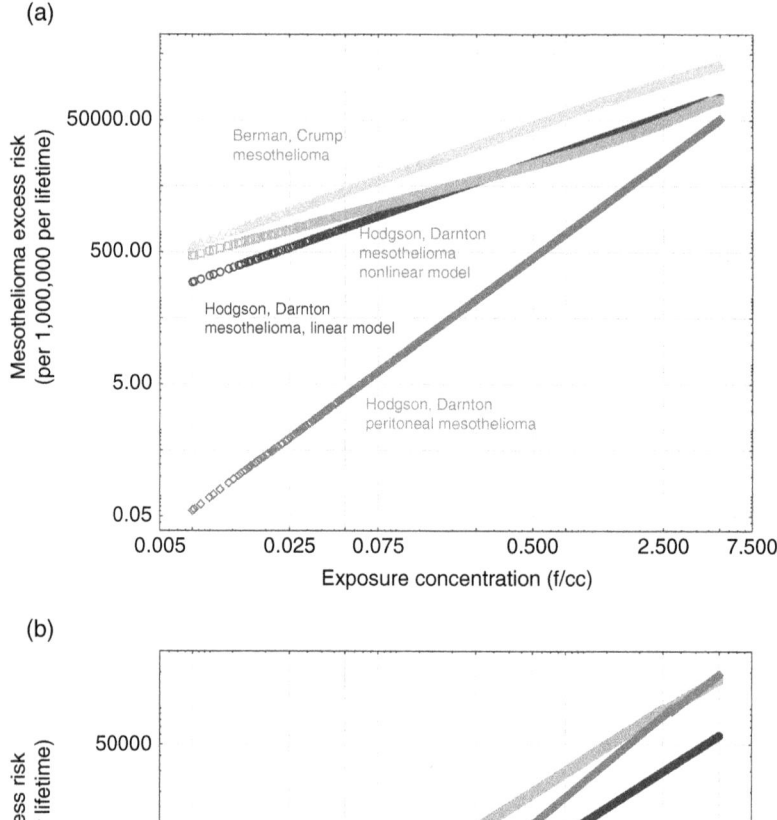

Figure 12.2 The behavior of dose–response models for mesothelioma (a) and lung cancer (b), assuming occupational exposure for 10 years to amosite, starting at the age of 20 (both axes log10-transformed).

Relationship Between Different Estimation of Mesothelioma and Lung Cancer Potency Factors | 373

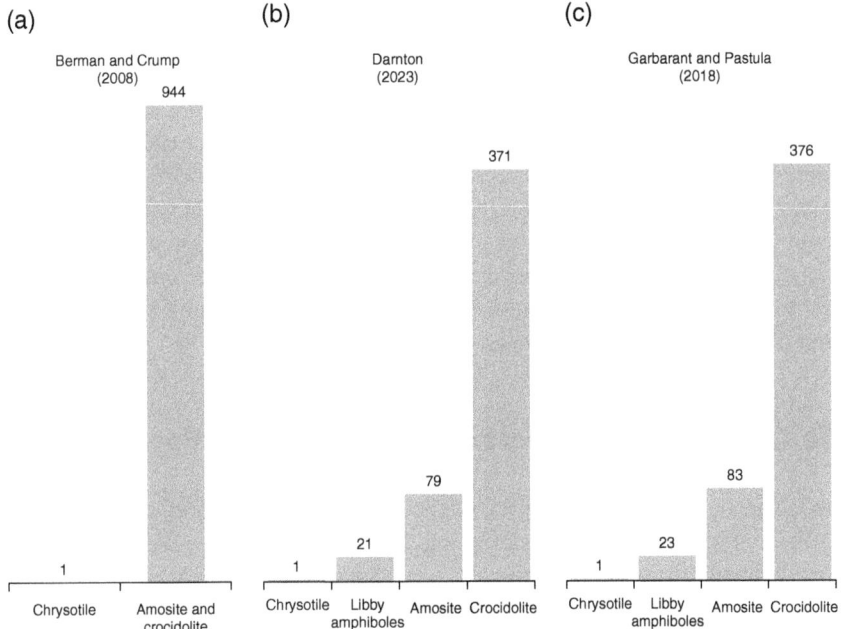

Figure 12.3 Comparison of the ratios of mesothelioma potency factors between amphibole asbestos types and chrysotile by three independent research groups. (a) Berman and Crump (2008). (b) Darnton (2023), and (c) Garabrant and Pastula (2018). The scale is different in plot (a) than for plots (b) and (c) for purposes of visualization.

the analysis of Berman and Crump (2008) combined crocidolite and amosite potency into one category, which increases the amosite potency factor compared with the Darnton (2023) updated value. The increased potency ratio between chrysotile and amphibole asbestos with amosite and crocidolite taken together in the Berman and Crump (2008) study can also be explained by Berman and Crump attempting to take into account the fraction of amphiboles in each of the epidemiological cohorts (see Chapter 7). This also allowed for consideration of chrysotile and amphibole fractions in mixed fiber cohorts in the mesothelioma potency estimates. Therefore, the mesothelioma chrysotile potency factor determined by Berman and Crump should be considered for relatively pure chrysotile while the mesothelioma potency factor for chrysotile based upon the Darnton and the Garabrant and Pastula work included the results from cohorts where chrysotile could be contaminated with other types of fibers (like tremolite and balangeroite). Our (Korchevskiy et al. 2019) empirical modeling of mesothelioma potency factors based on chemical composition and dimensions of mineral fibers also shows consistency with the ratio between chrysotile and amphiboles as reported by Darnton and by Garabrant and Pastula (see also Chapter 11).

Finally, additional comparisons between chrysotile and amphibole asbestos mesothelioma potency factors have been published. ECHA in 2021 determined a K_M value for a combination of amphibole asbestos of 7.95×10^{-8} based on only two studies and a K_M value for chrysotile of 0.017×10^{-8} based upon five studies, selecting only the highest quality cohort studies. The ECHA K_M values were only slightly updated from the DECOS (2010) published values when additional studies and updates of prior studies were considered. With the ECHA mesothelioma potency factors for mesothelioma, the ratio of amphibole potency to chrysotile potency is 468, between the Berman and Crump (2008) ratio for amosite and crocidolite combined and the Darnton (2023), and the Garabrant and Pastula (2018) ratios for amosite and crocidolite as shown in Figure 12.3.

Life Tables and Life Expectancy of the Exposed Population

Life tables are used to calculate the risk levels based on both Method I (the Peto model) and Method II (the Hodgson and Darnton model) (see section General Considerations and Model of Asbestos Dose-Response Assessment). However, the two methods treat the life table approach differently. In particular, application of the Peto model requires the direct use of life tables for mesothelioma risk, while Hodgson and Darnton potency factors for mesothelioma are in units of percent of disease per cumulative exposure (f/cc-years), with a factor of 0.7 applied to approximate the effect of the population "dying out." When age and duration are not far from age 30 and duration of 5 years, approximate mesothelioma risk can easily be estimated with the Darnton (2023) potency factors based on cumulative exposure and application of the correction factor (as we have calculated, this factor is usually less than or equal to 0.7). The correction factor was not applied by Hodgson and Darnton to lung cancer absolute or relative risk calculations. It should be noted, however, that changes in life tables (for example, an increase in general survivability of the population) are expected to increase the projected number of lung cancer cases for the same level of exposure.

Table 12.3 contains a comparison of estimated risk levels, depending on the choice of lifetables, for the occupational exposure to 0.5 f/cc of chrysotile asbestos, for 10 years, starting at the age of 20. The Peto method with Berman and Crump coefficients is used. The maximum age group is 85 and higher from the lifetables.

As we can see, the risk level for mesothelioma increased by about 22.2% and for lung cancer by 74% after 32 years. (This reflects the increase in both life expectancy for about 5 years and crude lung cancer mortality rate in the population by about 64%.) The lung cancer risk level is changed more dramatically because the lung cancer risk depends on the lung cancer baseline rate, and mesothelioma risk is just on the general mortality in the population.

Table 12.3 A comparison of estimated risk levels for amosite, depending on the choice of lifetables, for the exposure to 0.5 f/cc of chrysotile asbestos, for 10 years, starting at the age of 20 (cases per 1,000,000 per lifetime).

	Lung cancer	Mesothelioma
Lifetables 1977	485	36
Lifetables 2009	844	44

Also, the increase in mesothelioma and lung cancer rates in the exposed populations is expected to affect their surviving pattern and life expectancy (as happens with any cancers in general and with any carcinogenic agents). However, Korchevskiy and Korchevskiy (2023) demonstrated that as a rule respiratory cancers are much more frequent in the elder age groups of the population, and therefore the impact of respiratory carcinogens on the life expectancy at birth is limited. For example, it can be demonstrated that according to the Peto model with Berman and Crump coefficients, the above scenario (exposure to 0.5 f/cc of chrysotile asbestos, for 10 years, starting at the age of 20) would be associated with the overall loss of life expectancy at the level not exceeding 4 days, well below statistical significance.

Linearity and Nonlinearity of the Dose–Response Curves

As was demonstrated in Section 3.2, nonlinear and linear models can be fitted with different levels of certainty to the asbestos epidemiological dose–response data.

During the last decades, for several carcinogens, nonlinear dose–response curves were considered along with linear models. For example, Cox (2019) demonstrated that an inflammation-mediated two-stage clonal expansion model of respirable crystalline-silica-induced lung cancer may imply a sublinear shape of a dose–response curve. A sublinear relationship between the internal dose of hexavalent chromium and excess cancer risk was found (Haney Jr. 2015). At the same time, the US EPA (US EPA 2016) used a supralinear dose–response curve (as a two-piece spline model) for the assessment of a relationship between cumulative inhalation exposure to ethylene oxide and the risk of lymphoid and breast cancers though several subsequent sources criticized the approach of the US EPA in the interpretation of the data and argued for the linear dose–response relationship to be applied instead (Vincent et al. 2019; Hartwig et al. 2020).

Different explanations were proposed for the observed nonlinear dose–response curve in cancer. In particular, Stayner et al. (2003) listed various factors, causing attenuation of exposure–response curves in occupational cohorts at high levels of exposure (producing supralinear, or as it is called in calculus, "downward concavity," shape of the risk functions), including a bias introduced by the healthy worker survivor effect, a depletion of the number of susceptible people in the population at high exposure levels, a natural limit on the relative risk for diseases with a high background rate, mismeasurement or misclassification of exposures, the influence of other risk factors that vary by the level of the main exposure, and the saturation of key enzyme systems or other processes involved in the development of disease. Remarkably, Crump (2005) developed a mathematical model demonstrating that a lognormally distributed error in exposure measurements would cause a reduction of the exponent in a power function representing the dose–response relationship (in this case, for example, a linear relationship, for which the dose is in the power of $K = 1$, would look like having the dose in the power of $K < 1$, concaving the shape down).

Korchevskiy and Korchevskiy (2023) assessed two different hazard rate equations for mesothelioma – the Peto model and the two-stage clonal expansion (TSCE) model and proposed analytical formulas for lifetime risk. For both models, it was shown that mesothelioma lifetime risk changes supralinearly with duration; the exponent of the power function was ranging from 0.68 to 0.89. The dose–response curve for cumulative exposure tends to be linear shape with increasing exposure intensity. For the Peto model, the mesothelioma risk is linear with cumulative exposure if the duration is constant. The model has been tested for chrysotile asbestos cohorts, with good agreement demonstrated with published mesothelioma excess mortality ($R^2 = 0.733$, $p < 0.0041$). It was concluded that for pleural mesothelioma, observed deviation from linearity in the dose-response relationship can be potentially explained by the impact of a change in the duration of exposure. In a meta-analysis, this deviation can be eliminated by standardizing the mortality data for various cohorts by duration of exposure.

In general, the deviation from linear models in asbestos risk assessment seems to be of minor relevance compared to other sources of uncertainty. However, nonlinearity in some cases is important evidence for the mode and mechanisms of action for asbestos fibers in mesothelioma and lung cancer. Further studies are needed to explore the observed patterns in more detail.

Threshold and Benchmark Dose Response in Asbestos Risk Assessment

While some sources insist that there is no known asbestos dose which can be considered safe (e.g. OSHA "Asbestos: Overview"), this idea appears contradictory to general scientific approaches in toxicology. Dose–response relationships in general, and specifically the form of the relationship between the degree of response of the biological system and the amount of toxicant administered, is considered the most fundamental concept in toxicology (Casarett and Doull's Toxicology 2008). A dose–response relationship certainly was established for asbestos and cancer (IARC 2012). This means that lower exposure to asbestos is expected to produce lower effects as compared to higher exposure. It should be noted that a very low cancer risk is not measurable (Nguyen and Wu 2011). One reason for this is that the number of people in exposure groups is always limited (for example, risk increase in 1 case per 1,000 cannot be measured in a group of 100 people). Also, spontaneous cancers at some level make it impossible to distinguish between random malignancies and cancers produced by external agents. For asbestos, there is an established range of background concentrations that every average citizen of the United States is exposed to during a lifetime (ATSDR 2001). For both mesothelioma and lung cancer – the major malignancies that can be caused by asbestos exposure – the level of spontaneous or baseline exposure in unexposed populations was observed (Price and Ware 2009; Samet et al. 2009). The exposure corresponding to the cancer risk that is essentially unmeasurable (in particular, not distinguishable from baseline risk) should constitute a threshold, or generally safe level. It is also unacceptable to suggest that every, even miniscule, exposure to asbestos is contributory to the overall cancer risk. The exposure below the level of measurability is just a part of the background exposure. The clearance mechanisms of the human body efficiently remove significant fractions of fibers in comparatively short time periods, preventing the past exposures from providing input for the overall "exposure memory," especially for chrysotile (Korchevskiy and Wylie 2022a). Specifically, Pierce et al. (2016) identified the "no-observed adverse effect level" (NOAEL) for chrysotile asbestos at the "best estimate" threshold of 89–168 f/cc-years for lung cancer and 208–415 f/cc-years for mesothelioma for short-fiber chrysotile.

Benchmark dose modeling is one of the tools which demonstrated efficiency in establishing the NOAEL for various toxicants. Korchevskiy (2020) used a benchmark dose modeling technique with standard US EPA software to establish a left confidence limit for benchmark dose (BMDL) at the point of departure of 3% for amosite at 32.8 f/cc-years. It means that based on the US EPA benchmark dose methodology, the increase of the amosite exposure by 32.8 f/cc-years increases mesothelioma risk by 3%. It corresponds to potency factor R_M for amosite at $3\%/32.8\,\text{f/cc-years} = 0.09\%$ (close to the estimation given in Chapter 8 of this book).

If a similar approach is applied to chrysotile (based on data from Darnton 2023), a BMDL at a point of departure of 0.5% per 383 f/cc-years, corresponding to the mesothelioma potency factor of 0.0013% (also very close to the estimation in Chapter 8), would be found. The BMDL calculations in this case are made based on the epidemiological data described in Chapter 8.

Community and Occupational Risk Assessment

The US EPA recommended a coefficient of 3.042 to account for the exposure and risk differences of occupational (8 hours/5 days per week) vs. community (24 hours/7 days per week) populations (US EPA 2020). This factor not only includes exposure time differences, but differences in breathing rates while working and not working and while sleeping. To apply the coefficients R_M, R_L, K_M, and K_L for community risk assessment, the exposure values should be multiplied by this coefficient.

Peritoneal Mesothelioma

Peritoneal mesothelioma is a comparatively rare type of malignant mesothelioma comprising no more than 15%–20% of new cases. Among the hallmarks of peritoneal mesothelioma such factors are listed as relatively equal distribution among men and women (vs. pleural mesothelioma predominancy in men), the variability of tumor aggressiveness and disease progression, the tendency of the disease to stay within the peritoneal cavity, and other (Greenbaum and Alexander 2020).

In French males, the incidence of malignant peritoneal mesothelioma has been estimated at 0.07–0.10 per 100,000 person years although a peak of 0.16, the same units, occurred during the 2001–2003 period. In women, the incidence was 0.04–0.11, with the same units. The authors stated that they found "…almost equal incidence in men and women, with a ratio of men to women of 1.3 (195/154). The median age of onset was 66 years; there was no significant difference between men (67 years) and women (65 years)" (Le Stang et al. 2019).

In a Dutch study (Van Kooten et al. 2022), with 11,539 cases of malignant pleural mesothelioma and 629 cases of malignant peritoneal mesothelioma, the overall incidence of malignant peritoneal mesothelioma was 0.15–0.25 (CI −0.4, 0.4) with the male incidence at 0.2–0.5 and the female incidence at 0.0–0.2. Further, it was concluded that malignant pleural mesothelioma represented over 90% of mesothelioma cases.

As a rule, peritoneal mesothelioma is not observed in the pure chrysotile exposed cohorts. As described in Chapter 8, only two cases of peritoneal mesothelioma were reported for the cohorts with dominant chrysotile exposure: one of the three mesotheliomas in men working in the South Carolina textile factory and one of the two mesotheliomas in the Chongqing factory, where asbestos textile was manufactured (along with other products).

It should be mentioned that Jiang et al. (2018) reported elevated rates of peritoneal mesothelioma in a case–control study of hand-spinning asbestos textile workers in Southeastern China. The workers were apparently exposed to chrysotile with the level of tremolite contamination less than 0.3%. The average proportion of peritoneal mesothelioma among all mesothelioma cases in the county where the asbestos hand-spinning industry was widely spread is reported as 50%, highly unusual for this type of mesothelioma. Interestingly, Guo et al. (2017) demonstrated the ratio of peritoneal to pleural mesothelioma observed at the level of 3 : 1 in the province of Zhejiang in China. The asbestos textile industry was also prevalent in this region. Apparently, some unusual factors are in play in these parts of China, including potentially higher than normal cumulative chrysotile and tremolite exposures and perhaps different genetic predisposition (Hiltbruner et al. 2022) that makes the risk pattern so different from the Western world.

Hodgson and Darnton (2000) suggested that peritoneal mesothelioma rates appear to demonstrate a sublinear dose–response relationship by cumulative exposure. Indeed, as an exercise rather than for application, based on the available cohort data (see Chapter 8), the following nonlinear regression can be constructed for crocidolite utilizing numbers of peritoneal cases from applicable cohorts compared with total expected deaths from all causes to estimate numbers of individuals. It is a simpler approach than that used by Hodgson and Darnton in 2000, but it also demonstrates the exponential nature of the dose–response relationship for peritoneal mesothelioma:

$$\text{Peritoneal mesothelioma rate} = 0.000034 \times CE^2,$$
$$\left(R = 0.9999, R^2 = 0.99997, p < 0.0001\right)$$

where peritoneal mesothelioma rate is the number of peritoneal mesothelioma cases per age-adjusted total expected mortality and CE is the average cumulative exposure (f/cc-years). The age-adjusting of mortality in this calculation is the same as for the Hodgson and Darnton method described in Chapter 8.

For amosite (with the data from Chapter 8):

$$\text{Peritoneal mesothelioma rate} = 0.0000019 \times CE^{2.36},$$
$$\left(R = 0.996, R^2 = 0.992, p < 0.055\right)$$
$$\left(\text{or peritoneal mesothelioma rate} = 0.000009 \times CE^2,\right.$$
$$\left.\left(R = 0.991, R^2 = 0.982, p < 0.009\right)\right)$$

While there is only one available data point for Libby amphiboles exposure (also see Chapter 8), still with an assumption that the power by cumulative exposure is a value around 2, we can write

$$\text{Peritoneal mesothelioma rate} = 0.000000238 \times CE^2$$

The proportion between peritoneal mesothelioma rate coefficients for crocidolite, amosite, and Libby amphiboles, assuming the power of 2 in all regression

Table 12.4 Linear potency factors for peritoneal mesothelioma.

Mineral type/location	Peritoneal mesothelioma potency (R_{PerM}) (%)
Crocidolite, Australia	0.077
Crocidolite, South Africa	0.059
Cummingtonite-grunerite, asbestiform (amosite)	0.015
Fluoro-edenite, Italy, Biancavilla	0.00923
Libby amphiboles	0.002

equations, is 142 : 37 : 1. No peritoneal mesothelioma rates can be calculated for chrysotile, because of the rarity of peritoneal mesothelioma cases in predominantly chrysotile cohorts. Assuming the nonlinear model, crocidolite appears to be about 140 times more potent than Libby amphiboles (most probably, because of the significant fraction of nonasbestiform particles in the Libby variety of amphiboles).

Linear potency factors for peritoneal mesothelioma can also be calculated, as listed in Table 12.4. In this case,

$$R_{PerM} = 100 \frac{O_{PerM}}{\left(E_{Adj} \times X\right)} \quad (12.12)$$

where O_{PerM} is the number of peritoneal mesothelioma deaths, E_{Adj} the total expected deaths from all causes adjusted to the age of first exposure of 30, and X is the mean cumulative exposure in PCME fibers/cubic centimeter-years (f/cc-years).

In this instance, the proportion of linear potency factors for crocidolite, amosite, and Libby amphiboles is 34 : 7.5: 1.

The listed potency factors can be used for risk calculations similar to the approach described in section General Considerations and Model of Asbestos Dose-Response Assessment – for example, the coefficient 0.7 should be applied to account for the reduction of the population at risk with age.

Other Types of Cancer

Different types of cancers other than lung cancer and mesothelioma were associated with asbestos exposure in epidemiological studies. For example, IARC (2012) determined that asbestos inhalation exposure causes cancers of the larynx and ovary, with cancer of the pharynx, stomach, and colorectum expressing positive association with asbestos exposure, though for cancer of colorectum, this association is not strong enough to warrant classification as sufficient.

No quantitative meta-analysis has yet been performed to establish potency factors for these types of cancer. However, some estimations for the potency can be provided based on available studies.

For example, Musk et al. (2008) reported 13 cases of laryngeal cancer death in workers in the Wittenoom crocidolite asbestos cohort vs. 5 expected cases, with 222 mesotheliomas reported for this population. This means that the larynx cancer potency factor for crocidolite can potentially be projected as 0.019% per f/cc-years, about 4% of the mesothelioma risk in the cohort, and about 1% of excess combined lung cancer and mesothelioma risk for crocidolite.

However, no excess of laryngeal cancer for chrysotile was found by McDonald et al. (1980) in the Quebec asbestos miner cohort. Similarly, Hein et al. (2007) did not find elevated laryngeal cancer mortality in asbestos textile workers. However, Wang et al. (2012) reported 1.53 excess cases of laryngeal cancer deaths in asbestos textile workers compared with 2 cases of mesothelioma in the same cohort (corresponding to 17,508 person years, and 259 deaths from all causes). The proportion of laryngeal cancers in Chinese textile workers to mesothelioma seems too high and most probably does not reflect a statistically significant relationship.

The Finnish Institute of Occupational Health (2014) in its document "Asbestos, Asbestosis, and Cancer: Helsinki Criteria for Diagnosis and Attribution" suggested that a regression equation can be used to calculate the risk of laryngeal cancer based on a standardized incidence ratio (SIR) of lung cancer in the cohorts (where SIR is a ratio between the incidence of cancer in the cohort and in the reference population that usually is standardized by age and gender). The formula:

$$y = 0.5416x + 0.4925 \left(R^2 = 0.9468 \right) \quad (12.13)$$

was proposed, where x – SIR for lung cancer and y – SIR for laryngeal cancer. Remarkably, no regression with a standardized mortality ratio (SMR) was determined (SMR is the ratio between cancer mortality in the study cohort and reference population; usually this value is standardized by age and gender).

Similar logic can be applied to studying other types of cancer in relationship to asbestos. For example, the Helsinki Criteria document also proposed the following equation for the risk of ovarian cancer:

$$y = 0.6441x + 0.8959 \left(R^2 = 0.3504 \right) \quad (12.14)$$

where

x – SMR for lung cancer,
y – SMR for ovarian cancer.

It is important to consider that the correlation in this case is very low, and only about 35% of the variability of ovarian cancer rates among the cohorts can be explained by the variability of lung cancer rates.

Similarly, the formula for colorectal cancer risk was proposed by the Helsinki Criteria document as follows:

$$y = 0.2785x + 0.5419 \left(R^2 = 0.5895\right) \quad (12.15)$$

where

x – SMR for lung cancer,
y – SMR for colorectal cancer.

For stomach cancer, the following equation was found:

$$y = 0.4011x + 0.4125 \left(R^2 = 0.5895\right) \quad (12.16)$$

where

x – SMR for lung cancer,
y – SMR for stomach cancer.

The dose–response relationship that might be derived from the Helsinki Criteria can be illustrated for stomach cancer.

According to the Hodgson and Darnton model, the relative risk of lung cancer mortality for chrysotile can be estimated as

$$RR = 1 + 0.00078 CE \quad (12.17)$$

where RR – relative risk of lung cancer and CE – cumulative exposure to chrysotile asbestos (f/cc-years).

Assuming that SMR of lung cancer can be approximated by relative risk, and following the Helsinki Criteria document, we can re-write the equation for the stomach cancer risk as

$$y = 0.8136 + 0.000313 CE \quad (12.18)$$

(with notation as previously).

From here, the stomach cancer SMR will not exceed 1 (the baseline risk level) until cumulative exposure to chrysotile exceeds 595 f/cc-years. Similarly, the colorectal cancer will not be elevated above the baseline up to the exposure level of 829 f/cc-years.

It is not expected that the total risk of all types of cancer (including respiratory and other localizations) will be significantly different from a small percentage of the level of lung cancer and mesothelioma risk taken together. It also seems that chrysotile asbestos because of the high dissolution rate and low rigidity of fibers would not have sufficient potential to cause nonrespiratory cancers in humans.

Inhalation Unit Risk (IUR) for Asbestos Fibers

The IUR is one of the useful risk assessment metrics that was introduced into risk methodology and practice within the United States Environmental Protection Agency (US EPA) Integrated Risk Information System (IRIS).

IUR can be defined from the linear equation:

$$\text{Risk} = \text{IUR} \times f \times N \qquad (12.19)$$

where Risk is the predicted number of excess cancer cases per population, N,

f is the continuous lifetime mean exposure to the distinct EMP type.

Korchevskiy et al. (2020) published IURs for several mineral types of asbestos as related to lung cancer and mesothelioma risk. Science-based IURs are listed in Table 12.5. Also in 2020, the US EPA published its risk evaluation for chrysotile asbestos and proposed estimations for chrysotile IUR (US EPA 2020). The EPA also published the estimations of IURs for Libby amphiboles (US EPA 2014). The EPA values are listed in Table 12.5 along with other estimations.

The US EPA IUR for lung carcinoma is generally consistent with the values derived from the meta-analytical study. However, for mesothelioma, the EPA IUR exceeds the average estimation from the Peto model with the Berman and Crump coefficients, and the Hodgson and Darnton methods by a factor of about 50.

Table 12.6 contains the comparison of mesothelioma cases for major chrysotile cohorts (listed in Chapter 8) that would be projected by average science-based IURs vs. the values derived from US EPA estimations vs. observed cases of disease. The duration and onset age correction coefficients were used as in Chapter 7. Also, the duration of exposure for each of the cohorts was quoted as in Korchevskiy and Korchevskiy (2022). The exposure levels were recalculated from occupational to environmental equivalents to apply the IURs (as in this chapter, section Community and Occupational Risk Assessment).

For all combined chrysotile cohorts, the US EPA mesothelioma IUR would project about 1,650 cases of mesothelioma vs. just 55 observed. For both lung cancer and mesothelioma, the US EPA potency factor would project 1,885 cases vs. 557 observed.

The science-based IURs provide better estimation for the available cohort data.

Also, the analysis of IURs provides important insights into the difference between nontextile and textile chrysotile cohorts. The ratio between the mesothelioma risk calculated from the science-based IUR for chrysotile and actually

Table 12.5 Inhalation Unit Risk (IUR) values for different methods of calculation and various cancer endpoints.

Asbestos fiber type	Total cancers			Mesothelioma			Lung carcinoma		
	Hodgson and Darnton linear method	Berman and Crump method (based on the Peto model)	US EPA (upper bound)	Hodgson and Darnton linear method	Berman and Crump method (based on the Peto model)	US EPA	Hodgson and Darnton linear method	Berman and Crump method (based on the Peto model)	US EPA (upper bound)
Crocidolite	1.50	1.83	N/A	1.12	1.59	N/A	0.38	0.23	N/A
Amosite	0.60	1.83	N/A	0.22	1.59	N/A	0.38	0.23	N/A
Libby amphibole	0.13	0.13	0.169	0.065	0.09	0.122	0.068	0.040	0.047
Chrysotile	0.016	0.035	0.16	0.004	0.002	0.14	0.012	0.033	0.031

IUR per PCM f/cc (assuming environmental, 24/7 exposure)

Table 12.6 Observed and projected mesothelioma cases in the published chrysotile cohorts.

Cohort	Observed mesothelioma cases	Mesothelioma cases projected by average IUR from Korchevskiy et al. (2020)	Mesothelioma cases projected by US EPA IUR
Quebec	33	32	1,242
Balangero	7	4	174
Qinghai	0	0.25	10
New Orleans, plant 2	0	0.15	6
South Carolina textile women	0	0.33	13
South Carolina textile men	3	0.29	12
North Carolina (textile)	8	3	116
Chongqing (textile)	2	0.19	8
Connecticut	2	2	69

observed cases is 0.91 for nontextile cohorts and 0.29 for textile cohorts. Chrysotile in textile cohorts apparently behaves as a different mineral (Darnton 2023), and it may require considering a separate IUR for asbestos textile workers in contemporaneous or retrospective exposure assessment.

Asbestos Dose–Response and Tobacco Smoking

Both the Hodgson and Darnton and the Peto methods of risk assessment for the lung cancer endpoint assume that the relative risk for exposed populations is proportional to the baseline lung cancer risk. While the coefficient for the models was derived from apparently heavy smoking populations, the models are based on a hypothesis of the linear dependence of excess risk on the changing baseline value. This assumption means that tobacco smoking multiplicatively increases the risk of asbestos exposure. In the exposed population, the cases of lung cancers can be seen as a combination of four groups: baseline cases, cases caused by tobacco smoking, cases caused by asbestos exposure, and cases caused by the interaction between asbestos and tobacco smoking.

Hammond (1979) demonstrated a classic example of lung cancer risk in asbestos insulators, where the multiplicative effect was especially pronounced (Table 12.7).

Table 12.7 Lung cancer mortality ratios for asbestos insulation workers comparatively to CPS-I smoking study.

	Nonsmokers	Smokers
No asbestos exposure	1	11.3
Asbestos exposure	5.17	58.4

Source: Adapted from Hammond et al. (1979).

Table 12.8 Lung cancer mortality ratios for asbestos insulation workers compared with the CPS-I smoking study.

	Nonsmokers	Smokers
No asbestos exposure	1	10.3
Asbestos exposure	3.6	14.4

Source: Markowitz et al. (2013)/American Thoracic Society.

In the case of this study on insulators, the relative risk of lung cancer because of asbestos exposure in smokers and nonsmokers is virtually identical (5.17 : 1–58.4 : 11.3). However, the data observed in various sources seems to demonstrate deviations from a multiplicative model. Reanalyzing the data from the same insulator cohort, Markowitz et al. (2013) determined updated values for the mortality ratios in insulators (Table 12.8).

In this case, the relative risk of lung cancer in nonsmokers with asbestos exposure is 3.6, and in smokers 1.4, which represents less multiplicativity (or synergism than in the earlier Hammond et al. study).

Berry and Liddell (2004) proposed a metric to assess the interaction of asbestos with tobacco smoking (RAE_m), according to a formula:

$$RAE_m = \frac{RR_{nonsmokers} - 1}{RR_{smokers} - 1} \tag{12.20}$$

where

$RR_{nonsmokers}$ is an asbestos-related relative risk of lung cancer in nonsmokers and
$RR_{smokers}$ is an asbestos-related relative risk of lung cancer in smokers with the same asbestos exposure in both instances.

With $RAE_m > 1$, the effect of interaction between smoking and tobacco exposure is less than multiplicative. For example, if the carcinogenic effects between smoking and asbestos exposure were simply additive with no multiplicity, RAE_m would simply be 1.

In a general sense, RAE_m is not expected to be constant and/or independent of lung cancer relative risk in smokers, asbestos exposure level, and the smoking patterns of individuals.

For example, for Hammond (1979 data), $RAE_m = 1$; for Markowitz et al. (2013), $RAE_m = 2.39$.

Based on the data from the prospective Netherland cohort study on occupational asbestos exposure (Offermans et al. 2014), RAE_m would be calculated as 2.19. Based on combined data from seven cohorts, Berry and Liddell suggested that $RAE_m = 3.19$ (95% CI 1.67, 6.13).

Meta-analysis by Ngamwong et al. (2015) demonstrated, based on 10 case–control and 7 cohort studies, that $RAE_m = 1.29$ (slightly less than multiplicative).

Rasmuson, Korchevskiy, and Rasmuson (unpublished data) hypothesized in 2014 that RAE_m can depend on the mineral type of fibers. In particular, the average estimate of RAE_m for Quebec chrysotile miners was about 28.24 (Berry and Liddell 2004), while based on Liddell (2001), the average RAE_m for Wittenoom crocidolite miners would be 0.31.

For the risk calculations, it is advisable to calculate lung cancer risk for the average population baseline, assuming it to be representative for the risk of a typical exposure person. It should be noted, however, that smokers generally have higher absolute risk of asbestos-related lung cancers than nonsmokers. Further steps to reduce smoking rates in the population are likely to be especially beneficial if exposure to other carcinogens is present.

Other types of cancer should also be mentioned in assessing the relationship between risk and tobacco smoking. For example, based on the Helsinki Criteria document (2014), the laryngeal cancer expresses a multiplicative effect with both tobacco and alcohol consumption. The effects of asbestos on laryngeal cancer rates in persons consuming tobacco and/or alcohol can be significantly driven by those risk factors.

Other Factors Impacting the Dose–Response Relationship for Elongate Mineral Particles

It was demonstrated that dimensions of elongate mineral particles are a strong predictor of their potency for mesothelioma and lung cancer (Berman and Crump 2008; Wylie et al. 2020; Korchevskiy and Wylie 2021, 2022, 2023). The category of longer, thinner particles is the most potent. Apparently, fibers longer than 5 µm and thinner than 0.25 µm are mostly responsible for mesothelioma development (Baron/NIOSH 2016). For mesothelioma potency, the specific surface area of fiber (ratio of surface area to mass of fibers and inversely proportional to the diameter of fibers) is found to be a major predictor of potency, and for lung cancer, the aspect ratio is the strongest predictor (Korchevskiy and Wylie 2022). For chrysotile asbestos, the relationship with dimensional characteristics is similar to amphibole asbestos, but a reduction coefficient of about 139 for mesothelioma and 36 for lung cancer should be applied (Wylie and Korchevskiy 2023) for chrysotile vs. amphibole asbestos. The nonasbestiform category of elongate mineral particles (like cleavage fragments) has zero or much lower cancer potency than asbestiform fibers, in particular, because of the pronounced difference in size categories, tensile strength, and biopersistence (Mossman 2008; Wylie et al. 2022).

References

Agency for Toxic Substances and Disease Registry (ATSDR). (2001). *Toxicological Profile for Asbestos*. Atlanta, GA: U.S. Department of Health and Human Services, Public Health Service. https://wwwn.cdc.gov/TSP/ToxProfiles/ToxProfiles.aspx?id=30&tid=4.

Baron, P.A. (2016). Measurement of fibers. In: *NIOSH Manual of Analytical Methods (NMAM)*, 5th Edition (eds. K. Ashley and P.F. O'Connor), F1-1-F1-31 National Institute for Occupational Safety and Health.

Berman, D.W. and Crump, K.S. (2008). A meta-analysis of asbestos-related cancer risk that addresses fiber size and mineral type. *Critical Reviews in Toxicology* 38 Suppl 1: 49–73. https://doi.org/10.1080/10408440802273156.

Berry, G. and Liddell, F.D.K. (2004). The interaction of asbestos and smoking in lung cancer: a modified measure of effect. *Annals of Occupational Hygiene* 48 (5): 459–462. https://doi.org/10.1093/annhyg/meh023.

Cox, L.A.T., Jr. (2019). Risk analysis implications of dose-response thresholds for NLRP3 inflammasome-mediated diseases: respirable crystalline silica and lung cancer as an example. *Dose Response* 17(2): 1559325819836900. https://doi.org/10.1177/1559325819836900.

Crump, K.S. (2005). The effect of random error in exposure measurement upon the shape of exposure response. *Dose Response* 3(4): 456–464. https://doi.org/10.2203%2Fdose-response.003.04.002.

Darnton, L. (2023). Quantitative assessment of mesothelioma and lung cancer risk based on phase contrast microscopy (PCM) estimates of fibre exposure: an update of 2000 asbestos cohort data. *Environmental Research* 230: 114753. https://doi.org/10.1016/J.ENVRES.2022.114753.

DECOS. (2010). *Asbestos: Risks of Environmental and Occupational Exposure*. The Hague: Health Council of the Netherlands; Publication No. 2010/10E.

Finnish Institute of Occupational Health (FIOH). (2014). *Asbestos, Asbestosis, and Cancer: Helsinki Criteria for Diagnosis and Attribution*. Juvenes Print, Tampere: Helsinki.

Garabrant, D.H. and Pastula, S.T. (2018). A comparison of asbestos fiber potency and elongate mineral particle (EMP) potency for mesothelioma in humans. *Toxicology and Applied Pharmacology* 361:127–136. https://doi.org/10.1016/j.taap.2018.07.003.

Greenbaum, A. and Alexander, H.R. (2020). Peritoneal mesothelioma. *Translational Lung Cancer Research* 9(Suppl 1): S120–S132. https://doi.org/10.21037/tlcr.2019.12.15.

Guo, Z., Carbone, M., Zhang, X., et al. (2017). Improving the accuracy of mesothelioma diagnosis in China. *Journal of Thoracic Oncology* 12(4): 714–723. https://doi.org/10.1016/j.jtho.2016.12.006. Epub 2016 Dec 19. PMID: 28007630; PMCID: PMC5567857.

Hammond, E.C., Selikoff, I.J., and Seidman, H. (1979). Asbestos exposure, cigarette smoking, and death rates. *Annals of the New York Academy of Sciences* 330: 473–490. https://doi.org/10.1111/j.1749-6632.1979.tb18749.x.

Haney, J., Jr. (2015). Consideration of non-linear, non-threshold and threshold approaches for assessing the carcinogenicity of oral exposure to hexavalent chromium. *Regulatory Toxicology and Pharmacology* 73(3): 834–852. https://doi.org/10.1016/j.yrtph.2015.10.011.

Hartwig, A., Arand, M., Epe, B., et al. (2020). Mode of action-based risk assessment of genotoxic carcinogens. *Archives of Toxicology* 94(6): 1787–1877. https://doi.org/10.1007%2Fs00204-020-02733-2.

Hein, M.J., Stayner, L.T., Lehman, E., and Dement, J.M. (2007). Follow-up study of chrysotile textile workers: cohort mortality and exposure-response. *Occupational and Environmental Medicine* 64(9): 616–625. https://doi.org/10.1136%2Foem.2006.031005.

Hiltbruner, S., Fleischmann, Z., Sokol, E.S., et al. (2022). Genomic landscape of pleural and peritoneal mesothelioma. *British Journal of Cancer* 127: 1997–2005. https://doi.org/10.1038/s41416-022-01979-0.

Hodgson, J.T. and Darnton, A. (2000). The quantitative risks of mesothelioma and lung cancer in relation to asbestos exposure. *Annals of Occupational Hygiene* 44(8):565–601. PMID: 11108782.

IARC (2012). Monograph 100C: asbestos (chrysotile, amosite, crocidolite, tremolite, actinolite, anthophyllite), 219–309. https://monographs.iarc.fr/wp-content/uploads/2018/06/mono100C-11.pdf. Accessed January 24th, 2024.

Jiang, Z., Chen, T., Chen, J., et al. (2018). Hand-spinning chrysotile exposure and risk of malignant mesothelioma: a case-control study in Southeastern China. *International Journal of Cancer* 142(3):514–523. https://doi.org/10.1002/ijc.31077.

van Kooten, J.P, Belderbos, R.A., von der Thüsen, J.H. et al. (2022). Incidence, treatment and survival of malignant pleural and peritoneal mesothelioma: a population-based study. *Thorax* 77(12): 1260–1267. https://doi.org/10.1136%2Fthoraxjnl-2021-217709.

Korchevskiy, A.A. and Korchevskiy, A. (2022). Using life expectancy as a risk assessment metric: the case of respirable crystalline silica. *Computational Toxicology* 27: 100285. https://doi.org/10.1016/j.comtox.2023.100285.

Korchevskiy, A.A. and Wylie, A.G. (2021). Dimensional determinants for the carcinogenic potency of elongate amphibole particles. *Inhalation Toxicology* 33(6-8): 244–259. https://doi.org/10.1080/08958378.2021.1971340. Epub 2021 Oct 6. PMID: 34612763.

Korchevskiy, A. and Wylie, A.G. (2022a). Asbestos exposure, lung fiber burden, and mesothelioma rates: mechanistic modelling for risk assessment. *Computational Toxicology* 24: 100249. https://doi.org/10.1016/j.comtox.2022.100249.

Korchevskiy, A.A. and Wylie, A.G. (2022b). Dimensional characteristics of the major types of amphibole mineral particles and the implications for carcinogenic risk assessment. *Inhalation Toxicology* 34(1-2): 24–38. https://doi.org/10.1080/08958378.2021.2024304. Epub 2022 Jan 10. PMID: 35001771.

Korchevskiy, A., Rasmuson, J.O., and Rasmuson, E.J. (2019). Empirical model of mesothelioma potency factors for different mineral fibers based on their chemical composition and dimensionality. *Inhalation Toxicology* 31(5): 180–191. https://doi.org/10.1080/08958378.2019.1640320.

Korchevskiy, A., Rasmuson, J.O., Rasmuson, E.J., and Strode, R.D. (2020). Inhalation unit risk (IUR) of asbestos based on available science. *Inhalation Toxicology* 32(9–10): 372–374. https://doi.org/10.1080/08958378.2020.1829210.

Le Stang, N., Bouvier, V., Glehen, O., Villeneuve, L., FRANCIM Network, MESOPATH Referent National Center, et al. (2019). Incidence and survival of peritoneal malignant mesothelioma between 1989 and 2015: a population-based study. *Cancer Epidemiology* 60: 106–111. https://doi.org/10.1016/j.canep.2019.03.014.

Liddell, F.D. (2001). The interaction of asbestos and smoking in lung cancer. *The Annals of Occupational Hygiene* 45(5): 341–356. PMID: 11418084.

Markowitz, S.B., Levin, S.M., Miller, A., and Morabia, A. (2013). Asbestos, asbestosis, smoking, and lung cancer. New findings from the North American Insulator Cohort. *American Journal of Respiratory Critical Care Medicine* 188(1): 90–96. https://doi.org/10.1164/rccm.201308-1436LE.

McDonald, J.C., Liddell, F.D., Gibbs, G.W., et al. (1980). Dust exposure and mortality in chrysotile mining, 1910–75. *British Journal of Industrial Medicine* 37(1): 11–24. https://doi.org/10.1136%2Foem.37.1.11.

Moolgavkar, S.H., Turim, J., Alexander, D.D., et al. (2010). Potency factors for risk assessment at Libby, Montana. *Risk Analysis* 30(8): 1240–1248. https://doi.org/10.1111/j.1539-6924.2010.01411.x.

Mossman, B.T. (2008). Assessment of the pathogenic potential of asbestiform vs. nonasbestiform particulates (cleavage fragments) in in vitro (cell or organ culture) models and bioassays. *Regulatory Toxicology and Pharmacology* 52(1 Suppl): S200–S203. https://doi.org/10.1016/j.yrtph.2007.10.004.

Musk, A.W., de Klerk, N.H., Reid, A., et al. (2008). Mortality of former crocidolite (blue asbestos) miners and millers at Wittenoom. *Occupation and Environmental Medicine* 65(8): 541–543. https://doi.org/10.1136/oem.2007.034280.

Ngamwong, Y., Tangamornsuksan, W., Lohitnavy, O., et al. (2015). Additive Synergism between asbestos and smoking in lung cancer risk: a systematic review and meta-analysis. *PLoS One* 10(8): e0135798. https://doi.org/10.1371/journal.pone.0135798.

Nguyen, P.K. and Wu, J.C. (2011). Radiation exposure from imaging tests: is there an increased cancer risk? *Expert Review of Cardiovascular Therapy* 9(2): 177–83. https://doi.org/10.1586/erc.10.184.

Offermans, N.S.M., Vermeulen, R., Burdorf, A., et al. (2014). Occupational asbestos exposure and risk of pleural mesothelioma, lung cancer, and laryngeal cancer in the prospective Netherlands cohort study. *Journal of Occupational and Environmental Medicine* 56(1): 6–19. https://doi.org/10.1097/jom.0000000000000060.

Pierce, J.S., Ruestow, P.S., and Finley, B.L. (2016). An updated evaluation of reported no-observed adverse effect levels for chrysotile asbestos for lung cancer and mesothelioma. *Critical Reviews in Toxicology* 46(7): 561–586. https://doi.org/10.3109/10408444.2016.1150960.

Price, B. and Ware, A. (2009). Time trend of mesothelioma incidence in the United States and projection of future cases: an update based on SEER data for 1973 through 2005. *Critical Reviews in Toxicology* 39(7): 576–588. https://doi.org/10.1080/10408440903044928.

Samet, J.M., Avila-Tang, E., Boffetta, P., et al. (2009). Lung cancer in never smokers: clinical epidemiology and environmental risk factors. *Clinical Cancer Research* 15(18): 5626–5645. https://doi.org/10.1158%2F1078-0432.CCR-09-0376.

Stayner, L., Steenland, K., Dosemeci, M., and Hertz-Picciotto, I. (2003). Attenuation of exposure-response curves in occupational cohort studies at high exposure levels. *Scandinavian Journal of Work, Environment & Health* 29(4): 317–324. https://doi.org/10.5271/sjweh.737. Attended on January 24, 2024.

U.S. EPA (2014) *IRIS Toxicological Review of Libby Amphibole Asbestos* (Final Report). U.S. Environmental Protection Agency: Washington, DC, EPA/635/R-11/002F. Accessed January 24th, 2024

US EPA. (2016). Evaluation of the inhalation carcinogenicity of ethylene oxide (Final Report). U.S. Environmental Protection Agency, Washington, DC, EPA/635/R-16/350F. https://cfpub.epa.gov/ncea/iris_drafts/recordisplay.cfm?deid=329730#main-content.

US EPA. (2020). *Risk Evaluation for Asbestos. Part 1. Chrysotile Asbestos.* Washington, DC: U.S. Environmental Protection Agency. https://www.epa.gov/assessing-and-managing-chemicals-under-tsca/final-risk-evaluation-asbestos-part-1-chrysotile.

Vincent, M.J., Kozal, J.S., Thompson, W.J., et al. (2019). Ethylene oxide: cancer evidence integration and dose-response implications. *Dose Response* 17(4): 1–17. https://doi.org/10.1177/1559325819888317.

Wang, X.R., Yu, I.T.S, Qiu, H., et al. (2012). Cancer mortality among Chinese chrysotile asbestos textile workers. *Lung Cancer* 75(2): 151–155. https://doi.org/10.1016/j.lungcan.2011.06.013.

Wylie, A.G. and Korchevskiy, A.A. (2023). Dimensions of elongate mineral particles and cancer: a review. *Environmental Research* 230: 114688. https://doi.org/10.1016/J.ENVRES.2022.114688.

Wylie, A.G., Korchevskiy, A., Segrave, A.M., Duane, A. (2020) Modeling mesothelioma risk factors from amphibole fiber dimensionality: mineralogical and epidemiological perspective. *Journal of Applied Toxicology* 40(4): 515–524. https://doi.org/10.1002/jat.3923. Epub 2020 Feb 10. PMID: k32040984.

Wylie, A.G., Korchevskiy, A.A., Van Orden, D.R., Chatfield, E.J. (2022). Discriminant analysis of asbestiform and non-asbestiform amphibole particles and its implications for toxicological studies. *Computational Toxicology* 23: 100233, ISSN 2468-1113, https://doi.org/10.1016/j.comtox.2022.100233.

Part IV

Risk Characterization

13

Risk Characterization for Occupational and Environmental Exposure to Asbestos: Case Studies

James Rasmuson, Andrey Korchevskiy, and Eric Rasmuson

Chemistry & Industrial Hygiene, Inc., Lakewood, CO, USA

We will analyze and discuss several case studies on asbestos risk characterization and lay out a methodology that can be used for the assessment of cancer risks.

Case Study 1. Ms. X, an industrial plumber in the United Kingdom, was exposed to the average level of 0.009 f/cc of amosite asbestos for 5 years, 8 hours per day, 5 days per week, 50 weeks per year, starting at the age of 30 years (exposure data from Burdett and Bard 2007). We will estimate the possible predicted excess level of mesothelioma mortality per 10,000 per lifetime.

The Hodgson and Darnton (2000) method, updated by Darnton in 2023 (see Chapter 8 of this book), yields the linear mesothelioma potency factor for amosite at the level of 0.11% per unit of cumulative exposure. This value should be corrected if the onset age is different from 30 years or if the duration is different from 5 years. However, without correcting for duration, a simple estimation of mesothelioma risk can be performed by the following equation:

$$\text{Excess Malignant Mesothelioma Risk} = 0.009 \text{ f/cc} \times 40 \text{ years} \times 0.11/100$$
$$\times 10,000 \times 0.7$$
$$= 2.8 \left(\text{cases per 10,000 per lifetime}\right)$$

which is slightly above the National Institute of Occupational Health (NIOSH 2016) upper-bound recommended level for acceptable occupational risk. (In this example, the risk of lung cancer is ignored.) The factor of 0.7 is an approximate correction factor recommended by Hodgson and Darnton to correct the standardized R_M values for the start of exposure at age 30 with a duration of five years, to account for population decreases with age (see Chapter 12). For durations not far from five

Health Risk Assessment for Asbestos and Other Fibrous Minerals, First Edition.
Edited by Andrey Korchevskiy, James Rasmuson, and Eric Rasmuson.
© 2024 John Wiley & Sons, Inc. Published 2024 by John Wiley & Sons, Inc.

years, or from an age of first exposure not too far from 30 years, the methodology shown in the risk equation can give acceptable results depending on the application.

This calculation does not fully account, however, for the Peto model (Peto et al. 1982) of age of first exposure and duration-related mesothelioma mortality (see Chapter 7). In Table 13.1, correction factors, based on the Peto method, normalized to the age of first exposure equal to 30 and duration equal to five years (correction factor of 1 in the table) for specific ages of first asbestos exposure and duration are given. Even with utilization of Table 13.1, the 0.7 correction factor is required, if Hodgson and Darnton (2000) or Darnton (2023) potency factors for mesothelioma are used. Extrapolation, if required, can be utilized between the age of first exposure and/or duration to apply correction with sufficient accuracy in the excess risk calculation to any asbestos exposure situation with application of a specific asbestos fiber type potency factor. For example, application of a correction factor for the age of first exposure being 30 and duration of 40 years from Table 13.1 is 0.332. When the factor of 0.332 is applied to the excess risk equation for amosite (above), the calculated risk is reduced to 0.92 per 10,000 lifetime cases of mesothelioma, close to or less than the NIOSH criteria for occupational cancer risk (ignoring lung cancer risk for now). In addition, the ratio between the correction coefficients for different durations becomes more important with increasing age of first exposure (as is demonstrated for 0.1-year duration and the 45-year duration in the last column of Table 13.1).

With the correction factor from Table 13.1, the industrial plumber is on the borderline of risk acceptability if lung cancer is discounted, which is probably appropriate for such low-level exposures. The lung cancer risk for this case is expected to be very low, considering, for example, the Hodgson and Darnton superlinear lung cancer dose–response model (see Chapter 8), or based on the arguments that asbestosis correlates better with lung cancer risk than a linear no-threshold dose–response model. OSHA risk acceptability criterion is considered to be 1 case per 1,000 per lifetime, according to the US Supreme Court Benzene decision (US Supreme Court 1992). However, NIOSH (2016) recommends an upper-bound target management level for carcinogenic excess risk at the level of 1 per 10,000 cases. The risk acceptability criteria in the United Kingdom may be different, but usually they are not stricter than in the United States.

Case Study 2. Mr. X, an industrial plumber in the United Kingdom, was exposed to the average level of 0.009 f/cc of amosite asbestos for 20–45 years with a range of exposure concentration from 0.004 to 0.011 f/cc, 8 hours per day, 5 days per week, 50 weeks per year, starting at the age of 30 years (exposure data from Burdett and Bard 2007). We will estimate the possible predicted level of mesothelioma per 10,000 per lifetime.

Using Monte Carlo simulation and Hodgson and Darnton methods combined with Peto formulas for age correction, the distribution of excess risk levels can be

Table 13.1 Correction factors for Darnton (2023) and Hodgson and Darnton (2000) mesothelioma risk calculations from normalized mesothelioma risk at age 30 with five-year duration for both linear and nonlinear methods.

Age of first exposure (years)	Duration (years)										Ratio between 0.1-year duration and 45-year duration for same cumulative exposure
N/A	0.1	5	10	15	20	25	30	35	40	45	N/A
0	5.815	5.248	4.723	4.250	3.824	3.443	3.104	2.804	2.539	2.306	2.52
5	4.690	4.200	3.750	3.350	2.990	2.680	2.400	2.150	1.940	1.752	2.68
10	3.717	3.303	2.925	2.590	2.295	2.036	1.810	1.615	1.450	1.303	2.85
15	2.895	2.547	2.234	1.959	1.719	1.512	1.334	1.181	1.053	0.944	3.07
20	2.207	1.920	1.665	1.443	1.253	1.091	0.954	0.839	0.743	0.664	3.32
25	1.641	1.409	1.205	1.030	0.883	0.761	0.659	0.575	0.507	0.452	3.63
30	1.184	1.000	0.841	0.708	0.598	0.509	0.436	0.378	0.332	0.295	4.01
35	0.824	0.682	0.562	0.465	0.386	0.323	0.274	0.236	0.207	0.184	4.48
40	0.548	0.442	0.356	0.287	0.234	0.193	0.162	0.139	0.122	0.108	5.07
45	0.344	0.269	0.209	0.164	0.131	0.106	0.089	0.076	0.066	0.059	5.83
50	0.200	0.150	0.112	0.084	0.065	0.053	0.044	0.038	0.033	0.029	6.90
55	0.105	0.074	0.052	0.037	0.028	0.023	0.019	0.016	0.014	0.013	8.08
60	0.047	0.030	0.019	0.015	0.010	0.008	0.007	0.006	0.005	0.004	11.75

Source: Adapted from Darnton (2023), Hodgson and Darnton (2000).

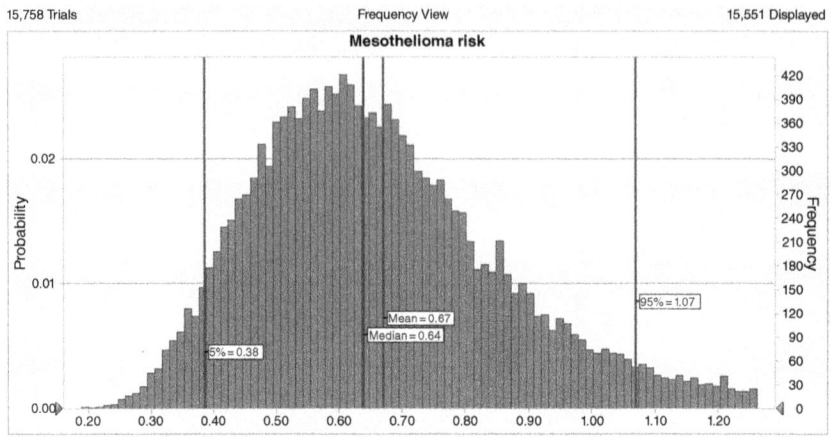

Figure 13.1 Excess mesothelioma risk for the industrial plumber (cases per 10,000 per lifetime). *Source:* Adapted from Darnton [2023] linear method.

derived as demonstrated in Figure 13.1. For the purposes of Monte Carlo simulation, an Excel spreadsheet was developed to estimate the correction factors for Hodgson and Darnton risk estimations according to the age-specific Peto model. (See Chapter 12 for details.)

The simulation confirmed the "borderline" acceptability of risk for the plumber's exposure (95% prediction interval of excess risk at 1.07 cases per 10,000, potentially requiring additional industrial hygiene interventions).

Case Study 3. Mr. Y worked in the construction industry for 15 years, 5 hours per day, 5 days per week, 50 weeks per year. He started working at the age of 20. The upper bound (95th confidence interval) of the daily eight-hour time-weighted average (TWA) of asbestos exposure was measured as 0.1 f/cc. The asbestos fiber type is chrysotile. What is the total excess cancer risk for Mr. Y per 10,000 workers (using the inhalation unit risk method)?

The inhalation unit risk (IUR) method for asbestos risk assessment was described in Chapter 12.

The IURs for different mineral types of fibers, averaged between "science based" IURs as calculated by both Hodgson and Darnton and the Peto method with Berman and Crump (2008b) coefficients, respectively, are listed in Table 13.2 (see Chapter 12).

The US EPA (2001) correction factors can be applied to account for the specific exposure duration and onset age as described in Chapters 7 and 12.

For convenience, in Table 13.3, we listed the US EPA values of the correction coefficients for various durations and onset age of exposure.

Table 13.2 The inhalation unit risk (IUR) values for different mineral types of asbestos.

	IUR per PCM f/cc (assuming environmental, 24/7 exposure)		
	Total cancers		
Asbestos Fiber type	Hodgson and Darnton linear method	Berman and Crump method (based on the Peto model)	Average
Crocidolite	1.50	1.83	1.665
Amosite	0.60	1.83	1.215
Libby amphibole	0.13	0.13	0.13
Chrysotile	0.016	0.035	0.026

To apply the US EPA method for workers, the exposure concentration should be recalculated to the continuous lifetime equivalent according to the formula:

$$\text{Lifetime average exposure (f/cc)} = C\,(\text{f/cc}) \times \text{Fraction of } \textbf{lifetime} \text{ exposed}$$
$$= \frac{C\,(\text{f/cc}) \times \text{Exposure duration (hours)}}{\text{Averaging time (hours per lifetime)}}$$

The fraction of the lifetime exposed at work is simple to calculate. For the numerator of the fraction in hours: 15 years × 50 weeks per year × 8 hours per day × 5 days per week = 30,000 hours.

For the denominator, the number of hours in a 70-year lifetime is 70 years × 365.25 days per year × 24 hours per day = 613,620 hours

So, the lifetime fraction of exposed time is equal to 30,000 hours/613,620 hours = 0.0489.

Thus, the average lifetime exposure is equal to 0.0489 × 0.1 f/cc = 0.00489 f/cc.

The correction coefficient based on age and duration of exposure for Mr. Y according to Table 13.3 will be 0.1983.

We can find, therefore,

$$\text{Excess risk (per 10,000)} = 0.00489 \text{ f/cc} \times 0.026\,(\text{cases per f/cc})$$
$$\times 0.1983 \times 10,000$$
$$= 0.25 \text{ lifetime cancer cases per 10,000.}$$

Based on the risk calculation, the exposure of Mr. Y to chrysotile would be fully acceptable based on NIOSH criteria.

Table 13.3 Correction coefficients for the inhalation unit risk for various durations and exposure onset age (proportionally to lifetime exposure from birth).

Age of first exposure (years)	Duration (years)									
N/A	0.1	5	10	15	20	25	30	35	40	45
0	0.0047	0.2107	0.3785	0.5121	0.6185	0.7032	0.7706	0.8243	0.8670	0.9010
5	0.0037	0.1661	0.2990	0.4053	0.4904	0.5584	0.6129	0.6564	0.6913	0.7192
10	0.0029	0.1309	0.2361	0.3205	0.3883	0.4427	0.4864	0.5215	0.5497	0.5723
15	0.0023	0.1030	0.1859	0.2527	0.3065	0.3498	0.3846	0.4127	0.4353	0.4535
20	0.0018	0.0807	0.1458	0.1983	0.2407	0.2749	0.3024	0.3247	0.3427	0.3572
25	0.0014	0.0627	0.1134	0.1544	0.1875	0.2143	0.2359	0.2534	0.2675	0.2789
30	0.0011	0.0483	0.0873	0.1189	0.1445	0.1652	0.1819	0.1954	0.2064	0.2153
35	0.0008	0.0366	0.0662	0.0901	0.1095	0.1253	0.1380	0.1483	0.1567	0.1634
40	0.0006	0.0271	0.0490	0.0668	0.0812	0.0928	0.1023	0.1100	0.1162	0.1212
45	0.0004	0.0193	0.0350	0.0477	0.0581	0.0664	0.0732	0.0787	0.0832	0.0868
50	0.0003	0.0131	0.0237	0.0323	0.0392	0.0449	0.0495	0.0532	0.0562	0.0587
55	0.0002	0.0079	0.0144	0.0196	0.0239	0.0273	0.0301	0.0324	0.0342	0.0357
60	0.0001	0.0038	0.0068	0.0093	0.0114	0.0130	0.0143	0.0154	0.0163	0.0170

If, however, Mr. Y would be exposed to Libby amphiboles, his excess risk would be

$$\text{Excess risk (per 10,000)} = 0.00489 \text{ f/cc} \times 0.13 \text{ (cases per f/cc)} \times 0.1983 \times 10,000$$
$$= 1.26 \text{ (cancer cases/10,000 exposed in this manner)}.$$

This value is slightly higher than the NIOSH criteria (1 case per 10,000).

Case Study 4. Mr. Z began doing drywall work with his father starting at age 14 in 1953 and continued for 12–16 years. He stated that they worked on three to four homes per month. He also stated that drywall work often took "at least part" of three to four days. He worked 12 months per year.

Later, he worked at an oil refinery during shutdowns where insulators were around him "every day" for a cumulative total of 12–18 months, 8 hours per day, 5 days per week, from age 30 to 36. Insulators worked at a distance of 10–30 ft from him.

Evaluate the lifetime and annual exposure level to chrysotile and amphibole fibers (f/cc). Estimate the excess risk of mesothelioma and lung cancer using the Darnton (2023) linear method and the Hodgson and Darnton (2000) nonlinear method as well as the Peto method with Berman and Crump (2008b) linear coefficients.

We will use the annual average worktime exposure to calculate the cumulative exposure of Mr. Z and also to utilize it for Excel spreadsheet calculations of lifetime risk.

Similar to average lifetime exposure, average annual worktime exposure is calculated in the following manner:

Worktime annual average exposure (f/cc) = C (f/cc) × Fraction of the **working year** exposed

$$= \frac{C(\text{f/cc}) \times \text{Exposure duration (hours)}}{\text{Averaging time (hours per working time)}}$$

$$= \frac{C(\text{f/cc}) \times \text{hours per day} \times \text{days per week} \times \text{weeks per year}}{8 \text{ hours per day} \times 5 \text{ days per week} \times 50 \text{ weeks}}$$

$$= \frac{TWA_8 (\text{f/cc}) \times \text{days per week} \times \text{weeks per year}}{5 \text{ days per week} \times 50 \text{ weeks}}.$$

Cumulative occupational exposure is calculated in this case as

Worktime annual average exposure (f/cc) × exposure durations (converted to units of full years) with overall cumulative exposure units in f/cc-years.

Table 13.4 Assumed typical TWA$_8$ exposure levels for some trades. (Often for these trades, travel and setup time reduce work on site to about 80% of a full workday.)

Shipyard insulator (amphibole typically amosite and chrysotile from thermal insulation)
- 1960s, ~7 f/cc (Cooper and Balzer 1968 for 1960s exposures)
- Earlier exposures, likely around 10–30 f/cc (based on Nicholson 1986 compiled data)
- 40–50% and possibly higher amphibole exposure in the asbestos of thermal insulation (Berman and Crump 2008b; Yarborough 2006)

Insulator in commercial and industrial applications (same fiber types as above)
- 1960s into early 1970s, ~3–6 f/cc (Cooper and Balzer 1968; Nicholson 1986)
- Earlier exposures, 10–15 f/cc (Nicholson 1986)
- 40–50% amphibole exposure (Berman and Crump 2008b; Yarborough 2006)

Drywall taper and sander (short-fiber chrysotile)
- Based on the analysis of much of all available historical and simulated data (Verma and Middleton 1981; Fischbein 1979; Rhodes and Ingalls 1975, 1976; and many others). Monte Carlo methods were used where phase contrast microscopy (PCM) analysis of chrysotile fibers in the air was available to estimate exposures for drywall tapers and sanders.
- Combining the mean and statistical asbestos exposure ranges and percent of workday for drywall tapers and sanders for application, mixing, sanding, and clean-up was typically in the 2–3 f/cc range (confidence interval), but mean values were closer to about 2 f/cc when premixed joint compound was used and closer to 3 f/cc when dry joint compound was mixed on site.
- A prediction interval range was found to be approximately 1–4 f/cc-years.

To evaluate the exposure concentration of Mr. Z, we will use exposure values from Table 13.4 where past typical exposure levels for US workers in some trades are tabulated:

Also, we will use Donovan's estimations of bystander factors varying with distance (Donovan et al. 2010), where

Exposure of bystander = Exposure of primary worker indoors × bystander factor.

Following Donovan's rule of thumb, the bystander at 1–5 ft is exposed to approximately 50% of the primary worker exposure, at 5–10 ft, 38%, at 10–30 ft, 10%, and more than 30 ft to 1–2%. (Obviously, ventilation and building geometries will alter these approximate values. More rigorous calculations of bystander factors for both indoor and outdoor scenarios are demonstrated in Chapter 6.)

The following assumptions can be made regarding the exposure of Mr. Z during his drywall job:

Exposure concentration (TWA$_8$): 4 f/cc × 80% = 3.2 f/cc (assuming 4 f/cc as Mr. Z's upper-bound exposure concentration while performing his task 80% of the time, as explained in Table 13.4).

Homes per month: 3–4 (assuming 3.5)
Days per home: 3–4 (assuming 3.5)
Duration: 12–16 years (assuming 14 years), starting at the age of 14 years
Fiber type: short-fiber chrysotile
The fraction of the year worked each year is as follows:

$$\frac{3.5\,(\text{homes/month}) \times 3.5\,(\text{days/home}) \times 12\,(\text{months/year})}{5\,\text{days/week} \times 50\,\text{weeks/year}}$$

$= 0.588\%$ or 58.8% of the work year.

The cumulative exposure can then be found as

3.2 f/cc × 0.588 × 14 years = 26.3 f/cc-years.

Sometimes, alternatively, to estimate the duration in calculating cumulative exposure, variable exposure times (hours, weeks, months, or years) are placed into an Excel Spreadsheet and simply added up and converted to years of exposure. Then, the number of years is multiplied by the mean or upper-bound exposure value to estimate the cumulative asbestos exposure in units of f/cc-years.

It should be noted that cumulative exposure is a useful metric that can be utilized for comparison between various exposure scenarios. One use is that the cumulative exposure can be compared to benchmark values to evaluate the risk acceptability. For example, it is useful to compare cumulative exposure values with levels that are not known to elevate the risk above acceptable levels such as cumulative asbestos exposure values that have historically occurred from ambient background airborne fiber concentrations for the general population on a fiber-specific basis. Another example is calculating cumulative asbestos exposure associated with one lifetime case per million or lower cancer risk, which would be considered to have zero or negligible risk by the FDA, the WHO, and the US EPA. Utilizing risk calculations can be helpful to estimate the cumulative asbestos exposure range to reach a relative risk of 2, above which it can be concluded that the causation of the particular disease from the exposure is more probable than not. To approximately calculate the relative risk, though, the assumed baseline unexposed disease rate must be known, which, within ranges, has been estimated in the general asbestos literature, some of which are listed here: Berry (1997); Price (1997); Price and Ware (2009); Roggli et al. (1997); Moore et al. (2008); British Thoracic Society (2007); Glynn et al. (2014).

Returning to the Mr. Z exposure estimation scenario, the following assumptions can be made regarding the exposure of Mr. Z during his bystander-to-insulators job:

- Primary worker (insulator) exposure concentration: 6 f/cc (assumed to be worst-case).

Table 13.5 Average annual worktime exposure and cumulative asbestos exposures for Case study 4.

Job duration and age of exposure onset	Fiber type	Average annual worktime exposure (f/cc)	Cumulative exposure (f/cc-years)
Taper and Sander, 14 years, starting at age 14	Short-fiber chrysotile	1.88	26.3
Bystander to insulator, 6 years, starting at age 30	Chrysotile Amosite	0.066 0.054	0.396 0.324

- Duration: 12–18 months over the course of 6 years; a "worst-case," top of range, 3 months per year is assumed; however, because of the likelihood of insulators not working a full day, the months per year is multiplied by 80%, reducing the assumed yearly duration to 2.4 months.
- Bystander factor: assumption of 10%, which reduces the exposure to the primary worker of 6 f/cc to an average exposure of 0.6 f/cc to Mr. Z.
- Duration of worktime: 6 years, starting at the age of 30 years.
- Assumed asbestos fiber types: 45% amosite, 55% chrysotile.
- Average annual worktime fraction can be found as

$$\frac{2.4 \text{ work months/year}}{12 \text{ months/year}} = 0.2$$

and 0.6 f/cc (mean exposure while being a bystander to an insulator) × 0.2 (yearly work fraction) = 0.12 f/cc mean asbestos exposure during the year, including average exposures of 0.054 f/cc to amosite and 0.066 f/cc to chrysotile.

The cumulative asbestos exposure of Mr. Z would therefore be

0.12 f/cc × 6 years = 0.72 f/cc-years, including 0.324 f/cc-years of amosite exposure and 0.396 f/cc-years of chrysotile exposure.

The exposure data for Mr. Z is summarized in Table 13.5.

With the use of the exposure values summarized in Table 13.5, the results of the risk assessment calculations based on the methods described in Chapters 7 and 8 are shown in Table 13.6.

For the Hodgson and Darnton nonlinear methods, the coefficients from Hodgson and Darnton (2000) were used for both chrysotile and amosite risk mesothelioma and lung cancer calculations. For the linear methods, the coefficients were used as in Darnton (2023) and in Berman and Crump (2008b). However, the 2023 Darnton publication notes that the meta-analysis of chrysotile heterogeneity

Table 13.6 Excess cancer risk for Mr. Z (cases per 10,000 per lifetime).

Method	Drywall (Taper and Sander), chrysotile risk	Chrysotile risk, bystander to insulator(s)	Amosite risk (bystander to insulators)	Total lifetime risk
Mesothelioma risk				
Berman and Crump (2008b), linear	2.94	0.02	15.52	18.48
Darnton (2023), linear	5.49	0.037	2.41	7.94
Hodgson and Darnton (2000), nonlinear	11.4	0.22	3.78	15.40
Lung cancer risk				
Berman and Crump (2008b), linear	44.39	0.66	3.81	48.86
Darnton (2023), linear	8.44	0.13	5.58	14.15
Hodgson and Darnton (2000), nonlinear	8.08	0.034	1.51	9.62
Total cancer risk for each method and exposure category				
Grand total cancer risks by method in column at right				
Berman and Crump (2008b), linear	47.33	0.68	19.33	67.34
Darnton (2023), linear	13.93	0.17	7.99	22.09
Hodgson and Darnton (2000), nonlinear	19.48	0.25	5.29	25.02

tests between textile plant cohorts and all other chrysotile cohorts failed. Therefore, three different potency factors are offered in the publication for the combined value along with separate chrysotile potency factors for the textile plants and also for all other cohorts. Further, also shown are the results of the risk calculation with the Peto model and the Berman and Crump (2008b) coefficients. It should be noted, however, that Berman and Crump did not distinguish between amosite and crocidolite when calculating the potency factors, combining the two fiber types into a single amphibole asbestos category in their 2008b publication. Calculated mesothelioma risk results for amosite appear to be biased somewhat high when the Berman and Crump method is utilized.

Although not considered in this exercise, it is important to note that joint compound typically utilizes Grade 7 chrysotile, with the vast majority of airborne fibers less than 5 µm. These shorter chrysotile fibers have much lower carcinogenicity (Phelka and Finley 2012; Bernstein 2022; Roggli 2015; Pierce et al. 2016) than the longer fibers in many of the chrysotile cohorts that Berman and Crump as well as Hodgson and Darnton studied. The combined value for chrysotile as determined by Darnton (2023) is utilized here for our calculations although Darnton separated out textile cohorts from nontextile cohorts. The textile cohorts have longer chrysotile fibers than other chrysotile cohorts with some evidence of the presence of amphibole asbestos exposure in the cohort members.

Case Study 5. Ms. Q lived in a county where naturally occurring asbestos (NOA) is found on a local trail. She arrived in the county at the age of 12 with her parents and stayed for four years. During a not-rainy period of the year (36 weeks), she was involved in a moderate-intensity activity pattern, including jogging/biking on the trail (two hours per day on average). Activity-based air samples were taken on the trail for jogging/biking, determining that the upper-bound level of phase contrast microscopy equivalent (PCME) fiber concentration was at 1 f/cc. The fibers were determined to be 50% chrysotile and 50% tremolite. We will assess exposure, risk, and its acceptability, utilizing the Darnton (2023) linear method and the Peto method with Berman and Crump (2008b) coefficients.

We will recalculate the exposure from the recreational activities of Ms. Q to the average annual worktime exposure equivalent and use the same potency factors as were developed for occupational exposure in the two cited methods.

The average annual worktime asbestos exposure (occupational equivalent) for Ms. Q can be calculated for this scenario as

$$\frac{1 \text{ f/cc} \times 2 \text{ (hours/day)} \times 7 \text{ (days/week)} \times 36 \text{ (weeks/yr)}}{8 \text{ hours/day} \times 5 \text{ days/week} \times 52 \text{ weeks/yr}} = 0.24 \text{ (f/cc)},$$

including 0.12 f/cc of chrysotile and 0.12 f/cc of tremolite.
Accordingly, with the four-year exposure duration considered, the cumulative exposure of Ms. Q can be estimated as 0.96 f/cc-years (chrysotile 0.48 f/cc-years, tremolite 0.48 f/cc-years).

The Hodgson and Darnton and the Peto models can be applied to occupational equivalent exposure data (if tremolite potency is approximated as equivalent to that of Libby amphiboles). The risk quantification results are listed in Table 13.7. To help the reader, the potency factors (risk calculation coefficients) are also shown in Table 13.7.

Table 13.7 Risk characterization results (based on the Darnton 2023 and the Peto methods).

Activity/duration/age of onset	Fiber type	Average "worktime" exposure equivalent (PCME f/cc)	Mesothelioma excess risk (cases per 1,000,000) (Darnton 2023 linear model)	Lung cancer excess risk (cases per 1,000,000) (Darnton 2023 linear model)	Mesothelioma excess risk (cases per 1,000,000) (Peto model)	Lung cancer excess risk (cases per 1,000,000) (Peto model)
Biker at the trail, four years, starting at the age of 12	Chrysotile	0.12	14	16	8	81
	Tremolite (as Libby amphiboles)	0.12	308	162	158	235
The potency factors used for calculations			For chrysotile: $R_M = 0.0014$ (%) For Libby amphiboles: $R_M = 0.03$ (%) (Darnton 2023)	For chrysotile: $R_L = 0.078$ (%) For Libby amphiboles: $R_L = 0.82$ (%) (Darnton 2023)	For chrysotile: $K_M = 0.009$ ($\times 10^{-8}$) (Berman and Crump 2008b) For Libby amphiboles: $R_M = 0.185$ (%) (US EPA 2014)	For chrysotile: $K_L = 0.2$ (%) (Berman and Crump 2008b) For Libby amphiboles: $K_L = 0.58$ (%) (US EPA 2014)

Source: The Darnton method is a Adapted from Darnton (2023).

The comparison between the Peto and the Darnton methods demonstrates that the estimations of lung cancer risk when using Berman and Crump coefficients for chrysotile from their dimensional-specific meta-analysis might be too conservative. (This can be seen from the comparison of nontextile chrysotile cohort data from Table 7.2 and R_L from Table 7.4, as shown in Chapter 7.) In any case, tremolite is expected to produce significantly higher input for the risk of Ms. Q from her recreational activities than chrysotile. Risk assessors can determine if this level of risk can be considered acceptable for the population. Potentially, tremolite at the trail can be analyzed for its dimensions and morphology that can significantly affect the estimations of risk (see Wylie and Korchevskiy 2023).

It should also be noted that some view lung cancer as a threshold disease rather than as a no-threshold linear response model with cumulative asbestos exposure. With the no-threshold linear model, lung cancer risks may be expressed too highly, because of evidence of lung cancer either having a threshold or an exponential dose–response as in the nonlinear Hodgson and Darnton lung cancer risk model. A cumulative asbestos exposure of about 25–100 f/cc-years is typically believed to be necessary to associate asbestos exposure with lung cancer, similar to the exposure criteria for asbestosis (Roggli et al. 2010; Sporn and Roggli 2014; Consensus Report-Helsinki Criteria 1997; FIOH 1997, 2014). There appears to be strong epidemiological evidence that the presence of clinical asbestosis or exposure required to produce clinical asbestosis correlates with the risk of asbestos-related lung cancer better than cumulative asbestos exposure by itself (Hughes and Weil 1991; Jones et al. 1996; Gibbs et al. 2007), which also indicates that the no-threshold linear model can produce estimated lung cancer risks that are too high at low asbestos cumulative exposure levels. However, for this example, we have applied the precautionary principle, using the no-threshold linear model for lung cancer as well as mesothelioma. There is a possibility that mesothelioma is also a threshold disease related to asbestos exposure.

References

Berman, D.W. and Crump, K.S. (2008b). A meta-analysis of asbestos-related cancer risk that addresses fiber size and mineral type. *Critical Reviews in Toxicology* 38 (Suppl 1): 49–73. https://doi.org/10.1080/10408440802273156.

Bernstein, D.M. (2022). The health effects of short fiber chrysotile and amphibole asbestos, *Critical Reviews in Toxicology* 52 (2): 89–112. https://doi.org/10.1080/10408444.2022.2056430.

Berry, M. (1997). Mesothelioma incidence and community asbestos exposure. *Environmental Research* 75 (1): 34–40. https://doi.org/10.1006/enrs.1997.3770.

[BTS] British Thoracic Society, Standards of Care Committee. (2007). BTS statement on malignant mesothelioma in the UK, 2007. *Thorax* 62 (Suppl 2): ii1–ii19. https://doi.org/10.1136/thx.2007.087619.

Burdett, G. and Bard, D. (2007). Exposure of UK industrial plumbers to asbestos, Part I: Monitoring of exposure using personal passive samplers. *The Annals of Occupational Hygiene* 51 (2): 121–130. https://doi.org/10.1093/annhyg/mel078.

Consensus Report-Helsinki Criteria (1997). "Asbestos, asbestosis, and cancer: the Helsinki criteria for diagnosis and attribution," *Scandinavian Journal of Work, Environment, and Health* 23 (4): 311–316.

Cooper, W.C. and Balzer, J.L. (1968). Evaluation and control of asbestos exposures in the insulating trade. *Prepared for Second International Conference on Biological Effects of Asbestos* (Dresden, Germany) April 22–25, 1968. (1965–1967 data, Light and Heavy Industry).

Darnton, L. (2023). Quantitative assessment of mesothelioma and lung cancer risk based on Phase Contrast Microscopy (PCM) estimates of fibre exposure: an update of 2000 asbestos cohort data. *Environmental Research* 230: 114753. https://doi.org/10.1016/j.envres.2022.114753.

Donovan, E.P., Donovan, B.L., Sahmel, J., Scott, P.K., and Paustenbach, D.J. (2010). Evaluation of bystander exposure to asbestos in occupational settings: a review of the literature and application of a simple eddy diffusion model. *Critical Reviews in Toxicology* 41 (1): 52–74. https://doi.org/10.3109/10408444.2010.506639.

[FIOH] Finnish Institute of Occupational Health. (1997). Asbestos, asbestosis, and cancer. *Proceedings of an International Expert Meeting* (Helsinki, Finland) January 20–22.

Finnish Institute of Occupational Health. (2014). Asbestos, asbestosis, and cancer – Helsinki criteria for diagnosis and attribution. http://www.ttl.fi/hcuasbestos.

Fischbein, A., Rohl, A.N., Langer, A.M., and Selikoff, I.J. (1979). Drywall construction and asbestos exposure. *American Industrial Hygiene Association Journal* 40 (5): 402–407. https://doi.org/10.1080/15298667991429750.

Gibbs, A., Attanoos, R.L., Churg, A., and Weill, H. (2007). The "Helsinki Criteria" for attribution of lung cancer to asbestos exposure: how robust are the criteria? *Archives of Pathology and Laboratory Medicine* 131 (2): 181–183. https://doi.org/10.5858/2007-131-181-thcfao.

Glynn, M., Gaffney, S., and Sahmel, J. (2014). Ambient Asbestos and Long-term Trends in Pleural Mesothelioma Incidence between Urban and Rural Areas in the United States. *Occupational and Environmental Epidemiology, American Industrial Hygiene Conference and Exhibition (AIHce)* (San Antonio, Texas) June 3, 2014, SR-116-04, PO 116.

Hodgson, J.T. and Darnton, A. (2000). The quantitative risks of mesothelioma and lung cancer in relation to asbestos exposure. *The Annals of Occupational Hygiene* 44(8): 565–601.

Hughes, J.M. and Weil, H. (1991). Asbestosis as a precursor of asbestos-related lung cancer: results of a prospective mortality study. *British Journal of Industrial Medicine* 48 (4): 229–233. https://doi.org/10.1136/oem.48.4.229

Jones, R.N., Hughes, J.M., and Weil, H. (1996). Asbestos exposure, asbestosis, and asbestos-attributable lung cancer. *Thorax* 51 (Suppl 2): S9–S15. https://www.researchgate.net/publication/14338140_Asbestos_exposure_asbestosis_and_asbestos-attributable_lung_cancer.

Moore, A.J., Parker, R.J., and Wiggins, J. (2008). Malignant mesothelioma. *Orphanet Journal of Rare Diseases* 3: 34. https://doi.org/10.1186/1750-1172-3-34.

[NIOSH] National Institute of Occupational Health. (2016). Current Intelligence Bulletin 68: NIOSH chemical carcinogen policy. By Whittaker, C., Rice, F., McKernan, L., Dankovic, D., Lentz, T.J., MacMahon, K., Kuempel, E., Zumwalde, R., Schulte, P., on behalf of the NIOSH Carcinogen and RELs Policy Update Committee. Cincinnati, OH: US Department of Health and Human Services, Centers for Disease Control and Prevention, National Institute for Occupational Safety and Health, DHHS (NIOSH) Publication No. 2017-100.

Nicholson, W. (1986). Airborne asbestos health assessment update. United States Environmental Protection Agency Report 600/884003F. Washington, DC, June, 1986. Document Display | NEPIS | US EPA

Peto, J., Seidman, H., and Selikoff, I.J. (1982). Mesothelioma mortality in asbestos workers: implications for models of carcinogenesis and risk assessment. *British Journal of Cancer* 45 (1): 124–135. https://doi.org/10.1038%2Fbjc.1982.15.

Phelka, A.D. and Finley, B.L. (2012). Potential health hazards associated with exposures to asbestos-containing drywall accessory products: A state-of-the-science assessment. *Critical Reviews in Toxicology* 42 (1): 1–27. https://doi.org/10.3109/10408444.2011.613067

Pierce, J.S., Ruestow, P.S., and Finley, B.L. (2016). An updated evaluation of reported no-observed adverse effect levels for chrysotile asbestos for lung cancer and mesothelioma. *Critical Reviews in Toxicology* 46 (7): 561–586. https://doi.org/10.3109/10408444.2016.1150960.

Price, B. (1997). Analysis of current trends in United States mesothelioma incidence. *American Journal of Epidemiology* 145 (3): 211–218. https://doi.org/10.1093/oxfordjournals.aje.a009093.

Price, B. and Ware, A. (2009). Time trend of mesothelioma incidence in the United States and projection of future cases: an update based on SEER data for 1973 through 2005. *Critical Reviews in Toxicology* 39 (7): 576–588. https://doi.org/10.1080/10408440903044928.

Rhodes, H.B. and Ingalls, B.L. (1975). Asbestos and silica dust in the drywall industry, Part I. Gypsum drywall contractors international. *Drywall* 21 (6–8): 30.

Rhodes, H.B. and Ingalls, B.L. (1976). Asbestos and silica dust in the drywall industry, Part II. Gypsum drywall contractors international. *Drywall* 22 (8–11): 30–31.

Roggli, V.L., Oury, T.D., and Moffatt, J.D. (1997). Malignant mesothelioma in women. *Anatomic Pathology* 2: 147–163. PMID: 9575374.

Roggli, V., Gibbs, A.R., Attanoos, R., Churg, A., et al. (2010). Pathology of asbestosis – an update of the diagnostic criteria. Report of the Asbestosis Committee of the College of American Pathologists and Pulmonary Pathology Society. *Archives of Pathology and Laboratory Medicine* 134 (3): 462–480. http://dx.doi.org/10.1043/1543-2165-134.3.462.

Roggli, V.L. (2015). The so-called short-fiber controversy. *Archives of Pathology and Laboratory Medicine* 139 (8): 1052–1057. https://doi.org/10.5858/arpa.2014-0466-ra.

Sporn, T.A. and Roggli, V.L. (2014). Asbestosis. In: *Pathology of Asbestos-Associated Diseases* (eds. T.D. Oury, T.A. Sporn, and V.L. Roggli), Third Edition. Springer.

US EPA. (2001). Framework for investigating asbestos-contaminated comprehensive environmental response, compensation and liability act sites. *Prepared by the Asbestos Committee of the Technical Review Workgroup of the Office of Land and Emergency Management U.S. Environmental Protection Agency*, OLEM Directive No. 9200.0-90. https://semspub.epa.gov/work/HQ/100002942.pdf

US EPA. (2014). Toxicological review of libby amphibole asbestos. December 2014. EPA/635/R-11/002F. www.epa.gov/iris.

US Supreme Court. 56 FR 64004, Dec. 6, 1991; 57 FR 29206, July 1, 1992.

Verma, D.K. and Middleton, C.G. (1981). Occupational exposure to asbestos in the ceiling and wall texture process. *Occupational Health and Safety Journal* 51: 21–24. https://www.researchgate.net/publication/236546302_Occupational_Exposure_to_Asbestos_in_the_Ceiling_and_wall_Texture_Process.

Wylie, A.G. and Korchevskiy, A.A. (2023). Dimensions of elongate mineral particles and cancer: a review. *Environmental Research* 230: 114688. https://doi.org/10.1016/j.envres.2022.114688.

Yarborough, C.M. (2006). Chrysotile as a cause of mesothelioma: an assessment. Based on epidemiology. *Critical Reviews in Toxicology*. 36: 165–187. https://doi.org/10.1080/10408440500534248.

14

Asbestos in Soil: Risk Characterization for Occupational and Environmental Exposures

Andrey Korchevskiy[1] and Robert Strode[2]

[1] Chemistry & Industrial Hygiene, Inc., Lakewood, CO, USA
[2] Summit Exposure and Risk Sciences LLC, Silverthorne, CO, USA

Asbestos contamination and its natural occurrence in soil are significant worldwide problems that should not be ignored, especially when industrial asbestos use is effectively prohibited in developed countries. Therefore, the focus of asbestos exposures in developed countries has shifted to focus on existing, in-place asbestos-containing products and secondary contamination problems, including waste disposal. At the same time, asbestos can be a natural component in soil, as in the serpentine or amphibole formations that can cover extensive areas and have worldwide distribution (Schreier 1989). In many cases, potential exposure scenarios are related to anthropogenic contamination of soils; however, *in situ* "naturally occurring asbestos" can also create significant concerns. It should be noted, however, that attempts of a direct correlation between geological content of minerals in the area and disease rates are often questioned from a methodological perspective (see, for example, Pan et al. 2005; Berman et al. 2013).

Several methods can be used to perform human health risk assessment for soils containing asbestos. Hazard identification should identify the pathways of the human exposure and susceptible populations and/or occupational groups. For exposure assessment, either activity-based or long-term air monitoring can be performed, as appropriate for duration and intensity of the exposures and risks being evaluated.

For risk prediction purposes, utilizing methods that estimate airborne concentrations based on the asbestos weight % in the soil may be useful. However, effective methods of converting soil to airborne concentrations require major assumptions, and further development in this area is necessary to accurately estimate airborne

---

Health Risk Assessment for Asbestos and Other Fibrous Minerals, First Edition.
Edited by Andrey Korchevskiy, James Rasmuson, and Eric Rasmuson.
© 2024 John Wiley & Sons, Inc. Published 2024 by John Wiley & Sons, Inc.

asbestos concentrations based on asbestos soil concentrations. A further complication is the need to assess the risks at the specific sites where asbestos fibers are present in soil as a part of the natural geology. It should be seen that a precise quantification of "asbestos content in soil" by itself is difficult, if not impossible, because this content would vary with a special location, as with the contamination depth and other characteristics. Nevertheless, some theoretical estimations can be made based on "ideal" models when asbestos content in soil is supposed to be uniformly distributed and equally available for release independent of the depth.

In particular, airborne asbestos concentrations can generally be approximated using airborne dust concentrations and a fraction of asbestos in dust. For example, Addison et al. (1988) used respirable dust data for calculation of airborne asbestos concentrations. The authors prepared artificial mixtures using three different soil types (clay, sand, and intermediate) with each of three asbestos types (chrysotile, amosite, and crocidolite) in concentrations of 1%, 0.1%, 0.01%, and 0.001% by weight. Airborne dust clouds were generated over periods of 4 hours from each mixture using a dust dispenser discharging into a 1.3 m³ test chamber. Airborne dust concentrations were measured for the full duration of the test using gravimetric dust sampling instruments. A sequence of membrane filter samples was collected for fiber counting by phase contrast optical microscopy (PCM). Airborne fiber concentrations were determined for each test as time-weighted averages.

To characterize the results of Addison's experiments, we derived the following formula relating the PCM concentrations of asbestos to the content of fibers in the soil mixtures and generated respirable dust concentrations:

$$C_{asbestos}(f/cc) = \text{Respirable dust concentration}(\mu g/m^3)/1{,}000 \times \lambda \\ \times \text{fraction of asbestos in soil}/100 \tag{14.1}$$

where

- λ is a coefficient with average values tabulated (see Table 14.1),
- $C_{asbestos}$ (f/cc) is the PCM airborne asbestos concentration resulting from the asbestos present in the soil mixtures, and
- fraction of asbestos in soil is the percent, weight-by-weight concentration of asbestos fibers in soil, assumed to be measurable and uniform.

In real-world scenarios, numerous factors would affect the propensity of asbestos fibers to be released from soil into the air. In particular, factors such as soil type, soil moisture content, and humidity and other weather parameters would impact the quantity of released fibers. As can be seen in Table 14.1, both the mineral type of asbestos and the type of soil affect its releasability. Addison et al. did not address the issue about the impact of other factors on release of asbestos from soil. For example, an asbestos-containing material (ACM) present in a matrix such as intact pieces of

Table 14.1 Asbestos soil to air coefficients.

Fiber type	Soil type	Coefficient λ (f/cc per 1 mg/m^3 of respirable dust per weight % asbestos in soil)
Amosite	Clay	212
	Intermediate	799
	Sand	1,088
Chrysotile	Clay	434
	Intermediate	316
	Sand	363
Crocidolite	Clay	809
	Intermediate	1,889
	Sand	498

Source: Addison et al. (1988)/U.S. Department of Energy

asbestos-cement material will significantly limit the potential of fibers to be released in comparison to a material that is already broken down into fibers or is readily friable such as paper-based or mineral-based thermal system insulation (TSI) materials, or asbestos present as free fibers from different types of environmental releases.

Various methods can be used to evaluate the soil-to-air coefficient (similar to λ in the formula above).

For example, Berman (2000) proposed the following formula for determining asbestos emissions from unpaved road dust:

$$C_{asbestos} = 1.7kn\left(\frac{s}{12}\right)\left(\frac{V}{48}\right)\left(\frac{W}{2.7}\right)^{0.7}\left(\frac{w}{4}\right)^{0.5}\left[\frac{365-p}{365}\right]\left[\frac{2}{(2\pi)^{0.5}\sigma_z U}\right]R_{a/d} \quad (14.2)$$

where

$C_{asbestos}$ – average airborne asbestos concentration (f/cc),
k – the aerodynamic particle size multiplier,
s – the silt content of the soil in wt%,
V – the velocity of the vehicle in km/h,
W – the weight of the vehicle,
W – the number of wheels in the vehicle,
p – the number of days per year when precipitation exceeds 0.254 mm,
n – the frequency with which vehicles pass the road of interest in veh/s,
σ_z – the vertical dispersion parameter (m),
U – the wind speed (m/s),
$R_{a/d}$ – the measured ratio of releasable asbestos structures to releasable dust (s/g).

In this case, $R_{a/d}$ is the measured approximation of the ratio (or coefficient), similar to λ above, between the respirable dust and airborne asbestos concentrations, but expressed in different units. Berman found measurable $R_{a/d}$ coefficients at the level from 1.9 to 3.6 f/cc per 1 mg/m^3 of respirable dust (size distribution of particles not reported) when the asbestos content in soil by polarized light microscopy (PLM) method ranged from 7.5% to 10%. This ratio would correspond to a λ ranging from approximately 25–36, a range that is significantly lower than the λ measured by Addison for soil emission.

To determine the $R_{a/d}$ coefficient, Berman and Kolk developed a method based on a customized elutriator device designed to imitate small-scale disturbances of soil material and to measure fibers in the airstream generated by the device (US EPA 1997). In addition to this aerosolization method, a fluidized bed asbestos segregator preparation technique can be used for determination of the airborne asbestos concentration generated from low-level asbestos concentrations in soil and other solid media (Januch et al. 2013).

In general, the approach of estimating airborne asbestos concentrations based on respirable dust concentrations using mathematical coefficients has been utilized in various exposure assessment studies for media other than soil. For example, Boelter et al. (2015) evaluated the potential asbestos exposure levels during sanding of joint compounds utilizing site observations, dispersion modeling, bench-test simulation, and Monte-Carlo stochastic analysis. In this publication, respirable dust concentrations (in mg/m^3) were derived from the Jones et al. (2011) study, where field measurements and two-zone dust exposure modeling had been combined to evaluate the overall exposure distribution of workers during joint compound sanding activities. To calculate asbestos fiber concentrations based on respirable dust concentrations, coefficients were derived from a bench-scale chamber study that utilized a ready mix (wet mix) product, and a dry mix joint compound product "manufactured" for the study. The joint compound composition was based on a chrysotile-containing calcium carbonate formula historically manufactured by Georgia-Pacific LLC (GP) predecessors according to the original 1960s formulations. Chrysotile 7RF3 grade with weight % of 5.5 was used in the material formulation (Brorby et al. 2008). During the bench test, a median value of the coefficient between respirable dust concentration and asbestos fibers concentrations was found at the level of 0.044 (95% CI 0.039–0.050) f/cc per mg/m^3 (Sheehan et al. 2011). It is noteworthy that the coefficient determined in the joint-compound study was several orders of magnitude lower than the coefficient λ derived from the Addison study for soil emission.

The difference in measured or estimated material-to-air or soil-to-air coefficients can be explained by various factors. In particular, it should be noted that the weight of one fiber, a function of length, width, and density of specific fiber types can affect the coefficients.

In simpler terms, airborne asbestos concentration can be found from the following general formula:

$$C_{asbestos}(f/cc) = \text{Dust concentration}(mg/m^3) \times \text{Fraction of asbestos in dust}(\%)$$
$$/\text{Weight of a fiber}(mg)/1,000,000 \text{ fibers} \quad (14.3)$$

In particular, for chrysotile (assuming a shape close to a half-cylinder), we can derive the following relationship:

$$C_{asbestos}(f/cc) = \frac{8 \times C_{respirable\ dust}\left(\frac{\mu g}{m^3}\right) \times \text{Asbestos Fraction in Respirable Dust}}{\pi \rho D^2 L} \quad (14.4)$$

where

L (μm) – fiber length,
D (μm) – fiber width,
ρ (g/cm^3) – fiber density.
$C_{asbestos}$ – calculated airborne concentration of asbestos (f/cc),
$C_{respirable\ dust}$ – concentration of respirable dust in the breathing zone (μg/m^3),
Asbestos fraction in respirable dust – the fraction of Phase-Contrast Microscopy Equivalent (PCME) fibers in respirable dust sample.

For amphibole fibers, a rectangular model better approximates the shape of fibers, where the "width" and "thickness" of fibers are not identical (Korchevskiy and Wylie 2022).

The formula, equivalent to Equation (14.4), for asbestiform amphiboles will be as follows:

$$C_{asbestos}(f/cc) = \frac{3 \times C_{respirable\ dust}\left(\frac{\mu g}{m^3}\right) \times \text{Asbestos Fraction in Respirable Dust}}{\rho D^{1.7} L} \quad (14.5)$$

For nonasbestiform amphiboles, the following formula can be derived:

$$C_{asbestos}(f/cc) = \frac{1.9 \times C_{respirable\ dust}\left(\frac{\mu g}{m^3}\right) \times \text{Asbestos Fraction in Respirable Dust}}{\rho D^2 L} \quad (14.6)$$

If the asbestos fraction in respirable dust, assumed to be emitted from soils, is unavailable from direct measurements, this parameter can be evaluated based on the level of asbestos in bulk soil samples:

$$\text{Asbestos fraction in respirable dust}(\%) = f(\text{Asbestos fraction in soil}), \quad (14.7)$$

where f is a function that theoretically should be monotonically increasing with its argument.

With many limitations involved, the fraction of asbestos in respirable dust can be approximated by the fraction of asbestos in soil from which the respirable dust is derived (in this case, the function f would be linear or even constant). This would also correspond to the assumption that asbestos is homogenously distributed throughout the soil layer, that the fraction of asbestos is the same across all particle size fractions in the soil, and that the probability of asbestos fibers becoming airborne is the same as for other respirable particulates in soil.

In order to evaluate the ratio between asbestos concentration in soil vs. asbestos concentration in the airborne respirable dust, we used computational fluid dynamic (CFD) modeling.[1] It was assumed that a 200 m^2 land parcel with flat terrain emits soil particulates at the rate of 0.4 µg/m^2/min. A wind speed of 5 mph was assumed, and the fraction of respirable particulates in soil was assumed to be 20%. It was also assumed that loose chrysotile fibers, with an average length of 20 µm and a width of 0.2 µm, comprised 1% of soil by weight, uniformly. The emissions are supposed to happen at the (flat) ground level.

The CFD modeling included estimating the airborne asbestos and respirable dust concentrations at various distances from the dust generation source within the land parcel. Figure 14.1 depicts the CFD modeled airborne asbestos concentrations at various distances from the closest perimeter boundary of the land parcel (in f/cc), demonstrating the order and trend of the airborne asbestos contamination.

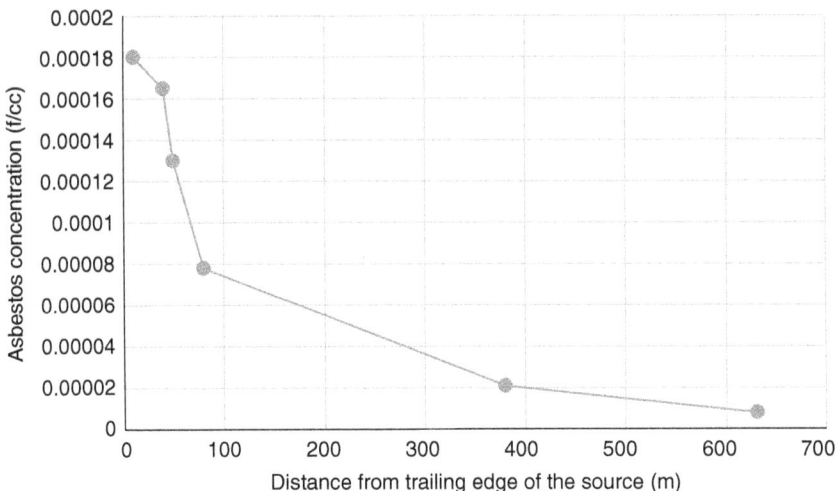

Figure 14.1 Asbestos concentration at different distances from the trailing edge of source, as modeled by CFD (f/cc).

---

1 Courtesy C. Strode and D. Hall of C&IH, Inc.

Using the CFD modeled results, we can estimate the relationship between the asbestos concentration in soil and the asbestos concentration in respirable dust (measured as weight %). The ratios between the modeled airborne asbestos mass concentrations and modeled respirable dust mass concentrations are plotted in Figure 14.2.

The CFD modeling results demonstrated that keeping the asbestos content in respirable dust equal to the asbestos content in soil (in this case, 1%) is reasonably conservative. Obviously, this relationship depends on numerous assumptions, some of which are described above.

For the maximum modeled airborne asbestos concentration (0.00018 f/cc), the corresponding respirable dust concentration was estimated by CFD at 0.159 µg/m^3, resulting in a λ coefficient of approximately 113 f/cc per 1 mg/m^3 per percent of asbestos in soil. This result is lower than Addison's estimations, but somewhere between Addison's λ and the λ calculated based on Berman's $R_{a/d}$ ratios. The average λ values calculated from the CFD modeling results at various distances are equal to approximately 112, a result that can be explained assuming similar distance-based decay ratios for asbestos dust and respirable dust.

The following case study is provided to demonstrate a quantitative human health risk assessment for exposure to soils contaminated with asbestos.

Let us assume that a site redevelopment project is planned for a former household and commercial waste landfill. Sampling and Transmission Electron Microscopy (TEM) laboratory testing for asbestos in bulk soil samples were performed. The results of the laboratory analyses are summarized in Table 14.2.

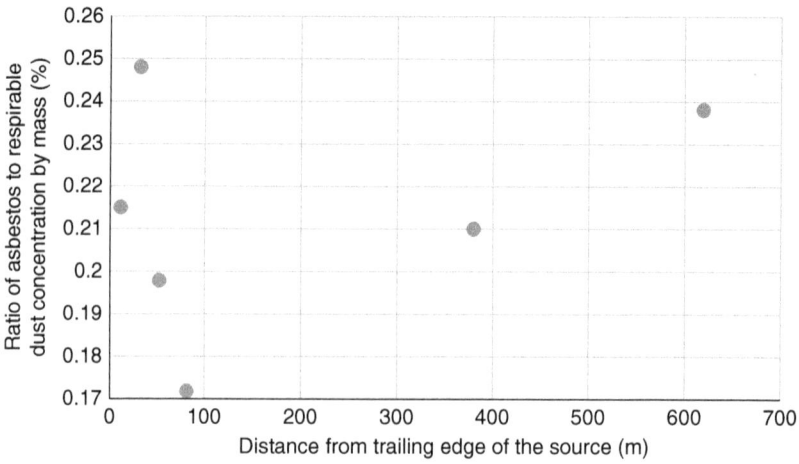

Figure 14.2 Airborne asbestos mass concentration with respect to total respirable mass concentrations (%).

Table 14.2 The results of soil sampling and analysis.

Sample #	Fiber type	Asbestos concentration (weight %)
1	NAD	
2	NAD	
3	NAD	
4	Chrysotile	<0.001
5	NAD	
6	NAD	
7	NAD	
8	Chrysotile	0.002
9	Amosite	<0.001
10	Chrysotile	<0.001
11	NAD	
12	Chrysotile	0.006
13	Chrysotile	0.001
14	NAD	
15	Chrysotile	0.003
16	Tremolite	<0.001
17	NAD	
18	NAD	
19	NAD	
20	Chrysotile	<0.001
21	Amosite	<0.001

NAD – No asbestos detected.

The asbestos analysis results for this case are heavily censored due to multiple nondetect data points. Different approaches could be used to address the censored data. Conservative methods would assume that some asbestos concentration less than the reported nondetect results, but greater than zero, is present. The most straightforward of these approaches is to replace the censored, "less than limit of quantification (LOQ)" sample results with the censored result divided by 2 (i.e. LOQ/2 where LOQ is equal to the reported limit of quantification). For the TEM method, we can assume the LOQ/2 = 0.0005%. We also assigned values of 0.00005% to the values where "no asbestos detected" was reported. It corresponds to the assumption that a level of 1 ppm of asbestos in soils can be theoretically detected by TEM if at least one fiber is present (HSE 2021).

Using the censored and uncensored data, a lognormal distribution can be applied to the chrysotile and amosite soil concentrations reported for the samples to determine the various statistical parameters associated with the soil concentrations using a stochastic approach. The resulting lognormal distributions for the chrysotile and amosite soil concentrations, derived from a standard statistical package (Statistica 13.0 and Crystal Ball), are illustrated in Figure 14.3.

Respirable dust concentrations can be modeled based on the available monitoring data; however, some assumptions are required about the respirable dust emitted from soil. For example, we can assume that the respirable dust concentration at the site would have a geometric mean of $12\,\mu g/m^3$ with a geometric standard deviation of 1.6. These assumptions result in a distribution similar to Figure 14.4. As previously, we would conditionally suppose that the area of interest is

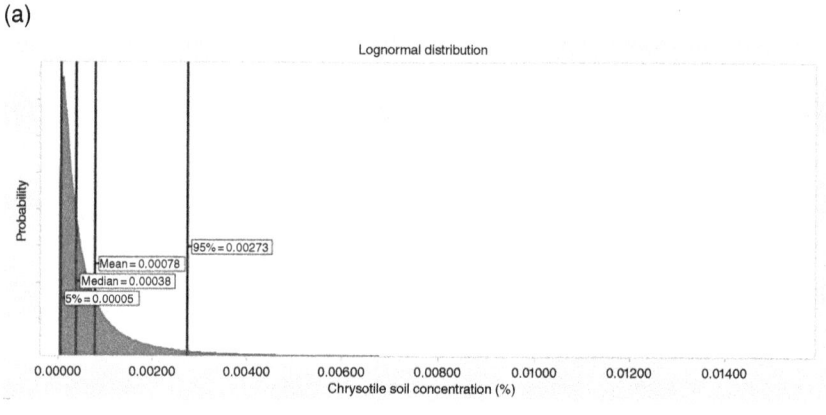

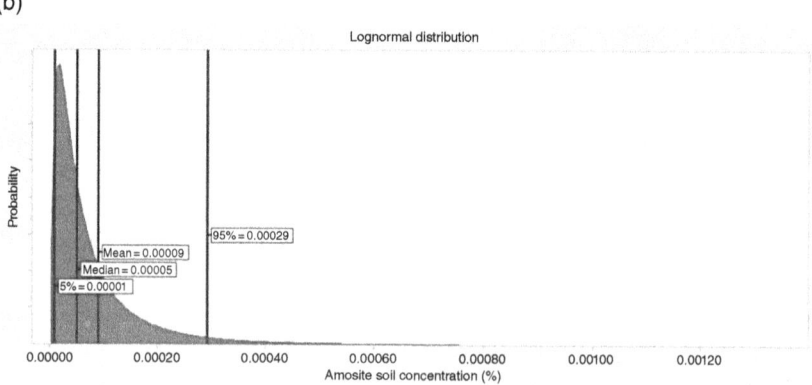

Figure 14.3 Distribution of the asbestos in soil sampling results (%). (a) Chrysotile concentration in soil (average 0.00078%, standard deviation 0.0014%), (b) amosite concentration in soil (average 0.00009%, standard deviation 0.00013%).

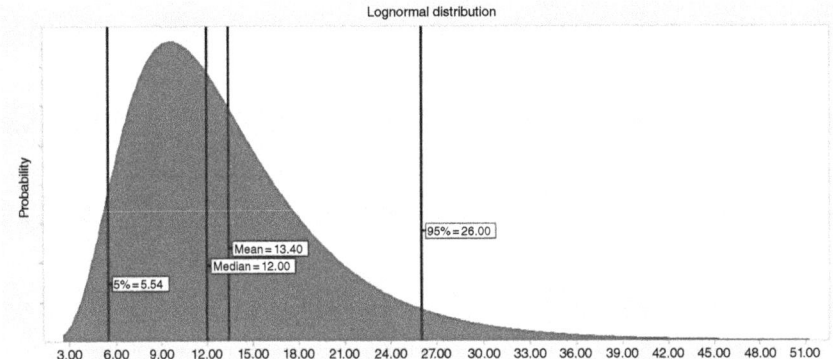

Figure 14.4 Distribution of the projected respirable dust concentrations ($\mu g/m^3$).

"separated" from the outside environment, and no other sources of asbestos emission are present other than the local soils.

Assuming that the dimensionality of the fibers was determined, we can estimate the potential airborne asbestos concentrations resulting from soil disturbance. For example, let us assume that for chrysotile, an average length of 30 μm was measured, with an average width of 0.35 μm. For amosite, an average length of 42 μm was measured, with an average width of 0.52 μm. The density of chrysotile was assumed to be equal to 2.53 g/cm^3, while the amosite density was assumed to be 3.2 g/cm^3. We will also conditionally assume that the fraction of asbestos fibers in soil and respirable dust is equivalent. In practice, we would potentially like to test the specific soil type for respirable fraction content and releasability of asbestos fibers.

Using the listed assumptions, we can apply Equations (14.4) and (14.5) to the available data to estimate the potential airborne asbestos concentrations for each type of asbestos fiber. Figures 14.5 and 14.6 demonstrate the projected empirical distributions of airborne concentrations for the site of interest for chrysotile and amosite, respectively.

With the potential airborne asbestos concentrations determined, we can utilize these data to assess the exposures and risks associated with various site use scenarios. For example, we can focus on the site redevelopment activities, with respirable dust concentrations related to specific disturbance scenarios impacting the soil at the site. We will assume that there is a potential exposure to on-site or off-site workers and other individuals present on or near the site. The duration of the project will be 2–4 years, with a daily exposure duration of 10–12 hours per day for 5 days per week. For a conservative estimation, we can assume age zero as the date of exposure onset (for children, potentially living close to the site). Also, we could conservatively assume that the distribution of the on-site airborne

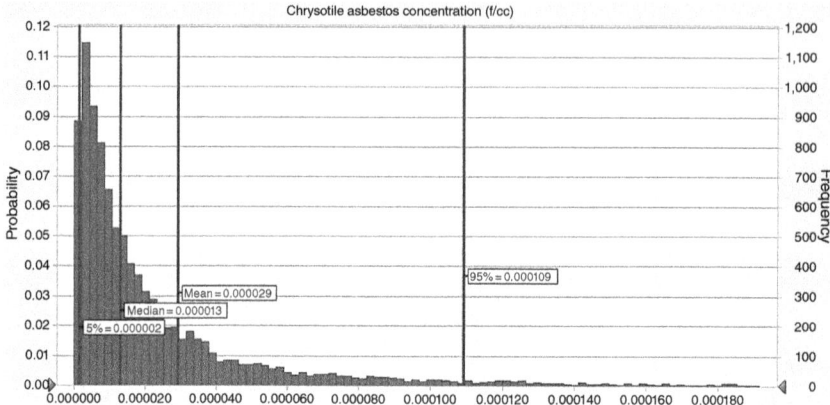

Figure 14.5 Projected airborne chrysotile asbestos concentrations (f/cc).

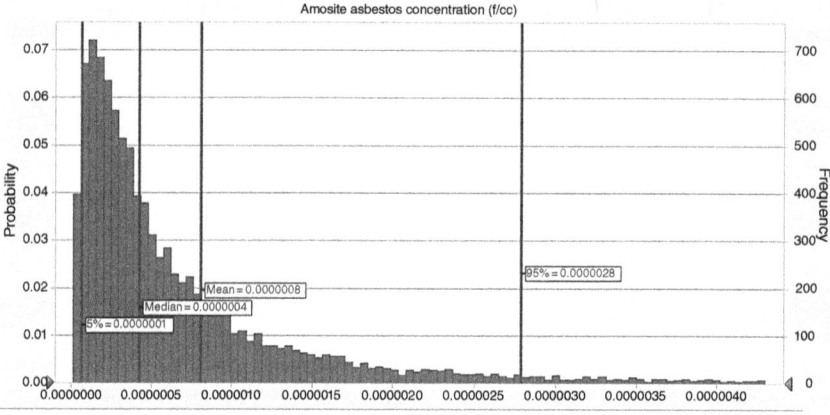

Figure 14.6 Projected airborne amosite asbestos concentrations (f/cc).

concentrations is the same as the off-site asbestos concentrations outdoors at any adjacent homes. We will ignore soil transfer to homes and other possible soil-related exposure routes in this example; however, these potential exposure pathways could be relevant under other scenarios (e.g. children playing on the site prior to or during development).

Using the on-site airborne asbestos statistics, we can calculate the excess mesothelioma mortality distribution using the Hodgson and Darnton linear method for the exposure scenario above (see Chapter 8 for the details of the method). The results of this analysis, as excess cases of mesothelioma per million per lifetime, are demonstrated in Figure 14.7.

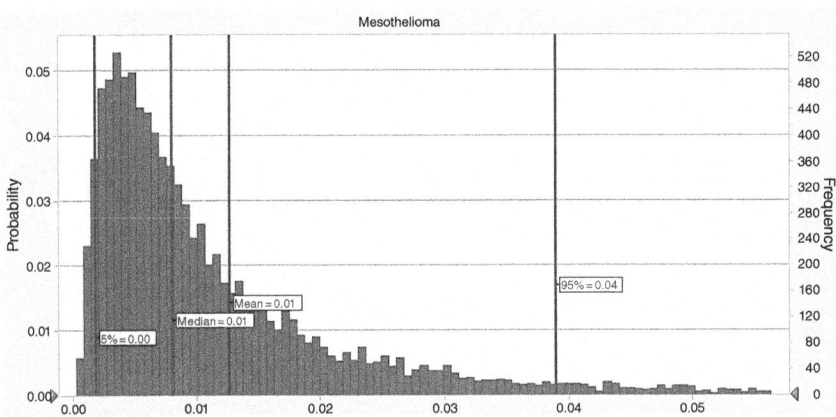

Figure 14.7 Excess mesothelioma mortality (per 1,000,000 per lifetime).

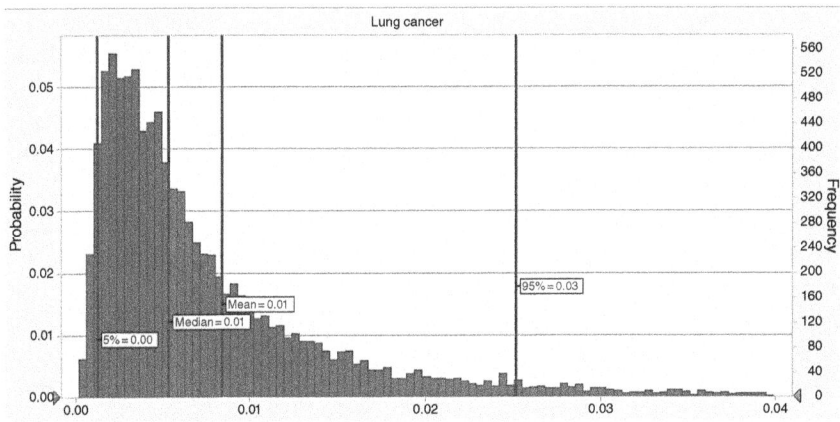

Figure 14.8 Excess lung cancer mortality (per 1,000,000 per lifetime).

An excess lung cancer mortality distribution, also calculated using the Hodgson and Darnton linear method for the same exposure scenario, is demonstrated in Figure 14.8.

The excess total asbestos cancer (i.e. lung cancer plus mesothelioma) mortality distribution, as calculated using the Hodgson and Darnton linear method for the same exposure scenario, is demonstrated in Figure 14.9.

It can be seen that total excess respiratory cancers attributable to asbestos for this scenario should not statistically exceed the threshold of one case of excess cancer per 1,000,000 per lifetime.

Our study on the risk modeling for asbestos in soil has uncertainties and limitations. One of them is related to potential flaws in the quantitative characterization

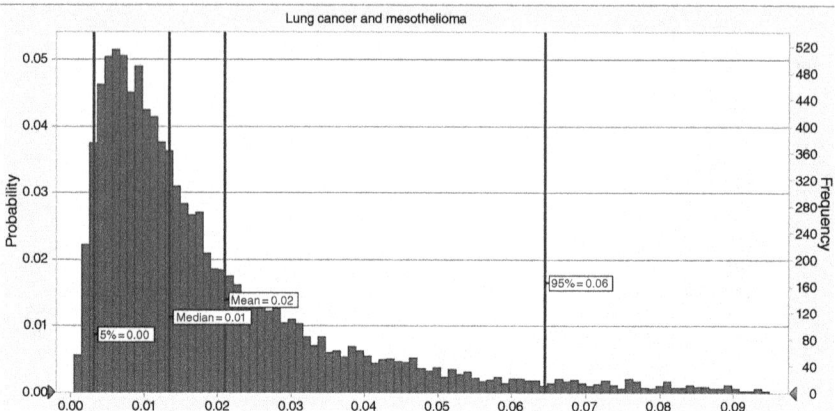

Figure 14.9 Excess lung cancer and mesothelioma mortality (per 1,000,000 per lifetime).

of asbestos content in soil in general. However, in spite of all analytical difficulties, asbestos soil content could still be assessed, theoretically or practically. For example, weight fraction of asbestos in soil is widely measured as a part of risk characterization in the United Kingdom (SoBRA 2020). The assumption of the homogeneity of asbestos in soil also seems to be an oversimplification of the reality. However, the content of asbestos in soil can be assumed from a theoretical angle: for example, the amount of asbestos migrating through the soil layers can be approximated from mass-balance considerations. Risk assessors can make reasonable estimations of the asbestos soil content that are not supposed to be less valid than lead (Shi et al. 2021) or per-and polyfluoroalkyl substances (PFAS) (Tang et al. 2020) content levels used for risk assessment purposes. We demonstrated various approaches for calculating the airborne asbestos concentration from the asbestos in soil fraction and related human health risks. This case study may be relevant to and potentially provides a road map for setting standards and designing monitoring programs and possible remediation measures.

References

Addison, J., Davies, LS.T., Robertson, A., et al. (1988). Release of dispersed asbestos fibres from soils. Technical Report # PB-89-170716/XAB; TM-88/14, United Kingdom. PB-89-170716/XAB; TM-88/14 Release of dispersed asbestos fibres from soils (Technical Report) | ETDEWEB (osti.gov).

Berman, D.W. (2000). Asbestos measurement in soils and bulk materials: sensitivity, precision, and interpretation – You can have it all. In: *Advances in Environmental*

Measurement Method for Asbestos (eds. M.E. Beard and H.I. Rook), 70–89. ASTM STP 1342, American Society for Testing and Materials. Advances in Environmental Measurement Methods for Asbestos – Michael E. Beard - Google Books.

Berman, D.W., Cox, L.A. Jr., and Popken, D.A. (2013) A cautionary tale: the characteristics of two-dimensional distributions and their effects on epidemiological studies employing an ecological design. *Critical Reviews in Toxicology* 43(Suppl 1): 1–25. https://doi.org/10.3109/10408444.2013.777688.

Boelter, F.W., Xia, Y., and Dell, L. (2015). Comparative risks of cancer from drywall finishing based on stochastic modeling of cumulative exposures to respirable dusts and chrysotile asbestos fibers. *Risk Analysis* 35(5): 859–871. https://doi.org/10.1111/risa.12297.

Brorby, G.P., Sheehan, P.J., Berman, D.W., et al. (2008). Re-creation of historical chrysotile-containing joint compounds. *Inhalation Toxicology* 20(11): 1043–1053. https://doi.org/10.1080%2F08958370802290595.

HSE (Health and Safety Executive). (2021). Asbestos: the analysts' guide. HSG248. May. UK. Asbestos: The Analysts' Guide - HSG248 (hse.gov.uk).

Januch, J., Brattin, W., Woodbury, L., et al. (2013). Evaluation of a fluidized bed asbestos segregator preparation method for the analysis of low levels of asbestos in soil and other solid media. *Analytical Methods* 5: 1658–1668. https://doi.org/10.1039/C3AY26254E.

Jones, R.M., Simmons, C., and Boelter, F. (2011). Development and evaluation of a semi-empirical two-zone dust exposure model for a dusty construction trade. *Journal of Occupational and Environmental Hygiene* 8(6): 337–348. https://doi.org/10.1080/15459624.2011.576330.

Korchevskiy, A.A. and Wylie, A.G. (2022). Dimensional characteristics of the major types of amphibole mineral particles and the implications for carcinogenic risk assessment. *Inhalation Toxicology* 34(1–2): 24–38. https://doi.org/10.1080/08958378.2021.2024304.

Pan, X. Day, H.W., Wang, W., et al. (2005). Residential proximity to naturally occurring asbestos and mesothelioma risk in California. *American Journal of Respiratory Critical Care Medicine* 172(8): 1019–1025. https://doi.org/10.1164%2Frccm.200412-1731OC.

Schreier, H. (1989). *Asbestos in the Natural Environment*, 1st Edition. Elsevier, Vancouver. 9780080874968.

Sheehan, P.J., Brorby, G.P., Berman, D.W., et al. (2011). Chamber for testing asbestos-containing products: validation and testing of a re-created chrysotile-containing joint compound. *Annals of* **Occupational Hygiene** 55(7): 797–809. https://doi.org/10.1093/annhyg/mer048. Epub 2011 Jul 26. PMID: 21795244.

Shi, J., Du, P., Luo, H., et al. (2021). Characteristics and risk assessment of soil polluted by lead around various metal mines in China. *International Journal of*

Environmental Research and Public Health 18(9): 4598. https://doi.org/10.3390%2Fijerph18094598.

SoBRA. (2020). The Distribution of Asbestos in Soil – what can the data mining of sample results held by UK laboratories tell us. Discussion Paper by the SoBRA asbestos sub-group, March 2020. The Distribution of Asbestos in Soil – what can the data mining of sample results held by UK laboratories tell us? – SoBRA.

Tang, L., Liu, X., Yang, G., et al. (2020). Spatial distribution, sources and risk assessment of perfluoroalkyl substances in surface soils of a representative densely urbanized and industrialized city of China. *CATENA* 198: 105059. https://doi.org/10.1016/j.catena.2020.105059.

US EPA (Berman and Kolk). (1997). Superfund method for the determination of asbestos in soils and bulk materials. Prepared for the Office of Solid Waste and Emergency Response. EPA 540-R-97-028. Document Display | NEPIS | US EPA.

15

Asbestos in Brakes: Risk Assessment for Exposure Patterns with Nonlinear Dynamics

Andrey Korchevskiy[1], Robert Strode[2], and Arseniy Korchevskiy[1]

[1] Chemistry & Industrial Hygiene, Inc., Lakewood, CO, USA
[2] Summit Exposure and Risk Sciences LLC, Silverthorne, CO, USA

Historically, asbestos-containing friction products such as vehicle brakes and clutches have been known to contribute to environmental (nonoccupational) asbestos exposure. As might be expected, the contribution of these anthropogenic sources to the otherwise ambient asbestos concentrations was generally much greater in areas with overall high traffic volume and/or high commercial vehicle use. As a result, there was greater concern for exposed urban populations vs. less populated areas.

Utilization of asbestos in vehicle brake linings and pads was widespread throughout the world during much of the twentieth century. It was not until the 1980s that practical alternatives to asbestos became widely available, with nonasbestos materials gradually replacing the asbestos components during the last few decades (Paustenbach et al. 2004). According to various sources, drum brake linings typically contained 30–70% chrysotile asbestos by weight, and disc brake pads contained 10–30% chrysotile asbestos by weight (Kauppinen and Korhonen 1987). Lemen (2004) noted that the asbestos used in brakes was almost exclusively chrysotile, as the amphibole asbestos type tended to be too harsh and tended to score the brake drums, making them wear much faster.

Different quantitative estimations have been made regarding the asbestos emissions related to asbestos-containing brakes. For example, brake wear has been recognized as one of the most important nonexhaust traffic-related particulate emission sources. Some authors have estimated that brake-wear particles account for 11–21% of the total traffic-related PM_{10} emissions (Grigoratos and Martini 2015). In addition, it has been reported that the traffic-generated emissions account for

Health Risk Assessment for Asbestos and Other Fibrous Minerals, First Edition.
Edited by Andrey Korchevskiy, James Rasmuson, and Eric Rasmuson.
© 2024 John Wiley & Sons, Inc. Published 2024 by John Wiley & Sons, Inc.

more than 50% of the PM_{10} emissions in the urban areas (Bathmanabhan and Madanayak 2010). At the same time, Williams and Muhlbaier (1982) indicated that the chrysotile asbestos fraction in dust from car brake operations was observed in the range from 0.001% to 0.19%, with an average of 0.03%, assuming that 99.9% of the original asbestos fiber mass was converted through the fosterite formation, or through other degradation processes, into nonfibrous magnesium silicates or other particles.

Considering the asbestos emission from brakes, it should be noted that the behavior of traffic-related emissions is complex and often can be described in terms of nonlinear dynamics. In particular, Lo and Cho (2005) noted that discrete dynamic traffic models can be described in terms of chaos theory.

The theory of chaos has become a promising approach applicable in the areas of modern thermodynamics, chemistry, medicine, and social sciences (Faber and Koppelaar 1994; Oestreicher 2007; Kondepudi et al. 2017). Although there are numerous definitions of chaotic systems, a behavior with sensitive dependence on initial conditions is a common description of a chaotic system (Effah-Poku et al. 2018). The usefulness of chaos theory has been demonstrated in various studies including models of chaotic behavior that were successfully applied to the infectious disease epidemiology (Philippe 1993).

In this chapter, we will propose two examples where the car emissions dynamics can be described in terms of the chaos theory. We will demonstrate the behavior of asbestos air contamination from the brake wear that consequently would also follow the nonlinear time trends.

Ambient Air Emissions from the Brakes in Street Canyons

Street canyons are one of the widespread examples of complicated air dynamics affecting the dispersion of the pollutants. In order to find the concentration of brake-related asbestos in a street canyon environment, we first separated concentration C (g/m^3) into two of its constituent parts (Berkowicz et al. 1997):

$$C = C_d + C_r$$

where C_d is the direct contribution from the brakes and C_r is the contribution due to recirculation. The Operational Street Pollution Model (OSPM) was used to quantify C_d and C_r. We began with calculating C_d, which requires us to first evaluate several parameters. In particular, the width of the street in meters, which we label W, was assumed to be 15 m. Another parameter, h_0, which is the initial dispersion height, was assumed to be 2 m. The proportionality constant, α, is assumed to be 0.1, since this value corresponds with typical levels of mechanically induced turbulence. Q is the emission level in the street, expressed in g/m/s (grams per meter per second). Wind speed, at the ground level, is labeled u_b (m/s) (we assumed it to be 1 m/s). Finally, the vertical turbulent velocity fluctuation was calculated, which is labeled σ_{wt} (Berkowicz et al. 1997). Now, to find the mechanical turbulence in the canyon, σ_w, we use the following equation:

$$\sigma_w = \left((\alpha u_b)^2 + \sigma_{wo}^2 \right)^{1/2}$$

Here, σ_{wo} is the traffic-related turbulence, which we have assumed to be 0.1 for the purposes of our analysis. Now, to calculate C_d, we used the following:

$$C_d = \sqrt{\frac{2}{\pi}} \frac{Q}{W\sigma_w} \ln \frac{h_0 + \left(\frac{\sigma_w}{u_b}\right)W}{h_0}$$

For simplicity, the wind direction perpendicular to the traffic was assumed. Next, we determined σ_{wt}, which is the canyon ventilation velocity, determined by turbulence at the top of the canyon. In order to do this, we need two more pieces of information. In the previous equation, u_b was the wind speed at the ground level, but in this case we need to use u_t, which is the wind speed at the top of the canyon, in meters per second (Wang et al. 2017). The complete equation to calculate the canyon ventilation velocity, σ_{wt}, is as follows:

$$\sigma_{wt} = \left((\zeta u_t)^2 + 0.4\sigma_{wo}^2 \right)^{1/2}$$

where ζ is the proportionality constant that we assume to be equal to $\alpha = 0.1$.
Finally,

$$C_r = \frac{Q}{\sigma_{wt} W}$$

(assuming the simple case when the vortex is totally immersed inside the canyon).

The following approach was used to quantify the asbestos emission source strength Q. The PM_{10} emission from brake wear from passenger cars was estimated as in US EPA (2014) at the level of 29.8 mg/mile per car. The fraction of passenger vehicles containing asbestos brakes in the United States was conservatively estimated at 13%, based on the data from the California Air Resources Board (CARB 2007). The chrysotile fraction from car operation dust was conservatively assumed to be 0.19%, as in Williams and Muhlbaier (1982). The fraction of fibers longer than 5 µm in the brake emissions was assumed to be about 1/300 of the total chrysotile emission (Lemen 2004).

We used the model of traffic density as proposed by Lo and Cho (2005):

$$\rho_{n+1} = \lambda \rho_n (1-\rho_n)$$

where

ρ_n – ratio between car density at the moment n and traffic jam density,
λ – ratio between the free flow speed and the actual average speed.

The free flow speed in the street canyon was estimated at 40 mph, jam density of 350 passenger car units (pcu) per mile, with an initial density of 10 pcu/mile.

The source strength Q (g/m/s) was found as a product of car density and chrysotile emissions per meter, divided by the time, in seconds, for the car to travel 1 m.

Figure 15.1 illustrates the dynamics of chrysotile concentrations (f/cc) in a street canyon assuming an average speed of 50 mph (a), 35 mph (b), and 10.5 mph (c).

The concentrations of chrysotile behave differently depending on the average speed of cars in the canyon. Concentrations decrease with average speeds close to the free flow speed: the increase starts at a speed of 35 mph; then, the behavior of concentrations becomes chaotic at the critical speed of 10.5 mph.

Figure 15.2 shows the dynamics of maximum and average concentrations of chrysotile depending on the average traffic speed.

For example, the average chrysotile concentration is estimated at the level of 3.8×10^{-7} f/cc with an average speed of 50 mph, while it is expected to reach 2.5×10^{-5} f/cc at the average speed of 13 mph, slightly decreasing with the chaotic trend to the level of 2.0×10^{-5} f/cc at 10.5 mph. At the same time, the highest concentrations are reached not at the starting point of the chaotic trend, but preceding it; the maximum concentration of 3.1×10^{-5} f/cc and highest average concentration of 2.9×10^{-5} f/cc are both reached at the average speed of 19.9 mph.

In chaotic systems, the parameters of interest depend significantly on the initial conditions. For example, Figure 15.3 demonstrates the dependence of the chrysotile concentrations on initial traffic density at the street canyon.

The risk assessment methods for asbestos allow us to project and evaluate the human health risks related to modeled concentrations of chrysotile fibers longer than 5 µm in street canyons. Assuming, for example, 35 years of residency at one of the buildings walling the street canyon, starting at birth, and conservatively assuming 24 hours/7 days per week frequency of exposure, we can relate the average concentration of 2.9×10^{-5} f/cc, with the excess risk of 0.084 cases of mesothelioma and 0.128 cases of lung cancer per 1,000,000 population per lifetime (according to the Hodgson and Darnton method) (see Chapter 8). This level is obviously negligible. The emission from the brakes of passenger cars does not significantly increase the risk of cancer in the population living by the street canyon under the given conditions. The level of risk would not be significant even if

Ambient Air Emissions from the Brakes in Street Canyons | 431

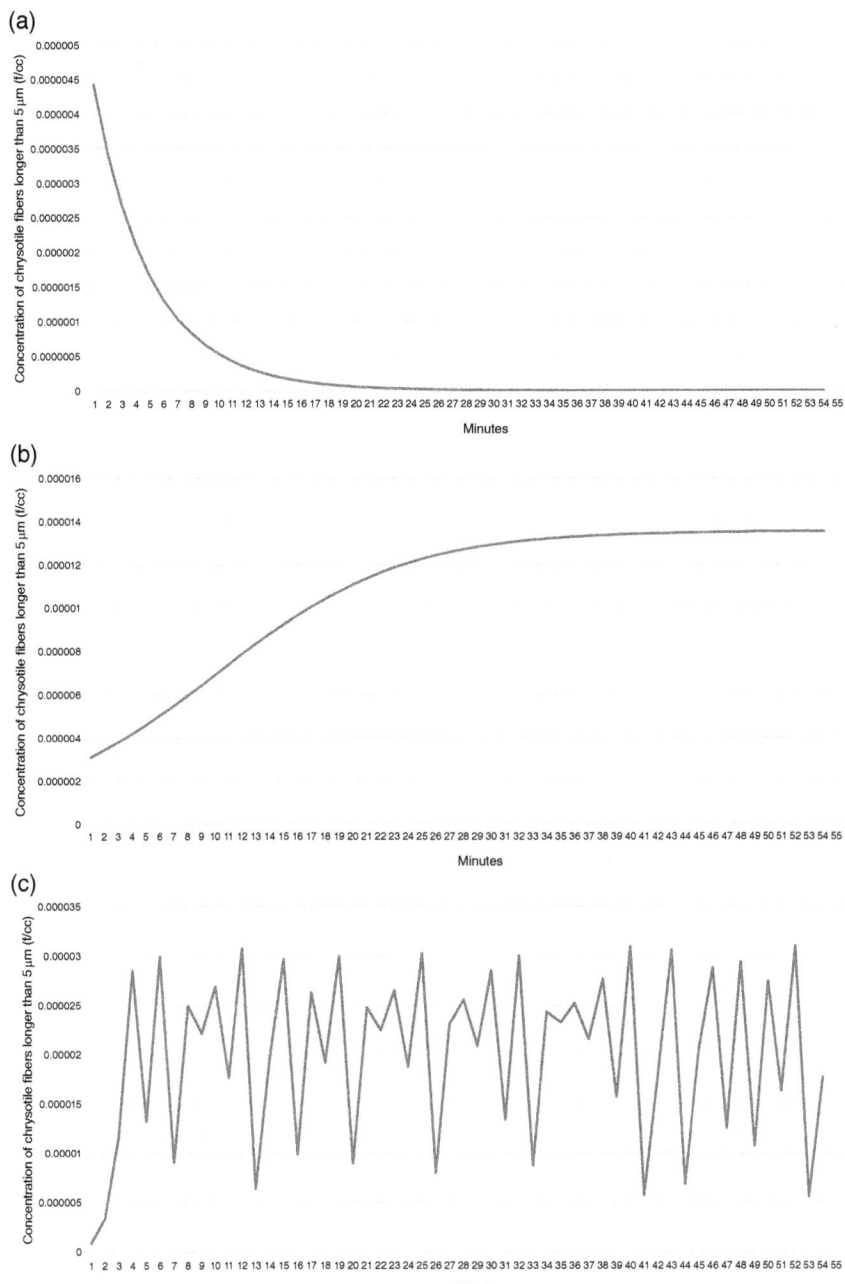

Figure 15.1 (a)–(c) Chrysotile concentrations in the street canyon depending on the average speed of vehicles (f/cc).

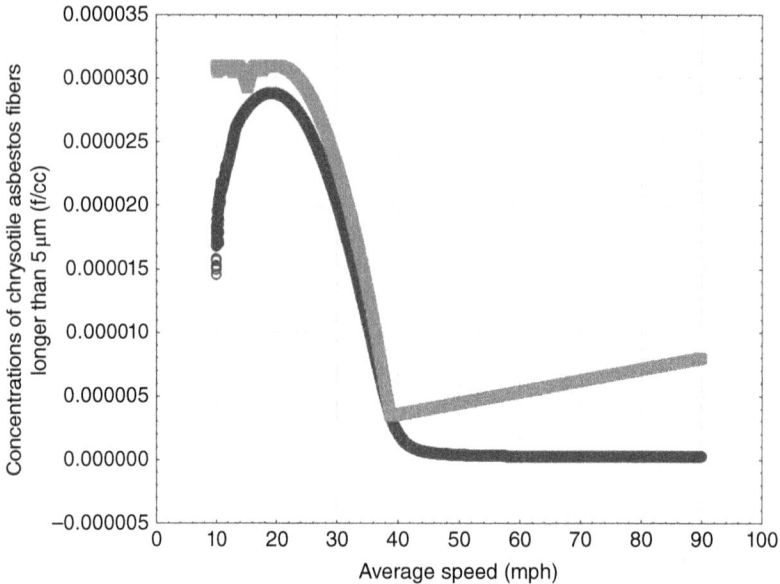

Figure 15.2 Chrysotile concentrations in the street canyon depending on the average traffic speed. The squares – maximum concentrations, the circles – average concentrations.

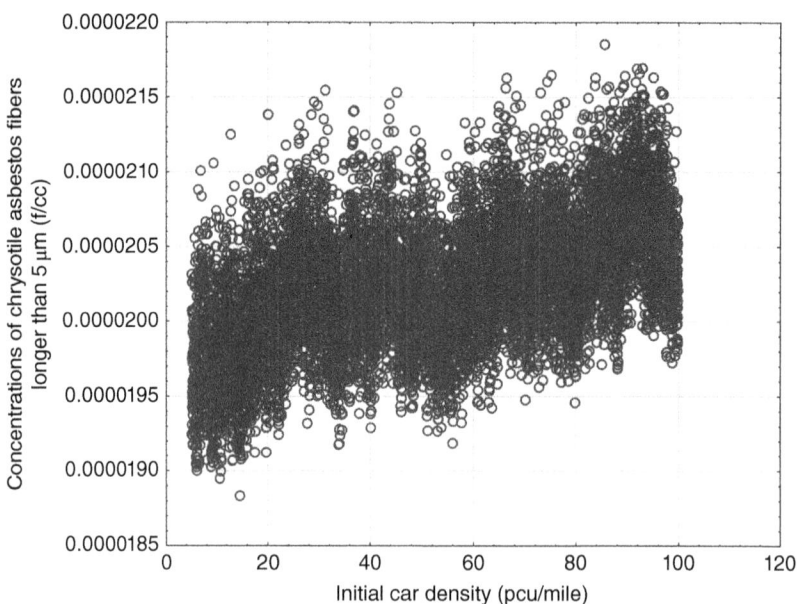

Figure 15.3 Chrysotile concentrations depending on initial traffic density at the street canyon (assuming an average speed of 10.5 mph).

we were to recreate the conditions where a majority of vehicles used chrysotile-containing brake materials.

Also, for developed countries, asbestos in brakes is mostly a problem of the past. However, asbestos is still used in brake manufacturing in various regions of the world. Additional attention is needed to evaluate the exposure and control it below acceptable risk levels.

Fibers in Car Brakes: Chaotic Behavior of Emissions in a Self-regulated Community

We can explore the following hypothetical example to demonstrate the contribution of individual car owners (and individuals in general) in determining risks, and how this can be used in decision-making management.

In 1994, Brian Arthur developed a statistical problem illustrating inductive reasoning in a group of people (Arthur 1994). The problem was inspired by a desire of a commercial establishment, the El Farol Restaurant in Santa Fe New Mexico, to maximize the enjoyment of its patrons during an Irish musical program offered on Thursday nights. However, space was limited, and the evening was supposed to be enjoyable as long as the establishment was not too crowded, and specifically, only if the occupancy was fewer than 60 people.

To solve this problem, it was assumed that N people independently decide to go to the establishment. Arthur used $N = 100$ for concreteness. Since there is no sure way to predict the numbers coming in advance, an individual that is aware of the 60-person enjoyment threshold (i.e. the tolerance threshold) will go to the establishment only if he expects fewer than 60 people to be present, but will not go if they expect 60 or more people to be present. The only information available is the occupancy in previous weeks. Arthur performed simulation modeling on the bar attendance (using at least six predictive models available for the population) and listed several important observations.

First of all, he noticed the dynamics of the attendance to be "chaotic" in the sense that there were "no persistent cycles" in the attendance data. At the same time, mean attendance always converged to 60, the tolerance threshold. Also, on average, 40% of the predictors forecasted occupancy greater than 60, and 60% forecasted occupancy less than the tolerance threshold (Arthur called this pattern the "ecology" of the system). Arthur also noted that "there is no deductively rational solution – no 'correct' expectational model." For example, if all believe few will go, all will go, invalidating this assumption as the tolerance threshold will always be exceeded. Arthur quoted one of Yogi Berra's famous ironic quotes: "Oh, that place. It's so crowded nobody goes there anymore."

It is easy to see that the El Farol problem can be generalized, with applications to market share price determination, traffic congestion, and other seemingly random problems (Marsili et al. 2000; Chen and Gostoli 2013). Various models were developed for the interpretation of the El Farol problem, especially in terms of game theory (Baccan and Macedo 2013). However, the model has not been applied to the situations of environmental risk assessment and management.

The following hypothetical is provided as an example of how one might utilize chaos theory in assessing environmental risks and how to manage those risks.

Let us assume that there are 102 vehicle owners in the town of Fibersville. Each car has brakes containing a small fraction of "fibrette," a hazardous fibrous mineral. When an individual vehicle is traveling on the road during the day, the concentration of airborne fibrette contributed to the ambient environment increases by three structures per cubic centimeter of air (s/cc). Also, without any additional vehicle emissions, the background fibrette concentration attains a steady-state concentration of 20 s/cc. The citizens of Fibersville consider 150 s/cc as the level of concern (i.e. the tolerance threshold per Arthur 1994) based on the published epidemiological studies indicating the potential to produce an uncurable tumor of the bronchi. It should be noted that all available epidemiologic data for fibrette cancer potency are based on occupational cohort cases, where people were exposed to at least 2,000 s/cc of fibrette on a daily basis.

Each morning the citizens of Fibersville read "The Fibersville Chronicles" to see the average exposure concentration trends in the city for the last 2 weeks. Based on this information, each citizen can predict the fibrette concentration for the current day.

Five different strategies to predict the fibrette exposure are used by some of the citizens:

1) Concentration from yesterday.
2) Concentration from 2 weeks ago.
3) Average concentration for the last 3 days.
4) Average concentration for the last 10 days.
5) Average concentration for the last 14 days, excluding the last 4 days.

Each citizen would use their car during the day if their individual predictions do not exceed a tolerance threshold of 150 s/cc. Otherwise, the individual would use a bicycle or walk during the day.

A simulation study was performed to calculate the fibrette concentration in Fibersville over time. The initial sequence of fibrette concentrations in the simulation was assumed to be 160, 140, 130, 170, 90, 160, 45, 180, 150, 155, 165, 55, 200, and 250 s/cc for the first through the 14th day.

The results for a simulation for the first 6 months of the year are demonstrated in Figure 15.4.

Fibers in Car Brakes: Chaotic Behavior of Emissions in a Self-regulated Community | **435**

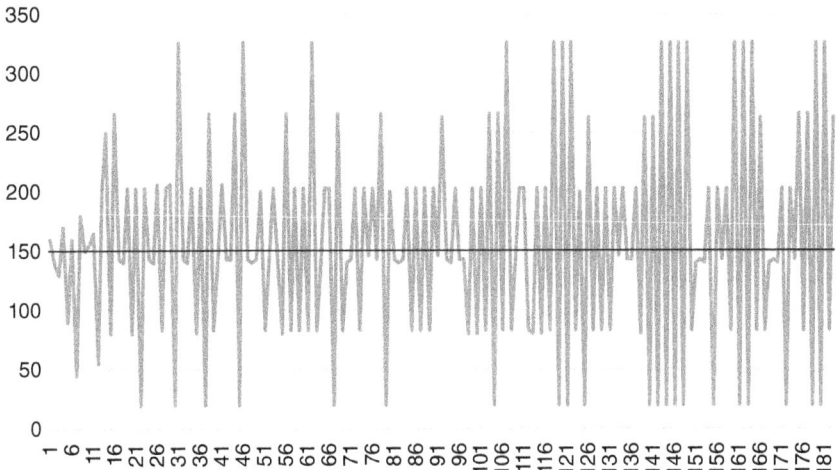

Figure 15.4 Simulated fibrette concentration in the air of Fibersville, as predicted by independent citizens (black line equals the tolerance threshold of 150 s/cc).

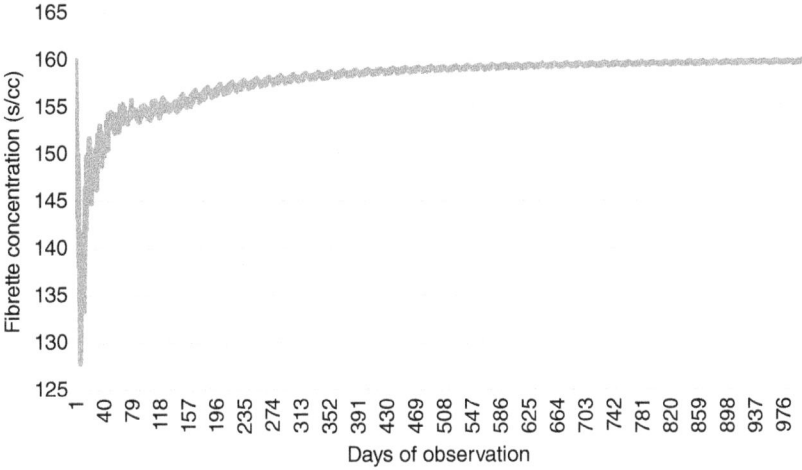

Figure 15.5 Moving average of the simulated concentration of fibrette in the air of Fibersville, as predicted by independent citizens.

The moving average for the first 1,000 days is illustrated in Figure 15.5.

The simulation generally confirms the observations of Arthur in 1994, but with some differences. The average concentration of fibrette in the air for the first 1,000 days of simulation is close to the level of concern (tolerance threshold), but slightly greater (average 159.9 s/cc, standard deviation 98.17 s/cc). Our observation is that the moving average continues to slowly increase with time, eventually leveling out at approximately 160 s/cc. Approximately 46% of residential predictions were below 150 s/cc, 54% above 150 s/cc, basically the inverse of the original El Farol data where 40% of the population forecasted occupancy greater than 60 and 60% forecasted occupancy less than 60. On average, the citizen predictions for the El Farol occupancy were correct in 54.5% of the cases and wrong in 45.5% of the cases. The actual airborne fibrette concentration was above 150 s/cc in 41.8% of the cases and below 150 s/cc in 58.2% of the cases. If no predictions were applied, and no one chose alternative transportation, the airborne fibrette concentration could reach the level of 326 s/cc 100% of time (i.e. 102 vehicles times 3 s/cc/vehicle, plus 20 s/cc background).

It is possible that the difference between the fibrette simulation and the El Farol simulation is related to the inclusion of a background fibrette level for the transportation prediction, whereas there was no background level of patrons in the El Farol occupancy decision (i.e. the El Farol occupancy predictions assumed no steady-state occupancy and occupancy would always include several people not participating in the overall predictive process).

Not submerging too deep into the mathematical metrics of chaos, we can generally observe chaotic behavior in the trend of fibrette airborne concentrations. There is no self-repeating cycle that would reveal itself, though the set of possible concentration values is generally limited. It is very interesting that in the simulation, the concentrations of fibrette are not random; they are fully and deterministically defined by the fixed algorithm methods of forecasting used by each of the citizens. However, the behavior of the airborne fibrette concentration imitates a random process. The trend of air contamination in Fibersville, being predicted by independent and ineffectively communicating citizens, is unstable, limiting our ability to predict the risk based on the initial conditions. As Prigogine (1989) noticed, "there is no risk in a deterministic universe." At the same time, the concept of instability provides us with the key to understand chaotic processes which can be understood as deterministic systems behaving in random-like patterns (i.e. the behavior of a chaotic system appears random, but the chaotic system is generated by simple, nonrandom, deterministic processes) (Rickles et al. 2007).

Chaotic processes are considered to be very sensitive to the initial conditions. When one observes the fibrette concentration on day 1,000 of the simulation, the fibrette concentration is 83 s/cc (Figure 15.6a). Note that if we change just one of the data inputs, the value from day 14, from 250 to 240 s/cc, the concentration on day 1,000 will change to 326 s/cc (Figure 15.6b). However, if we change the day 14 concentration to 241 s/cc, the result again would be 83 s/cc (Figure 15.6c).

(a)

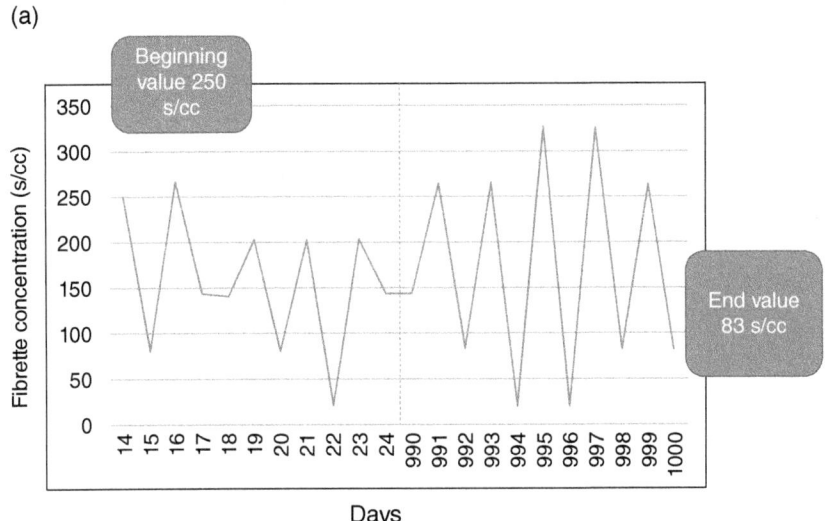

(b)

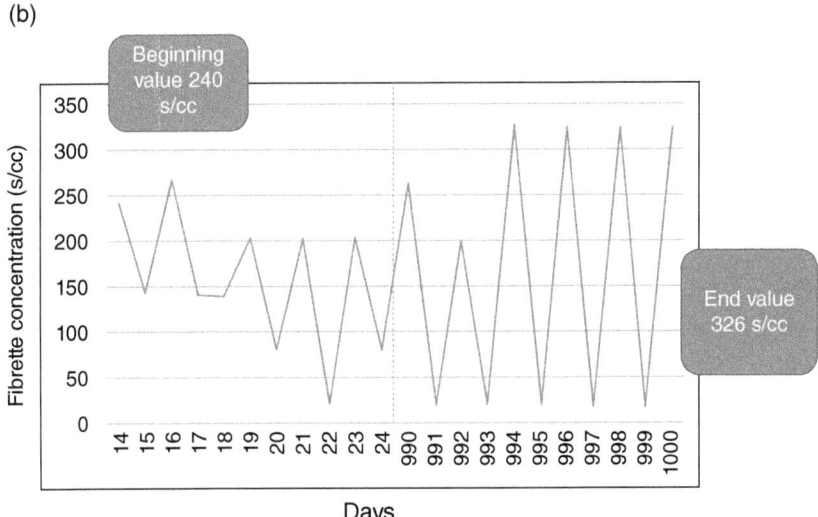

Figure 15.6 Concentration of fibrette at day 1,000 depending on the changing concentration at day 14. (a) Original simulation of the fibrette concentration.
(b) The simulation with the concentration for day 14 changes from 250 to 240 s/cc.
(c) The simulation with the concentration for day 14 changes from 250 to 241 s/cc.

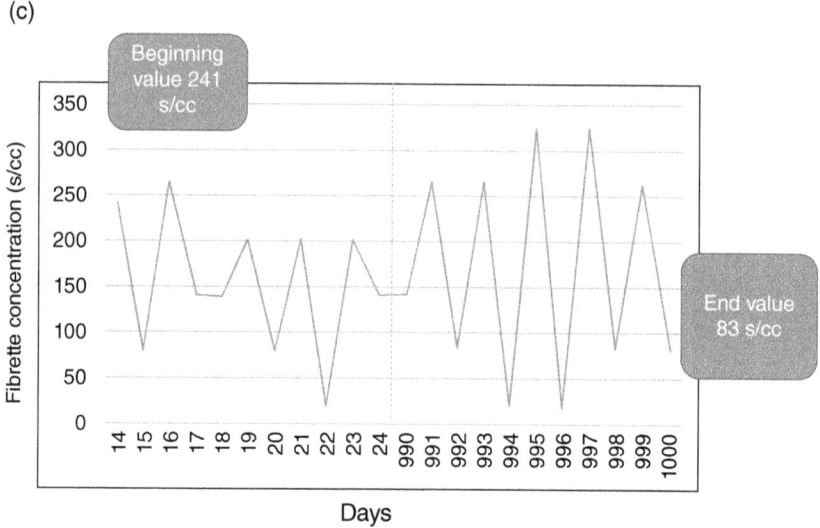

Figure 15.6 (Continued)

The following general observations can be made based on the provided simulation:

The cumulative exposure of the Fibersville's population was kept at the level of approximately 160 s/cc-year per each year of exposure, only slightly exceeding the tolerance threshold of 150 s/cc. However, the individual car owners, in spite of their relative success, were not able to reach their goal of reducing the exposure below an average of 150 s/cc for each day of exposure. The measure of the effectiveness of their efforts depends on our understanding of the "level of concern" or "tolerance threshold" for the fibrette emissions. Based on Rappaport and Kupper (2008), traditional occupational exposure limits (OELs) are intended to keep average exposure (i.e. the time-weighted average) below the defined OEL; however, they are not intended to require every instantaneous measurement collected throughout the weighted time interval to be under the limit. The same also would be true for environmental exposure limits based on an average (i.e. daily or annual) target concentration.

The citizens of Fibersville, however, would prefer to keep the fibrette concentration under the 150 s/cc tolerance threshold all the time. In this case, it is apparent that a more coordinated effort is necessary, and individuals would need to make collective decisions, and abide by them, on how to meet the tolerance threshold. If the collective decision-making cannot meet the tolerance threshold, centralized regulatory measures can be effective in managing the risk. This could be accomplished by the local transportation administration or another authority of Fibersville imposing a maximum daily (or other time interval) vehicle use limitation. For example, a personal car policy could be created wherein only cars with odd or even license plate numbers would be allowed to be driven during a specified time interval, such as normal business hours, on any day of the week when the tolerance threshold has historically been exceeded.

Alternatively, the local authorities could regulate the types of brake systems sold or installed in their community, including the stricter measures to limit or prohibit fibrette-containing brake lining.

The chaotic behavior of the airborne concentrations trend in Fibersville demonstrates a case where exposure assessment performed by collecting multiple air samples would fail in many situations since a reasonably efficient monitoring protocol would likely require the assumption of randomness. If the monitoring protocol assumes the measured parameter is not random, the monitoring could easily be cost prohibitive due to the number of samples required to evaluate the nonrandom conditions.

Also, the relative failure of individual predictions to adequately self-regulate the airborne fibrette concentration indicates that it is problematic to rely on individuals in a wider set of situations related to environmental air quality. For example, sometimes companies are brought into litigation for not self-regulating the potential exposures to agents which are not regulated by governments and/or for which no consensus standards exist. Many areas of current personal injury litigation law are based on historical exposures to chemical or physical agents wherein the concentration of the agents to which the individuals were exposed was well below the OELs applicable at the time of the exposures. In summary, attempts to self-regulate environmental and occupational health requirements, without due coordination with peers and with the government, can lead to chaotic, rather than fully predictable, consequences.

Diagnosing the Chaotic Trends

We have used several metrics to evaluate chaotic behavior in the airborne concentrations of fibers.

In particular, we utilized the metric of Approximated Entropy (ApEn), maximum Lyapunov exponent, Hurst exponent, and detrended fluctuation analysis (Delgado-Bonal and Marshak 2019) using Python 3.8.3 software to calculate these values.

While there are no definite criteria to distinguish between deterministic and chaotic trends, it is important to know that

The maximum Lyapunov exponent should be positive for chaotic systems;
ApEn is typically higher for chaotic trends;
The Hurst exponent is equal to 0.5 for "white noise" (random system), with more memory associated with higher or lower values of the parameter;
The result of a detrended fluctuation analysis is less than 1 for stationary processes; if the value is equal to 0.5, it indicates that there is no memory in the trend.

The results of the analysis for car traffic and "fibrette" emissions are demonstrated in Table 15.1.

The results demonstrate the nonlinear, chaotic trend for both car traffic emissions, and Fibrette concentrations, as demonstrated by the positive values of Lyapunov coefficients (Fell et al. 1993). The approximate entropy levels can be compared to the corresponding level for a fully random binary consequence that should have ApEn at the level of ln2 ($\approx$0.693) (Pincus and Kalman 1997). In this case, traffic data is less chaotic than the fully random consequence, but fibrette concentrations are more chaotic than random numbers. Similarly, traffic data (created by the mechanical process of car flow through the canyon) is less chaotic than "white noise," while the fibrette concentration trend (as a result of conscious, but not-coordinated, self-regulation) seems to be more chaotic.

Both the Hurst exponent and the detrended fluctuation analysis show values less than 0.5 for the explored processes, revealing the negative feedback from previous time moments. The airborne concentrations have "memory" of the previous levels, even if the trends are highly nonlinear.

Table 15.1 Characteristics of the chaotic behavior.

Metric	Traffic data (an average speed of 10.5 mph)	Fibrette concentrations
Maximum Lyapunov exponent (should be positive for chaotic systems)	0.05	0.03
Approximate entropy (ApEn) (higher for chaotic systems)	0.35	0.86
Hurst exponent (0.5 for white noise; further from 0.5, more "memory")	0.289	0.259
Detrended fluctuation analysis (<1 for the stationary process; 0.5 – no memory for previous values)	0.285	0.259

References

Arthur, W.B. (1994). Inductive reasoning and bounded rationality. *The American Economic Review* 84(2): 406–411.

Baccan, D.D. and Macedo, L. (2013). Revisiting the El Farol problem: a cognitive modeling approach. In: *Multi-Agent-Based Simulation XIII. MABS 2012* (eds. F. Giardini and F. Amblard). *Lecture Notes in Computer Science*, Vol. 7838. Berlin, Heidelberg: Springer. https://doi.org/10.1007/978-3-642-38859-0_5.

Bathmanabhan, S. and Madanayak, S. (2010). Analysis and interpretation of particular matter – PM10, PM2.5, and PM1 emissions from the heterogeneous traffic near an urban roadway. *Atmospheric Pollution Research* 1(3): 184–194. https://doi.org/10.5094/APR.2010.024.

Berkowicz, R., Hertel, O., Larsen, S.E., et al. (1997). *Modelling Traffic Pollution in Streets*. Ministry of Environment and Energy, National Environmental Research Institute.

California Air Resources Board (CARB). Wall, S., and Wagner, J. (2007). Determination of asbestos content of current automotive dry friction materials, and the potential contribution of asbestos to particular matter derived from brake wear. Final Report. CARB #01-333.

Chen, S.H. and Gostoli, U. (2013). Coordination in the El Farol Bar problem: The role of social preferences and social networks, Economics Discussion Papers 2013-20, Kiel Institute for the World Economy (IfW Kiel).

Delgado-Bonal, A. and Marshak, A. (2019). Approximate entropy and sample entropy: a comprehensive tutorial. *Entropy (Basel)* 21(6): 541. https://doi.org/10.3390/e21060541.

Effah-Poku, S., Obeng-Denteh, W., and Dontwi, I.K. (2018). A study of chaos in dynamical systems. *Journal of Mathematics* 2018: 1808953. https://doi.org/10.1155/2018/1808953.

Faber, J. and Koppelaar, H. (1994). Chaos theory and social science: a methodological analysis. *Quality & Quantity* 28(4): 421–433. https://doi.org/10.1007/BF01097019.

Fell, J., Röschke, J., and Beckmann, P. (1993). Deterministic chaos and the first positive Lyapunov exponent: a nonlinear analysis of the human electroencephalograph during sleep. *Biological Cybernetics* 69(2): 139–146. https://doi.org/10.1007/bf00226197.

Grigoratos, T. and Martini, G. (2015). Brake wear particle emissions: a review. *Environmental Science and Pollution Research* 22: 2491–2504. http://dx.doi.org/10.1007/s11356-014-3696-8.

Kauppinen, T. and Korhonen, K. (1987). Exposure to asbestos during brake maintenance of automotive vehicles by different methods. *American Industrial Hygiene Association Journal* 48(5): 499–504. https://doi.org/10.1080/15298668791385101.

Kondepudi, D., Petrosky, T., and Pojman, J.A. (2017). Dissipative structures and irreversibility in nature: celebrating 100th birth anniversary of Ilya Prigogine (1917–2003). *Chaos* 27(10): 104501. https://doi.org/10.1063/1.5008858.

Lemen, R.A. (2004). Asbestos in brakes: exposure and risk of disease. *American Journal of Industrial Medicine* 45: 229–237.

Lo, S.-C. and Cho, H.-J. (2005). Chaos and control of discrete dynamic traffic model. *Journal of the Franklin Institute* 342(7): 839–851. https://doi.org/10.1016/j.jfranklin.2005.06.002.

Marsili, M., Challet, D., and Zecchina, R. (2000). Exact solution of a modified El Farol's bar problem: efficiency and the role of market impact. *Physica A: Statistical Mechanics and its Applications* 280(3–4): 522–553. https://doi.org/10.1016/S0378-4371(99)00610-X.

Oestreicher C. (2007). A history of chaos theory. *Dialogues in Clinical Neuroscience* 9(3): 279–289. https://doi.org/10.31887/dcns.2007.9.3/coestreicher.

Paustenbach, D., Finley, B.L., Lu, E.T., et al. (2004). Environmental and occupational health hazards associated with the presence of asbestos in brake linings and pads (1900 to present): a "State-of-the-art" review. *Journal of Toxicology and Environmental Health, Part B* 7(1): 25–80. http://dx.doi.org/10.1080/10937400490231494.

Philippe, P. (1993). Chaos, population biology, and epidemiology: some research implications. *Human Biology* 65(4): 525–546. PMID: 8406405.

Pincus, S. and Kalman, R.E. (1997). Not all (possibly) "random" sequences are created equal. *Proceedings of the National Academy of Sciences of the United States of America* 94(8): 3513–3518. https://doi.org/10.1073/pnas.94.8.3513.

Prigogine, I. (1989). The philosophy of instability. *Futures* 21(4): 396–400. https://doi.org/10.1016/S0016-3287(89)80009-6.

Rappaport, S.M. and Kupper, L.L. (2008). *Quantitative Exposure Assessment*. El Cerrito, CA: S. Rappaport.

Rickles, D., Hawe, P., and Shiell, A. (2007). A simple guide to chaos and complexity. *Journal of Epidemiology Community Health* 61(11): 933–937. https://doi.org/10.1136%2Fjech.2006.054254.

US EPA. (2014). Brake and tire wear emissions from onroad vehicles in MOVES2014. EPA-420-R-14-013. MOVES Onroad Technical Reports | US EPA.

Wang, Y., Huang, Z., Liu, Y., et al. (2017). Back-calculation of traffic-related PM_{10} emission factors based on roadside concentration measurements. *MDPI* 8(6): 99. https://doi.org/10.3390/atmos8060099.

Williams, R.L. and Muhlbaier, J.L. (1982). Asbestos brake emissions. *Environmental Research* 29(1): 70–82. https://doi.org/10.1016/0013-9351(82)90008-1.

Index

a

aerodynamic diameter   5, 26–28, 37, 128, 131, 141, 189, 300
airborne asbestos   6, 116, 119, 140, 174, 177, 182, 183, 185, 278, 295, 413–418, 421–422, 424
ambient air emissions   428–433
American Meteorological Society/Environmental Protection Agency Regulatory Model (AERMOD) atmospheric dispersion model   154, 157, 163–164, 188
   asbestos cement pipe manufacturing plant   191, 193–194
      annual average asbestos concentration   195, 196
      meteorology set-up   191, 193
amosite cohorts   239–240, 246, 249, 250, 254, 265, 345
   Patterson, New Jersey   314
   South African amosite mining   311, 313–314
   Tyler, Texas   315
   Uxbridge   315–316

amphibole asbestos, lung burdens of   279, 281–282
amphiboles   4–5, 7, 8, 10, 12, 15–16, 20, 22–25, 28, 30–32, 35–37, 39–40, 64, 67, 70–72, 82–87, 93, 118, 123, 216, 219, 221–227, 234, 237, 241, 242, 246, 249–251, 254, 256, 259–262, 264, 272, 279, 289–290, 294–296, 306, 313, 327, 328, 345, 355–360, 371, 373–374, 379, 380, 388, 406, 416, 427
AP-1 transcription factor   69
Armitage–Doll age-distribution cancer model   203–205, 208, 210
asbestiform habit   3, 4, 8, 14, 24, 31, 82, 345
asbestos   3, 13, 82
   aerosols   39–40
   bodies   21
   dissolution rates   19
   effusion   84
   exposures   24
   fibrils   3–4, 23–24
   lung burden of   4
   physical properties of   3

Health Risk Assessment for Asbestos and Other Fibrous Minerals, First Edition.
Edited by Andrey Korchevskiy, James Rasmuson, and Eric Rasmuson.
© 2024 John Wiley & Sons, Inc. Published 2024 by John Wiley & Sons, Inc.

asbestos (cont'd)
    stiffness 28
    tensile strength 28–30
    traffic-related emissions 428
    vehicle brake 427
asbestos cement pipe manufacturing plant
    AERMOD model 191, 193
        annual average asbestos concentration 195, 196
        meteorology set-up 191, 193
        results 194
        set-up 193
    baghouse source emission rates 182–183
    CALMET 189
    CALPUFF model
        annual average asbestos concentration 196
        parameters 189
        results 191–193
    CFD model
        annual average asbestos concentration 196
        domain 177
        primary parameters 186, 187
        results 186–188
    crusher source emission rate 183–184
    EPA outdoor dispersion models 188
    fugitive emission building sources generation rates 180, 181
    geophysical set-up 188–189
    meteorology 184–186
    pipe storage and shipping yard source emission rate 183
    receptor descriptions 177–179
    source descriptions 177–178, 180, 184

asbestos-containing material (ACM) 164, 413–414
asbestos dose–response assessment
    community and occupational risk assessment 378
    cumulative exposure 366
    Hodgson and Darnton method 367
    IUR (Inhalation Unit Risk) for asbestos fibers 383–385
    linear model 367
    mesothelioma and lung cancer potency factors 214, 233–235, 265, 332, 348, 355, 367, 371–374, 387
    nonlinear model 368–371
    other types of cancer 380–383
    peritoneal mesothelioma 378–380
    Peto method 366–367
    tobacco smoking 385–387
asbestos exposure, health outcomes of
    malignant diseases 92–102
        asbestos-related lung cancer 94–96
        general comments 92–94
        mesothelioma 96–102
    nonmalignant change in structure or function 83–84
    nonmalignant diseases 84–92
        lung disease 89–92
        pleural disease 84–89
asbestos exposure measurements
    early sampling strategies 135
    fiber counting 136–140
    gravimetric methods 131–132
    historic methods 131
    microscopic particle counting
        direct reading instruments 134–135
        impaction sampling 132
        impinger sampling 132–133

thermal precipitator sampling   133–134
variability of MF-PCM index over time   140–146
asbestos fibers   3–4, 23–24, 28, 31, 36, 52, 55, 56, 58, 63–65, 68–72, 83, 85, 89, 95, 118, 120, 131, 134, 137–138, 140, 143, 144, 169–170, 180, 182, 189, 216, 221–224, 226, 227, 257, 274, 287, 292–296, 300, 313, 325, 327, 356–360, 369–370, 376, 383–385, 396, 398, 404, 413, 415, 417, 421, 424, 428
asbestos in soil
 airborne asbestos concentrations   412–413, 415, 416, 418, 421–422
 airborne respirable dust   413, 417, 420–422
 for amphibole fibers   416
 asbestos soil to air coefficients   413–415
 for chrysotile   416, 422
 contamination problems   412
 hazard identification   412
 lung cancer mortality   422–424
 mesothelioma mortality   422–424
 PCM concentrations   413
 results of soil sampling and analysis   418–419
 risk modeling   423–424
asbestosis   40, 52, 84, 86, 88–92, 94–95, 128–129, 134, 136, 301, 313, 315–316, 320, 322, 327, 330, 381, 396, 408
Asbestosis Research Council (ARC)   128, 136, 144, 304
asbestos-related diseases (ARD)   4–6, 39–40, 82–89, 129, 146

asbestos-related lung cancer   93–96, 235, 237, 408
asbestos-related lung disease   128–130
asbestos worker cohorts
 air concentration measurements   287
 amosite cohort exposures
  Patterson, New Jersey   314
  South African amosite mining   311, 313–314
  Tyler, Texas   315
  Uxbridge   315–316
 chrysotile mining and milling cohort exposures
  Balangero, Italy   318–319
  Qinghai, China   319–321
  Quebec, Canada   318
  Uralasbest, Russia   321–322
 chrysotile textiles   322
  Chongqing chrysotile cohort   327–328
  North Carolina textile workers   325–327
  South Carolina textile workers   324–325
 crocidolite cohort exposures
  Massachusetts cigarette filter manufacturing   309–310
  mesothelioma rate   311, 312
  South African mines and mills   306–309
  UK gas mask workers   310
  Wittenoom environmental   305–306
  Wittenoom occupational   302–305
 MF-PCM fiber counts
  conversion from gravimetric measurement   299–302
  conversion from impinger counts   297–299

asbestos worker cohorts (*cont'd*)
  conversion from particle counting methods 299
    TEM fiber-size-distribution 287
    from manufacturing cohorts 289–292
    mines and mills 288–289
aspect ratios (ARs) 3, 6–8, 22, 23, 27, 30, 31, 36, 55, 118, 129, 140, 143, 144, 287, 290, 291, 293, 319, 347, 353, 388

b

Balangero, Italy 16, 242–244, 250, 254, 262, 265, 295, 317–319, 322
benchmark dose modeling 377
"best subset" procedure model 347, 348
biodurability 8, 17, 18, 70
biopersistence 5, 8, 17, 18, 20, 37, 72, 357–359, 371, 388
Bolivian crocidolite 239, 305, 309, 355
brake dust 62–64
brake-related operations
  asbestos concentration drop-off *vs.* distance 165
British mesothelioma projections 272–273
bronchoalveolar lavage fluid (BALF) 66
bystander exposure factor 121, 174
  defined 164
  rule-of-thumb diagram 165, 166

c

California Puff (CALPUFF) atmospheric dispersion model 154, 156, 163–164, 188, 194
  asbestos cement pipe manufacturing plant

  annual average asbestos concentration 196
    parameters 189
    results 191–193
  CALMET 189
cancer
  Armitage–Doll age-distribution cancer model 203–205
  probability formula 204
  risks 203
    for amosite and chrysotile exposure levels 226, 227
  types of 380–383
carcinoembryonic antigen (CEA) 99–100
chaos theory 428
  chaotic trends 439–440
  self-regulated community 433–439
chaotic behavior 428, 433–440
Chongqing, China 238, 242, 244, 245, 249, 254, 292, 294, 295, 301, 327–328, 330
chrysotile 406
  asbestos 62–64
  cohorts 242–245
chrysotile mining and milling cohort exposures
  Balangero, Italy 318–319
  Qinghai, China 319–321
  Quebec, Canada 318
  summary 322, 323
  Uralasbest, Russia 321–322
chrysotile textiles 322
  Chongqing chrysotile cohort 327–328
  North Carolina textile workers 325–327
  South Carolina textile workers 324–325
cleavage fragments 4, 5, 8, 11, 13, 23, 25, 29, 35–37, 39, 66, 69, 388

colorectal cancer   382
community and occupational risk assessment   378
computational fluid dynamics (CFD)   154, 417–418
  air dispersion modeling
    application   155–157
    validation   155–157
  bystander exposure estimation, in outdoor setting (*see* asbestos cement pipe manufacturing plant)
  indoor modeling   164–176
  simulation   157
    geometry creation and set-up   159–160
    mesh creation   160, 161
    parameter set-up   160
    post-processing   162–163
    steady-state simulations   162, 197
    transient simulations   162
crocidolite   8, 10, 16, 24
  asbestos   62–64, 68–71, 96, 264, 294, 295, 302, 313, 356, 381
  cohorts   238–239
crocidolite cohort exposures
  Massachusetts cigarette filter manufacturing   309–310
  mesothelioma rate   311, 312
  South African mines and mills   306–309
  UK gas mask workers   310
  Wittenoom
    environmental   305–306
    occupational   302–305
  cumulative occupational exposure   401, 404

d
diffuse pleural thickening   84, 86–89

e
electron microscopy (EM)   32–33, 69, 128, 137–139, 287, 294
elongate mineral particles (EMPs)   3
  amphiboles and serpentines   4
  bioreactivity of mineral surfaces   17–24
  carcinogenic potential   5–6, 38–39
  categorization of   5
  chemical and physical properties   3, 5, 8–10
  chemical composition   15–16
  chemical factors   17–24
  dimension   11–13, 30–32, 38–39
  exposure measurement   6
  fragmentation   5
  frequency distributions of length and width   35–38
  lung burden   37
  measurement protocols   32–39
  mineral intergrowths and associations   16–17
  nomenclature   6–8
  parallel sides   5
  pathogenic potential   5
  structural groupings of   13–15
  surfaces, chain silicates and zeolites   23–24
  types and characteristics   13, 14
endothelioma   96–98
energy-dispersive X-ray (EDX)   9, 15, 119, 138
energy dispersive X-ray analyzer (EDXA)   137, 138
enthalpy of formation   26, 345
EPA outdoor dispersion models   188

erionite   4–5, 17, 21–22, 24, 40, 55, 65, 67–69, 71, 72, 86, 120, 345–347, 352, 354–357, 359–360
exposure assessment for EMPs
　analytical methodologies 120–123
　　censored data   122
　　lung burden analysis   122–123
　　proximity to emission source 121–122
　NRC/NAS risk assessment process   112
　principles and methods   113–120
　　evaluating quality of data 114–116
　　gathering information   113–114
　　measurement techniques 116–120

f

fiber   7, 8
　counting   136–140
　　direct reading instruments 134–135
　　limitations of current indices of exposure assessment   139–140
　　MF-PCM   136–137
　　MF sampling and EM analysis 137–139
　definition   140, 143–144
Fiber Half-Life in Lungs   358
fiberization   25
Fiber Potential Toxicity Index (FPTI)   70
fibrils   3–5, 7–10, 17, 21, 23–25, 28–29, 31, 35, 37, 39–40, 137
fibrosis   17, 18, 24, 52, 54–55, 62–65, 71, 84–85, 89–92, 128, 134, 221, 296

fibrous habit   8, 10
Field Emission Gun (FEG-SEM) 138, 333
Field Emission Scanning Electron Microscopy (FESEM)   33
fragment   4–6, 8, 11–13, 23, 25, 29, 31, 35–37, 39, 66, 69, 144
frustrated phagocytosis   69

g

GATA binding protein 3 (GATA3)   101
genotoxicity   67
Great Britain, mesothelioma mortality   270, 272, 282

h

habit   3–5, 8–11, 14, 24, 26, 31, 35, 39, 70, 82, 100, 143, 146, 212, 265, 345, 360, 371
Hodgson and Darnton methods   367, 369–371, 374, 396, 397, 404
　meta-analyses   237–238
　　amosite cohorts   239–240
　　amphiboles   241
　　chrysotile cohorts   242–245
　　crocidolite cohorts   238–239
　　lung cancer   250–256
　　mesothelioma   245–250
　metrics and data requirements
　　data issues   236–237
　　lung cancer   235–236
　　mesothelioma   236
　nonlinear exposure–response relationship   256–257
　peritoneal mesothelioma 259–260
　pleural mesothelioma   257–259
　overview of   234–235
　risk assessment   262–264

hot strip steel mill, indoor CFD
    modeling
  CFD results
    bystander factor distribution, for
      ACH 174–176
    horizontal breathing zone airflow
      distribution 172, 173
    three-dimensional 5 to 10-foot
      bystander zone volume 173
    vertical airflow distribution,
      turbulent nature of 172
  indoor modeling domain 168
  model parameters 169, 172
  pipe insulation removal source 171
  reheat furnace brick removal source
    170–171
  source description 170
  ventilation 168–169
^{3}H-thymidine 62
hydroxyproline 62

i

impaction sampling 132
impinger sampling 132–133
indoor CFD modeling 164–176
  asbestos concentration drop-off *vs.*
    distance, for brake-related
    operations 165
  bystander exposure factor
    defined 164
    hot strip steel mill (*see* hot strip
      steel mill)
    preliminary outdoor CFD wind
      simulation 165–168
    rule-of-thumb diagram 165, 166
inhalation unit risk (IUR) 214,
  215, 226
  for asbestos fibers 383–385
  method 398–400

inosilicates 15, 16, 24
International Mineralogical Association
  (IMA) 7, 8
International Standards Organization
  (ISO) 119, 128, 138–139
intraclass correlation coefficients
  (ICC) 122
intraperitoneal injection model 55

k

konimeters 132, 135, 299, 305, 307–308

l

laryngeal cancer 295, 380–381, 387
less than limit of quantification (LOQ)
  sample results 419
Libby amphibole (LA) 55, 65–66, 71,
  82, 86, 225, 227, 237, 249, 251,
  262, 345, 354–355, 358, 379–380,
  383, 401, 406
Libby Amphibole Asbestos (LAA)
  86–88, 345, 428
lifetime average exposure 399
light microscopy 122, 133
linear model 218, 262, 263, 367, 408
localized mesothelioma 98
localized pleural thickening (LPT)
  84–88
long-running thermal precipitator
  (LRTP) 134, 135, 286,
  302, 304–306
lung burden 4, 22, 31, 34, 37, 53, 62,
  221, 259, 279–282, 284, 294, 296,
  305, 309, 310, 316
lung burden analysis (LBA) 122–123,
  296, 428, 429
lung cancers 52, 54, 67, 71, 112, 401,
  405, 408
  asbestos-related 94–96, 102

lung cancers (*cont'd*)
  chrysotile *vs.* amphibole potency factors for   219
  dose–response coefficients   219, 220, 224
  erionite potency   354
  histological types of   95–96
  meta-analysis   250–256
  modeled and published   353
  mortality, time-and age-dependent patterns   203
  nonlinear exposure-response relationship   260–262
  Peto model for   209, 211, 212
  potency factors   346
  and smoking   95
  Standardized Mortality Ratios for   235–236

m

malignant diseases   92–102
  asbestos-related lung cancer   94–96
  mesothelioma   96–102
malignant mesotheliomas (MMs)   52, 54, 63–65, 68–71, 96, 97, 101, 212, 354, 378
malignant pleural mesothelioma (MPM)   101, 271–272, 378
man-made synthetic vitreous fibers   64
Massachusetts cigarette filter manufacturing   239, 309–311
membrane filter (MF) sampling   127, 137–139, 287, 413
membrane filter sampling and phase contrast microscopy (MF-PCM)   127–128, 132–135, 287–288, 296–303, 306, 308–310, 320, 321, 324–325, 327, 332
  adoption of   136
  counting procedures and performance   144–145
  effect of changes   145–146
  fiber counting   136–137
  fiber definition   143–144
  index of exposure   137, 139, 140
  microscope equipment and set-up   142–143
  sample preparation   141–142
  sampling method   140–141
mesothelioma   10, 96–102, 112, 236
  asbestos exposure   97
  cumulative risk *vs.* time since first exposure   206, 207
  death rate   205, 206
  derived models for   349
  diagnostic revolution   100–102
  dose–response coefficients   221, 224
  exposure to   96–97
  half-life for fibers   359
  histologic classification of   98
  immunohistochemistry   99–102
  and lung cancer potency factors   371–374
    behavior of dose–response models   371, 372
    comparison ratios of   371, 373–374
    life tables and life expectancy of exposed population   374–375
    linearity and nonlinearity   375–376
    threshold and benchmark dose response   376–377
  meta-analyses   245–250
  microscopic characteristics   98
  misdiagnosis   97–98
  modeled and published   351
  mortality

annual population exposure
profiles   278
fitted model parameters   276
Great Britain   272, 282
projections of   270
original data set   351
peritoneal   259–260
Peto model for   208–209, 211, 212
pleural   257–259
pleural neoplasms   98
potency factors   345–347
primary endothelioma of pleura   97
risk   395, 398, 401, 404, 405
risk among American insulators   206, 207
time-and age-dependent patterns   203
meta-analysis   30, 128, 140, 235, 297, 299, 387
  Berman and Crump   218–227
  Hodgson and Darnton approach   237–238, 264–265, 309, 330, 354
    amosite cohorts   239–240
    amphiboles   241
    chrysotile cohorts   242–245
    crocidolite cohorts   238–239
    lung cancer   250–256
    mesothelioma   245–250
metal-oxide nanoparticles   344–345
MF-PCM fiber counts
  gravimetric measurement   299–302
  impinger counts   297–299
  particle counting methods   299
mineral fiber   5–8, 10, 16, 18–19, 21, 30, 39, 237, 238, 250, 264, 287, 330, 345, 352, 373
mineral fibers, toxicology of
  defined   52
  induced diseases, systems biology approach   71–72

properties   68–71
rodent models   53–66
  with asbestos fibers   54–62
  inhalation studies   53
  intraperitoneal injection studies   54
  intrapleural injection studies   54
  intratracheal instillation and oropharyngeal aspiration studies   54
in vitro models   66–68
  advantages and disadvantages   52, 66–67
  cytotoxicity and carcinogenesis   67–68
  historical use of   52
mineral name   7, 8
mineral's habit   8
mineral surfaces
  chemical reactivity of   17
  coatings   21–22
  dissolution rates   19, 20
  quartz   17–18
  ROS, formation of   20–21
  solubility   18–20
  specificity of   17–18
  surface charge   22–23
modeling software tools. *see also* computational fluid dynamics (CFD)
  AERMOD   163–164
  CALPUFF   163–164
models, defined   153
Monte Carlo simulation   396, 398
mutagenesis   67

n

National Academy of Sciences (NAS)   111

National Institute of Occupational
    Safety and Health (NIOSH)
    5–7, 32, 86, 91, 111–112, 223,
    241, 396, 399, 401
National Research Council (NRC)
    111, 297
naturally occurring asbestos (NOA)
    4, 406, 412
near-infrared instruments (NIR)   120
NIOSH 7400 method   116–119,
    136, 145
NIOSH 7402 method   32, 119
nitrilotriacetate (NTA)   356
nonasbestiform   5, 29, 31, 35, 39, 82,
    120, 344–345, 380, 388, 416
nonlinear exposure–response
        relationship   256–257
    lung cancer   260–262
    peritoneal mesothelioma   259–260
    pleural mesothelioma   257–259
nonlinear model   257, 259, 262, 264,
    368–371, 380
nonmalignant diseases   84–92
    lung asbestosis   89–92
    pleural
        asbestos effusion   84
        diffuse pleural thickening   88–89
        pleural plaques and LPT   84–88
        rounded atelectasis   89
"no-observed adverse effect level"
    (NOAEL)   377
North Carolina textile workers
    249, 325–327

o

occupational and environmental
    exposure to asbestos, risk
    characterization, case
    studies   395–408

occupational exposure limits (OELs)
    332, 333, 438, 439
Operational Street Pollution Model
    (OSPM)   428
optical microscopy   32, 33, 223, 226
optical phase contrast microscopy
    (PCM)   127, 138
optimum exposure index   222
ovarian cancer   381–382

p

Patterson, New Jersey   311, 314,
    315, 317
PCME fibers   223, 380, 406, 416
per-and polyfluoroalkyl substances
    (PFAS)   424
peritoneal mesothelioma   98, 256–257,
    262, 265, 356, 368, 378–380
    nonlinear exposure-response
        relationship   259–260
Peto method   366–368, 374, 385, 396,
    398, 401, 406, 407
Peto model   205, 371, 375
    Berman and Crump meta-analysis
        based on   218–227
    for American insulators' mesothelioma
        mortality data   211
    for lung cancer   209, 211, 212
    for mesothelioma   208–209, 211, 212
    US EPA utilization of   212–218
phagocytosis   22, 69, 128
phase contrast microscopy (PCM)   31,
    120–121, 127, 287, 413
phase contrast microscopy equivalent
    (PCME)   32, 121, 223, 225, 226,
    380, 406, 416
phyllosilicates or sheet silicates
    7, 15, 355
pleural fibrosis   64–65, 84–85

pleural mesothelioma 98–99, 101, 210, 236, 244, 257–259, 265, 271, 295, 316, 347, 368, 376, 378–379
pleural plaques 84–89, 97
pleural tuberculosis 97
polarized light microscopy (PLM) 4, 32, 119, 415
polymorphs 7
potassium octatitanate (PO) 55
Pott hypothesis 129
progressive interstitial fibrosis 63
pulmonary dysfunction 86–89, 92
pulmonary fibrosis (asbestosis) 52, 54, 62–64, 71, 221
pyroxene 4–5, 23

q

Qinghai, China 237–238, 242, 243, 245, 254, 255, 294, 301, 317–322, 330
quantitative structure–activity relationship (QSAR) modeling 344, 345, 356–357, 360
  methodology and principles 347
quartz 9, 15, 17–18, 22–23
Quebec, Canada 94, 102, 218, 222, 224, 242–246, 250, 254–256, 258, 262, 265, 289, 290, 294, 298, 299, 317, 318, 322, 325, 328, 332, 352, 354, 355, 381, 387

r

reactive oxygen species (ROS) 16, 17, 20–21, 71, 72
refractive index (RI) 140, 142, 287
relevance 115
reliability 115–116, 122, 287
retention 8, 296
retrospective exposure assessment (REA) 122, 123, 385
rounded atelectasis 89

s

sarcomatoid mesotheliomas 101
scanning electron microscopy (SEM) 11, 25, 33, 34, 99, 100, 119, 122, 138, 287, 288, 295, 333
selected area electron diffraction (SAED) 119
SEM size distributions 292–293
silicosis 17, 20, 85, 113, 131
solid solution series 7, 10
South African 10, 25, 63, 96, 97, 131, 141, 233, 238–240, 288, 354, 355
  amosite mining 311, 313–314
  mines and mills 306–309
South Carolina textile workers 221, 242, 290, 324–325
specific gravity (SG) 26–27, 170, 191
specific surface area 24–25, 30, 39, 70, 347
standard heat of formation 26
Standardized Mortality Ratios (SMRs) 235–237, 322, 381–382
stratified counting 34, 289
street canyons 428–433
  chrysotile concentrations in 430–432
Surveillance, Epidemiology, and End Results (SEER) 210, 272

t

tectosilicates 15, 16, 18, 24
TEM fiber-size-distribution 287
  from manufacturing cohorts 289–292
  mines and mills 288–289
textile workers 214, 244, 378, 381, 385
  North Carolina 325–327
  South Carolina 40, 218, 221, 242, 290, 324–325

thermal precipitator (TP) sampling   133–134
thermal system insulation (TSI) materials   414
thermodynamic properties of silicates   26
Threshold Limit Value (TLV)   298
time-weighted average (TWA) exposure   112, 114, 135, 141, 304, 398, 413, 438
tobacco smoking   385–387
trace metals   9
transmission electron microscopy (TEM)   11, 119, 120, 137, 287, 418
  direct transfer method   34–35
  indirect transfer method   34, 35
tumorigenicity   54, 64, 65, 69
two-stage clonal expansion (TSCE) model   209–210, 375, 376
  for American insulators' mesothelioma mortality data   211
Tyler, Texas   237, 240, 311, 314–317

u

UK gas mask workers   310

Uralasbest, Russia   265, 301, 321–322, 331
Uxbridge   240, 293, 311, 315–316

v

vermiculite miners   88, 241
visceral pleural thickening   63, 85

w

Weather research forecasting (WRF) model meteorological data   184–185, 189, 191
Wittenoom crocidolite miners cohort   218, 222, 224, 237–239, 265, 302–306, 309–311, 381, 387
wollastonite   20, 55, 65, 70, 71
woolly erionite   4, 5, 24
worktime exposure   401, 404, 406

x

X-ray diffraction patterns   9

y

Young's modulus   28

z

zeolites   15, 23–24, 65

Printed in the USA
CPSIA information can be obtained
at www.ICGtesting.com
JSHW010759151124
73655JS00001B/25